Mathematical Conversations

Selections from *The Mathematical Intelligencer*

Springer Science+Business Media, LLC

Mathematical Conversations

Selections *from* The Mathematical Intelligencer

Compiled by

Robin Wilson and Jeremy Gray

With 231 Illustrations

Springer

Robin Wilson
Jeremy Gray
Faculty of Mathematics and Computing
The Open University
Walton Hall
Milton Keynes MK7 6AA, UK

Library of Congress Cataloging-in-Publication Data
Mathematical conversations: selections from The mathematical intelligencer / compiled by Jeremy
  Gray, Robin Wilson.
      p. c.m.
    Includes bibliographical references.
    ISBN 978-1-4612-6556-6   ISBN 978-1-4613-0195-0 (eBook)
    DOI 10.1007/978-1-4613-0195-0
    1. Mathematics.   I. Gray, Jeremy, 1947–   .   II. Wilson, Robin J.
  III. Mathematical intelligencer.
  QA7.C57 1999
  510—dc21   98-43867

Printed on acid-free paper.

© 2001 Springer Science+Business Media New York
Originally published by Springer-Verlag New York, Inc. in 2001
Softcover reprint of the hardcover 1st edition 2001

Production coordinated by Impressions Book and Journal Services, Inc., and managed by Lesley Poliner; manufacturing supervised by Joe Quatela.
Typeset by Impressions Book and Journal Services, Inc., Madison, WI.

9 8 7 6 5 4 3 2 1

ISBN 978-1-4612-6556-6   SPIN 10706527

Contents

Part Six

GEOMETRY AND TOPOLOGY

Part Seven

HISTORY OF MATHEMATICS

Introduction

Jeremy Gray and Robin Wilson

Over the past twenty years *The Mathematical Intelligencer* has put before a general mathematical audience a variety of articles that are by turns exciting, important, diverting, fun, profound, and surprising. Reading these articles again has given us insights into what animates mathematicians as they shared their enthusiasm with non-specialists, including experts in fields other than their own.

One passion that they share is for hard, significant problems—and discovering their solutions. *The Intelligencer* has carried a considerable number of articles that report on such breakthroughs as Faltings' affirmative answer to the Mordell conjecture (for which he won a Fields Medal in 1986), Vaughan Jones' work on knots (part of the work that earned him a Fields Medal in 1990), the solution of the Bieberbach conjecture, and progress in the fast-moving area of work on hyperbolic three-manifolds, solutions, instantons, and string theory.

Another passion is for the sheer joy of doing mathematics. Can you imagine Boy's surface or think geometrically about differential equations? Calculating, whether in algebra, analysis, combinatorics, or geometry, is not just a mathematician's bread and butter; it has its own appeal and excitement, and some of that is conveyed in these pages.

A third passion is for the sensuous beauty of the subject. From space-filling curves to the newly discovered minimal surfaces, from the delights of Celtic knots and quasi-crystals to the patterns on a Florentine pavement, much of what mathematicians discover is visually lovely. Since its start, *The Intelligencer* has been able to carry many illustrations, thus conveying to its readers another part of mathematical life that many specialist journals have perforce to leave out.

Mathematics is an intrinsically human activity, and this aspect has also been well represented by *The Intelligencer*. It has carried a steady stream of articles in the history of mathematics, some biographical, some analytical, and some showing the links with the wider world that mathematicians have made in their lives. It has presented a number of articles in which mathematicians have gone back to the past and found new things to say as a result, thereby exemplifying the remarkable way in which the longevity of the subject continues to animate it. A number of interviews with living mathematicians have also appeared in its pages; these enrich our understanding of how research is done today.

At a time when mathematics has to fight harder than ever for its funding, there is a need for journals that display in an accessible way what mathematics is and what it can do. Popularization wins friends for the subject. It brings in students, it informs colleagues in other fields, and it helps to assure politicians and taxpayers that public money is being well and responsibly spent; it can refresh professional mathematicians who necessarily dig deep into their own studies but still wish to know what their neighbors are doing, and it helps to locate and affirm the place of mathematics in the collective cultural activity of humanity. We are pleased to present a selection of forty articles from *The Mathematical Intelligencer*'s first eighteen years of work in this field.

1. Interviews and Reminiscences

The Mathematical Intelligencer has mined a rich seam of addresses and reflective pieces by distinguished mathematicians of the past and present and has also been able to obtain a number of interviews with leading figures of today. The essay by Steve Smale captures something of the turbulent times of the International Congresses of Mathematicians during the Cold War as well as of his own work. There is a moving portrait of Julia Robinson by her collaborator Yuri Matijasevich; she had a solution to one of Hilbert's problems to her credit and died in 1996—see the article about her [1] by her sister. Bourbaki is represented here by Jean-Pierre Serre and Michael Atiyah, who both speak interestingly on how they went about learning and doing mathematics. Atiyah's interests ran more towards mathematical physics as his career proceeded, and he considered himself most inspired by Hermann Weyl; the mathematicians around him have been an important bridge between Ed Witten and the mathematics community (see Ronald J. Stern's article in Part 6). In view of this connection, the interview with C. N. Yang acquires particular interest for indicating how the theory of fiber bundles looked to a leading physicist.

Lastly we mention a small fraction of the humorous side of mathematics as collected by Steven G. Krantz. *The Intelligencer* is distinguished for the accuracy of its presentations—no easier, we suggest, when the author is popularizing than when writing for research journals. So it is with Krantz's article until, as he admits, one or two of his stories become almost too good to be true. But then, knowing mathematicians . . .

2. Algebra and Number Theory

One of the great successes of the 1980s was Gerd Faltings' proof of the Mordell conjecture, which asserts that a polynomial with rational coefficients in two variables, and of genus greater than 1, has at most finitely many rational points; in fact, Faltings proved three results in a row, of which the Mordell conjecture was the last. The mathematical community's interest in this achievement was intense, not least because an effective bound on the number of solutions would have opened up the way to an attack on Fermat's Last Theorem, then still unresolved. The challenge posed to a popular expositor was tough because significant results in such a difficult area as modern algebraic number theory are forbiddingly technical. Spencer Bloch's account remains a gem; it is not easy, but is clear, accurate, and remarkably informative.

Number theory of the analytic kind is the subject of R. C. Vaughan's article, which connects such seemingly disparate themes as the Riemann Hypothesis and Fermat's Last Theorem (again) via investigations into Goldbach's conjecture that every even number greater than 2 is the sum of two primes. The fact of mathematical life that seemingly simple problems can still be intractable is the theme of Tien-Yien Li's article, which addresses an old topic so natural, and so unexpectedly difficult, that a lot of time is spent making sure that students never see it. The sad truth is that solving systems of polynomial equations is far from routine, although the problem could be put before a teenager. Readers seeking an update in the era of high-speed computers can consult Cox, Little, and O'Shea [2].

Beneath the surface of the simple questions, "How long is the repeating block in the decimal expansion of $1/p$, where p is a prime?" and "How long is the block in base n?" lurks another fascinating topic that is the subject of M. Ram Murty's article. As he explains, the problem concerns generators of the multiplicative group of integers mod p and Artin's still-unresolved conjecture claims that for each integer a (other than -1, 0, or 1), there are infinitely many primes p for which a is a generator mod p. Part of the thrill is discovering that we are led to the generalized Riemann Hypothesis and back to the article by Vaughan.

The role of group theory in mathematics needs no explanation, but it is often mediated by representation theory. Charles W. Curtis' article is one of many in this volume that shows how deeply leaders in the field today have drawn on the work of their predecessors. In the space of a few pages, he takes us through the work of Frobenius, Schur, Burnside, Emmy Noether, and Brauer, ending his account in effect where his own research begins.

3. Analysis

As Florin N. Diacu explains, Painlevé's conjecture, which survived for ninety-one years, asserts the existence of solutions to the so-called n-body problem in Newtonian gravitation (with $n > 3$) in which one body would, to put it loosely, go off to infinity in finite time. Such solutions were discovered by Xia in 1988, building on the work of numerous predecessors. The implications, if only for the philosophy of mathematics, may be even more disconcerting. The time-reversed process has objects arriving from infinity in finite time, too.

Counter-intuitive properties are a prominent feature of analysis. They might be thought of as the mathematical equivalent of snakes: not something you would expect and not always agreeable at first sight, but lovely in their own way and important not to ignore. Among the rich collection of "snakes," W. F. Osgood's example of a continuous curve with no crossings but having area has an honorable position. As Hans Sagan explains, Osgood showed that his class of curves has positive outer measure. Sagan then goes on to show that Osgood's curves are Lebesgue measurable with positive measure.

The article by Nail H. Ibragimov takes us from integration to differentiation and to the revival (in which he has played a prominent part) of Sophus Lie's geometrical theory of differential equations. Lie had hoped that the jungle of equations, ordinary and partial, and the ad hoc character of the solution methods then available, would yield to a group-theoretic approach by analogy with the Galois theory of polynomial equations. He was only partially successful, but in each generation the return to his original papers has led to new results with further implications for the study of mathematical physics.

4. Applied Mathematics

The core of applied mathematics remains physics, and it continues to stimulate profound work, as two articles by leading Russian mathematicians attest. Yu. I. Manin's was one of the first (and remains one of the most instructive) attempts to explain string theory, which is one way in to quantum field theory. Whatever the final verdict of physicists will be, this theory has already generated some splendid mathematics and garnered a Fields Medal for Ed Witten in 1990. S. P. Novikov's article is both classical and modern. He considers general dynamical systems and asks when they are integrable, and he takes the topic of solutions (non-linear travelling waves, often visible in canals) and shows how their theory involves the mathematics of quantum scattering and the theory of Riemann surfaces.

In the past twenty years, significant applications of mathematics have cropped up in many other areas; for example, Steven Strogatz describes the relevance of topology to the behavior of yeast. Such examples combine two of the features that contribute to the current revival of applied mathematics: an interesting topic and new kinds of mathematics.

5. Arrangements and Patterns

The mathematics of design has charmed people in every culture. Two-dimensional patterns, commonly called wallpaper patterns, tilings, or tesselations (from the Latin word for a marble chip), are to be found in Egyptian remains, numerous Islamic sites, in many places in China and Japan, and in pavements and walls around us, while their more serious younger brother, three-dimensional patterns, is the mathematical mainstay of crystallography and solid-state physics.

A tricky question for mathematicians and designers alike is the sense in which a pattern repeats or goes on for ever. It is quite easy to prove under mild restrictions that no such tiling can contain a regular pentagon and that no crystal can be dodecahedral. But tilings with pentagons are easy to find in the Islamic world, and in the 1980s Penrose created much excitement by showing how they could be introduced into mathematical tilings. Soon thereafter, dodecahedral crystals turned up; this is a nice scientific case of nature imitating art. All of this is discussed in the article by Marjorie Senechal and Jean Taylor on quasi-crystals.

The rich mixture of mathematical, artistic, and cultural themes is developed in an architectural setting in Kim Williams' article. An interesting complication of these investigations is the question of how the builders of the day could have carried out the actual constructions with the mathematics they knew. The patterns on a strip or frieze can also be intricate, so much more so than patterns on a line, so that friezes are sometimes said to have one and a half dimensions. As Peter Cromwell lucidly explains, they were much appreciated by Celtic designers; he also speculates on the question of classification.

Finally, there are articles on the third dimension. Here readers can enjoy a modern encounter between artists and mathematicians in which H. S. M. Coxeter, the doyen of "polyhedronists," describes the work of two sculptors using hollow triangles. There is a lot of life in these ancient traditions.

6. Geometry and Topology

Poincaré's conjecture ranks among the most famous conjectures in mathematics. It asserts that if a topological space has the same homotopy groups as a sphere, then it is homeomorphic to a sphere. Surprisingly, advances in algebraic topology made it possible in the mid-1960s to show that the conjecture is true in dimensions five and above. It is also trivially true in dimension two, but in dimensions three and four nothing was known.

Two results changed all that, opening up new fields for research as a result; they are described in Ronald J. Stern's article, which remains a fine introduction to the whole subject. First, Freedman characterized simply connected four-manifolds in terms of an algebraic object attached to their second homology group. Second, using techniques from gauge theory, Donaldson showed that if the simply connected four-manifold is smooth, then the associated algebraic object is very special. At once a series of results followed. Smooth four-manifolds could be classified in terms of what became known as their Donaldson invariants, a topic recently and powerfully redescribed by Ed Witten using quantum field theory (see [3]). In particular, the Poincaré conjecture is true for smooth four-manifolds. For their work, Freedman and Donaldson received Fields Medals in 1986.

The study of three-dimensional manifolds was revitalized in the 1970s by Bill Thurston, who developed a program (some of which still lacks proof) for imposing geometrical structures on them. The largest class of three-manifolds is the set of hyperbolic ones (ones that are modeled on non-Euclidean three-dimensional space), and the article by Linda Keen provides an introduction to this still-expanding subject. The prospect that this program will do for three-manifolds what Poincaré's insights in the 1880s eventually did for Riemann surfaces is an exciting one; one consequence would be a proof of the Poincaré conjecture in dimension three—but that remains tantalizingly out of reach.

In a different area, the classic subject of minimal surfaces also saw renewed interest in the 1980s, and *The Intelligencer* was pleased to carry one of the fruits of this, David Hoffman's use of high-quality computer graphics to suggest a proof that there are new, embedded minimal surfaces of genus greater than 1. The pictures themselves are things of beauty.

7. History of Mathematics

This history of mathematics has been a mainstay of *The Mathematical Intelligencer*. A number of interesting biographies have appeared, alongside articles drawing on the experience of producing editions of collected works (here, Kurt Gödel's) and on the microclimate of mathematics—life in departments and editorial boards. The frog-and-mouse story that D. van Dalen tells so vividly may not place Hilbert and Brouwer in the warmest of lights, but it does illuminate one of the fiercest struggles for influence over the foundations of mathematics that this century has seen. Meanwhile, in a very few pages, Jean-Michel Kantor provides a panorama of what Hilbert's famous twenty-three problems of 1900 have inspired.

The theme of community is prominent in O. M. Fomenko and G. V. Kuz'mina's article on the resolution of the Bieberbach conjecture, which does much more than explain what this famous result in geometrical complex function theory says and how it was finally proved after seventy years. It is also a vivid account of how research is done, of the interactions between researchers, of the role of suggestions and simplifications, of human psychology, and lastly of a fruitful international collaboration.

The Mathematical Intelligencer's success and attractiveness over the past twenty years is a tribute to the far-sightedness of its begetters Klaus and Alice Peters and others involved with its origination and continued publication. In particular, it is here that we have placed our memorial to Walter Kaufmann-Bühler, who saw the journal through to a secure future, only to die of an asthma attack. The journal has been fortunate in its editors, but no-one has done more for it than he did.

References

1. C. Reid, Being Julia Robinson's sister, *Notices Amer. Math. Soc.* 43 (1996), 1486–92.
2. D. Cox, J. Little, and D. O'Shea, *Ideals, Varieties, and Algorithms*, Springer-Verlag, New York, 1992.
3. J. W. Morgan, *The Seiberg-Witten Equations and Applications to the Topology of Smooth Four-Manifolds*, Mathematical Notes 44, Princeton University Press, 1996.

Part One

INTERVIEWS AND REMINISCENCES

1

An Interview with Michael Atiyah

Roberto Minio

Michael Atiyah was born in 1929 and received his B.A. and Ph.D. from Trinity College, Cambridge (1952, 1955). During his career he has been Savilian Professor of Geometry at Oxford (1963–69) and Professor of Mathematics at the Institute for Advanced Study in Princeton (1969–72); he is currently a Royal Society Research Professor of Mathematics at Oxford University. Since this article appeared he has been Master of Trinity College, Cambridge (1990–97), President of the Royal Society (1990–95), and Director of the Isaac Newton Institute, Cambridge.

FIGURE 1-1 *Michael Atiyah*

Among other honors Professor Atiyah is a Fellow of the Royal Society and a member of the National Academies of France, Sweden, and the USA. He received the Fields Medal at the International Congress of Mathematicians held in Moscow, 1966. His research interests span a broad area of mathematics including topology, geometry, differential equations, and mathematical physics.

The following is an edited version of an interview in Oxford with Roberto Minio, former editor of The Intelligencer.

MINIO: *I think some information about your background might be [valuable]. When did you start getting interested in mathematics? How early?*

ATIYAH: I think I was always interested in maths from a very young age. But there was a stage—I was about fifteen—when I got very interested in chemistry, and I thought that would be a great thing; after about a year of advanced chemistry I decided that it wasn't what I wanted to do and I went back to maths. I never seriously considered doing anything else.

MINIO: *And that was already noticeable very early?*

ATIYAH: Yes, I think so. My parents always thought that I was cut out to be a mathematician from a very young age, all the way through.

MINIO: *But they weren't mathematicians?*

ATIYAH: They weren't mathematicians, no.

Volume 6, No. 1 (Winter 1984), 9–19

MINIO: *And were you helped at school? Were your teachers reasonably good with you?*

ATIYAH: Well, I think I had good teachers and my relations with them were good. I was at school first in Egypt and I went to quite a good school there.

MINIO: *Were you born there?*

ATIYAH: No, I was born in England, but we lived in the Middle East—my father worked in the Sudan—so I went to school in Egypt for my main secondary schooling. I did a couple of years in England when I came over after the war—that was a good school. There were a lot of good pupils there. Then I went to Cambridge, and there were many good students around.

I don't think I was particularly influenced by one person. But I had a good education, and I had plenty of opportunity to meet good mathematicians; I had a good background in that sense.

MINIO: *Did you largely work on your own at Cambridge?*

ATIYAH: Well, I came to Cambridge after I had two years of National Service. It was a marvellous contrast. Actually, I went to Cambridge a little early in the academic year. I had a summer term there and the lovely weather and beautiful surrounding made a tremendous impression on me. I used to enjoy just going into the library to read, being surrounded by all of those books. It was an impressive atmosphere; it captured my imagination.

There were a lot of very bright students there and I got reasonable help from the teachers. I don't think any of the teachers was particularly inspiring; some of the lectures were good and some were not so good.

MINIO: *One of your early papers was with Hodge, is that right?*

ATIYAH: Yes, that was part of my thesis really. He was my research supervisor, and that was very important for me—working with him. I'd come up to Cambridge at a time when the emphasis in geometry was on classical projective algebraic geometry of the old-fashioned type, which I thoroughly enjoyed. I would have gone on working in that area except that Hodge represented a more modern point of view—differential geometry in relation to topology; I recognized that. It was a very important decision for me. I could have worked in more traditional things, but I think that it was a wise choice, and by working with him I got much more involved with modern ideas. He gave me good advice and at one stage we collaborated together. There was some recent work in France at the time on sheaf theory. I got interested in it, he got interested in it, and we worked together and wrote a joint paper which was part of my thesis. That was very beneficial for me.

MINIO: *One thing that's noticeable is that you worked quite a lot with other people—with Singer, with Hirzebruch, with Bott.*

ATIYAH: Yes, that's right, I work a lot with people, and I think that that's my style. There are various reasons, one of which is that I dabble in a number of different areas. My interest is in the fact that things in different subjects interact; it's very helpful to work with other people who know a bit more about something else and complement your interest. I find it very stimulating to exchange ideas with other people.

I've collaborated with many people, some of them—many of them—on an extended basis for many years. It's partly my personality, the way I think and interact with people, and partly because of the kind of mathematics I like doing, which is rather broad and therefore hard to be completely expert in. It is very helpful to have someone else who knows a bit more about something different. So when I work with Singer, for example, he's much stronger on the analysis side, where I was rather weak, and I know more about the algebraic geometry and topology.

MINIO: *Do you then separate out the problems?*

ATIYAH: No, no. The collaboration is completely intermingled; we merge our interests and learn about each other's techniques. After a while we're more on an equal footing in most parts of the subjects. Our interests are very close; it's just that our backgrounds are a bit different.

MINIO: *How do you select a problem to study?*

ATIYAH: I think that presupposes an answer. I don't think that's the way I work at all. Some people may sit back and say, "I want to solve this problem" and they sit down and say, "How do I solve this problem?" I don't. I just move around in the mathematical waters, thinking about things, being curious, interested, talking to people, stirring up ideas; things emerge and I follow them up. Or I see something which connects up with something else I know about, and I try to put them together and things develop. I have practically never started off with any idea of what I'm going to be doing or where it's going to go. I'm interested in mathematics; I talk, I learn, I discuss and then interesting questions simply emerge. I have never started off with a particular goal, except the goal of understanding mathematics.

MINIO: *Is that how K-Theory emerged?*

ATIYAH: Yes. It was very much an accident in some ways. I was interested in what Grothendieck had been doing in algebraic geometry. Having gone to Bonn, I was interested in learning some topology. I was interested in some of the questions that Ioan James had been studying on topological problems related to projective spaces. I found that by using Grothendieck's formulas these things could be explained, one got nice results. There was Bott's work on the periodicity theorem; I knew him and his work. Using this I found that one could solve some interesting problems. It seemed necessary to develop some machinery to make this formal, and *K*-Theory grew out of that.

You can't develop completely new ideas or theories by predicting them in advance. Inherently, they have to emerge by intelligently looking at a collection of problems. But different people must work in different ways. Some people decide that there is a fundamental problem that they want to solve, such as the resolution of singularities or the classification of finite simple groups. They spend a large part of their life devoted to working towards this end. I've never done that, partly because that requires a single-minded devotion to one topic which is a tremendous gamble.

It also requires a single-minded approach, by direct onslaught, which means you have to be tremendously expert at using technical tools. Now some people are very good at that; I'm not really. My expertise is to skirt the problem, to go around the problem, behind the problem . . . and so the problem disappears.

MINIO: *Do you feel there are mainstream topics in mathematics? Are some subjects more important than others?*

ATIYAH: Yes, well, I think that is true. I strongly disagree with the view that mathematics is simply a collection of separate subjects, that you can invent a new branch of mathematics by writing down axioms 1, 2, 3 and going away and working on your own. Mathematics is much more of an organic development. It has a long history of connections with the past and connections with other subjects.

Hardcore mathematics is, in some sense, the same as it has always been. It is concerned with problems that have arisen from the actual physical world and other problems inside mathematics having to do with numbers and basic calculations, solving equations. This has always been the main part of mathematics. Any development that sheds light on these topics is an important part of mathematics.

FIGURE 1-2 *Pictured from left to right: Montgomery, Spencer, de Rham, Mrs. Garding, Lars Garding, Chandra-sekharan, Bott, and Atiyah. Bombay, 1963.*

Parts that go off and are very far removed from these, and don't shed much light on the essentials of mathematics, are unlikely to be important. It may be that a new branch grows on its own and can eventually cast light on other things, but if it goes too far away and gets cut off, then it really isn't very significant in mathematical terms. There are really original ideas which may, for a while, open up new things, but still they are connected with other important parts of mathematics and interact. The importance of a part of mathematics is something one can judge roughly by the amount of interaction it has with other parts of the subject. It is a kind of self-consistent definition of importance.

MINIO: *Isn't it possible though that something in fact has no impact for quite a while, but then many years later is taken up?*

ATIYAH: Well, I think it is true that somebody may have a mathematical idea which is in advance of its time, and it may be that someone makes a clever suggestion and people don't see its significance for a long period of time. It obviously does happen.

I wasn't thinking quite so much of things like that. I was thinking more of the tendency today for people to develop whole areas of mathematics on their own, in a rather abstract fashion. They just go on beavering away. If you ask what is it all for, what is its significance, what does it connect with, you find that they don't know.

MINIO: *Do you feel like giving an example?*

ATIYAH: There are some examples in all parts of modern mathematics: some parts of abstract algebra, some parts of functional analysis, some parts of general topology—those parts where one sees the axiomatic method at its worst.

Axioms are designed to temporarily isolate a class of problems for which you can then develop techniques of solution. Some people think of axioms as a way of defining a whole area of mathematics that is self-contained. That I think is wrong. The narrower the axioms, the more you cut out.

When you abstract something in mathematics, you separate out what you want to concentrate on and what you regard as irrelevant. Now that may be convenient for a while; it concentrates the mind. But by definition, you have cut away a lot of things you said you're not interested in and, in the long run, that has cut a lot of roots. If you can develop something axiomatically, then at some stage you should return it to its origin, merging and producing cross-fertilization. That's healthy.

You will find views like these expressed by von Neumann and Hermann Weyl, some thirty years ago. They worried about the way mathematics might be going; if it goes too far away from its sources then it might become sterile. I think that is fundamentally correct.

MINIO: *It's clear that you have a strong feeling for the unity of mathematics. How much do you think that is a result of the way you work and your own personal involvement in mathematics?*

ATIYAH: It is very hard to separate your personality from what you think about mathematics. I believe that it is very important that mathematics should be thought of as a unity. And the way I work reflects that; which comes first is difficult to say. I find the interactions between the different parts of mathematics interesting. The richness of the subject comes from this complexity, not from the pure strand and isolated specialization.

But there are philosophical and social arguments as well. Why do we do mathematics? We mainly do mathematics because we *enjoy* doing mathematics. But in a deeper sense, why should we be paid to do mathematics? If one asks for the justification for that, then I think one has to take the view that mathematics is part of the general scientific culture. We are contributing to a whole, organic collection of ideas, even if the part of mathematics which I'm doing now is not of direct relevance and usefulness to other people. If mathematics is an integrated body of thought, and every part is potentially useful to every other part, then we are all contributing to a common objective.

If mathematics is to be thought of as fragmented specializations, all going off independently and justifying themselves, then it is very hard to argue why people should be paid to do this. We are not entertainers, like tennis players. The only justification is that it is a real contribution to human thought. Even if I'm not directly working in applied mathematics, I feel that I'm contributing to the sort of mathematics that can and will be useful for people who are interested in applying mathematics to other things.

Everybody has to try to justify his life philosophically, to himself at least. If you are teaching you can say, "Well, my job is to teach, I turn out educated young people and I am paid for that. Research I do in my spare time and they allow me to do that out of generosity." But if you're a full-time researcher, then you've got to think much harder about justifying your work.

In some sense, I still do mathematics because I enjoy doing it. I'm glad that people pay me to do what I enjoy. But I try to feel that there is a serious side to it which provides a justification.

MINIO: *What do you think about statements like "pure mathematics isn't very useful and within five years everybody will be just computing"?*

ATIYAH: There's always a danger in such a point of view. If pure mathematicians take an ivory tower attitude, don't think about their relationship with other subjects, there is the danger that people will eventually turn around and say, "We don't really need you—you are a luxury—and we will employ people doing much more practical things." I think that is a danger that is always there and becomes much more serious in times of financial difficulty, such as we are going through now. And I think the message is beginning to come through.

Certainly in the last five or ten years there has been quite a growing appreciation among pure mathematicians that they have to justify themselves a bit more. But I still think with many people this hasn't come naturally; they have only done it under pressure. I think it would be healthy if pure mathematicians in general were more self-critical.

MINIO: *Back to the mathematics that you do. Is there a theorem that you are most happy to have proved?*

ATIYAH: I think so. The Index Theorem I proved with Singer is in many ways the clearest single thing I've done. I really think the Index Theorem is a nice, clear theorem which one can point to. Most of my work centers around it in one form or another.

It started from work in topology and algebraic geometry, but then it has had quite an impact on functional analysis; over the last ten years this aspect has been developed by many people. And also it is now being realized that it has interesting connections with mathematical physics. So it's still developing and active in many ways. It symbolizes, in a way, my main interest which is the interactions and connections between all parts of mathematics. It is an area in which algebraic topology and analysis come together in a very natural way, along with differential equations in various forms.

MINIO: *Did you anticipate the recent revived interest in mathematical physics among mathematicians?*

ATIYAH: Not really. I have had an interest in mathematical physics for quite a long time. Not very deep—I tried to understand quantum mechanics and related topics. But what has happened in the last five years—the interest of mathematicians in Gauge theories—was unexpected for me. I didn't know enough physics to know it was likely. Quantum Field Theory was one of those big mystic phrases as far as I was concerned.

I think the physicists themselves were surprised. The fact that the geometrical aspect became significant and dominant wasn't predicted by many of them (and is still disputed by some!). The main problems looked as though they were different—analytical questions, algebraic problems. Some people like Roger Penrose weren't really surprised. They had been working in it from their own point of view for a long time anyway. But I think it's a nice example: if you do interesting, basic mathematical work in the mainstream, then you shouldn't be surprised when others find it a useful tool. It justifies one's belief in the unity of mathematics, including physics.

MINIO: *How far would you go with that statement?*

ATIYAH: The more I've learned about physics, the more convinced I am that physics provides, in a sense, the deepest applications of mathematics. The mathematical problems that have been solved or techniques that have arisen out of physics in the past have been the lifeblood of mathematics. And it's still true. The problems that physicists tackle are extremely interesting, difficult, challenging problems from a mathematical point of view. I think more mathematicians ought to be involved in and try to learn about some parts of physics; they should try to bring new mathematical techniques into conjunction with physical problems.

Physics is very sophisticated. It is tremendously mathematical and the combination of physical insight, on the one hand, and mathematical technique, on the other, is a very deep connection between the subjects. Newer applications of mathematics, say in the social sciences, economics, computing, are important. It is important that we turn out students with this view of applied mathematics because this is what is required in the commercial world; thousands and thousands of students require this.

On the other hand, from the point of view of the depth of mathematics involved, there is no comparison. Although there are interesting questions in, say, economics and statistics, broadly speaking, the depth of mathematics involved is very shallow. The really deep questions are still in the physical sciences. For the health of mathematics at its research level, I think it is very important to maintain that link as much as possible.

MINIO: *You are obviously interested in education. On the other hand, as far as your job is concerned, you are very clearly a research mathematician. How do you explain this?*

ATIYAH: My reasons for an interest in education are the same as my reasons for an interest in the unity of mathematics. Universities are institutions that are educational *and* involved in research. I think that is very important—there should be unity in the university and unity in the whole social structure that attempts to keep a broad balance between mathematical research and mathematical education. And when universities give courses for educational purposes, they should be sure that they are performing the right task for the students, not just giving courses in (say) advanced topology because they are interested in turning out research students. That's a disastrous mistake.

Universities must try to balance two activities. They ought to know what's useful for students to learn, bearing in mind what they are going to be doing later on. At the same time, they ought to foster research. Some people will be doing all research and some people will be mostly teaching, and mainly people will be in between. Although I am only involved with the research end of it, I live in the university, I have colleagues in the university, I know what they are involved with, so I am concerned to see that a proper balance is struck between the different functions of the university.

MINIO: *Do you think there was too much expansion of the universities in Britain during the last twenty years?*

ATIYAH: I don't think there was too much expansion. By comparison with other countries, particularly America, it is clear that the number of people going on to higher education was really very small and ought to have been increased; fundamentally it ought to continue to expand. I can't believe that for the next century the proportion of people getting higher education will stay the same as it is now. It is bound to change.

When you have a period of rapid expansion, which was necessary after the war, it does produce some problems. It produces a discontinuity. You recruit a lot of people to teach in the universities, and when the expansion stops you have filled up all of the positions—you can't appoint any young people. One can criticize the overenthusiasm or lack of caution which the universities had in not foreseeing some of the ultimate difficulties which were going to emerge. For example, unlike American universities, English universities in this period of expansion gave tenure appointments immediately after the Ph.D. because the universities were competing with each other. I think that was a mistake and they are paying for it now.

It would have been wiser not to have given people lifetime positions from the moment they arrived but to have had some more flexible system which enabled them to adapt. Now we are getting a very sharp discontinuity and it's going to produce crisis and confrontations. Perhaps people in the universities ought to be a bit more cautious.

MINIO: *Back for a moment to teaching and research. You talked about both being essential parts of the university life, but you talk about them still quite separately. There has been a growth of research institutes—Bonn, Warwick, Princeton—at which there is no teaching. Do you find that a healthy development?*

ATIYAH: I think the first thing to say about these kinds of institutions is that they have either no permanent staff or they have a very small permanent staff. Most of the people who go there are going there on refresher courses. They are going from the universities to spend a term or a year there and then they go back. So these are a kind of generalized conference center, where people get together to exchange ideas and then go back to carry on their work. They are simply helping to keep people in universities actively interested in their research—that is their main function.

On the other hand, if you had a system like Eastern Europe, where they have big research institutions employing large numbers of people on a permanent basis, siphoning off from the universities a significant percentage of their staff—that raises different problems. Then you really are separating the university from research in a big way. But in mathematics the number of these centers is so small, the number of staff is minute, and the people going and coming back are simply strengthening the university system. I think that is quite healthy.

They can also serve another purpose which is to help to orient or steer or guide people into profitable areas of mathematical thought. In addition to going to a center to boost your own current work, you can go to a center of this type as a young man in order to be led into productive areas of research.

The institute at Princeton, which is where I went after I took my Ph.D., serves that purpose very well. I had done my Ph.D., written my thesis, but I was still looking around for a niche in life mathematically. I didn't know where I was going, what I was going to be doing. I went to this large center where there were lots of very able young men, and older men, from different parts of the world with lots of different ideas. After a year or so there, I came away full of new ideas and new directions. That had an enormous bearing on my subsequent mathematical development.

MINIO: *Who was it at Princeton who influenced you most?*

ATIYAH: Well, I think it wasn't so much the permanent staff. I went in 1955 and many of the people there were probably slightly older than the comparable people who go now.

I met Hirzebruch, Serre, Bott, Singer . . . I met them all when I was at the institute. Kodaira and Spencer were there too. That whole group of people—I got to know them, and I was influenced by their mathematics. It's not an accident that I've collaborated subsequently with these same people I met at the institute.

I think there is another aspect. Not only do you alter your point of view and your work, but it puts you in contact with other active people and you keep up these personal contacts; they are very important in maintaining your mathematical development afterwards. Meeting people from different countries is important—mathematics is very international and these centers provide an opportunity which is very hard to get otherwise.

MINIO: *Conferences also provide an opportunity for meeting people, but maybe less of an opportunity for working together and really learning things?*

ATIYAH: Yes, conferences are very useful, but probably not so useful for the young person starting off. They are useful for the person who is established. If you know other people well already, and you are active, then in a very short time you can benefit from a rapid exchange of ideas. If you are a young student, or post-doctoral, you can't really talk to a lot of people because you don't really know them, you are inhibited, and also you don't understand enough to follow what they are saying. Then you need much longer exposure, I think. You need a year or so to absorb things slowly, to get to know people well. So I think conferences fulfill different functions.

MINIO: *What about the International Congress?*

ATIYAH: Well, I think the International Congress is totally different. I've been to every International Congress since 1954, I think; the benefit I have derived from those is very mixed.

The first one, which I attended as a young student, was great. I had a chance to hear Hermann Weyl give a talk and it was a tremendous psychological boost. I felt I was one of a large community of several thousand mathematicians. I didn't understand most of the talks. I'd go to them and be lost. I don't think I gained anything in terms of concrete mathematical understanding, but the psychological boost was substantial.

Now as I get older, the International Congress is of little value. I go out of a sense of duty—I have functions to perform—to talk to people, to give lectures. I don't really benefit because there are so many people. Some lectures I quite enjoy; I think the International Congress has some benefit, but not very much.

Besides the benefit they give to young people, giving them some sense of international identity, their main function is probably to help people from countries outside the small circuit of very active mathematical countries. If you come from Western Europe or the United States, they are probably not very essential. But if you come from Africa or Asia or Eastern Europe, where the opportunity for travelling and meeting people is much less, then I think it is the one chance you have of seeing what is going on. I suspect that is its main justification.

FIGURE 1-3 *M. Atiyah (right) talks to F. Hirzebruch (center). Michael Artin is pictured on left.*

MINIO: *Do you think that the Fields Medals serve a useful function?*

ATIYAH: Well, I suppose in some minor way. I think it's a good thing that Fields Medals are not like the Nobel prizes. The Nobel prizes distort science very badly, especially physics. The prestige that goes with the Nobel prizes, and the hooplah that goes with them, and the way universities buy up Nobel prizemen—that is terribly discontinuous. The difference between someone getting a prize and not getting one is a toss-up—it is a very artificial distinction. Yet, if you get the Nobel prize and I don't, then you get twice the salary and your university builds you a big lab; I think that is very unfortunate.

But in mathematics the Fields Medals don't have any effect at all, so they don't have a negative effect. They are given to young people and are meant to be a form of encouragement to them and to the mathematical world as a whole.

I was encouraged by getting a Fields Medal. It helped my self-confidence, my morale. I don't know whether if I hadn't got a medal it would have been any different, but certainly getting it at that stage gave me encouragement and made me enthusiastic. So I think in that sense it can help.

I found out that in a few countries the medals have a lot of prestige—for example, Japan. Getting a Fields Medal in Japan is like getting a Nobel prize. So when I go to Japan and am introduced, I feel like a Nobel prize winner. But in this country, nobody notices at all.

MINIO: *Do you find that mathematicians generally are treated significantly differently in different countries?*

ATIYAH: Well, mathematics can mean slightly different things in different countries, of course. Particularly the division between mathematics, applied mathematics, and physics is quite different in this country; in most other countries, pure mathematics is much more separated. This probably has a general effect on what people think about mathematicians; they don't identify them quite so narrowly here with pure mathematics as they do in America, where mathematician means pure mathematician.

Otherwise, I suppose it is true in France that mathematicians have a higher status, traditionally. That is because France has a tradition of assigning higher status to philosophy, literature, and the arts, and mathematics belongs to that group, whereas in this country, they never attach much weight to these things. In Germany also professors traditionally had a higher status, though that is now rapidly changing.

I think there are obvious national differences, about how people view mathematics or universities. But that is changing—the distinctions between different cultures are blurring.

MINIO: *I have a few questions here about how you work. For example, what sort of mental images do you use?*

ATIYAH: I'm not sure I know the answer. I think I do have a visual picture in my mind sometimes, some schematic diagram. But whether that is really a picture or whether it is purely symbolic I don't know. I think it is a very difficult question, more to do with the general nature of psychology than mathematics.

MINIO: *I suppose the question was meant to draw a distinction between geometrical intuition and algebraic manipulation.*

ATIYAH: Yes, there are differences there. I suspect that dichotomy is quite real in the brain. I work with things that are more geometrical but I'm not the sort of person like Thurston who sees complicated, multidimensional geometry in that way. My geometry is rather more formal. But I'm not an algebraist either—I don't just enjoy manipulations. Perhaps I am not sufficiently extremal for the psychology to be evident; I'm sort of the ordinary man in the middle.

If you ask Thurston, perhaps he says that he does see complicated pictures in his mind and all he has to do is draw it on paper to get the proof. Ask Thompson how he sees a group; I don't know what his answer is. There are differences. It is a complicated question, but it is three-quarters psychology and only one-quarter mathematics.

MINIO: *How important is memory for your work?*

ATIYAH: I mentioned that when I was fifteen I was very keen on doing chemistry. I did a whole year of chemistry, then I gave it up for the simple reason that in chemistry you have to memorize vast amounts of facts. There were big books I had in inorganic chemistry and I simply had to remember that you produced different chemicals out of different substances by various processes. The amount of structure that helped you to remember these things was infinitesimal. Organic chemistry was a bit better. Compared with this, in mathematics you needed practically no memory at all. You didn't need to memorize facts; all you needed to do was to have an understanding of the way the whole thing fitted together. So I think, in that sense, mathematicians don't really need the sort of memory that scientists or medical students do.

Memory is important in mathematics in a different way. I will be thinking about something and suddenly it will dawn on me that this is related to something else I heard about last week, last month, talking to somebody. Much of my work has come that way. I go around shopping, talking to people, I get their ideas, half understood, pigeon-holed in the back of my mind, I have this vast card-index of bits of mathematics from all of those areas. So I think memory plays a role in mathematics, but it is a different sort of memory than you would use in other areas.

MINIO: *When you're working do you know if a result is true even if you don't have a proof?*

ATIYAH: To answer that question I should first point out that I don't work by trying to solve problems. If I'm interested in some topic, then I just try to understand it; I just go on thinking about it and trying to dig down deeper and deeper. If I understand it, then I know what is right and what is not right.

Of course it is also possible that your understanding has been faulty, and you thought you understood it but it turns out eventually that you were wrong. Broadly speaking, once you really feel that you understand something and you have enough experience with that type of question through lots of examples and through connections with other things, you get a feeling for what is going on and what ought to be right. And then the question is: How do you actually prove it? That may take a long time.

The Index Theorem, for example, was formulated and we knew it should be true. But it took us a couple of years to get a proof. That was partly because different techniques were involved and I had to learn some new things to get the proof, in that case several proofs. I don't pay very much attention to the importance of proofs. I think it is more important to understand something.

MINIO: *Then what is the importance of a proof?*

ATIYAH: Well, a proof is important as a check on your understanding. I may think I understand, but the proof is the check that I have understood, that's all. It is the last stage in the operation—an ultimate check—but it isn't the primary thing at all.

I remember one theorem that I proved and yet I really couldn't see why it was true. It worried me for years and years. It had to do with the relationship between *K*-Theory and representations of finite groups. To prove the theorem I had to break up the group into solvable groups and cyclic groups; there were lots and lots of inductions and there were various bits on the way. In order for the proof to work, every single thing had to go just right—you had to be remarkably lucky, so to speak. I was staggered that it all worked and I kept thinking that if any one link of this chain were to snap, if there was some flaw in the argument, the whole thing would collapse. Because I didn't understand it, it might not be true at all. I kept worrying about it, and five or six years later I understood why it had to be true. Then I got an entirely different proof by going from finite groups to compact groups. Using quite different techniques, it was quite clear why it had to be true.

MINIO: *Do you see some way of communicating this understanding to someone without the proofs?*

ATIYAH: Well, I think ideally as you are trying to communicate mathematics, you ought to be trying to communicate understanding. It is relatively easy to do this in conversation. When I collaborate with people, we exchange ideas at this level of understanding—we understand topics and we cling to our intuition.

If I give talks, I try always to convey the essential ingredients of a topic. When it comes to writing papers or books, however, then it is much more difficult. I don't tend to write books. In papers I try to do as much as I can in writing an account and an introduction which gives the ideas. But you are committed to writing a proof in a paper, so you have to do that.

Most books nowadays tend to be too formal most of the time, they give too much in the way of formal proofs, and not nearly enough in the way of motivation and ideas. Of course it is difficult to do that—to give motivation and ideas.

There are some exceptions. I think the Russians are an exception. I think the Russian tradition in mathematics has been less formalized and structured than the Western tradition, which is under the influence of French mathematics. French mathematics has been dominant and has led to a very formal school. I think it is very unfortunate that most books tend to be written in this overly abstract way and don't try to communicate understanding.

But it is hard to communicate understanding because that is something you get by living with a problem for a long time. You study it, perhaps for years, you get the feel of it and it is in your bones. You can't convey that to anybody else. Having studied the problem for five years, you may be able to present it in such a way that it would take somebody else less time to get to that point than it took you, but if they haven't struggled with the problem and seen all the pitfalls, then they haven't really understood it.

MINIO: *Where do you get your ideas for what you are doing? Do you just sit down and say, "All right, I'm going to do mathematics for two hours"?*

ATIYAH: I think that if you are actively working in mathematical research, then the mathematics is always with you. When you are thinking about problems, they are always there. When I get up in the morning and shave, I'm thinking about mathematics. When I have my breakfast, I am still thinking about my problems. When I am driving my car, I am still thinking about my problems. All in various degrees of concentration.

Sometimes you wonder whether it is worthwhile thinking about it while you are doing these things, whether it really helps. You are just turning it over idly in your mind.

There are occasions when you sit down in the morning and start to concentrate very hard on something. That kind of acute concentration is very difficult for a long period of time and not always very successful. Sometimes you will get past your problem with careful thought. But the really interesting ideas occur at times when you have a flash of inspiration. Those are more haphazard by their nature; they may occur just in casual conversation. You will be talking with somebody and he mentions something and you think, "Good God, yes, that is just what I need . . . it explains what I was thinking about last week." And you put two things together, you fuse them and something comes out of it. Putting two things together, like a jigsaw puzzle, is in some sense random. But you have to have these things constantly turning over in your mind so that you can maximize the opportunities for random interaction. I think Poincaré said something like that. It is a kind of probabilistic effect: ideas spin around in your mind and the fruitful interactions arise out of some random, fortunate mutation. The skill is to maximize this degree of randomness so that you increase the chances of a fruitful interaction.

From my point of view, the more I talk with different types of people, the more I think about different bits of mathematics, the greater the chance that I am going to get a fresh idea from someone else that is going to connect up with something I know.

For example, the Index Theorem was partly a matter of chance. Singer and I happened to be working at Oxford on things related to the Riemann-Roch Theorem, coming out of Hirzebruch's work. We were playing around and we had the idea of looking for a formula with the Dirac operator. And then Smale passed through and we talked to him. He told us that he had read a paper just the other day by Gel'fand, which had been about the general question of the index of operators, and he suggested that it might have some relation to what we were doing. I found this paper very difficult to understand, but it was the general formulation of the problem and we were looking at an important special case. Then we realized that we had to generalize what we were doing and that led to the whole thing. But it was Smale passing through that put us on the right track.

Another example is the work I did on instantons. That was also a bit of a chance. I knew that Roger Penrose and his group were working on geometrical aspects of physics and one of them named Richard Ward was doing some nice things. He was giving a seminar and I asked myself, "Should I go to it or not, would it be a bit of a bore? Well, all right." So I went to the seminar. It was a very clear seminar; I understood what he was doing, and I came away saying, "Gee, that's really good." I went back and spent three days thinking very hard about it, and suddenly it dawned on me what was going on, how it related to algebraic geometry. From then on, the thing took off. I could easily have not gone to that seminar and the subject would still be where it was. The gap between the mathematicians and physicists was very big; I doubt whether the ideas would have been taken up nearly so rapidly. But of course, a lot of time you go to seminars and don't get ideas.

MINIO: *Do you have a favorite theorem or problem?*

ATIYAH: That's not such a serious question because I don't really believe in theorems, per se. I believe in mathematics as sort of an entity; a theorem is just a staging post. I know lots of nice nuggets, nice facts, nice things, but I don't attach much importance to them individually. I guess the same goes for problems.

I don't want to give the impression that I think of mathematics as simply abstract theory with no body to it. A theory is interesting because it solves lots of special problems and puts them in proper context; it enables you to understand them all. Quite often a theory evolves because somebody solved some very hard problem first and then you try to understand what is going on—you build a super-structure around it. Soft theory which has no hard problems in it is of no use.

MINIO: *What is your feeling about the classification of finite simple groups?*

ATIYAH: Well, I have slightly mixed feelings about that. First of all, it takes so many pages to prove; it seems to me the degree of understanding must be pretty limited if that is the only way it can be done. One would hope that a much better understanding of all this would emerge. Maybe I am wrong in this, but I believe that if it is going to emerge, it is going to come from people who look at groups from an extroverted rather than introverted way.

As groups arise in nature, they are things that move things around, they are transformations or permutations. In the abstract version, you think of the group as an internal structure with its own multiplication—that's a very introverted point of view. If you only allow yourself to use the intro-verted point of view, you have a very limited array of techniques. But if you think of the group through its manifestations, from the outside world, then you have the whole of the outside world to help you. And you have got, or ought to have, a much more powerful understanding. My idea, my dream, is that one should prove these deep theorems about groups by using the fact that the group occurs in some natural context as a group of transformations, and then out of this the structure should become transparent.

Also, I am not entirely sure how important this whole result is. Some people will say that in mathematics the most important thing is to set up an axiom system with axioms 1, 2, 3. There are objects—groups, spaces. The problem is then to classify all such objects. I don't think that is the right point of view. The goal is to understand the nature of these things and to use them; classification sim-ply gives you an indication of the scope of the theory.

For example, the classification of Lie groups is a bit peculiar. You have this list of groups, both classical and exceptional. But for most practical purposes, you just use the classical groups. The exceptional Lie groups are just there to show you that the theory is a bit bigger; it is pretty rare that

FIGURE 1-4 *Michael Atiyah (left) and Laurent Schwartz (right) in Japan.*

they ever turn up. And the theory of Lie groups would not be very different if the classification had been vastly more complicated, if there had been infinitely many more of these exceptional groups.

So I don't think it makes much difference to mathematics to know that there are different kinds of simple groups or not. It is a nice intellectual endpoint, but I don't think it has any fundamental importance.

MINIO: *But if there were another way of doing it, some extroverted point of view, would that have more of an impact?*

ATIYAH: That would have an impact in the sense that it would show people that one can do things in some other way. But I don't think the result has fundamental importance; it doesn't compare with, say, the theory of group representations.

The classification point of view can be greatly exaggerated. It is a focus for a while, it points out nice problems and challenges. But if it takes a lot of effort, one suspects that there ought to be better ways to do it. Getting a better way to do it may in itself be interesting and show new ideas and new techniques at work. You see this result, it looks nice, you've got this long complicated proof, it is a challenge to find better ways to prove it. The search may be beneficial, but the benefits come more from the new ideas than from the fact that you've got the new proof.

George Mackey once said to me something that I think is very true. In a given area of mathematics, the things that are important are quite often not the most technically difficult parts—the hardest things to prove. They are quite often the more elementary things because those are the parts that have the widest interaction with other fields and other areas—they have the widest impact.

There are many things about group theory which are tremendously important and occur in all sorts of mathematics everywhere. Those tend to be the more elementary things: basic ideas about groups, about homomorphisms, about representations. General features, general techniques—those are the things that are really important.

The same is true in analysis. There are some very fine points in proving exactly under which conditions a Fourier series may converge; these are technically demanding, very interesting. But for the rest of mathematicians who use Fourier analysis, they are not really important. The specialists in a field, of course, get enamored of the hard technical problems. But from the point of view of mathematicians as a whole, while they admire them, they don't use them.

MINIO: *Who is your most admired mathematician?*

ATIYAH: Well, I think that is rather easy. The person I admire most is Hermann Weyl. I have found that in almost everything I have ever done in mathematics, Hermann Weyl was there first. Most of the areas I have worked in were areas where he worked and did pioneering, very deep work himself—except topology, of course, which came after his time. But he had interests in group theory, representation theory, differential equations, spectral properties of differential equations, differential geometry, theoretical physics; nearly everything I have done is very much in the spirit of the sort of things he worked in. And I entirely agree with his conceptions about mathematics and his view about what are the interesting things in mathematics.

I heard him at the International Congress in Amsterdam. He gave the Fields Medals there, to Serre and Kodaira. Then I went to the Institute at Princeton, but he was in Zurich at the time and he died there. I never saw him at Princeton, I only saw him that one time. So it wasn't personal contact that makes me admire him.

For many years whenever I got into a different topic I found out who was behind the scene, and sure enough, it was Hermann Weyl. I feel my center of gravity is in the same place as his. Hilbert was more algebraical; I don't think he had quite the same geometrical insights. Von Neumann was more analytical and worked more in applied areas. I think Hermann Weyl is clearly the person I identify with most in terms of mathematical philosophy and mathematical interests.

2

On the Steps
of Moscow University

Steve Smale

From the front page of the *New York Times*, Saturday, August 27, 1966:

Moscow Silences a Critical American
By Raymond H. Anderson

Special to the *New York Times*

MOSCOW. Aug. 26—A University of California mathematics professor was taken for a fast and unscheduled automobile ride through the streets of Moscow, questioned and then released today after he had criticized both the Soviet Union and the United States at an informal news conference.

Speaking on the steps of the University of Moscow, the professor, Dr. Stephen Smale . . .

August 15, 1966, was a typical hot day in Athens and I was alone at the airport. My wife and children had just left by car for the northern beaches. My plane to Moscow was to depart shortly.

I had arrived in Paris from Berkeley at the end of May. It was a great pleasure to see Laurent Schwartz again. Though tall and imposing, Schwartz was soft in voice and demeanor. He was a man of many achievements. As a noted collector of butterflies, he had visited jungles throughout the world; Schwartz was an outstanding mathematician; he was also a leader of the French left. With Jean-Paul Sartre, he had been in the forefront of intellectuals opposing the French war in Algeria. He was accused of disloyalty, and his apartment bombed. More recently, Schwartz and I had been in frequent correspondence about internationalizing the protest against the Vietnam War. He had asked me to speak at "Six Hours for Vietnam" which he and other French radicals had organized.

The Salle de la Mutualité—I remembered that great old hall on the left bank for the political rallies I had attended there fifteen years earlier. Now there were several thousand exuberant young people in the audience, and I was at the microphone. My French was poor and since my talk could be translated, I decided to speak in English. I still wasn't at ease giving non-mathematical talks; even though I had scribbled out my brief talk on scratch paper, I was nervous.

Volume 6, No. 2 (Spring 1984), 21–27

There was a creative tension in that atmosphere that inspired me to communicate my feelings about the United States in Vietnam. As I was interrupted with applause my emotional state barely permitted me to give the closing lines; ". . . As an American, I feel very ashamed of my country now, and I appreciate very much your organizing and attending meetings like this. Thank you." Then I was led across the stage to M. Vanh Bo, the North Vietnamese representative in Paris; we embraced and the applause reached its peak.

The next day I eagerly read the news story in *Le Monde*; and Joseph Barry, "datelined Paris", in the *Village Voice* June 2, 1966, gave an account of the meeting:

> . . . The dirty war was one's own country's and several thousand came to the Mutualité to condemn it. "Six Hours for Vietnam" it was titled. Familiar names, such as Sartre, Ricoeur, Schwartz, Vidal Naquet, were appended to the call. And it was organized by the keepers of the French conscience: French teachers and students.
>
> There was some comfort in the presence of American professors—from Princeton, Boston, and Berkeley. Indeed Professor Smale of the University of California got almost as big a hand as Monsieur Vanh Bo of North Vietnam. Such is the sentimentality of the left. Even the chant that followed Smale's speech—"U.S. go home! U.S. go home!"—was obviously a left-handed tribute. . . .

After that I very quickly turned to the world of mathematics. I remember spending the next evening at dinner with René and Suzanne Thom. René Thom was a deeply original mathematician who was to become known to the public for his "catastrophe theory." A few days later René and I drove together to Geneva to a mathematics conference, and I to a rendezvous with my wife, Clara, and young children, Laura and Nat.

Both Schwartz and Thom had won the Fields Medal. Two—sometimes four—of these medals are awarded at each International Congress of Mathematics (held every four years). It is the most esteemed prize in mathematics, and I had been greatly disappointed in not getting the Fields Medal at the previous International Congress in Stockholm in 1962. That disappointment led me to deemphasize the importance of the Fields Medal; I was (and I still am) conscious of the considerations of mathematical politics in its choice. Thus I showed little concern about getting the medal at the Congress that summer in Moscow. Nevertheless, when René Thom told me during the ride to Geneva that I was to be a Fields medalist at the Moscow Congress, it was a great thrill. Thom had been on the awarding committee and was giving me informal advance notice. Georges de Rham notified me officially in Geneva a few days later.

The time in Geneva was beautiful. It included intense mathematical activity, seeing old friends, and Alpine hikes with my family.

Medal or no medal, I had been planning to go to the Moscow conference where I was to give an hour-long address. Clara and I had ordered a Volkswagen camper to be delivered in Europe, and we drove it from Geneva to Greece via Yugoslavia, camping with our kids on the route. The plan was for Clara and the kids to stay in Greece while I was in Moscow. They would meet me at the airport on my return. We would go by auto-ferry to Istanbul, where we had reservations at the Hilton Hotel (a change from camping!). The return was to feature an overnight roadside stop in the region of Dracula in Transylvania.

It was our first visit to Greece and we enjoyed the beaches, visited monasteries at Meteora, and traveled to the isle of Mykonos. I was especially moved by the home of the oracle at Delphi, I knew it well from stories my father had read to me when I was a child.

Now at the Athens airport, I reviewed those beautiful memories; tomorrow I would be receiving the Fields Medal in front of thousands of mathematicians gathered in Moscow from all over the world.

At first I was only slightly annoyed when the customs official stopped me, objecting to something about my passport. I knew my passport was OK; what was this little hassle about? Slowly I began to understand. When we had come across the Greek border, customs had marked in my passport that we were bringing in a car. (The government was concerned that we would sell the car without paying taxes.) Now the Greek officials were not letting me leave the country without that car. The customs officials were adamant. Unfortunately, the car was out of reach for the next ten days. The plane left the ground—it was the only one to Moscow until the next day. My heart sank. There was no possibility of getting to Moscow in time to get my medal.

The American embassy was closed, but after some frantic efforts, I managed to find a helpful consular official, Mr. Paul Sadler. He became convinced of my story, waived the protocol of the embassy, and wrote to the Director of Customs:

> Dear Sir:
>
> As explained in our telephone conversation with Mr. Psilopoulos. . .
>
> The Embassy is aware of the customs regulations which require an automobile imported by a tourist to be sealed in bond if the tourist leaves the country without exporting the automobile. Unfortunately, Mr. Smale's wife has departed Athens for Thessaloniki with the car and Mr. Smale is unable to deliver the car to the customs.
>
> Since it is of such urgency that Mr. Smale be in Moscow no later than the afternoon of August 16, the Embassy would greatly appreciate your approval to permit Mr. Smale to leave Greece while his car remains in the Country. The Embassy will guarantee to the Director of Customs any customs duties which may be due on the automobile in the event Mr. Smale does not return to Greece and export the car by the end of August, 1966. . . .

I was able to take the plane the next day. At a stop in Budapest, Paul Erdős, a Hungarian mathematician I knew, boarded the plane. Erdős gave me the startling news that the House Un-American Activities Committee had issued a subpoena for me to testify in Congressional hearings. I guessed immediately it was because of my activity on the Vietnam Day Committee in Berkeley; I had been co-chairman with Jerry Rubin. We had organized Vietnam Day and made an attempt to stop troop trains. Most of all, there were the days of protest—the "International Days of Protest."

I arrived late in Moscow and rushed from the airport to the Kremlin where I was to receive the Fields Medal at the opening ceremonies of the International Congress. Without a registration badge the guards at the gate refused me admission into the palace. Finally, through the efforts of a Soviet mathematician who knew me, I obtained entrance and found a rear seat. René Thom was speaking about me and my work:

> . . . si les oeuvres de Smale ne possedent peut-etre pas la perfection formelle du travail definitif, c'est que Smale est un pionnier qui prend ses risques avec un courage tranquille; . . .

After Thom's talk I found a letter from my friend, Serge Lang, waiting for me. He wrote that the *San Francisco Examiner* (August 5) had written a slanderous article about me and added ". . . I hope you sue them for everything they are worth . . . I promise you 2,000 dollars as help in a court fight . . ." The article started as follows:

UC Prof Dodges Subpoena, Skips U.S. for Moscow

Stephen Smale, Supporter of VDC-FSM

By ED MONTGOMERY, Examiner Staff Writer

Dr. Stephen Smale, University of California professor and backer of the Vietnam Day Committee and old Free Speech Movement, is either on his way or is in Moscow, The Examiner learned today.

In leaving the country, he has dodged a subpoena directing him to appear before the House Committee on Un-American Activities in Washington.

One of a number of subpoenaed Berkeley anti-war activist[s], Dr. Smale took a leave of absence from UC and leased his home there before his trip abroad.

And it went on to quote a University of California spokesman:

According to press reports the committee has subpoenaed several persons. No Berkeley *student* [my emphasis] was included . . .

Not only was I under fire from HUAC, but my University, it seemed, wouldn't even acknowledge me. On the other hand, my friends and the many other mathematicians came to my defense. Serge, who was in Berkeley at the time, and my department chairman, Leon Henkin, acted quickly to clear my name. Supportive petitions were circulated in Berkeley and Moscow. An article in the *San Francisco Chronicle* (August 6) which followed the *Examiner* was terrific. (See Figure 2-1.)

Next, I received a letter from Beverly Axelrod, which further lifted my spirit. (See Figure 2-2.) I took special pleasure at her addressing me as "Provocateur Extraordinaire"; and that preceded my most provocative act by a week!

The HUAC hearings in Washington were the most tumultuous ever. Jerry Rubin appeared in costume as an American Revolutionary soldier; members of the Maoist Progressive Labor party proudly proclaimed themselves communists. In the *New York Times,* August 18 (1966), Beverly Axelrod was quoted as saying that she felt in physical danger staying after lawyer Arthur Kinoy was dragged out of the hearings by court police.

Another *Times* article *that day* was captioned:

AWARD IN MOSCOW BALKS HOUSE
WEST COAST MATHEMATICIAN UNABLE TO TESTIFY

A year earlier I would have welcomed the chance to testify in Washington. Now, politics was playing a secondary role to mathematics, and so it was just as well to have missed the subpoena. I had the best of both worlds, the "prestige" of having been called, but not the problem of going.

In Moscow the scientific aspects of the Congress were quite exciting. I renewed friendships with the Russians, Anosov, Arnold and Sinai, whom I had met in Moscow in 1961. The four of us were working in the fast-developing branch of mathematics called dynamical systems and had much to discuss in a short time.

I became caught up in the social life of a group of non-establishment Muscovites, visiting their homes and attending their parties. These Russians were scornful of communism, sarcastic about their government, and, in the privacy of their own company, free and uninhibited. This was quite a change from the serious, cautious, apolitical manner of Soviet mathematicians to which I had become accustomed. Considering that dissent was being punished at that time by prison (recall Daniel and Sinavsky), it was surprising to find how open they were in my presence.

I loved the Russians in that crowd. Nevertheless, we argued heatedly about Vietnam. Some of those dissidents were so anti-communist that they wished for an American victory in Vietnam. Chinese communism was akin to Stalinism and they hoped that the Americans could stop both Soviet and Chinese expansion.

There was political agitation among the 5,000 mathematicians in Moscow on two fronts. One aim was to obtain condemnation of the United States intervention in Vietnam, and I was quite active in that movement. The other aim was to pressure the International Congress to condemn HUAC's attempt to subpoena me. A *Times* story, datelined Moscow, August 21, led off with:

> · SCIENTISTS URGED TO CONDEMN U.S. WAR ROLE
> AND HOUSE INQUIRY TARGETS AT WORLD PARLEY

UC Prof's Subpoena May Boomerang on Probers

By Jack Smith

The attempt to subpoena a brilliant University of California mathematician for anti-war activities may boomerang on the House Un-American Activities Committee, The Chronicle was told yesterday.

"American [sic] is going to be the laughing stock of the world for treating such a way [sic] in that way," his colleague Dr. Serge Lang declared.

The highly respected mathematician, Professor Stephen Smale, was named along with seven others in the Bay Area to appear before the Congressional committee on August 16 because of anti-Vietnam war activities.

Subpoenas were served on a number of those involved Thursday, but Professor Smale, a one-time Vietnam Day Committee leader, was out of the country.

REPORT

Refuting a somewhat exaggerated published report that the 36-year-old professor had "dodged" the subpoena, colleagues noted that he had applied for a leave of absence at least six months ago. He left here in June and spent two months at the University of Geneva in Switzerland.

At present, according to Dr. Lang, a visiting professor of mathematics at UC from Columbia University, Dr. Smale is driving with his wife and two children from Geneva to Moscow for the International Congress on Mathematics, scheduled August 16 through 26.

During the congress, Dr. Smale will be given the Field[s] Medal, the highest honor in mathematics and comparable to the Nobel Prize.

Smale was one of the organizing members of Berkeley's Vietnam Day Committee, which began as the sponsor of a 36-hour "teach-in" in May, 1965. Its anti-war activities continued afterward, and Smale was one of the leaders of demonstrations seeking to halt troop trains last fall and took part in campus rallies.

FIGURE 2-1

BEVERLY AXELROD, ATTORNEY AT LAW

345 FRANKLIN STREET   TELEPHONE
SAN FRANCISCO 2, CALIFORNIA   MARKET 1-3968

August 14, 1966

Dr. Steven Smale
Provocateur Extraordinaire
Moscow, U.S.S.R.

Dear Steve:

THIS lawsuit is not just another legal exercise; it offers the possibility of really deal-ing a death blow to HUAC. Most, if not all, the others who have been subpoenaed will be parties to this action, and we intend to get "prestige" names to be added afterwards. Yours would be the most important name, for obvious reasons. We hope you will agree and authorize the use of your name as plaintiff.

Also, as I explained to Leon Henkin, we would like to be able to use the names of as many of your colleagues as are willing to participate. The theory is that everyone suffers an injury from the activities of the Committee, and particularly those who, because of their profession, feel a kinship with you.

I'm working on this in association with a number of New York lawyers, including Kuntsler & Kinoy, and national A.C.L.U., which is footing the bill.

Please read the enclosed complaint, and, as soon as possible, at any hour, phone me, or Arthur Kinoy, or William Kunstler, at the Congressional Hotel, Wash. D.C. phone 202 LI 6-6611. (Collect)

We can take telephone authorization for use of names, to be followed up by a letter similar to the one enclosed.

Hope your stay in Moscow is a great experience—I envy you.

Best regards,
Beverly

FIGURE 2-2

The *New York Times* story caught the attention of the State Department which became concerned enough to initiate an investigation. The National Academy of Sciences, Office of the Foreign Secre-tary, cooperated by contacting the official delegates at the Congress in Moscow:

... we would appreciate any comments you may be able to make to us about the political sub-jects raised in the context of the Congress which we in turn could pass on to the Department of State ...

Indeed, the writing and circulation of a petition was in progress:

... We the undersigned mathematicians from all parts of the world express our support for the Vietnamese people and their right to self-determination, and our solidarity with those American colleagues who oppose the war which is dishonoring their country.

FIGURE 2-3 *Steve Smale in Moscow, 1967.*

This activity brought me together with Laurent Schwartz and Chandler Davis once more. Davis had served a prison term stemming from his HUAC appearance back in Michigan. Unable to find a regular job in the United States, he had become professor of mathematics at the University of Toronto. I had been friends with Chandler as a student at Ann Arbor and had been present at his HUAC hearings.

The three of us were invited by four Vietnamese to a banquet on Tuesday, August 23, in appreciation for our opposition to the war. Without any assistance from the Russians, these Hanoi mathematicians put a great amount of energy into preparing a fine Vietnamese feast under difficult conditions; the banquet took place in a dormitory. I gained a convincing picture of the struggle that these particular Vietnamese were going through, continuing to do their mathematics while American bombs fell on their capital city. And for the first time I could see directly the meaning of the American anti-war movement for the Vietnamese. Even though our language and culture were different, much warmth flowed between us.

At that banquet, I was asked to give an interview to a reporter named Hoang Thinh from Hanoi. I didn't know what to say and struggled with the problem for the next day. I felt a great debt and obligation to the Vietnamese—after all, it was my country that was causing them so much pain. It was my tax money that was supporting the U.S. Air Force, paying for the napalm and cluster bombs. On the other hand, I was a mathematician, with compelling geometrical ideas to be translated into theorems. There was a limit to my ability to survive as a scientist and weather further political storms. I was conscious of the problems that could develop for me from a widely publicized anti-U.S. interview given to a Hanoi reporter in Moscow. In particular, I knew that what I said might come out quite differently in the North Vietnamese newspaper, and even more so when translated back into the U.S. press.

This was the background for my rather unusual course of action. On the one hand, I would give the interview; on the other hand, I would ask the American reporters in Moscow to be present so that my statements could be reported more directly. I would give a press conference on Friday morning, August 26.

The mechanics for holding such a press conference in Moscow were not simple. I surely wasn't going to become dependent on the Russians by asking for a room. Since Moscow University was the location for the Congress, it was natural to hold the interview on the steps of this university. Out of respect for my hosts, I invited the Soviet press. There was obviously a provocative aspect to what I was doing, so for some protection, I asked a number of friends to be present, including Chandler Davis, Laurent Schwartz, Leon Henkin, and a close friend and colleague, Moe Hirsch. I was already scheduled to leave for Greece early the following morning.

The interview with Hoang Thinh was now set up under my conditions. Then late Thursday I received word that Thinh wouldn't be present! His questions were given me in writing and I was to return my answers in writing. I began to draft a statement.

As I wrote down my words attacking the United States from Moscow, I felt that I had to censure the Soviet Union as well. This would increase my jeopardy, but, having just received the Fields Medal and being the center of much additional attention because of HUAC, I was as secure as anybody. If *I* couldn't make a sharp anti-war statement in Moscow and criticize the Soviets, who could?

The Congress was coming to a close and my rendezvous with the journalists was close at hand. Amidst my friends and all the mathematicians I felt a certain loneliness. No one was giving me encouragement for what I was doing now. Chandler showed some sympathy, and Thursday night the two of us walked together locked in discussion and debate.

That morning, Friday, August 26, as the hour arrived, the main question on my mind was: How would the Soviets react to my press conference? Everything so far seemed to be going smoothly as the participants assembled on those broad steps with the enormous main building of the University behind. A woman from a Soviet news agency asked if she could interview me personally, alone. I said yes, but afterwards. Then I read the following statement:

> This meeting was prompted by an invitation to an interview by the North Vietnamese Press. After much thought, I accepted, never having refused an interview before. At the same time, I invited representatives from Tass and the American Press, as well as a few friends.
>
> I would like to say a few words first. Afterwards I will answer questions.
>
> I believe the American Military Intervention in Vietnam is horrible and becomes more horrible every day. I have great sympathy for the victims of this intervention, the Vietnamese people. However, in Moscow today, one cannot help but remember that it was only 10 years ago that Russian troops were brutally intervening in Hungary and that many courageous Hungarians died fighting for their independence. Never could I see justification for Military Intervention, ten years ago in Hungary or now in the much more dangerous and brutal American Intervention in Vietnam.
>
> There is a real danger of a new McCarthyism in America, as evidenced in the actions of the House Un-American Activities Committee. These actions are a serious threat to the right of protest, both in the hearings and in the legislation they are proposing. Again saying this in the Soviet Union, I feel I must add that what I have seen here in the discontent of the intellectuals on the Sinavsky-Daniel trial and their lack of means of expressing this discontent, shows indeed a sad state of affairs. Even the most basic means of protest are lacking here. In all countries it is important to defend and expand the freedoms of speech and the press.

After I had read my prepared remarks, a woman, whom I didn't recognize, approached me to say that I was wanted immediately by Karmonov, the organizing secretary of the Congress. Keeping cool, I replied, "OK, but wait till I finish here." There were a few more questions from the American correspondents, which I answered. Now there were the two remaining pieces of business, which seemed to be converging. The woman from the Congress Committee reappeared to lead me to Karmonov, and the Soviet reporter was content to follow along. Sensing a story, the American press followed, as did Moe Hirsch and some other friends.

Karmanov was very friendly; no mention of politics or the press conference came up. It was an unhurried conversation about my impressions of Moscow, inconsistent with the urgency with which I had been called. Karmonov was quite solicitous, and gave me a big picture book about the treasures of the Kremlin (in German!). He next wished to accommodate me in seeing the museums and other touristic aspects of Moscow. A car and guide were put at my disposal. At that point I had no interest whatsoever in sightseeing. However, I had promised the woman from the Soviet press a per-

FIGURE 2-4 *Steve Smale speaks at the Chicago National Convention, 1968.*

sonal interview and reluctantly acceded to go with her by car, not at all clear who was going, where we were going or why. I felt pressured and a little scared. But all the while I was treated not just politely, but like a dignitary. It was hard to resist.

As several of us left Karmonov's office, the American reporters were waiting, and asked me what was happening. I really didn't know; in any case, I had no chance to answer as I was rushed out into a waiting car. The newsmen were pushed aside by Russians on each side of me. Moe yelled out, "Are you all right, Steve?" I only had time for a hurried "I think so," before the car was driven off at high speed with the American press in hot pursuit. I was to read later in the newspapers how most, but not all, of the American reporters were eluded. The car stopped at the headquarters of the Soviet news agency, Novosti, and I was taken inside. The same kind of red carpet treatment continued. The top administrators showed me around, but there was no interview. My notes were copied, but otherwise my visit to Novosti seemed to have no purpose. We were just passing the time of day, and, finally, I insisted that I be returned to the Congress activities.

I felt an enormous release of tension as I rejoined my colleagues at the closing reception and ceremonies of the International meeting. There had been a lot of concern for me, and friends told me: "Stay in crowds." With trepidation I returned alone after midnight to my room at the Hotel Ukraine. The phone was ringing. I. Petrovskii, the president of the Congress, wanted to see me the first thing next morning. That was impossible since I was leaving by air at 7:00 A.M. The phone rang again. This time it was the American Embassy. Was I all right? Did I need any assistance? I replied that I probably didn't need any help.

Indeed, I was on that 7:00 A.M. plane. Clara, Nat and Laura had been enjoying the sea and sun during this time and were unaware of the mix-up with customs and the events in Moscow. It certainly was a welcome sight to see them at the Athens airport.

3

An Interview with Jean-Pierre Serre

C. T. Chong and Y. K. Leong

Jean-Pierre Serre was born in 1926 and studied at the École Normale Supérieure in Paris. He was awarded a Fields Medal in 1954 and has been Professor of Algebra and Geometry at the Collège de France since 1956.

Professor Serre visited the Department of Mathematics, National University of Singapore, during February 1985 under the French-Singapore Academic Exchange Program. In addition to giving several lectures organized by the Department of Mathematics and the Singapore Mathematical Society, he was also interviewed by C. T. Chong and Y. K. Leong on 14 February 1985.

Q: *What made you take up mathematics as your career?*

A: I remember that I began to like mathematics when I was perhaps seven or eight. In high school I used to do problems for more advanced classes. I was then in a boarding house in Nîmes, staying with children older than I was, and they used to bully me. So to pacify them, I used to do their mathematics homework. It was as good a training as any.

My mother was a pharmacist (as was my father), and she liked mathematics. When she was a pharmacy student, at the University of Montpellier, she had taken a first-year course in calculus, just for fun, and passed the exam. And she had carefully kept her calculus books (by Fabry and Vogt, if I remember correctly). When I was fourteen or fifteen, I used to look at these books, and study them. This is how I learned about derivatives, integrals, series and such (I did that in a purely formal manner—Euler's style so to speak: I did not like, and did not understand, epsilons and deltas). At that time, I had no idea one could make a living by being a mathematician. It is only later I discovered one could get paid for doing mathematics! What I thought at first was that I would become a high school teacher. This looked natural to me. Then, when I was nineteen, I took the competition to enter the École Normale Supérieure, and I succeeded. Once I was at "l'École," it became clear that it was not a high school teacher I wanted to be, but a research mathematician.

Q: *Did other subjects ever interest you, subjects like physics or chemistry?*

A: Physics not much, but chemistry yes. As I said, my parents were pharmacists, so they had plenty of chemical products and test tubes. I played with them a lot when I was about fifteen or sixteen besides doing mathematics. And I read my father's chemistry books (I still have one of them, a fascinating one, "Les Colloïdes" by Jacques Duclaux). However, when I learned more chemistry, I got

Volume 8, No. 4 (Fall 1986), 8–13

disappointed by its almost mathematical aspect: there are long series of organic compounds like CH_4, C_2H_6, etc. all looking more or less the same. I thought, if you have to have series, you might as well do mathematics! So, I quit chemistry—but not entirely: I ended up marrying a chemist.

Q: *Were you influenced by any school teacher in doing mathematics?*

A: I had only one very good teacher. This was in my last year in high school (1943–1944), in Nîmes. He was nicknamed "Le Barbu": Beards were rare at the time. He was very clear, and strict; he demanded that every formula and proof be written neatly. And he gave me a thorough training for the mathematics national competition called "Concours Général,"

FIGURE 3-1 *Jean-Pierre Serre.*

where I eventually got first prize.

Speaking of Concours Général, I also tried my hand at the one in physics the same year (1944). The problem we were asked to solve was based entirely on some physical law I was supposed to know, but did not. Fortunately, only one formula seemed to me possible for that law. I assumed it was correct, and managed to do the whole six-hour problem on that basis. I even thought I would get a prize. Unfortunately, my formula was wrong, and I got nothing—as I deserved!

Q: *How important is inspiration in the discovery of theorems?*

A: I don't know what "inspiration" really means. Theorems, and theories, come up in funny ways. Sometimes, you are just not satisfied with existing proofs, and you look for better ones, which can be applied in different situations. A typical example for me was when I worked on the Riemann-Roch theorem (circa 1953), which I viewed as a "Euler-Poincaré" formula (I did not know then that Kodaira-Spencer had had the same idea). My first objective was to prove it for algebraic curves—a case which was known for about a century! But I wanted a proof in a special style; and when I managed to find it, I remember it did not take me more than a minute or two to go from there to the two-dimensional case (which had just been done by Kodaira). Six months later, the full result was established by Hirzebruch, and published in his well-known Habilitationsschrift.

Quite often, you don't really try to solve a specific question by a head-on attack. Rather you have some ideas in mind, which you feel should be useful, but you don't know exactly for what they are useful. So, you look round, and try to apply them. It's like having a bunch of keys, and trying them on several doors.

Q: *Have you ever had the experience where you found a problem to be impossible to solve, and then after putting it aside for some time, an idea suddenly occurred leading to the solution?*

A: Yes, of course this happens quite often. For instance, when I was working on homotopy groups (around 1950), I convinced myself that, for a given space X, there should exist a fiber space E, with base X, which is contractible; such a space would indeed allow me (using Leray's methods) to do lots of computations on homotopy groups and Eilenberg-MacLane cohomology. But how to find it? It took me several weeks (a very long time, at the age I was then . . .) to realize that the space of "paths" on X had all the necessary properties—if only I dared call it a "fiber space," which I did. This was the starting point of the loop space method in algebraic topology; many results followed quickly.

Q: *Do you usually work on only one problem at a time or several problems at the same time?*

A: Mostly one problem at a time, but not always. And I work often at night (in half sleep), where the fact that you don't have to write anything down gives to the mind a much greater concentration, and makes changing topics easier.

Q: *In physics, there are a lot of discoveries which were made by accident, like X-rays, cosmic background radiation, and so on. Did that happen to you in mathematics?*

A: A genuine accident is rare. But sometimes you get a surprise because some argument you made for one purpose happens to solve a question in a different direction; however, one can hardly call this an "accident."

Q: *What are the central problems in algebraic geometry or number theory?*

A: I can't answer that. You see, some mathematicians have clear and far-ranging "programs." For instance, Grothendieck had such a program for algebraic geometry; now Langlands has one for representation theory, in relation to modular forms and arithmetic. I never had such a program, not even a small size one. I just work on things which happen to interest me at the moment. (Presently, the topic which amuses me most is counting points on algebraic curves over finite fields. It is a kind of applied mathematics. You try to use any tool in algebraic geometry and number theory that you know of . . . and you don't quite succeed!)

Q: *What would you consider to be the greatest developments in algebraic geometry or number theory within the past five years?*

A: This is easier to answer. Faltings' proof of the Mordell conjecture, and of the Tate conjecture, is the first thing which comes to mind. I would also mention Gross-Zagier's work on the class number problem for quadratic fields (based on a previous theorem of Goldfeld), and Mazur-Wiles' theorem on Iwasawa's theory, using modular curves. (The applications of modular curves and modular functions to number theory are especially exciting: You use GL_2 to study GL_1, so to speak! There is clearly a lot more to come from that direction . . . maybe even a proof of the Riemann Hypothesis some day?)

Q: *Some scientists have done fundamental work in one field and then quickly moved on to another field. You worked for three years in topology, then took up something else. How did this happen?*

A: It was a continuous path, not a discrete change. In 1952, after my thesis on homotopy groups, I went to Princeton, where I lectured on it (and on its continuation: "C-theory"), and attended the celebrated Artin-Tate seminar on class field theory.

Then I returned to Paris, where the Cartan seminar was discussing functions of several complex variables, and Stein manifolds. It turned out that the recent results of Cartan-Oka could be expressed much more efficiently (and proved in a simpler way) using cohomology and sheaves. This was quite exciting, and I worked for a short while on that topic, making applications of Cartan theory to Stein manifolds. However, a very interesting part of several complex variables is the study of projective varieties (as opposed to affine ones—which are somewhat pathological for a geometer); so, I began working on these complex projective varieties, using sheaves. That's how I came to the circle of ideas around Riemann-Roch, in 1953. But projective varieties are algebraic (Chow's theorem), and it is a bit unnatural to study these algebraic objects using analytic functions, which may well have lots of essential singularities. Clearly, rational functions should be enough—and indeed they are. This made me go (around 1954) into "abstract" algebraic geometry, over any algebraically closed field. But why assume the field is algebraically closed? Finite fields are more exciting, with Weil conjectures and such. And from there to number fields it is a natural enough transition. . . . This is more or less the path I followed.

Another direction of work came from my collaboration (and friendship) with Armand Borel. He told me about Lie groups, which he knows like nobody else. The connections of these groups with topology, algebraic geometry, number theory . . . are fascinating. Let me give you just one such example (of which I became aware about 1968):

Consider the most obvious discrete subgroup of $SL_2(\mathbf{R})$, namely $\Gamma = SL_2(\mathbf{Z})$. One can compute its "Euler-Poincaré characteristic" $\chi(\Gamma)$, which turns out to be $-1/12$ (it is not an integer; this is because Γ has torsion). Now $-1/12$ happens to be the value $\zeta(-1)$ of Riemann's zeta-function at the point $s = -1$ (a result known already to Euler). And this is not a coincidence! It extends to any totally real number field K, and can be used to study the denominator of $\zeta_K(-1)$. (Better results can be obtained by using modular forms, as was found later.) Such questions are not group theory, nor topology, nor number theory: They are just mathematics.

Q: *What are the prospects of achieving some unification of the diverse fields of mathematics?*

A: I would say that this has been achieved already. I have given above a typical example where Lie groups, number theory, etc. come together and cannot be separated from each other. Let me give you another such example (it would be easy to add many more):

There is a beautiful theorem proved recently by S. Donaldson on four-dimensional compact differentiable manifolds. It states that the quadratic form (on H^2) of such a manifold is severely restricted; if it is positive definite, it is a sum of squares. And the crux of the proof is to construct some auxiliary manifold (a "cobordism") as the set of solutions of some partial differential equation (non-linear, of course)! This is a completely new application of analysis to differential topology. And what makes it even more remarkable is that, if the differentiability assumption is dropped, the situation becomes quite different: By a theorem of M. Freedman the H^2-quadratic form can then be almost anything.

Q: *How does one keep up with the explosion in mathematical knowledge?*

A: You don't really have to keep up. When you are interested in a specific question, you find that very little of what is being done has any relevance to you; and if something does have relevance, then you learn it much faster, since you have an application in mind. It is also a good habit to look regularly at *Math. Reviews* (especially the collected volumes on number theory, group theory, etc). And you learn a lot from your friends, too: it is easier to have a proof explained to you at the blackboard, than to read it.

A more serious problem is the one of the "big theorems" which are both very useful and too long to check (unless you spend on them a sizable part of your lifetime . . .). A typical example is the Feit-Thompson Theorem: groups of odd order are solvable. (Chevalley once tried to take this as the topic of a seminar, with the idea of giving a complete account of the proof. After two years, he had to give up.) What should one do with such theorems, if one has to use them? Accept them on faith? Probably. But it is not a very comfortable situation.

I am also uneasy with some topics, mainly in differential topology, where the author draws a complicated picture (in two dimensions), and asks you to accept it as a proof of something taking place in five dimensions or more. Only the experts can "see" whether such a proof is correct or not—if you can call this a proof.

Q: *What do you think will be the impact of computers on the development of mathematics?*

A: Computers have already done a lot of good in some parts of mathematics. In number theory, for instance, they are used in a variety of ways. First, of course, to suggest conjectures, or questions. But also to check general theorems on numerical examples—which helps a lot with finding possible mistakes.

They are also very useful when there is a large search to be made (for instance, if you have to check 10^6 or 10^7 cases). A notorious example is the proof of the Four Color theorem. There is however a problem there, somewhat similar to the one with Feit-Thompson: such a proof cannot be checked by hand; you need a computer (and a very subtle program). This is not very comfortable either.

Q: *How could we encourage young people to take up mathematics, especially in the schools?*

A: I have a theory on this, which is that one should first *discourage* people from doing mathematics; there is no need for too many mathematicians. But, if after that, they still insist on doing mathematics, then one should indeed encourage them, and help them.

As for high school students, the main point is to make them understand that mathematics *exists*, that it is not dead (they have a tendency to believe that only physics, or biology, has open questions). The defect in the traditional way of teaching mathematics is that the teacher never mentions these questions. It is a pity. There are many such, for instance in number theory, that teenagers could very well understand: Fermat of course, but also Goldbach, and the existence of infinitely many primes of the form $n^2 + 1$. And one should also feel free to state theorems without proving them (for instance Dirichlet's theorem on primes in arithmetic progressions).

Q: *Would you say that the development of mathematics in the past thirty years was faster than that in the previous thirty years?*

A: I am not sure this is true. The style is different. In the 50s and 60s, the emphasis was quite often on general methods: distributions, cohomology, and the like. These methods were very successful, but nowadays people work on more specific questions (often, some quite old ones: for instance the classification of algebraic curves in three-dimensional projective space!). They *apply* the tools which were made before; this is quite nice. (And they also make new tools: microlocal analysis, supervarieties, intersection cohomology . . .).

Q: *In view of this explosion of mathematics, do you think that a beginning graduate student could absorb this large amount of mathematics in four, five, or six years and begin original work immediately after that?*

FIGURE 3-2 *Manin—Serre—Arnold, Moscow, 1984.*

A: Why not? For a given problem, you don't need to know that much, usually—and, besides, very simple ideas will often work.

Some theories get simplified. Some just drop out of sight. For instance, in 1949, I remember I was depressed because every issue of the *Annals of Mathematics* would contain another paper on topology which was more difficult to understand than the previous ones. But nobody looks at these papers any more; they are forgotten (and deservedly so: I don't think they contained anything deep . . .). Forgetting is a very healthy activity.

Still, it is true that some topics need much more training than some others, because of the heavy technique which is used. Algebraic geometry is such a case; and also representation theory.

Anyway, it is not obvious that one should say "I am going to work in algebraic geometry," or anything like that. For some people, it is better to just follow seminars, read things, and ask questions to oneself; and then learn the amount of theory which is needed for these questions.

Q: *In other words, one should aim at a problem first and then learn whatever tools that are necessary for the problem.*

A: Something like that. But since I know I cannot give good advice to myself, I should not give advice to others. I don't have a ready-made technique for working.

Q: *You mentioned papers which have been forgotten. What percentage of the papers published do you think will survive?*

A: A non-zero percentage, I believe. After all, we still read with pleasure papers by Hurwitz, or Eisenstein, or even Gauss.

Q: *Do you think that you will ever be interested in the history of mathematics?*

A: I am already interested. But it is not easy; I do not have the linguistic ability in Latin or Greek, for instance. And I can see that it takes more time to write a paper on the history of mathematics than in mathematics itself. Still, history is very interesting; it puts things in the proper perspective.

Q: *Do you believe in the classification of finite simple groups?*

A: More or less—and rather more than less. I would be amused if a new sporadic group were discovered, but I am afraid this will not happen.

More seriously, this classification theorem is a splendid thing. One may now check many properties by just going through the list of all groups (typical example: the classification of n-transitive groups, for $n > 4$).

Q: *What do you think of life after the classification of finite simple groups?*

A: You are alluding to the fact that some finite group theorists were demoralized by the classification; they said (or so I was told) "there will be nothing more to do after that." I find this ridiculous. Of course there would be plenty to do! First, of course, simplifying the proof (that's what Gorenstein calls "revisionism"). But also finding applications to other parts of mathematics; for instance, there have been very curious discoveries relating the Griess-Fischer monster group to modular forms (the so-called "Moonshine").

It is just like asking whether Faltings' proof of the Mordell conjecture killed the theory of rational points on curves. No! It is merely a starting point. Many questions remain open.

(Still, it is true that sometimes a theory can be killed. A well-known example is Hilbert's fifth problem: to prove that every locally Euclidean topological group is a Lie group. When I was a young topologist, that was the problem I really wanted to solve—but I could get nowhere. It was Gleason, and Montgomery-Zippin, who solved it, and their solution all but killed the problem. What else is there to find in this direction? I can only think of one question: Can the group of p-adic integers act effectively on a manifold? This seems quite hard—but a solution would have no application whatsoever, as far as I can see.)

Q: *But one would assume that most problems in mathematics are like these, namely that the problems themselves may be difficult and challenging, but after their solutions they become useless. In fact there are very few problems like the Riemann Hypothesis where even before its solution, people already know many of its consequences.*

A: Yes, the Riemann Hypothesis is a very nice case. It implies lots of things (including purely numerical inequalities, for instance on discriminants of number fields). But there are other such examples: Hironaka's desingularization theorem is one; and of course also the classification of finite simple groups we discussed before.

Sometimes, it is the method used in the proof which has lots of applications: I am confident this will happen with Faltings. And sometimes, it is true, the problems are not meant to have applications; they are a kind of test on the existing theories; they force us to look further.

Q: *Do you still go back to problems in topology?*

A: No. I have not kept track of the recent techniques, and I don't know the latest computations of the homotopy groups of spheres $\pi_{n+k}(S_n)$ (I guess people have reached up to $k = 40$ or 50. I used to know them up to $k = 10$ or so.)

But I still use ideas from topology in a broad sense, such a cohomology, obstructions, Stiefel-Whitney classes, etc.

Q: *What has been the influence of Bourbaki on mathematics?*

A: A very good one. I know it is fashionable to blame Bourbaki for everything ("New Math" for instance), but this is unfair. Bourbaki is not responsible. People just misused his books; they were never meant for university teaching, even less high school teaching.

Q: *Maybe a warning sign should have been given?*

A: Such a sign was indeed given by Bourbaki: It is the séminaire Bourbaki. The séminaire is not at all formal like the books; it includes all sorts of mathematics, and even some physics. If you combine the séminaire and the books, you get a much more balanced view.

Q: *Do you see a decreasing influence of Bourbaki on mathematics?*

A: The influence is different from what it was. Forty years ago, Bourbaki had a point to make; he had to prove that an organized and systematic account of mathematics was possible. Now the point is made and Bourbaki has won. As a consequence, his books now have only technical interest; the question is just whether they give a good exposition of the topic they are on. Sometimes they do (the one on "root systems" has become the standard reference in the field); sometimes they don't (I won't give an example: It is too much a matter of taste).

Q: *Speaking of taste, can you say what kind of style (for books or papers) you like most?*

A: Precision combined with informality! That is the ideal, just as it is for lectures. You find this happy blend in authors like Atiyah or Milnor, and a few others. But it is hard to achieve. For instance, I find many of the French (myself included) a bit too formal, and some of the Russians a bit too imprecise . . .

A further point I want to make is that papers should include more side remarks, open questions, and such. Very often, these are more interesting than the theorems actually proved. Alas, most people are afraid to admit that they don't know the answer to some question, and as a consequence they refrain from mentioning the question, even if it is a very natural one. What a pity! As for myself, I enjoy saying "I do not know."

4

Mathematical Anecdotes

Steven G. Krantz

In any field of human endeavor, the "great" participants are distinguished from everyone else by the arcana and apocrypha that surround them. Stories about Wolfgang Amadeus Mozart abound, yet there are few stories about his musical contemporaries. Mozart had the *je ne sais quoi* that made people *want* to tell stories about him.

And so it goes with mathematicians. Over the years I have collected dozens of anecdotes about famous mathematicians (a necessary condition for being the subject of legend is fame; it is by no means sufficient). These stories are of several types: (i) incidents to which I have been witness (there are few of these); (ii) incidents related to me by someone who witnessed them (on statistical grounds, one expects a greater number of these); and (iii) incidents that have been passed down through iterated tellings and are therefore unverifiable. I shall not consistently classify the stories that I will relate here. In many cases I cannot remember which of the three types they are, and actually knowing would generally spoil the fun. In any event, I must bear the ultimate responsibility for the stories.

In writing this article, I am running the risk that readers will think me flip, disrespectful, or (worse) that I am attacking people who cannot fight back. Let me set the record straight once and for all: To me, the mathematicians described here are among the gods of twentieth-century mathematics. Much of what we know, and certainly much of my own work, follows from their insights. The enormous scholarly reputations of these men sometimes cause their humanity to be forgotten. Bergman, Besicovitch, Gödel, Lefschetz, and Wiener were not merely collections of theorems masquerading as people; they had feet of clay like the rest of us. In telling stories about them, we bring them back to life and celebrate their careers.

Bergman

Stefan Bergman (1898–1977) was a native of Poland. He began his career in the United States at Brown University. It is said that shortly after Bergman and his mistress arrived in the United States, he took her aside and told her, "Now we are in the United States where customs are different. When we are with other people, you should call me 'Stefan.' But at home you should continue to call me 'Professor Doktor Bergman.' " Others who knew Bergman will say that he was not the sort of man who would have had a mistress. It is more likely that the man in question was von Mises (Bergman's sponsor); there is general agreement that the woman was Hilda Geiringer. In fact another story holds that Norbert Wiener (more on him later) went to D. C. Spencer around this time and said, "I think that we should call the FBI (Federal Bureau of Investigation)." Puzzled, Spencer asked why. "Because von Mises has a mistress," was the serious reply.

Volume 12, No. 4 (Fall 1990), 32–38

After a few years Bergman moved to Harvard and then to Stanford, where he spent most of his career. Supported almost always on grants and other soft money, Bergman rarely taught. This fact may have contributed to the general murkiness of his verbal and written communications. Murkiness aside, Bergman was proud of his ability to express himself in many tongues. Said he, "I speak twelve languages—English the bestest."

In fact Bergman had a stammer and was sometimes difficult to understand in any language. Once he was talking to Antoni Zygmund, another celebrated Polish analyst, in their native tongue. After a bit Zygmund said, "Please let's speak English. It's more comfortable for me."

Although Bergman had many fine theorems to his credit (including the invention of a version of the Šilov boundary), the crowning achievement of his mathematical work was the invention of the *kernel function,* now known as the Bergman kernel. He spent most of his life developing properties and applications of the Bergman kernel and the associated Bergman metric. It must have been a special source of pride and pleasure for him when, near the end of Bergman's life, Charles Fefferman (1974) found a profound application of the Bergman theory to the study of biholomorphic mappings. Fefferman's discoveries, coupled with related ideas of J. J. Kohn and Norberto Kerzman, created a renaissance in the study of the Bergman kernel. Indeed, a major conference in several complex variables was held in Williamstown in 1975 in which many of the principal lectures mentioned or discussed the Bergman kernel.

Bergman had always felt that the value of his ideas was not sufficiently appreciated. He attended the conference and commented to several people how pleased he was that his wife (also present) could see his work finally being recognized. I sat next to him at most of the principal lectures. In each of these, he listened carefully for the phrase "and in 1922 Stefan Bergman invented the kernel function." Bergman would then dutifully record this fact in his notes—and nothing more. I must have seen him do this twenty times during the three-week conference.

There was a rather poignant moment at the conference. In the middle of one of the many lectures on biholomorphic mappings, Bergman stood up and said, "I think you people should be looking at representative coordinates (also one of Bergman's inventions)." Most of us did not know what he was talking about, and we ignored him. He repeated the comment a few more times, with the same reaction. Five years later S. Webster, S. Bell, and E. Ligocka found astonishing simplifications and extensions of the known results about holomorphic mappings using—guess what?—representative coordinates.

Bergman was an extraordinarily kind and gentle man. He went out of his way to help many young people begin their careers, and he made great efforts on behalf of Polish Jews during the Nazi terror. He is remembered fondly by all who knew him. But he was a shark when it came to his mathematics. When he attended a lecture about a theorem he liked, he often went to the lecturer afterwards and said "I really like your theorem. It reminds me of my studies of the kernel function. Consider complex two-space . . ." And Bergman was off and running on his favorite topic. On another occasion a young mathematician gave Bergman a manuscript he had just written. Bergman read it and said "I like your result. Let's make it a joint paper, and I'll write the next one."

Whenever someone proved a new theorem about the Bergman kernel or Bergman metric, Bergman made a point of inviting the mathematician to his house for supper. Bergman and his wife were a gracious host and hostess and made their guest feel welcome. However, after supper the guest had to pay the piper by giving an impromptu lecture about the importance of the Bergman kernel.

Bergman's wife Edy was very devoted to him, but life with Stefan was sometimes trying. When they first got married, Bergman had just completed a difficult job search. In the days immediately following World War II, jobs were scarce, and Bergman wanted a position with no teaching. After a long period of disappointment, Shiffer got Bergman a position at Stanford; so the mood was high at the Bergmans' wedding reception. The reception took place in New York City, and Bergman was delighted that one of the guests was a mathematician from New York University with whom he had many mutual interests. They got involved in a passionate mathematical discussion and after a while Bergman announced to the guests that he would be back in a few hours: He had to go to NYU to discuss mathematics. On hearing this, Shiffer turned to Bergman and said "I got you your job at Stanford; if you leave this reception, I will take it away." Bergman stayed.

Bergman thought intensely about mathematics and cared passionately about his work. One day, during the 1950 International Congress of Mathematicians in Cambridge, Bergman had a luncheon date with two Italian friends. Right on schedule they appeared at Bergman's office: the distinguished elder Italian mathematician Piccone (bearing a bouquet of flowers for Bergman!) and his younger colleague Sichera. This was Piccone's first visit to the United States, and he spoke no English; Sichera acted as interpreter. After greetings were exchanged, Bergman asked Sichera whether he had read Bergman's latest paper. Sichera allowed that he had, and that he thought it was very interesting. However, he said that he felt that certain additional differentiability assumptions were required. Bergman said "No, no, you don't understand," and proceeded to explain on the blackboard. Piccone, understanding none of this, waited patiently. After the explanation, Bergman asked Sichera whether he now understood. Sichera said that he did, but he still thought that some differentiability hypotheses were required in a certain step. Bergman became adamant and a heated argument ensued—Piccone comprehending none of it. After some time, Sichera said "Well, let's forget it and go to lunch." Bergman cried "No differentiability—no lunch!" and he remained in his office while the two Italians went to lunch. Piccone gave the flowers to the waitress.

There is considerable evidence that Bergman thought about mathematics constantly. Once he phoned a student, at the student's home number, at 2:00 A.M. and said "Are you in the library? I want you to look something up for me!"

On another occasion, when Bergman was at Brown, one of Bergman's graduate students got married. The student planned to attend a conference on the West Coast, so he and his new bride decided to take a bus to California as a sort of makeshift honeymoon. There was a method in their madness: the student knew that Bergman would attend the conference but that he liked to get where he was going in a hurry. The bus seemed the least likely mode of transportation for Bergman. But when Bergman heard about the impending bus trip, he thought it a charming idea and purchased a bus ticket for himself. The student protested that this trip was to be part of his honeymoon, and that he could not talk mathematics on the bus. Bergman promised to behave. When the bus took off, Bergman was at the back of the bus and, just to be safe, Bergman's student took a window seat near the front with his wife in the adjacent aisle seat. But after about ten minutes Bergman got a great idea, wandered up the aisle, leaned across the scowling bride, and began to discuss mathematics. It wasn't long before the wife was in the back of the bus and Bergman next to his student—and so it remained for the rest of the bus trip! The story has a happy ending: The couple is still married, has a son who became a famous mathematician, and several grandchildren.

Presumably it was his preoccupation with mathematics that caused Bergman to appear to be out of touch with reality at times. For example, one day he went to the beach in northern California with a group of people, including a friend of mine who told me this yarn. Northern California beaches are cold, so when Bergman came out of the water, he decided that he'd better put on his street clothes. As he wandered off to the parking lot, his friends noticed that he was heading in the wrong direction; but they were used to his sort of behavior and paid him no mind. In a while, Bergman returned—clothed—exclaiming "You know, there's the most unfriendly woman in our car!"

Bergman was a prolific writer. Of course he worked in the days before the advent of word processors. His method of writing was this. First, he would write a manuscript in longhand and give it to the secretary. When she had it typed up, he would begin revising, stapling strips of paper over the portions that he wished to change. Strips would be stapled over strips, and then again and again, until parts of the manuscript would become so thick that the stapler could no longer penetrate. Then the manuscript would be returned to the secretary for a retype and the whole cycle would begin again.

FIGURE 4-1 *Stefan Bergman (top) and Abram S. Besicovitch.*

Sometimes it would repeat ad infinitum. Bergman once told a student that "a mathematician's most important tool is the stapler."

Bergman had a self-conscious sense of humor and a loud laugh. He once walked into a secretary's office and, while he spoke to her, inadvertently stood on her white glove that had fallen on the floor. After a bit she said "Professor Bergman, you're standing on my glove." He acted embarrassed and exclaimed "Oh, I thought it was a mousy." (It should be mentioned here that a number of wildly exaggerated versions of this story are in circulation, but I got this version from a primary source.)

Besicovitch

Abram S. Besicovitch (1891–1970) was a geometric analyst of extraordinary power. He became world-famous for his solution of the Kakeya needle problem. The problem was to find the planar region of least area with the property that a segment of unit length lying in the region can be moved through all direction angles $\theta, 0 \leq \theta \leq 2\pi$, within the region. Besicovitch's surprising answer was that for any $\epsilon > 0$, there is such a region with area less than ϵ.

Besicovitch, a Russian by birth, was a creature of the old world. After leaving Russia (a prudent move on account of his rumored black market dealings during World War I), Besicovitch ended up at Cambridge University in England. A dinner was given in his honor, at which the main course was some sort of game bird. In his thick Russian accent, Besicovitch asked the name of the tasty food that they were eating. When he heard the reply, he exclaimed, "In Russia, we are not allowed to eat the peasants!"

Besicovitch was a smart man, so he quickly became proficient at English. But it was never perfect. He adhered to the Russian paradigm of never using articles before nouns. One day, during his lecture, the class chuckled at his fractured English. Besicovitch turned to the audience and said "Gentlemen, there are fifty million Englishmen speak English you speak; there are two hundred million Russians speak English I speak". The chuckling ceased.

In another lecture series, on approximation theory, he announced "zere is no t in ze name Chebyshóv." Two weeks later he said "Ve now introduce ze class of T-polynomials. Zey are called T-polynomials because T is ze first letter of ze name Chebyshóv."

Besicovitch, in spite of his apparent powers, was modest. On his thirty-sixth birthday, he convinced himself that his best and most intense years of research were over. He said "I have had four-fifths of my life." Twenty-three years later, when in 1950 he was awarded the Rouse Ball Chair of Mathematics at Cambridge, someone reminded him of this frivolous remark. He replied, "Numerator was correct."

In the 1960s, the Mathematical Association of America made a series of delightful one-hour films in each of which a great mathematician gave a lecture, for a general mathematical audience, about one of his achievements. One of these films starred Besicovitch, and he explained his solution of the Kakeya needle problem. Besicovitch was a natty dresser under any circumstances, and he wore to this lecture an attractive light beige suit. However the lights were hot and, after a while he removed his jacket, revealing bright red suspenders! The producers were most surprised (this was thirty-five years ago, and nobody but firemen wore red suspenders), but the filming continued and the suspenders can be seen today.

At one point during the filming of Besicovitch, the aged professor had to blow his nose. He drew a large white handkerchief from his pocket and did so—loudly. Later, when Besicovitch viewed the finished product, he objected to the noseblowing scene as undignified—he wanted it removed. The producers were able to replace the offending video segment, but it was decided that the sound should remain. As a result, if you view the film today, there comes a point in the action where the camera abruptly leaves Besicovitch and focuses on the side of the room—and you can *hear* Besicovitch blow his nose.

Gödel

Kurt Gödel (1906–1978) was one of the most original mathematicians of the twentieth century. Any thesaurus links "originality" with "eccentricity," and Gödel had his fair share of both. Toward the end of his life, Gödel became convinced that he was being poisoned, and he ended up starving himself to death. However, years before that, his peculiar point of view exhibited itself in other ways.

Einstein was Gödel's closest personal friend in Princeton. For several years Einstein, Gödel, and Einstein's assistant Ernst Straus (who later became a well-known combinatorial theorist) would lunch together. During lunch they discussed non-mathematical topics—frequently politics. One notable discussion took place the day after Douglas MacArthur was given a ticker-tape parade down Madison Avenue upon his return from Korea. Gödel came to lunch in an agitated state, insisting that the man in the picture on the front page of the *New York Times* was not MacArthur but an imposter. The proof? Gödel had an earlier photo of MacArthur and a ruler. He compared the ratio of the length of the nose to the distance from the tip of the nose to the point of the chin in each picture. These were different: Q.E.D.

Gödel spent a significant part of his career trying to decide whether the Continuum Hypothesis (CH) is independent of the Axiom of Choice (AC). In the early 1960s, a brash, young, and extremely

FIGURE 4-2 *Kurt Gödel (top) and Solomon Lefschetz.*

brilliant Fourier analyst (student of the aforementioned Zygmund) named Paul J. Cohen (people who knew him in high school and college assure me that he was always brash and brilliant) chatted with a group of colleagues at Stanford about whether he would become more famous by solving a certain Hilbert problem or by proving that CH is independent of AC. This (informal) committee decided that the latter problem was the ticket. [To be fair, Cohen had been interested in logic and recursive functions for several years; he may have conducted this seance just for fun.] Cohen went off and learned the necessary logic and, in less than a year, had proved the independence. This is certainly one of the most amazing intellectual achievements of the twentieth century. Cohen's technique of "forcing" has become a major tool of modern logic, and Cohen was awarded the Fields Medal for the work. But there is more.

Proof in hand, Cohen flew off to the Institute for Advanced Study to have his result checked by Kurt Gödel. Gödel was naturally skeptical, as Cohen was not the first person to claim to have solved the problem, and Cohen was not even a logician! Gödel was also, at this time, beginning his phobic period. When Cohen went to Gödel's house and knocked on the door, it was opened six inches and a hoary hand snatched the manuscript and slammed the door. Perplexed, Cohen departed. However, two days later Cohen received an invitation for tea at Gödel's home. His proof was correct: The master had certified it.

Lefschetz

The story goes that Solomon Lefschetz (1884–1972) was trained to be an engineer. This was in the days, near the turn of the century, when engineering was part carpentry, part alchemy, and part luck (the pre–von Kármán era). In any event, Lefschetz had the misfortune to lose both his hands in a laboratory accident. This mishap was lucky for us, for he subsequently, at the age of thirty-six, became a mathematician.

Lefschetz had two prostheses in place of his hands—they looked like hands, loosely clenched, but they did not move or function in any way. Over each he wore a shiny black glove. A friend of mine was a graduate student of Lefschetz; he tells me that one of his daily duties was to push a piece of chalk into Lefschetz's hand each morning and to remove it at the end of the day.

Lefschetz starred in one of the MAA films. He gave a lovely lecture, punctuated by a cacophony of squeaky chalk, about his celebrated fixed point theorem. His feelings about the film were mixed. At one point he says on film "I hope this is clear; it's probably about as clear as mud." After his lecture comes a filmed round table discussion including John Moore, Lefschetz, and a few others. For ten or fifteen minutes they reminisce about the old days at Princeton. One person reminds Lefschetz that in the late 1940s, during the heyday of the development of algebraic topology, they were on a train together. Lefschetz was asked the difference between algebra and topology. He is reported to have said "If it's just turning the crank, it's algebra; but if there is an idea present, then it's topology." When Lefschetz was reminded of this story in the film, he became most embarrassed and said "I couldn't have said anything like that."

With his artificial hands, Lefschetz could not operate a doorknob, so his office door was equipped with a lever. Presumably he had difficulty with other routine daily matters, too—dialing a phone, turning on a light, etc. By the time I was a graduate student at Princeton, Lefschetz was 87. He was still mathematically sharp but he had trouble getting around. In those days Fine Hall, the mathematics building in Princeton, was having constant trouble with the elevators: Push the button for the fifth floor and you're shot to the penthouse, down to the basement, and ejected on seven; or variations on that theme. The receptionist kept a log of complaints so that she could report them to the person who came to repair the elevator. One day Lefschetz got into the elevator and it delivered him to the fourth floor "machine room"; this room houses the air conditioning equipment and is ordinarily only accessible with a janitor's key. Poor Lefschetz unwittingly wandered out into the room, only to have the elevator door shut behind him before he realized what was going on. He was trapped in total darkness, could not summon the elevator (no key), could not turn the doorknob to use the stairwell, and could not find a telephone (which, even had he found, he probably could not have dialed). The members of the mathematics department rode that elevator for several hours, not realizing that Lefschetz was missing, before someone finally heard Lefschetz's shouts and understood what was going on. Fortunately Lefschetz survived the incident unharmed.

Speaking of the elevators at Princeton, one of my earliest memories as a graduate student was of the elevator emergency stop alarm going off three or four times a day. Especially puzzling was that everyone ignored it. Bear in mind that this alarm only sounds if someone inside the elevator sets it off. It is sometimes used by janitors to hold the elevator at a certain floor; but the janitors never used it during the day. After I had asked around for some time, someone finally told me the secret. When the mathematics department moved from old Fine Hall to new Fine Hall (sometimes called "Finer Hall," overlooking "Steenrod Square"), Ralph Fox, the famous topologist, was annoyed that there was no men's room on his floor. So, whenever he had to use the facilities, he would take the elevator to the next floor, set the emergency stop alarm, do what needed to be done, and then return to his floor. Now I knew why everyone smiled when the alarm went off.

So much for boyhood memories; back to Lefschetz.

Lefschetz was famous for his aggressive self-confidence. He could terrorize most other mathematicians easily. At committee meetings he would pound his fist on the table with terrifying effect. So it is with pleasurable surprise that one hears of exceptions. The one I have in mind is a certain unflappable graduate student at the time of the student's qualifying examination. The qualifying exams at Princeton are administered as one long oral exam: three professors and one graduate student locked in an office for about three and one-half hours. The student is examined on real analysis, complex analysis, algebra, and two advanced topics of the student's choosing (subject to the approval of the Director of Graduate Studies). Our confident student had Lefschetz on his committee. Lefschetz was famous for, among other things, profound generalizations of Picard's theorems in function theory to several complex variables. So it came as no surprise when Lefschetz asked the student "Can you prove Picard's Great Theorem?" Came the reply "No, can you?" Lefschetz had to admit that he could not remember, and the exam moved on to another topic.

Lefschetz was one of those mathematicians, of whom we all know at least one, who would sleep during lectures and then wake up at the end with a brilliant question. At one colloquium, the speaker got stuck on a point about twenty minutes into his talk. A silence of several minutes ensued. This threw off Lefschetz's rhythm. He woke up, said "Are there any questions? Thank you very much," and the seminar was ended with a round of applause.

The "roasting" of an individual is a peculiarly American custom. A group of close friends holds a fancy dinner in honor of the victim, after which they stand up one by one and make a collection of (humorously delivered) insulting remarks about him. Some anecdotes are in the nature of a roast. Here is an example. In the fifties, it was said in Princeton that there were four definitions of the word "obvious." If something is obvious in the sense of Beckenbach, then it is true and you can see it immediately. If something is obvious in the sense of Chevalley, then it is true and it will take you several weeks to see it. If something is obvious in the sense of Bochner, then it is false and it will take you several weeks to see it. If something is obvious in the sense of Lefschetz, then it is false and you can see it immediately.

This last item reminds me of the old concept of "true in the sense of Henri Cartan." In the 1930s and 1940s, a theorem was "true in the sense of Cartan" if Grauert could not find a counter-example in the space of an hour.

The discussion of "truth" and "obviousness" raises the issue of standards. Perhaps the least delightful arena in which we all wrestle with standards is that of referees' reports. The *Annals of Mathematics*, Princeton's journal, has very high standards and exhorts its referees to be tough-minded. Lefschetz was instrumental in establishing the pre-eminence of the *Annals*. But I doubt that even he could have anticipated the following event. Many years ago, Gerhard Hochschild (who sets high standards for himself and everyone else) submitted a paper to the *Annals*. The referee's report said "Good enough for the *Annals*. Not good enough for Hochschild. Rejected."

Wiener

The brilliant analyst Norbert Wiener (1894–1964) is a favorite subject of anecdotes. He is just modern enough that many living mathematicians knew him and was just eccentric enough to be a never-ending object of stories and pranks.

Born the son of a distinguished professor of languages, Wiener became one of America's first internationally recognized mathematicians. Because of anti-Semitism in the American mathematical establishment, Wiener spent the early years of his career working in England. The story goes that

FIGURE 4-3 *Norbert Wiener.*

when he met Littlewood, he said, "Oh, so you really exist. I thought that 'Littlewood,' was just a pseudonym that Hardy put on his weaker papers." Poor Wiener was so chagrined by this story that he denied it vehemently in his autobiography, thus inadvertently fueling belief in its validity. [In fairness to Wiener I should point out that another popular version of the story involves Edmund Landau: Landau so doubted the existence of Littlewood that he made a special trip to Great Britain to see the man with his own eyes.]

After Wiener left Britain, he moved to MIT where he stayed for more than twenty-five years. He developed a reputation all over campus as a brilliant scientist and a bit of a character. He was always working—either thinking or writing or reading. When he walked the halls of MIT, he invariably read a book, running his finger along the wall to keep track of where he was going. One day, engaged in this activity, Wiener passed a classroom where a class was in session. It was a hot day and the door had been left open. But of course Wiener was unaware of these details—he followed his finger through the door, into the classroom, around the walls (right past the lecturer) and out the door again.

People who knew Wiener tell me—and this comes through clearly in his autobiography as well—that he struggled all his life with feelings of inferiority. These feelings applied to non-mathematical as well as to mathematical activities. Thus, when he played bridge at lunch with a group of friends, he would invariably say, every time he bid or played, "Did I do the right thing? Was that a good play?" His partner, Norman Levinson, would patiently reassure him each time that he couldn't have done any better.

It is not a well-known fact that Wiener wrote a novel. The villain in the novel was a thinly disguised version of R. Courant. The hero was a thinly disguised version of Wiener himself. Friends were successful in discouraging him from publication. [Another version of the story is that the villain was Osgood. In the book, he proves a theorem that is a thinly disguised version of a celebrated theorem of Osgood, but in a different branch of mathematics; he ends up dying in China.]

Students liked to play pranks on Wiener. He read the newspaper every day at the same time in a certain lounge at MIT. As Wiener sat with the newspaper spread open before him, a student would sneak up and set the bottom of the paper afire. The results were spectacular, and the joke was repeated again and again.

And sometimes Wiener played jokes on his students, though he did not realize that he was doing so. On one occasion, a student asked him how to solve a certain problem. Wiener thought for a moment and wrote down the answer. The student hadn't really wanted the answer but wanted the method to be explained (this really was a long time ago!). So he said "But isn't there some other way?" Wiener thought for another moment, smiled, and said "Yes there is"—and he wrote down the answer a second time.

Probably the most famous Wiener story concerns a day when the Wiener family was moving to a new home. Wiener's wife knew Norbert only too well. So on the night before, as well as the morning of, the moving day, she reminded him over and over that they were moving. She wrote the new address for him on a slip of paper (the new house was just a few blocks away), gave him the new keys, and took away his old keys. Wiener dutifully put the new address and keys into his pocket and left for work. During the course of the day, Wiener's thoughts were elsewhere. At one point somebody asked him a mathematical question, and Wiener gave him the answer on the back of the slip of paper his wife had given him. So much for the new address! At the end of the day Wiener, as was his habit, walked home—to his old house. He was puzzled to find nobody home. Looking through the window, he could see no furnishings. Panic took over when he discovered that his key would not fit the lock. Wild-eyed, he began alternately to bang on the door and to run around in the yard. Then he spotted a child coming down the street. He ran up to her and cried "Little girl, I'm very upset. My family has disappeared and my key won't fit in the lock." She replied, "Yes, daddy. Mommy sent me for you."

My final Wiener story, indeed my final story, does not seem to be well known. Even inveterate Wienerologists proclaim it too good to be true. But it's not too good for this article. I believe that I heard it when I was a graduate student at Princeton. As I've mentioned, Wiener was quite a celebrated figure on the MIT campus. Therefore, when one of his students spied Wiener in the post office, the student wanted to introduce himself to the famous professor. After all, how many MIT students could say that they had actually shaken the hand of Norbert Wiener? However, the student wasn't sure how to approach the man. The problem was aggravated by the fact that Wiener was pacing back and forth, deeply lost in thought. Were the student to interrupt Wiener, who knows what profound idea might be lost? Still, the student screwed up his courage and approached the great man. "Good morning, Professor Wiener," he said. The professor looked up, struck his forehead, and said "That's it: Wiener!"

5

My Collaboration with Julia Robinson

Yuri Matijasevich

The name of Julia Robinson cannot be separated from Hilbert's tenth problem. This is one of the twenty-three problems stated by David Hilbert in 1900. The section of his famous address [4] devoted to the tenth problem is so short that it can be cited here in full:

10. DETERMINATION OF THE SOLVABILITY OF A DIOPHANTINE EQUATION

Given a Diophantine equation with any number of unknown quantities and with rational integral numerical coefficients: To devise a process according to which it can be determined by a finite number of operations whether the equation is solvable in rational integers.

The tenth problem is the only one of the twenty-three problems that is (in today's terminology) a *decision problem;* i.e., a problem consisting of infinitely many individual problems, each of which requires a definite answer: yes or no. The heart of a decision problem is the requirement to find a single method that will give an answer to any individual subproblem. Since Diophantus' time, number-theorists have found solutions for a large number of Diophantine equations and also have established the unsolvability of a large number of other equations. Unfortunately, for different classes of equations and even for different individual equations, it was necessary to invent different specific methods. In the tenth problem, Hilbert asks for a *universal* method for deciding the solvability of Diophantine equations.

A decision problem can be solved in a positive or in a negative sense, by discovering a proper algorithm or by showing that none exists. The general mathematical notion of algorithm was developed by A. Church, K. Gödel, A. Turing, E. Post, and other logicians only thirty years later, but, in his lecture [4], Hilbert foresaw the possibility of negative solutions to some mathematical problems.

I have to start the story of my collaboration with Julia Robinson by telling about my own involvement in the study of Hilbert's tenth problem. I heard about it for the first time at the end of 1965 when I was a sophomore in the Department of Mathematics and Mechanics of Leningrad State University. At that time I had already obtained my first results concerning Post's canonical systems, and I asked my scientific adviser, Sergei Maslov (see [3]), what to do next. He answered: "Try to prove the algorithmic unsolvability of Diophantine equations. This problem is known as Hilbert's tenth problem, but that does not matter to you."—"But I haven't learned any proof of the unsolvability of any decision problem."—"That also does not matter. Unsolvability is nowadays usually proved by reduc-

Volume 14, No. 4 (Fall 1992), 38–45.

ing a problem already known to be unsolvable to the problem whose unsolvability one needs to establish, and you understand the technique of reduction well enough."—"What should I read in advance?"—"Well, there are some papers by American mathematicians about Hilbert's tenth problem, but you need not study them."—"Why not?"—"So far the Americans have not succeeded, so their approach is most likely inadequate."

Maslov was not unique in underestimating the role of the previous work on Hilbert's tenth problem. One of these papers was by Martin Davis, Hilary Putnam, and Julia Robinson [2], and even the reviewer of it for *Mathematical Reviews* stated:

> These results are superficially related to Hilbert's tenth Problem on (ordinary, i.e., non-exponential) Diophantine equations. The proof of the authors' result, though very elegant, does not use recondite facts in the theory of numbers nor in the theory of r.e. [recursively enumerable] sets, and so it is likely that the present result is not closely connected with Hilbert's tenth Problem. Also it is not altogether plausible that all (ordinary) Diophantine problems are uniformly reducible to those in a fixed number of variables of fixed degree, which would be the case if all r.e. sets were Diophantine.

The reviewer's skepticism arose because the authors of [2] had considered not ordinary Diophantine equations (i.e., equations of the form

$$P(x_1, x_2, \ldots, x_m) = 0, \tag{1}$$

where P is a polynomial with integer coefficients) but a wider class of so-called *exponential Diophantine equations*. These are equations of the form

$$E_1(x_1, x_2, \ldots, x_m) = E_2(x_1, x_2, \ldots, x_m), \tag{2}$$

where E_1 and E_2 are expressions constructed from $x_1, x_2, \ldots, x_m$ and particular natural numbers by addition, multiplication, and exponentiation. (In contrast to the formulation of the problem as given by Hilbert, we assume that all the variables range over the natural numbers, but this is a minor technical alteration.)

Besides single equations, one can also consider parametric families of equations, either Diophantine or exponential Diophantine. Such a family

$$Q(a_1, \ldots, a_n, x_1, \ldots, x_m) = 0 \tag{3}$$

determines a relation between the *parameters* $a_1, \ldots, a_n$ which holds if and only if the equation has a solution in the remaining variables, called *unknowns*. Relations that can be defined in this way are called *Diophantine* or *exponential Diophantine* according to the equation used. Similarly, a set $\mathcal{M}$ of n-tuples of natural numbers is called (exponential) Diophantine if the relation "to belong to $\mathcal{M}$" is (exponential) Diophantine. Also a function is called (exponential) Diophantine if its graph is so.

Thus, in 1965 I did not encounter even the name of Julia Robinson. Instead of suggesting that I first study her pioneer works, Maslov proposed that I try to prove the unsolvability of so-called *word equations* (or *equations in a free semi-group*) because they can be reduced to Diophantine equations. Today we know that this approach was misleading, because in 1977 Gennadii Makanin found a decision procedure for word equations. I started my investigations on Hilbert's tenth problem by showing that a broader class of word equations with additional conditions on the lengths of words is also reducible to Diophantine equations. In 1968, I published three notes on this subject.

I failed to prove the algorithmic unsolvability of such extended word equations (this is still an open problem), so I then proceeded to read "the papers by some American mathematicians" on Hilbert's tenth problem. (Sergei Adjan had initiated and edited translations into Russian of the most important papers on this subject; they were published in a single issue of *Математика*, a journal dedicated to translated papers.) After the paper by Davis, Putnam, and Robinson mentioned above, all that was needed to solve Hilbert's tenth problem in the negative sense was to show that exponentiaton is Diophantine; i.e., to find a particular Diophantine equation

$$A(a, b, c, x_1, \ldots, x_m) = 0 \tag{4}$$

which for given values of the parameters a, b, and c, (4) has a solution in $x_1, \ldots, x_m$ if and only if $a = b^c$. With the aid of such an equation, one can easily transform an arbitrary exponential Diophantine equation into an equivalent Diophantine equation with additional unknowns.

As it happens, this same problem had been tackled by Julia Robinson at the beginning of the 1950s. According to "The Autobiography of Julia Robinson," an article written by her sister Constance Reid [11], Julia Robinson's interest was originally stimulated by her teacher, Alfred Tarski, who suspected that even the set of all powers of 2 is *not* Diophantine. Julia Robinson, however, found a sufficient condition for the existence of a Diophantine representation (4) for exponentiation; namely, to construct such an A, it is sufficient to have an equation

$$B(a, b, x_1, \ldots, x_m) = 0 \tag{5}$$

which defines a relation $J(a, b)$ with the following properties:

for any a and b, $J(a, b)$ implies that $a < b^b$;
for any k, there exist a and b such that $J(a, b)$ and $a > b^k$.

Julia Robinson called a relation J with these two properties *a relation of exponential growth*; today such relations are also known as *Julia Robinson predicates*.

My first impression of the notion of a relation of exponential growth was "what an unnatural notion," but I soon realized its important role for Hilbert's tenth problem. I decided to organize a seminar on Hilbert's tenth problem. The first meeting where I gave a survey of known results was attended by five logicians and five number-theorists, but then the numbers of participants decreased exponentially and soon I was left alone.

I was spending almost all my free time trying to find a Diophantine relation of exponential growth. There was nothing wrong when a sophomore tried to tackle a famous problem, but it looked ridiculous when I continued my attempts for years in vain. One professor began to laugh at me. Each time we met he would ask; "Have you proved the unsolvability of Hilbert's tenth problem? Not yet? But then you will not be able to graduate from the university!"

Nevertheless I did graduate in 1969. My thesis consisted of my two early works on Post canonical systems because I had not done anything better in the meantime. That same year I became a postgraduate student at the Steklov Institute of Mathematics of the Academy of Sciences of the USSR (Leningrad Branch, LOMI). Of course, the subject of my study could no longer be Hilbert's tenth problem.

One day in the autumn of 1969, some of my colleagues told me, "Rush to the library. In the recent issue of the *Proceedings of the American Mathematical Society* there is a new paper by Julia Robinson!" But I was firm in putting Hilbert's tenth problem aside. I told myself, "It's nice that Julia Robinson goes on with the problem, but I cannot waste my time on it any longer." So I did not rush to the library.

Somewhere in the Mathematical Heavens there must have been a God or Goddess of Mathematics who would not let me fail to read Julia Robinson's new paper [15]. Because of my early publications on the subject, I was considered a specialist on it, and so the paper was sent to me to review for *Рефативный журнал Математика*, the Soviet counter-part of *Mathematical Reviews*. Thus, I was forced to read Julia Robinson's paper, and on December 11, I presented it to our logic seminar at LOMI.

Hilbert's tenth problem captured me again. I saw at once that Julia Robinson had a fresh and wonderful new idea. It was connected with the special form of Pell's equation

$$x^2 - (a^2 - 1)y^2 = 1. \tag{6}$$

Solutions $\langle \chi_0, \psi_0 \rangle, \langle \chi_1, \psi_1 \rangle, \ldots, \langle \chi_n, \psi_n \rangle, \ldots$ of this equation listed in the order of growth satisfy the recurrence relations

$$\chi_{n+1} = 2a\chi_n - \chi_{n-1}, \quad \psi_{n+1} = 2a\psi_n - \psi_{n-1}. \tag{7}$$

It is easy to see that for any m the sequences $\chi_0, \chi_1, \ldots, \psi_0, \psi_1, \ldots$ are purely periodic modulo m and hence so are their linear combinations. Further, it is easy to check by induction that the period of the sequence

$$\psi_0, \psi_1, \ldots, \psi_n, \ldots \quad (\bmod\ a - 1) \tag{8}$$

is

$$0, 1, 2, \ldots, a - 2, \tag{9}$$

whereas the period of the sequence

$$\chi_0 - (a - 2)\psi_0, \chi_1 - (a - 2)\psi_1, \ldots, \chi_n - (a - 2)\psi_n, \ldots \quad (\bmod\ 4a - 5) \tag{10}$$

begins with

$$2^0, 2^1, 2^2, \ldots. \tag{11}$$

The main new idea of Julia Robinson was to synchronize the two sequences by imposing a condition $G(a)$ which would guarantee that

the length of the period of (8) is a multiple of the length of the period of (10). (12)

If such a condition is Diophantine and is valid for infinitely many values of a, then one can easily show that the relation $a = 2^c$ is Diophantine. Julia Robinson, however, was unable to find such a G and, even today, we have no direct method for finding one.

I liked the idea of synchronization very much and tried to implement it in a slightly different situation. When, in 1966, I had started my investigations on Hilbert's tenth problem, I had begun to use Fibonacci numbers and had discovered (for myself) the equation

$$x^2 - xy - y^2 = \pm 1 \tag{13}$$

which plays a role similar to that of the above Pell equation; namely, Fibonacci numbers ϕ_n and only they are solutions of (13). The arithmetical properties of the sequences ψ_n and ϕ_n are very similar. In particular, the sequence

$$0, 1, 3, 8, 21, \ldots \tag{14}$$

of Fibonacci numbers with even indices satisfies the recurrence relation

$$\phi_{n+1} = 3\phi_n - \phi_{n-1} \tag{15}$$

similar to (7). This sequence grows like $[(3 + \sqrt{5})/2]^n$ and can be used instead of (11) for constructing a relation of exponential growth. The role of (10) can be played by the sequence

$$\psi_0, \psi_1, \ldots, \psi_n, \ldots \quad (\text{mod } a - 3) \tag{16}$$

because it begins like (14). Moreover, for special values of a the period can be determined explicitly; namely, if

$$a = \phi_{2k} + \phi_{2k+2}, \tag{17}$$

then the period of (16) is exactly

$$0, 1, 3, \ldots, \phi_{2k}, -\phi_{2k}, \ldots, -3, -1. \tag{18}$$

The simple structure of the period looked very promising.

I was thinking intensively in this direction, even on the night of New Year's Eve of 1970, and contributed to the stories about absentminded mathematicians by leaving my uncle's home on New Year's Day wearing his coat. On the morning of January 3, I believed I had found a polynomial B as in (5), but by the end of that day I had discovered a flaw in my work. But the next morning I managed to mend the construction.

What was to be done next? As a student I had had a bad experience when once I had claimed to have proved unsolvability of Hilbert's tenth problem but during my talk found a mistake. I did not want to repeat such an embarrassment, and something in my new proof seemed rather suspicious to me. I thought at first that I had just managed to implement Julia Robinson's idea in a slightly different situation; however, in her construction an essential role was played by a special equation that implied one variable was exponentially greater than another. My supposed proof did not need to use such an equation at all, and that was strange. Later I realized that my construction was a dual of Julia Robinson's. In fact, I had found a Diophantine condition $H(a)$ which implied that

the length of the period of (16) is a multiple of the length of the period of (8).   (19)

This H, however, could not play the role of Julia Robinson's G, which resulted in an essentially different construction.

I wrote out a detailed proof without finding any mistake and asked Sergei Maslov and Vladimir Lifshits to check it but not to say anything about it to anyone else. Earlier, I had planned to spend the winter holidays with my bride at a ski camp, so I left Leningrad before I got the verdict from Maslov and Lifshits. For a fortnight I was skiing, simplifying the proof, and writing the paper [6]. I tried to convey the impact of Julia Robinson's paper [15] on my work by a rather poetic Russian word навеять, which seems to have no direct counter-part in English, and the later English translator used plain "suggested."

On my return to Leningrad I received confirmation that my proof was correct, and it was no longer secret. Several other mathematicians also checked the proof, including D. K. Faddeev and A. A. Markov, both of whom were famous for their ability to find errors.

On 29 January 1970 at LOMI I gave my first public lecture on the solution of Hilbert's tenth problem. Among my listeners was Grigorii Tseitin, who shortly afterward attended a conference in Novosibirsk. He took a copy of my manuscript along and asked my permission to present the proof in Novosibirsk. (It was probably due to this talk that the English translation of [6] erroneously gives the Siberian Branch instead of the Leningrad Branch as my address.) Among those who heard Tseitin's talk in Novosibirsk was John McCarthy. In "The Autobiography" [11], Julia Robinson recalls that on his return to the United States, McCarthy sent her his notes on the talk. This was how Julia Robinson learned of my example of a Diophantine relation of exponential growth. Later, at my request, she sent me a copy of McCarthy's notes. They consisted of only a few main equations and lemmas, and I believe that only a person like Julia, who had already spent a lot of time intensively thinking in the same direction, would have been able to reconstruct the whole proof from these notes as she did.

In fact, Julia herself was very near to completing the proof of the unsolvability of Hilbert's tenth problem. The question sometimes asked is *why she did not* (this question is also touched upon in [11]). In fact, several authors (see [7] for further references) showed that ψs can be used instead of ϕs for constructing a Diophantine relation of exponential growth. My shift from (12) to (19) redistributed the difficulty in the entire construction. The path from a Diophantine H to a Diophantine relation of exponential growth is not as straightforward as the path from Julia Robinson's G would have been. On the other hand, it turned out that to construct an H is much easier than to construct a G. In [6], I used for this purpose a lemma stating that

$$\phi_n^2 | \phi_m \Rightarrow \phi_n | m. \tag{20}$$

It is not difficult to prove this remarkable property of Fibonacci numbers *after* it has been stated, but it seems that this beautiful fact was not discovered until 1969. My original proof of (20) was based on a theorem proved by the Soviet mathematician Nikolai Vorob'ev in 1942 but published only in the third augmented edition of his popular book [18]. (So the translator of my paper [6] made a misleading error by changing in the references the year of publication of [18] from 1969 to 1964, the year of the second edition.) I studied the new edition of Vorob'ev's book in the summer of 1969 and that theorem attracted my attention at once. I did not deduce (20) at that time, but after I read Julia Robinson's paper [15], I immediately saw that Vorob'ev's theorem could be very useful. Julia Robinson did not see the third edition of [18] until she received a copy from me in 1970. Who can tell what would have happened if Vorob'ev had included his theorem in the first edition of his book? Perhaps, Hilbert's tenth problem would have been "unsolved" a decade earlier!

The Diophantine definition of the relation of exponential growth in [6] had fourteen unknowns. Later I was able to reduce the number of unknowns to five. In October 1970, Julia sent me a letter with another definition also in only five unknowns. Having examined this construction, I realized that she had used a different method for reducing the number of unknowns and we could combine our ideas to get a definition in just three unknowns!

This was the beginning of our collaboration. It was conducted almost entirely by correspondence. At that time there was no electronic mail anywhere and it took three weeks for a letter to cross the ocean. One of my letters was lost in the mail and I had to rewrite eleven pages (copying machines were not available to me.) On the other hand, this situation had its own advantage: Today I have the pleasure of rereading a collection of letters written in Julia's hand. Citations from these letters are incorporated into this paper.

One of the corollaries of the negative solution of Hilbert's tenth problem (implausible to the reviewer for *Mathematical Reviews*) is that *there is a constant N such that, given a Diophantine equation with any number of parameters and in any number of unknowns, one can effectively transform this equation into another with the same parameters but in only N unknowns such that both equations are solvable or unsolvable for the same values of the parameters.* In my lecture at the Nice International Congress of Mathematicians in 1970, I reported that this N could be taken equal to 200. This estimate was very rough. Julia and her husband, Raphael, were interested in getting a smaller value of N, and in the above-mentioned letter Julia wrote that they had obtained $N = 35$. Our new joint construction of a Diophantine relation of exponential growth with three (instead of five) unknowns automatically reduced N to 33. Julia commented, "I consider it in the range of 'practical' number theory, since Davenport once wrote a paper on cubic forms in 33 variables."

Julia sent me a detailed proof of this reduction, and it became the basis for our further work. We were exchanging letters and ideas and gradually reducing the value of N further. In February 1971, I sent a new improvement that reduced N to 26 and commented that now we could write equations in Latin characters without subscripts for unknowns. Julia called it "breaking the 'alphabetical' barrier."

In August 1971, I reported to the IV International Congress on Logic, Methodology and Philosophy of Science in Bucharest on our latest result: *Any Diophantine equation can be reduced to an equation in only 14 unknowns* [7]. At that Congress Julia and I met for the first time. After the Congress I had the pleasure of meeting Julia and Raphael in my native city of Leningrad.

"With just 14 variables we ought to be able to know every variable personally and why it has to be there," Julia once wrote to me. However, in March 1972 the minimal number of unknowns unexpectedly jumped up to 15 when she found a mistake in my count of the number of variables! I would like to give the readers an idea of some of the techniques used for reducing the number of unknowns and to explain the nature of my mistake. Actually, we were constructing not a single equation but a system of equations in a small number of unknowns. (Clearly, a system $A = B = \cdots = D = 0$ can be compressed into a single equation $A^2 + B^2 + \cdots + D^2 = 0$.) Some of the equations used in our reduction were Pell equations:

$$x_1^2 - d_1 y_1^2 = 1, \quad x_2^2 - d_2 y_2^2 = 1. \tag{21}$$

We can replace these two Diophantine equations by a single one:

$$\prod \left(x \pm \sqrt{(1 + d_1 y_1^2)} \pm (1 + d_1 y_1^2)\sqrt{(1 + d_2 y_2^2)} \right) = 0, \tag{22}$$

where the product is over all the four choices of signs $\pm$. In the remaining equations, we substitute $\sqrt{(1 + d_1 y_1^2)}$ for x_1, $\sqrt{(1 + d_2 y_2^2)}$ for x_2, and eliminate the square roots by squaring. Thus, we reduce the total number of unknowns by 1 by introducing x but eliminating x_1 and x_2. It was in the count of variables, introduced and eliminated, that I made my error.

The situation was rather embarrassing for us because the result had been announced publicly. I tried to save the claimed result, but having no new ideas, I was unable to reduce the number of unknowns back to 14.

Soon I got a new letter from Julia. She tried to console me: "I think mistakes in reasoning are much worse than arithmetical ones which are sort of funny." But more important, she came up with new ideas and managed to reduce the number of unknowns to 14 again, thus saving the situation.

We discussed for some time the proper place for publishing our joint paper. I suggested the Soviet journal *Известия*. The idea of having a paper published in Russian was attractive to Julia. (Her paper [16] had been published in the USSR in English in spite of what is said in *Mathematical Reviews.*) On the other hand, she wanted to attract the attention of specialists in number theory to the essentially number-theoretical results obtained by logicians, so she suggested *Acta Arithmetica.* Finally, we decided that we had enough material for several papers and would publish our first joint paper in Russian in *Известия* and our second one somewhere else in English.

We found writing out a paper when we were half a world apart quite an ordeal. Later Julia wrote to me, "It seems to me that we had little trouble in collaborating mathematically on four-week turn-around time but it is hopeless when it comes to writing the results up. Namely, by the time you could answer a question, it was no longer relevant." We decided that one of us would write the whole manuscript, which was then to be subject to the other's criticism. Because the first paper was to be in Russian, I wrote the first draft (more than 60 typewritten pages) and sent it to Julia in autumn 1972. Of course, she found a number of misprints and small errors but, in general, she approved it.

The reader, however, need not search the literature for a reference to this paper because that manuscript has never been published! In May 1973, I found "a mistake in reasoning."

The mistake was the use of the incorrect implication

$$a \equiv b \pmod{q} \Rightarrow \binom{a}{c} \equiv \binom{b}{c} \pmod{q}. \tag{23}$$

The entire construction collapsed. I informed Julia and she replied:

I was completely flabbergasted by your letter of May 11. I wanted to crawl under a rock and hide from myself! Somehow I had never questioned that

$$\binom{a}{c} \equiv \binom{b}{c} \pmod{a-b}$$

I usually know enough not to divide by zero. I had even mentioned (asserted) it to Raphael several times, and he had not objected. He said he would have said 'No' if I had asked if it were true. I guess I would have myself if I had asked!

Earlier, we had discussed a similar situation, and in 1971 Julia had written to me, "Almost all mathematical mistakes come about from not writing out proofs and especially making changes after the proof is written out." But that was not the case this time. The mistake was present from an early stage and was not detected either when one mathematician (myself) wrote out a detailed proof or when another mathematician (Julia) carefully read it.

Luckily, this time I was able to repair the proof on the spot. Julia wrote, "I am very glad you sent a way around the mistake at the same time you told me about it!" However, the manuscript had to be completely rewritten.

In 1973, the prominent Soviet mathematician A. A. Markov celebrated his seventieth birthday. His colleagues from the Computing Center of the Academy of Sciences of the USSR decided to publish a collection of papers in his honor. I was invited to contribute to the collection. I suggested a joint paper with Julia Robinson, and the editors agreed. Because of the imminent deadline, we had no time to discuss the manuscript. I just asked Julia to authorize me to write the paper and to send it to the editors without her approval. Later I would incorporate her suggestions on the proof sheets. She agreed.

So our first joint publication [9] appeared, and it was in Russian. The paper was a by-product of our main investigations on reduction of the number of unknowns in Diophantine equations. The first theorem stated that given a parametric Diophantine equation (3) we can effectively find polynomials with integer coefficients $P_1, D_1, Q_1, \ldots, P_k, D_k, Q_k$ such that the Diophantine relation defined by (3) is also defined by the formula

$$\exists x \exists y \underset{i=1}{\overset{k}{\&}} \exists z \left[P_i(a, x, y) < D_i(a, x, y)z < Q_i(a, x, y) \right]. \tag{24}$$

While k can be a particular large fixed number, each inequality involves only 3 unknowns.

The second theorem states that we can also find polynomials F and W such that the same relation is defined by the formula

$$\exists x \exists y \forall z \left[z \le F(a, x, y) \Rightarrow W(a, x, y, z) > 0 \right]. \tag{25}$$

This formula also has only three quantifiers, but the third is a (bounded) universal one. Such representations have a close connection to equations because the main technical result of [2] is a method for eliminating a single bounded quantifier at the cost of introducing several extra existential quantifiers and allowing exponentiation to come into the resulting purely existential formula.

One of Julia's requests in regard to this paper was that her first name should be given. She had good reason for that. I had been the Russian translator of one of the fundamental papers on automatic theorem-proving by John A. Robinson [12]. When the translation appeared in 1970 in a collection of important papers on that subject, Soviet readers saw the names of Дж. Робинсон as the author of a paper translated by Ю. Матиясевич and М. Дэвис as the author of another fundamental paper on automatic theorem-proving. In the minds of many, these three names were associated with the recent solution of Hilbert's tenth problem, so a number of people got the idea that it was Julia Robinson who had invented the resolution principle, the main tool from [12]. To add to the confusion, John Robinson in his paper thanked George Robinson, whose name in Russian translation also becomes Дж. Робинсон.

As a student I had made "a mistake of the second kind": I did not identify J. Robinson, the author of a theorem in game theory with J. Robinson, the author of important investigations on Hilbert's tenth problem. (In fact, Julia's significant paper [13] was her only publication on game theory.)

Julia's request was agreed to by the editors, and as a result our joint paper [9] is the only Russian publication where *my* first name is given in full.

This short paper was a by-product of our main investigation, which was still to be published. As it had been decided beforehand that our second publication should be in English, Julia wrote the new paper about the reduction of the number of unknowns. Now we were able to eliminate one more variable and so had only "a baker's dozen" of unknowns.

The second paper [10] was published in *Acta Arithmetica*. We had a special reason for this choice because the whole volume was dedicated to the memory of the prominent Soviet mathematician Yu. V. Linnik, whom we both had known personally. I was introduced to him soon after showing Hilbert's tenth problem to be unsolvable. Someone had told Linnik the news beginning with one of the corollaries, "Matijasevich can construct a polynomial with integer coefficients such that the set of all natural number values assumed by this polynomial for natural number values of the variables is exactly the set of all primes." "That's wonderful," Linnik replied. "Most likely we soon shall learn a lot of new things about primes." Then it was explained to him that the main result is in fact much more general: Such a polynomial can be constructed for every recursively enumerable set, i.e., a set the elements of which can be listed in some order by an algorithm. "It's a pity," Linnik said. "Most likely, we shall not learn anything new about the primes."

Since there was some interest in our forthcoming paper with the proof of a long-announced result becoming at last accessible to other researchers, numerous copies were circulated. We had exhausted our ideas but there was a chance that someone with a fresh view of the subject might improve our result. "Of course there is the possibility that someone will make a breakthrough and supersede our paper too," Julia wrote, "but we should think of that as being good for mathematics!" Raphael, on the other hand, believed that thirteen unknowns would remain the best result for decades. Actually, the record fell even before our paper appeared.

The required "new idea" turned out, as so often happens, to be an old one that had been forgotten. In this case, it was the following nice result by E. E. Kummer: *The greatest power of a prime p which divides the binomial coefficient $\begin{pmatrix} a+b \\ a \end{pmatrix}$ is p^c, where c is the number of carries needed when adding a and b written to base p.* This old result was rediscovered and reproved a number of times and I was lucky to learn it from the review of [17] in *Рефтивный журнал Математика*. Kummer's theorem turned out to be an extremely powerful tool for constructing Diophantine equations with special properties. (Julia once called it "a gold mine.") It would be too technical to explain all the applications, but one of them can be given here.

Let p be a fixed prime and let f be a map from $\{0, 1, \ldots, p-1\}$ into itself such that $f(0) = 0$. Such an f generates a function F defined by

$$F(\overline{a_n a_{n-1} \cdots a_0}) = \overline{f(a_n) f(a_{n-1}) \cdots f(a_0)}, \tag{26}$$

where $a_n a_{n-1} \cdots a_0$ is the number with digits $a_n, a_{n-1}, \ldots, a_0$ to base p. Now we can easily prove that F is an exponential Diophantine function. Namely, $b = F(a)$ if and only if there are natural numbers $c_0, \ldots, c_{p-1}, d_0, \ldots, d_{p-1}, k, s, u, w_0, \ldots, w_{p-1}, v_0, \ldots, v_{p-1}$ such that

$$a = 0 * d_0 + 1 * d_1 + \cdots + (p-1) * d_{p-1}, \tag{27}$$

$$b = f(0) * d_0 + f(1) * d_1 + \cdots + f(p-1) * d_{p-1}, \tag{28}$$

$$s = d_0 + d_1 + \cdots + d_{p-1}, \tag{29}$$

$$s = (p^{k+1} - 1)/(p-1), \tag{30}$$

$$u = 2^{s+1}, \tag{31}$$

$$(u+1)^s = w_i u^{d_i+1} + c_i u^{d_i} + v_i, \tag{32}$$

$$v_i < u^{d_i}, \tag{33}$$

$$c_i < u, \tag{34}$$

$$p \nmid c_i. \tag{35}$$

This system has a solution with

$$d_i = \sum_{l=0}^{k} \delta_i(a_l) p^l, \tag{36}$$

where δ_i is the delta-function: $\delta_i(i) = 1$, otherwise $\delta_i(j) = 0$. In this solution

$$w_i = \sum_{k=d_i+1}^{s} \binom{s}{k} u^k, \tag{37}$$

$$c_i = \binom{s}{d_i}, \tag{38}$$

$$v_i = \sum_{k=0}^{d_i-1} \binom{s}{k} u^k; \tag{39}$$

and for any given value of k that solution is in fact unique.

Kummer's theorem serves as a bridge between number theory and logic because it enables one to work with numbers as sequences of indefinite length consisting of symbols from a finite alphabet. Application of Kummer's theorem to reducing the number of unknowns resulted in a real break-through and, in one jump, that number dropped from 13 to 9. I wrote out a sketch of the new construction and sent it to Julia. When we met for the second time in London, Ontario, during the V International Congress on Logic, Methodology and Philosophy of Science, she confirmed that the proof was correct, so I dared to present the result in my talk [8]. We hoped to be able to publish it as an addendum to our paper in *Acta Arithmetica*, but it turned out to be too late.

In 1974, the American Mathematical Society organized a symposium on "Mathematical Developments Arising From Hilbert's Problems" at DeKalb, Illinois. I was invited to speak about the tenth problem, but my participation in the meeting did not get the necessary approval in my country, so Julia became the speaker on the problem; however, she suggested that the paper for the *Proceedings* of the meeting be a joint one by Martin Davis and the two of us. Again we had the problem of an approaching deadline. So we first discussed by phone what topics each of us would cover. Of course, Julia and Martin had much more communication with each other than with me. The final difficult work of combining our three contributions into a coherent exposition [1] was done by Martin. I believe that this paper turned out to be one about which Julia had thought for a long time: a non-technical introduction to many results obtained by logicians in connection with Hilbert's tenth problem.

Writing the paper for the *Proceedings* prevented me from immediately writing a paper about the new reduction to 9 unknowns (clearly it was my turn to write it up). Unfortunately, Julia firmly refused to be a coauthor. She wrote, "I do not want to be a joint author on the 9 unknowns paper— I have told everyone that it is your improvement and in fact I would feel silly to have my name on it. If I could make some contribution it would be different."

I am sure that without Julia's contribution to [10] and without her inspiration I would never have reduced N to 9. I was not inclined to publish the proof by myself, and so the result announced in [9] did not appear in print with a full proof for a long time. At last James P. Jones of the University of Calgary spent half a year in Berkeley, where Julia and Raphael lived. He studied my sketch and Julia's comments on it, and made the proof available to everybody in [5].

The photo of Julia Robinson, Martin Davis, and myself accompanying this article was taken in Calgary at the end of 1982 when I spent three months in Canada collaborating with James as part of a scientific exchange program between the Steklov Institute of Mathematics and Queen's University at Kingston, Ontario. Julia at that time was very much occupied with her new duties as president of the American Mathematical Society and was not very active in mathematical research, but she visited Calgary on her way to a meeting of the society. Martin also came to Calgary for a few days.

I conclude these reminiscences with yet another citation from Julia's letters with which I completely agree, "Actually I am very pleased that working together (thousands of miles apart) we are obviously making more progress than either one of us could alone."

References

1. Martin Davis, Yuri Matijasevich, and Julia Robinson, Hilbert's tenth problem. Diophantine equations: positive aspects of a negative solution, *Proc. Symp. Pure Math.* 28 (1976), 323–378.
2. Martin Davis, Hilary Putnam, and Julia Robinson, The decision problem for exponential Diophantine equations, *Ann. Math.* (2) 74 (1961), 425–436.
3. G. V. Davydov, Yu. V. Matijasevich, G. E. Mints, V. P. Orevkov, A. O. Slisenko, A. V. Sochilina and N. A. Shanin, "Sergei Yur'evich Maslov" (obituary), *Russian Math. Surveys* 39(2) (1984), 133–135 [translated from *Uspekhi Mat. Nauk* 39(2) (1984), 129–130].
4. David Hilbert, Mathematische Probleme. Vortrag, gehalten auf dem intenationalen Mathematiker Kongress zu Paris 1900, *Nachr. K. Ges. Wiss., Göttingen, Math.-Phys. Kl.* (1900), 253–297.
5. James P. Jones, Universal diophantine equation, *J. Symbolic Logic* 47 (1928), 549–571.
6. Ju. V. Matijasevich, Enumerable sets are Diophantine, *Soviet Math. Doklady* 11(20) (1970), 354–357.
7. Yuri Matijasevich, On recursive unsolvability of Hilbert's tenth problem, *Proceedings of the Fourth International Congress on Logic, Methodology and Philosophy of Science, Bucharest, 1971,* Amsterdam: North-Holland (1973), 89–110.
8. Yuri Matijasevich, Some purely mathematical results inspired by mathematical logic, *Proceedings of the Fifth International Congress on Logic, Methodology and Philosophy of Science, London, Ontario, 1975,* Dordrecht: Reidel (1977), 121–127.
9. Yuri Matijasevich, *AH CCCP* (1974), 112–123.
10. Yuri Matijasevich and Julia Robinson, Reduction of an arbitrary Diophantine equation to one in 13 unknowns, *Acta Arith.* 27 (1975), 521–553.
11. Constance Reid, The autobiography of Julia Robinson, *College Math. J.* 17 (1986), 3–21.
12. John A. Robinson, A machine-oriented logic based on the resolution principle, *J. Assoc. Comput. Mach.* 12 (1965), 23–41.
13. Julia Robinson, An iterative method of solving a game, *Ann. Math.* (2) 54 (1951), 296–301.
14. Julia Robinson, Existential definability in arithmetic, *Trans. Amer. Math. Soc.* 72 (1952), 437–449.
15. Julia Robinson, Unsolvable diophantine problems, *Proc. Amer. Math. Soc.* 22 (1969), 534–538.
16. Julia Robinson, Axioms for number theoretic functions, *Selected Questions of Algebra and Logic (Collection Dedicated to the Memory of A. I. Mal'cev),* Novosibirsk: Nauka (1973), 253–263; MR 48#8224.
17. D. Singmaster, Notes on binomial coefficients, *J. London Math. Soc.* 8 (1974), 545–548.
18. N. N. Vorob'ev, *Fibonacci numbers,* 2nd ed., Moscow: Nauka, 1964; 3rd ed., 1969.

6

C. N. Yang and Contemporary Mathematics

D. Z. Zhang

C. N. Yang, one of the twentieth century's great theoretical physicists, shared the Nobel Prize in physics with T. D. Lee in 1957 for their joint contribution to parity non-conservation. Mathematicians, however, know Yang best for the Yang–Mills theory and the Yang–Baxter equation. After Einstein and Dirac, Yang is perhaps the twentieth-century physicist who has had the greatest impact on the development of mathematics. I interviewed Dr. Yang in 1991; this article is based on my notes of the interviews together with his published papers and books.

Early Interactions between Yang and Chern

Yang was born in 1922 in Hefei, a mid-sized city in eastern China. His father, K. C. Yang (Yang Ko-Chuen, also known as W. C. Yang), served as a professor of mathematics at Tsinghua (now Qinghua) University in Peking (now Beijing) and later at Fudan University in Shanghai. The elder Yang had received his Ph.D. in number theory from the University of Chicago in 1928 under L. E. Dickson. One of the first to introduce modern mathematics into China, K. C. Yang taught many talented students. Among them, two later became famous: Loo-Keng Hwa and S. S. Chern.

ZHANG: *When did you first meet Professor Chern?*

YANG: I do not remember whether I had met him when he was a graduate student of Qinghua University in Beijing, 1930–1934, where my father was a mathematics professor. But I do remember when and how I first met Mrs. Chern. It was in early October 1929. Her father, Professor Tsen, had been a professor of mathematics in Qinghua University for a number of years, and the Yangs were newcomers that fall. I was seven years old and was going to elementary school. The Tsens invited us to their house for dinner and that was when I first made the acquaintance of "big sister Tsen." The Tsen and Yang families were very close, and it was a great joy for my parents to have been among the "introducers" in the marriage of the Cherns in Kunming in 1939.

ZHANG: *As a student of the Department of Physics at Qinghua University in 1938–1942, did you learn mathematics from Chern?*

YANG: When Chern came back to China to teach in 1937, Qinghua had, because of the Japanese invasion, combined with Beijing University and Nankai University to form the wartime National

Volume 15, No. 4 (Fall 1993), 13–21.

FIGURE 6-1 *C. N. Yang and his parents.*

Southwest Associated University in Kunming. Chern taught at this university for six years, 1937–1943. He was a brilliant and popular professor. I was first an undergraduate, later a graduate student, at the same university. I have very fond memories of my student years on that campus and am deeply grateful for the excellent education I received there.

I probably audited several of Professor Chern's courses in mathematics, but a transcript of my records, which I still have today, shows that I had taken only one course with him, Differential Geometry. That was in the fall of 1940 when I was a junior.

ZHANG: *This course benefited you, didn't it?*

YANG: Of course. But I do not remember that course today very distinctly. Only one thing sticks in my mind: how to prove that every two-dimensional surface is conformal to the plane. I knew how to transform the metric into the form $A^2 du^2 + B^2 dv^2$ but for a long time had not been able to make further progress. When Chern told me to use complex variables and write $C\,dz = A\,du + iB\,dv$, it was like a bolt of lightning which I never forgot.

ZHANG: *When did you arrive in the United States?*

YANG: In November 1945. I had hoped to study with Fermi or Wigner after I got to the United States. However, I did not find Fermi at Columbia, where he had been before 1942. I went to Princeton and discovered to my deep regret that Wigner would be mostly unavailable to students for the next year. Fortunately, I learned Fermi would join a newly established Institute at Chicago. That is why I went to the University of Chicago for my Ph.D.

ZHANG: *Chern was a Professor at the University of Chicago for a long time.*

YANG: Yes, but only after I had left Chicago in 1949. Chern and I, however, have met frequently in Princeton, Chicago, and Berkeley, ever since he came back to the United States in early 1949.

ZHANG: *Did you discuss fiber bundles?*

YANG: Not until the 1970s. Our earlier contacts were social. We did discuss mathematicians but not mathematics.

The 1954 Yang–Mills Paper

While a graduate student in Kunming and Chicago, Yang was impressed with the fact that gauge invariance determined all electromagnetic interactions. This was known from the work in the years 1918–1929 of Weyl, Fock, and London, and through later review papers of Pauli. But by the 1940s and early 1950s, it played only a minor and technical role in physics. In Chicago, Yang tried to generalize the concept of gauge invariance to non-Abelian groups [the gauge for electromagnetism being the Abelian $U(1)$]. In analogy with Maxwell's equations, he tried

$$F_{\mu\nu} = \frac{\partial B_\mu}{\partial x_\nu} - \frac{\partial B_\nu}{\partial x_\mu} \qquad (*)$$

FIGURE 6-2 *Robert Mills.*

which had appeared to him to be a natural generalization of Maxwell's equations. "This led to a mess, and I had to give up" [1, p. 19].

In 1954, as a visiting physicist at Brookhaven National Laboratory on Long Island, New York, Yang returned once again to the idea of generalizing gauge invariance. His officemate was R. L. Mills, who was about to finish his Ph.D. degree at Columbia University. Yang introduced the idea of non-Abelian gauge field to Mills, and they decided to add a quadratic term to the right side of (*). That cleared up the "mess" and led to a beautiful new field theory. They submitted a paper in the summer of 1954 to the *Physical Review*, which was published in October of that year as "Conservation of Isotopic Spin and Isotopic Gauge Invariance." [2] Mills wrote later about this period:

> During the academic year 1953–1954, Yang was a visitor to Brookhaven National Laboratory . . . I was at Brookhaven also . . . and was assigned to the same office as Yang. Yang, who has demonstrated on a number of occasions his generosity to young physicists beginning their careers, told me about his idea of generalizing gauge invariance and we discussed it at some length. . . . I was able to contribute something to the discussions, especially with regard to the quantization procedures, and to a small degree in working out the formalism; however, the key ideas were Yang's" [3, p. 495].

ZHANG: *I read that Mills was in England: "In 1954, Yang in the United States and Mills in England constructed a non-linear version of Maxwell Equations that incorporated a non-Abelian group" [4, p. 463].*

YANG: That was wrong. Mills was in the United States in 1954. He later did visit England many times, but not in 1954.

ZHANG: *M. E. Mayer said in 1977: "A reading of the Yang–Mills paper shows that the geometric meaning of the gauge potentials must have been clear to the authors, since they use the gauge-invariant derivative and the curvature form of the connection, and indeed the basic equations in that paper will coincide with the ones derived from a more geometric approach . . ." [5, p. 2]. Is that correct?*

YANG: Totally false. What Mills and I were doing in 1954 was generalizing Maxwell's theory. We knew of no geometrical meaning of Maxwell's theory, and we were not looking in that direction. To a physicist, gauge potential is a concept *rooted* in our description of the electromagnetic field. Connection is a geometrical concept which I only learned around 1970. That Maxwell's equations have deep geometrical meaning was a surprising revelation to the physicists.

ZHANG: *An interesting question is whether you understood in 1954 the tremendous importance of your original paper on non-Abelian gauge theory.*

YANG: No. In the 1940s we felt our work was elegant. I realized its importance in the 1960s and its great importance to physics in the 1970s. Its relationship to deep mathematics became clear to me only after 1974.

ZHANG: *As is well known, H. Weyl initiated the idea of Abelian gauge theory. Why was Weyl's work not mentioned in your paper?*

YANG: In the 1940s and 1950s, physicists knew that Weyl had introduced the Abelian gauge idea but always referred to Pauli's review papers [6, 7]. In fact, I did not read any of Weyl's papers at that time.

ZHANG: *Did you meet Weyl in Princeton?*

YANG: Of course. I shall show you what I said about this matter in my talk at Zurich in 1985 in celebration of the centenary of Weyl's birth:

> I had met Weyl in 1949 when I went to the Institute for Advanced Study in Princeton as a young "member". I saw him from time to time in the next years, 1949–1955. He was very approachable, but I don't remember having discussed physics and mathematics with him at any time. His continued interest in the idea of gauge fields was not known among the physicists. Neither Oppenheimer nor Pauli ever mentioned it. I suspect they also did not tell Weyl of the 1954 papers of Mills and mine. Had they done that, or had Weyl somehow come across our paper, I imagine he would have been pleased and excited, for we had put together two things that were very close to his heart: gauge invariance and non-Abelian Lie groups. [8, pp. 19–20].

ZHANG: *I read in this beautiful article of yours on Weyl that he had originated the two-component theory of the neutrino.*

YANG: That is correct. He wrote about this theory in 1929 and pointed out that it did not observe right–left symmetry, and therefore would not be realized in nature. About thirty years later, in 1956–1957, when it was found that right–left symmetry is not strictly observed anyway, Weyl's theory was revived. It is still the correct theory of the neutrino today.

By the way, we bought Weyl's house in Princeton two years after Weyl died. We lived in it for nine years: 1957–1966.

ZHANG: *What was Weyl's reaction to the news that his neutrino theory had become reality?*

YANG: Weyl unfortunately died two years before the great excitement in physics in 1957. Early that year it was announced that right–left symmetry is not strictly observed, that is, parity is not strictly conserved. Then the Weyl theory was revived. It fitted beautifully the experiments on mu-decay. There followed six months of great confusion about beta-decay, which is related to the question whether the Weyl neutrino is right-handed or left-handed. Then in the fall there was the V–A proposal for the structure of beta-decay. In December there was an ingenious experiment which clarified everything, including the finding that the Weyl neutrino is left-handed.

Weyl was thirty-seven years older than Yang. They belonged to different academic generations, came from different countries, were in different disciplines. Could it be said that Weyl was a mathematician who deeply appreciated physics, and Yang is a physicist who deeply appreciates mathematics?

Yang–Mills Theory and Geometry

After the original paper written by Yang and Mills, a large number of papers were devoted to the quantization and renormalization of gauge theories and to finding solutions of Yang–Mills equations. Relatively few people paid attention to the geometric and topological aspects of gauge theories. Among those who did, however, were S. Mandelstam (1962), E. Lubkin (1963), and H. G. Loos (1967). In addition, R. Hermann published a series of mathematical books for physicists, some of which were on this subject. None of these seemed to have left much impact. I asked Yang about his own experience in realizing the relationship between gauge theory and geometry.

ZHANG: *Did you study gauge theory continuously after 1954?*

YANG: Yes, I did. Although non-Abelian gauge theory was not used in any practical way in physics in the 1950s and 1960s, the elegance of it was more and more appreciated as time went on. For example, in 1964, D. Ivanienko published a collection of Russian translations of twelve articles on gauge theory by Yang and Mills, Lee and Yang, Sakurai, Gell-Mann, and so on. I myself continued to work on various aspects of gauge fields throughout the 1950s, although I did not obtain many useful results.

In the late 1960s, I began a new formulation of gauge fields, through the approach of non-integrable phase factors. It happened that one semester I was teaching general relativity, and I noticed that the formula in gauge theory

$$F_{\mu\nu} = \frac{\partial B_\mu}{\partial x_\nu} - \frac{\partial B_\nu}{\partial x_\mu} + i\epsilon(B_\mu B_\nu - B_\nu B_\mu) \tag{1}$$

and the formula in Riemann geometry

$$R^l_{ijk} = \frac{\partial}{\partial x^j}\begin{Bmatrix} l \\ ik \end{Bmatrix} - \frac{\partial}{\partial x^k}\begin{Bmatrix} l \\ ij \end{Bmatrix} + \begin{Bmatrix} m \\ ik \end{Bmatrix}\begin{Bmatrix} l \\ mj \end{Bmatrix} - \begin{Bmatrix} m \\ ij \end{Bmatrix}\begin{Bmatrix} l \\ mk \end{Bmatrix} \tag{2}$$

are not just similar—they are, in fact, the same if one makes the right identification of symbols! It is hard to describe the thrill I felt at understanding this point.

ZHANG: *Is that the first time you realized the relation between gauge theory and differential geometry?*

YANG: I had noticed the similarity between Levi-Civita's parallel displacement and non-integrable phase factors in gauge fields. But the exact relationship was appreciated by me only when I realized that (1) and (2) are the same.

With an appreciation of the geometrical meaning of gauge theory, I consulted Jim Simons, a distinguished geometer, who was then the chairman of the Mathematics Department at Stony Brook. He said gauge theory must be related to connections on fiber bundles. I then tried to understand fiber-bundle theory from such books as Steenrod's *The Topology of Fibre Bundles*, but learned nothing. The language of modern mathematics is too cold and abstract for a physicist.

ZHANG: *I suppose only mathematicians appreciate the mathematical language of today.*

YANG: I can tell you a relevant story. About ten years ago, I gave a talk on physics in Seoul, South Korea. I joked, "There exist only two kinds of modern mathematics books: one which you cannot read beyond the first page and one which you cannot read beyond the first sentence." *The Mathematical Intelligencer* later reprinted this joke of mine. But I suspect many mathematicians themselves agree with me.

ZHANG: *When did you understand bundle theory?*

YANG: In early 1975, I invited Jim Simons to give us a series of luncheon lectures on differential forms and bundle theory. He kindly accepted the invitation, and we learned about de Rham's theorem, differential forms, patching, and so on. That was very useful and allowed us to understand the mathematical meaning of the Aharonov–Bohm experiment and of the Dirac quantization rule of electric and magnetic monopoles. H. S. Tsao and I later also understood the profound and very general Chern–Weil theorem. In retrospect, it was these lectures that taught me the concept of manifold, which I had appreciated only very vaguely.

Yang–Singer–Atiyah

Simons' lectures helped T. T. Wu and Yang to write a famous paper: "Concept of non-integrable Phase Factors and Global Formulation of Gauge Fields" [9]. In this paper, they analyzed the intrinsic meaning of electromagnetism, emphasizing especially its global topological connections. They discussed the mathematical meaning of the Aharonov–Bohm experiment and of the Dirac magnetic monopole. They exhibited a dictionary as presented in Table 6-1.

Table 6-1. **Translation of Terminologies.** [9]

GAUGE FIELD TERMINOLOGY	BUNDLE TERMINOLOGY
gauge (or global gauge)	principal coordinate bundle
gauge type	principal fiber bundle
gauge potential b_μ^k	connection on a principal fiber bundle
S_{ba}	transition function
phase factor Φ_{QP}	parallel displacement
field strength $f_{\mu\nu}^k$	curvature
source J_μ^K	?
electromagnetism	connection on a $U_1(1)$ bundle
isotopic spin gauge field	connection on a SU_2 bundle
Dirac's monopole quantization	classification of $U_1(1)$ bundle according to first Chern class
electromagnetism without monopole	connection on a trivial $U_1(1)$ bundle
electromagnetism with monopole	connection on a non-trivial $U_1(1)$ bundle

Half a year later, in the summer of 1976, I. M. Singer of MIT visited Stony Brook and discussed these matters at length with Yang. Singer had been an undergraduate student in physics and a graduate student in mathematics in the 1940s. He wrote in 1985:

> Thirty years later, I found myself lecturing on gauge theories at Oxford, beginning with the Wu and Yang dictionary and ending with instantons, i.e., self-dual connections. I would be inaccurate to say that after studying mathematics for thirty years, I felt prepared to return to physics. [10, p. 200].

In order to explain the developments of the past decade, Singer reproduced the Wu–Yang dictionary in this paper of 1985.

In April–May of 1977, an Oxford–Berkeley–MIT preprint by M. F. Atiyah, N. J. Hitchin, and I. M. Singer [11] was circulated. It applied the Atiyah–Singer Index Theorem to the problem of self-dual gauge fields. Thus began the interest of many mathematicians in gauge fields.

In 1979, Atiyah published a monograph called *Geometry of Yang–Mills Fields* [12]. Volume 5 of his *Collected Works* has the subtitle "Gauge Theories." I found a copy of this volume signed by Atiyah on the bookshelf in Yang's office at Stony Brook. By way of introducing the volume, Atiyah wrote:

From 1977 onward my interests moved in the direction of gauge theories and the interactions between geometry and physics. I had for many years had a mild interest in theoretical physics, stimulated on many occasions by lengthy discussions with George Mackey. However, the stimulus in 1977 came from two other sources. On the one hand, Singer told me about the Yang–Mills equations, which through the influence of Yang were just beginning to percolate into mathematical circles. During his stay in Oxford in early 1977, Singer, Hitchin and I took a serious look at the self-duality equations. We found that a simple application of the index theorem gave the formula for the number of instanton parameters . . . The other stimulus came from the presence in Oxford of Roger Penrose and his group. [13]

ZHANG: *Why did you leave a question mark in the middle of your dictionary?*

YANG: Because mathematicians had not investigated the concept, so familiar and important to physicists, called source, usually denoted by J. It was a key concept in Maxwell's formulation of Coulomb's and of Ampere's laws. In modern mathematical notations, it is

$$ {}^*D\,{}^*f = J = \text{source.} $$

The sourceless case satisfies

$$ D\,{}^*f = 0 $$

which is fulfilled if $f = \pm\,{}^*f$, and that is what led both physicists and mathematicians to the study of self-dual gauge fields.

ZHANG: *This is an extremely interesting story. The study of self-dual gauge fields led later to much beautiful mathematics, including Donaldson's result which won a Fields Medal [see below].*

YANG: Yes. The story supplies a modern example of how mathematicians can derive concepts from physics, which was prevalent in earlier centuries, but unfortunately rare now.

ZHANG: *How about ideas in mathematics becoming important for physics? We may recall Einstein was advised to pay attention to tensor analysis. Is that similar to your getting help from Simons?*

YANG: Einstein's profound depth and stunning insight were such that no mortals should be compared with him in any way. As to the entry of mathematics into general relativity and into gauge theory, the processes were quite different. In the former, Einstein could not formulate his ideas without Riemannian geometry, while in the latter, the equations were written down, but an intrinsic overall understanding of them was later supplied by mathematics.

ZHANG: *There were many scholars who had pointed out earlier that gauge theory is related to bundle theory. Why did their papers not exert the same degree of influence in mathematical circles as your paper?*

YANG: There may be many factors. The work may have been so formal that physicists were unable to understand what it said. It may have seemed trivial to mathematicians because the physical content may not have been clarified. As to the paper Wu and I wrote in 1975, our discussion of the Aharonov–Bohm experiment and the Dirac monopole helped to draw people's attention to it. Also the dictionary helped.

ZHANG: *Have you had scientific correspondence with Singer and Atiyah?*

YANG: I have met them from time to time, but there has been no research cooperation.

Yang–Baxter Equation

The other mathematical structure Yang contributed to mathematics is the Yang–Baxter equation, which arose from his work on statistical mechanics.

In 1967, Yang tried to find the eigenfunctions of a one-dimensional fermion gas with delta function interaction [14]. This was a rather difficult problem. He solved it and showed that a crucial identity in the intermediate steps was a matrix equation:

$$A(u)B(u + v)A(v) = B(v)A(u + v)B(u). \tag{**}$$

A few years later, R. J. Baxter in his solution of another problem in physics, the eight-vertex model [15], again used equation (**).

Both lines of development were later pursued in several centers of research, especially in the USSR, where the largest efforts were concentrated. In 1980, Faddeev coined the term "Yang–Baxter relation" or "Yang–Baxter equation" and that has become the generally accepted name today. In the last six or seven years, a number of exciting developments in physics and mathematics have led to the conclusion that the Yang–Baxter equation is a fundamental mathematical structure with connections to various subfields of mathematics, such as knot and braid theory, operator theory, the theory of Hopf Algebra, quantum group theory, the topology of three-manifolds, the monodromy of differential equations, and so on. There has been an explosion of literature on these subjects [10–12].

ZHANG: *The YBE (Yang–Baxter equation) is just a simple matrix equation. Why does it have such great importance?*

YANG: In the simplest situation, the YBE has the form

$$ABA = BAB.$$

This is the fundamental equation of Artin for the braid group. The braid is, of course, a record of the history of permutations. It is not difficult to understand that the history of permutations is relevant to many problems in mathematics and physics.

Looking at the developments of the last six or seven years, I got the feeling that the YBE is the next pervasive algebraic equation after the Jacobi identity

$$C_{ab}^i C_{ic}^j + C_{ca}^i C_{ib}^j + C_{bc}^i C_{ia}^j = 0.$$

The study of the Jacobi identity has, of course, led to the whole of Lie algebra and its relationship to Lie groups.

ZHANG: *The influences of YBE upon mathematics seems to be stronger than upon physics.*

YANG: This is true right now. In fact, some physicists think the YBE is pure mathematics. But I think that will change. The YBE is a fundamental structure. Even if a physicist does not like it, he/she may have to use it eventually. In the 1920s, many physicists called group theory "group pest." That attitude persisted well into the 1930s, but disappeared later.

FIGURE 6-3   *R. J. Baxter.*

Fields Medals of 1986 and 1990

Yang–Mills theory and the Yang–Baxter equation both figure prominently in today's core mathematics. One can see this by the Fields Medals awarded in 1986 and 1990.

Simon Donaldson was awarded a Fields Medal at the ICM held in Berkeley in 1986. M. F. Atiyah spoke on Donaldson's work:

> Together with the important work of Michael Freedman (another Fields Medal winner in 1986), Donaldson's result implied that there exist "exotic" four-dimensional spaces which are topologically but not differentially equivalent to the standard Euclidean four-space $\mathbf{R}^4$ ... Donaldson's results are derived for the Yang–Mills equations of theoretical physics, which are non-linear generalizations of Maxwell's equations. In the Euclidean case, the solution to the Yang–Mills equations giving the absolute minimum are of special interest and called instantons. [19].

There were four Fields Medalists in 1990: V. Drinfeld, V. F. R. Jones, S. Mori, and E. Witten. The work of three of them was related to Yang–Mills theory and/or the Yang–Baxter equation. The following quotes are from reports on the Kyoto Conference:

> We should mention Drinfeld's pioneering work with Manin on the construction of instantons. These are solutions to the Yang–Mills equations which can be thought of as having particle-like properties of localization and size. . . . Drinfeld's interests in physics continued with his investigation of the Yang–Baxter Equation. [20]

> Jones opened a whole new direction upon realizing that under certain conditions solutions of the Yang–Baxter equation could be used for constructing invariants of links . . . The theory of quantum groups, non-commutative Hopf algebras, was devised by Jimbo and Drinfeld to produce solutions of Yang–Baxter equations. [21]

> Witten described in these terms the invariant of Donaldson and Floer (extending the earlier ideas of Atiyah) and generalized the Jones knot polynomial to the case of an arbitrary ambient 3-manifold. [22]

We note with amusement that there were complaints that the plenary lectures at the ICM-1990 (Kyoto) were heavily slanted toward the topics of mathematical physics:

> Everywhere we heard quantum group, quantum group, quantum group! [23]

Mathematics and Physics

ZHANG: *Why did your work in physics produce such a great impact in mathematics?*

YANG: This is, of course, very difficult to answer. Luck is a factor. Beyond that, two points may be relevant. First, if one chooses to look into simple problems, one has a bigger chance of coming close to fundamental structures in mathematics. Second, one must have a certain appreciation of the value judgment of mathematics.

ZHANG: *Please say more about the first point.*

YANG: Most papers in theoretical physics are produced in the following way: A publishes a paper about his theory. B says he can improve on it. Then C points out that B is wrong, and so forth. Most of the time, it turns out that the original idea of A is totally wrong or irrelevant.

ZHANG: *In mathematical circles, too, one has this situation.*

YANG: No, no. It is very different. Mathematical theorems are proved, or supposed to be proved. In theoretical physics, we are pursuing instead a guessing game, and guesses are mostly wrong.

ZHANG: *It is, however, necessary to read the newest publications.*

YANG: Of course. It is important to know what other research workers in one's field are thinking about. But to make real progress, one must face original simple physical problems, not other people's guesses.

ZHANG: *Was that what you were doing with Mills in 1954?*

YANG: Yes. We asked, "Could we generalize Maxwell's equations so as to obtain general guiding rules for interactions between particles?"

ZHANG: *What about the Yang–Baxter equation? You were not treating in 1967 a basic important problem in physics.*

YANG: This is correct. But I was looking at one of the simplest mathematical problems in quantum mechanics: A fermion system in one dimension with the simplest interaction possible.

ZHANG: *Why do you emphasize "simplest"?*

YANG: Because the simpler the problem, the more the analysis is likely to be close to some basic mathematical structure. I can illustrate this with the following observation: If there is a mathematics-based winning strategy in the game of chess or in Wei-qi (known in the United States by the later Japanese name of "go"), then it must be in Wei-qi, because Wei-qi is a simpler, more basic game.

ZHANG: *Please talk about the second point.*

YANG: Many theoretical physicists are, in some ways, antagonistic to mathematics, or at least have a tendency to downplay the value of mathematics. I do not agree with these attitudes. I have written:

> Perhaps because of my father's influence, I appreciate mathematics more. I appreciate the value judgement of the mathematician, and I admire the beauty and power of mathematics. There are ingenuity and intricacy in tactical maneuvers, and breathtaking sweeps in strategic campaigns. And, of course, miracle of miracles, some concepts in mathematics turn out to provide the fundamental structures that govern the physical universe! [1, p. 74]

ZHANG: *What were your father's mathematical influences on you?*

YANG: To give one example: I was exposed to the rudiments of group theory by my father when I was a high school student, and I had always been fascinated by the beautiful diagrams in the book by A. Speiser on finite groups that he had on his bookshelf. When I worked on my Bachelor's degree thesis, he suggested that I should learn about group representations from a small book called *Modern Algebraic Theory* by L. E. Dickson, which presented in a short chapter of twenty pages the essentials of the theory of characters. The elegance and potency of the chapter introduced me to the incredible beauty and power of group theory.

ZHANG: *It is said that you had been a mathematics teacher and that your wife was a student of yours.*

YANG: Yes. I spent the year 1944–1945 teaching mathematics at a high school in Kunming, and she was in one of my classes. But we did not know each other well. Several years later, I accidentally ran into her in Princeton. It was an interesting experience, teaching high school mathematics. But that has nothing to do with my attitude toward mathematics.

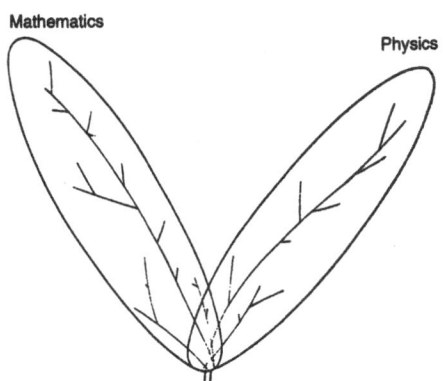

FIGURE 6-4 *S. S. Chern and W. Z. Yang, Geneva, 1964.*

FIGURE 6-5 *Mathematics and physics.*

ZHANG: *Is it important for a physicist to learn a lot of mathematics?*

YANG: No. If a physicist learns too much mathematics, he or she is likely to be seduced by the value judgment of mathematics and may lose his or her physical intuition. I have likened the relationship between physics and mathematics to a pair of leaves (Figure 6-5). They share a small common part at the base, but mostly they are separate:

> They have their own aims and distinctly different value judgements, and they have different traditions. At the fundamental conceptual level they amazingly share some concepts, but even there, the life force of each discipline runs along its own veins. [1]

ZHANG: *For a physicist, experimental results are more important to learn?*

YANG: This is right.

ZHANG: *Did you have much exchange with mathematicians?*

YANG: Some. When T. D. Lee and I were working in 1951 on what was later called the "unit circle theorem," von Neumann and Selberg had suggested to us to read *Inequalities* by Hardy, Littlewood, and Pólya, and H. Whitney taught my brother, C. P. Yang, and me in 1965 the topological concept of index. For the method of solving Wiener–Hopf equations, M. Kac referred us to M. G. Krein's long review on this subject. In the 1970s, I collaborated with a mathematics group under C. H. Gu at Fudan University in Shanghai, China. In addition to all these and Simons' lectures, I had benefited from many interactions with A. Borel in Princeton and with my mathematics colleagues in Stony Brook, R. Douglas, M. Gromov, I. Kra, B. Lawson, C. H. Sah, and others.

ZHANG: *Did you have much interaction with Chern?*

YANG: As mentioned earlier, I had taken a course on differential geometry with him in my junior year in China and had probably audited some other courses with him. We talked to each other often in 1949 and subsequent years, but we did not go into any real mathematics. I had heard of the great importance of the Chern class, I think in the 1950s, but did not know what it was.

It was only in 1975, when Simons gave a series of talks to us at the Institute for Theoretical Physics at Stony Brook, that I finally understood the basic ideas of fiber bundles and connections on

FIGURE 6-6 *S. S. Chern and C. N. Yang (right) in 1985, when Chern received an honorary DSci at Stony Brook.*

fiber bundles. After some struggles, I also finally understood the very general Chern–Weil theorem. It is hard to describe the joy I had in understanding this profoundly beautiful theorem. I would say the joy even surpassed what I had experienced upon learning, in the 1960s, Weyl's powerful method of computing characters for the representations of the classical groups, or upon learning the beautiful Peter–Weyl theorem. Why? Perhaps because the Chern–Weil theorem is more geometrical.

But it was not just joy. There was something more, something deeper: After all, what could be more mysterious, what could be more awe-inspiring, than to find that the basic structure of the physical world is intimately tied to deep mathematical concepts, concepts which were developed out of considerations rooted only in logic and in the beauty of form? On this feeling I had written:

> In 1975, impressed with the fact that gauge fields are connections on fiber bundles, I drove to the house of S. S. Chern in El Cerrito, near Berkeley . . . I said I found it amazing that gauge theory are exactly connections on fibre bundles, which the mathematicians developed without reference to the physical world. I added "this is both thrilling and puzzling, since you mathematicians dreamed up these concepts out of nowhere." He immediately protested: "No, no. These concepts were not dreamed up. They were natural and real." [1, p. 567]

References

1. C. N. Yang, *Selected Papers, 1945–1980, with Commentary,* W. H. Freeman and Company, San Francisco, 1983.
2. C. N. Yang and R. L. Mills, "Conservation of isotopic spin and isotopic gauge invariance," *Phys. Rev.* 96 (1954), 191–195.
3. R. Mills, "Gauge fields," *Ann. J. Phys.* 57 (1989), 493–507.
4. P. A. Griffith, "Mathematical sciences: A unifying and dynamical resource—Report of the Panel on Mathematical Sciences, initiated by the National Research Council," *Notices AMS* 33 (1986), 463.
5. M. E. Mayer, *Fibre Bundle Technique in Gauge Theories,* Lecture Notes in Physics No. 67, Springer-Verlag, Berlin, 1977, p. 2.
6. W. Pauli, *Handbuch der physik,* 2nd ed., Geiger and Scheel, 1933, Vol. 24(1), p. 83.
7. W. Pauli, *Reviews of Modern Physics* 13 (1941), 203.

8. C. N. Yang, "Herman Weyl's contributions to physics," in *Herman Weyl (1885– 1955)*. Springer-Verlag, Berlin, 1985.

9. T. T. Wu and C. N. Yang, "Concept of non-integrable phase factors and global formulation of gauge fields," *Phys. Rev. D* 12 (1975), 3845–3857.

10. I. M. Singer, "Some problems in the quantization of gauge theories and string theories," *Proc. Symposia in Pure Math.* 48 (1988), 198–216.

11. M. F. Atiyah, N. J. Hitchin, and I. M. Singer, "Self-duality in four dimensional Riemann geometry," *Proc. Roy. Soc. London Ser. A,* 362 (1978), 425–461.

12. M. F. Atiyah, *Geometry of Yang–Mills Fields,* Scuola Normale Superiore, Pisa, 1977.

13. M. F. Atiyah, *Collected Works, Vol. 5. Gauge Theories.* Cambridge University Press, Cambridge, England, 1988, p. 1.

14. C. N. Yang, "Some exact results for the many-body problem in one dimension with repulsive delta-function interaction," *Phys. Rev. Lett.* 19 (1967), 1312–1315.

15. R. J. Baxter, "Partition function of the eight-vertex lattice model," *Ann. Phys.* 70 (1972), 193–228.

16. M. Barber and P. Pearce, eds., *Yang–Baxter Equations, Conformal Invariance and Integrability in Statistical Mechanics and Field Theory,* World Scientific, Singapore, 1990.

17. M. Jimbo, ed., *Yang–Baxter Equation in Integrable Systems,* World Scientific, Singapore, 1990.

18. C. N. Yang and M. L. Ge, eds., *Braid Group, Knot Theory and Statistical Mechanics,* World Scientific, Singapore, 1989.

19. M. F. Atiyah, "The work of Donaldson," *Notices AMS* 33 (1986), 900.

20. A. Jaffe and B. Mazur, "Vladimir Drinfeld," *Notices AMS* 37 (1990), 1210.

21. R. H. Hermann, "Vaughan F. R. Jones," *Notices AMS* 37 (1990), 1211.

22. K. Galwedzki and C. Soule, "Edward Witten," *Notices AMS* 37 (1990), 1214.

23. *Mathematical Intelligencer,* vol. 9 (1991), no. 2, p. 7.

ALGEBRA AND NUMBER THEORY

7

The Proof of the Mordell Conjecture

Spencer Bloch

Andre Weil has frequently criticized the use of "conjecture" in mathematics:

> Sans cesse le mathématicien se dit: "Ce serait bien beau" (ou: "Ce serait bien commode") si telle ou telle chose était vrai. Parfois il le vérifie sans trop de pein; d'autres fois il ne tarde pas à se détromper. Si son intuition a résisté quelque temps à ses efforts, il tend à parler de "conjecture", même si la chose a peu d'importance en soi. Le plus souvent c'est prématuré. [17]

In this connection he refers to the Mordell conjecture:

> Nous sommes moins avancés a l'égard de la "conjecture de Mordell". Il s'agit là d'une question qu'un arithmeticien ne peut guère manquer de se poser; on n'aperçoit d'ailleurs aucun motif sérieux de parier pour ou contre.

Probably most mathematicians would have agreed with Weil (certainly I would have), until earlier last year, when a German mathematician, Gerd Faltings, proved the Mordell conjecture, opening thereby a new chapter in number theory. In fact, his paper also establishes two other important conjectures, due to Tate and Shafarevich, and these achievements may well prove to be equally significant. I want to sketch briefly what these three conjectures say and what the ingredients are in Faltings' proof. To maintain the "G" rating appropriate to a (mathematical) "family magazine" for as long as possible, there will of course be no question of giving details. I will also not enter into the history of the conjecture other than to note it was first posed by Mordell in [7] and that the analogue for curves over a function field was proved independently by Manin in the USSR and by Grauert [6], [12].

The Mordell Conjecture

As originally formulated, the conjecture said that any irreducible polynomial in two variables with rational coefficients, having "genus" greater than or equal to two, has at most a finite number of solutions. Writing $f(x, y)$ for the polynomial and $\mathbf{Q}$ for the rational numbers, the conjecture said that there exist at most a finite number of pairs $x_i, y_i \in \mathbf{Q}$ with $f(x_i, y_i) = 0$. If f has degree d, the genus is a number $\leq (d - 1)(d - 2)/2$. (I will be slightly more precise in a moment.) For example, before Faltings it was not known that the equation $y^2 = x^5 + a$ has only finitely many solutions in $\mathbf{Q}$ for every non-zero integer a.

Volume 6, No. 2 (Spring 1984), 41–47.

Subsequent generations have broadened the conjecture to include polynomials defined over any number field (indeed any "global field") and have with the advent of abstract algebraic geometry tended to rephrase it in terms of algebraic curves. What Faltings actually proves, then, is that any algebraic curve of genus ≥ 2 defined over a number field K has at most a finite number of K-points.

Roughly speaking, the algebraic curve is the set of all solutions of $f(x, y) = 0$ with values in any field containing K. In fact, this statement has to be amended on several counts. For one thing, the solution set misses "points at infinity." To avoid having some fiend stash all the goodies out at infinity where we cannot get at them (cf. *The Hitchhiker's Guide to the Galaxy*), we introduce projective space. Projective n-space over a field k is the set of lines through the origin in a vector space of dimension $n + 1$. If we fix a basis of the vector space, we can specify a line, and hence a point of projective space, by the coordinates $(x_0, \ldots, x_n)$ of a non-zero point on the line. If F is a homogeneous polynomial in $X_0, X_1, \ldots, X_n$, we say F vanishes at the projective point if it vanishes along the corresponding line.

For example, let $F(X, Y, Z)$ be the unique homogeneous polynomial of degree d, where d is the degree of $f(x, y)$, such that $F(x, y, 1) = f(x, y)$. The locus of zeros of F in projective 2-space will then contain the solutions of $f(x, y) = 0$, but there will be other points corresponding to lines in the plane $Z = 0$ along which $F = 0$. For example, if $f(x, y) = y^2 - x^3 - 1$, then $F(X, Y, Z) = Y^2Z - X^3 - Z^3$, and there is one extra point at infinity, namely $X = Z = 0$, $Y = 1$ (see Figure 7-1).

We can give a more precise formula for the genus of f:

$$\text{genus} = (d - 1)(d - 2)/2 - \Sigma\, v_p$$

where the sum is taken over all projective points with

$$\frac{\partial F}{\partial X}(p) = \frac{\partial F}{\partial Y}(p) = \frac{\partial F}{\partial Z}(p) = 0$$

and $v_p \geq 1$ for all p. Such points are called singular points of the zero locus. For example, the Fermat polynomial $x^n + y^n - 1$ has no singular points; its genus is $(n - 1)(n - 2)/2$.

On the other hand, $y^2 = x^2(x + 1)$ has a double point at the origin, and the corresponding projective curve $F = Y^2Z - X^2(X + Z) = 0$ has these partial derivatives:

$$\frac{\partial F}{\partial X} = -3X^2 - 2XZ, \quad \frac{\partial F}{\partial Y} = 2YZ, \quad \frac{\partial F}{\partial Z} = Y^2 - X^2,$$

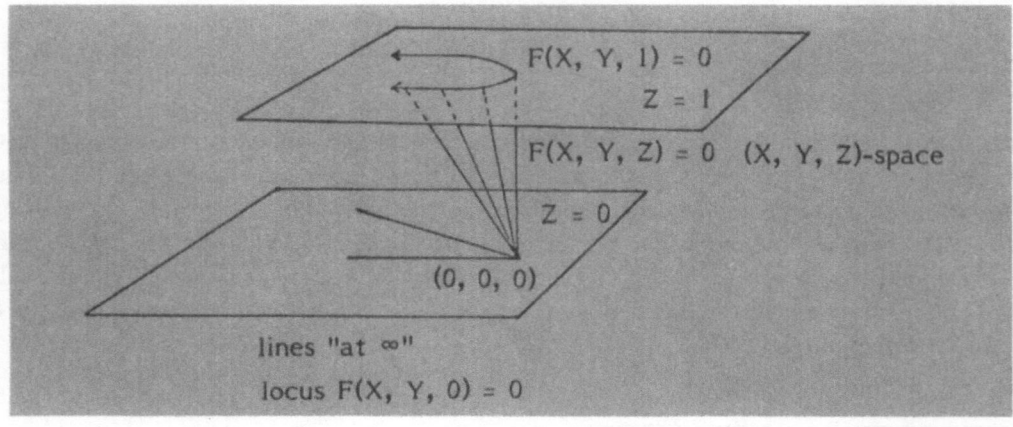

FIGURE 7-1

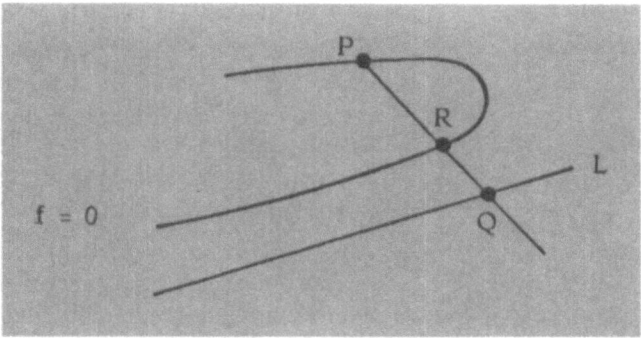

FIGURE 7-2

which all vanish at $[0, 0, 1]$. So the genus of this curve is not 1 but 0. Singular curves, like $y^2 = x^2 (x + 1)$, have smaller genus than their degree would suggest. Since Mordell's conjecture is formulated in terms of genus, it will be necessary to keep track of how singular are the curves that we introduce.

Curves of Genus 0 and 1

What happens when degree $f \leq 3$, so the conjecture doesn't apply? For $d = 1, f = ax + by + c$ clearly has an infinite number of solutions. For $d = 2, f$ may have no solutions (e.g., $x^2 + y^2 + 1$ when $K = \mathbf{Q}$), but if it has a solution, it will have an infinite number. To see this, we argue geometrically. Let P be a point in the solution set of f, and let L be a line not passing through P (see Figure 7-2). For Q a point on L with coordinates in our given field K, the line PQ usually meets the solution set in a second point R. As Q runs through the infinite set of K-points of L, the collection of points R is an infinite set of K-solutions of f. For example, applying this procedure to the equation $x^2 + y^2 - 1$ gives the familiar parametrization

$$x = \frac{t^2 - 1}{t^2 + 1} \quad y = \frac{2t}{t^2 + 1} \, .$$

When F is non-singular of degree 3, the set of solutions forms a group, a so-called *elliptic curve*. The line through points P and Q meets the zero locus in a third point R which equals $-P - Q$ for the group law. If one fixes a suitable origin P_0, the procedure for multiplying a point R by 2 is given in Figure 7-2. Assuming the original point was not of finite order for the group law, iteration will yield an infinite set of solutions. One consequence of Faltings' work is that for F non-singular of degree ≥ 4, no such geometric method of generating points is possible. However, there are higher dimensional algebraic varieties, called abelian varieties, which *do* have similar group laws on them. It is the study of these abelian varieties that form the heart of Faltings' proof.

How Shafarevich's Conjecture Comes In

One key step in the proof was taken by A. N. Parshin [10], who showed that the Mordell conjecture would follow from a conjecture of Shafarevich about good reduction for curves. Suppose for example that our homogeneous polynomial F is non-singular (i.e., has no singular points) and $K = \mathbf{Q}$. By clearing denominators, we may suppose the coefficients of F are integers with no common factors, and we may then consider the equation mod p for some prime number p. If the partial derivatives have no common zero mod p, then the equation $F(X, Y, Z) = 0$ (mod p) gives a non-singular curve over the field $\mathbf{Z}/p\mathbf{Z}$, and the original curve is said to have good reduction at p. For example the

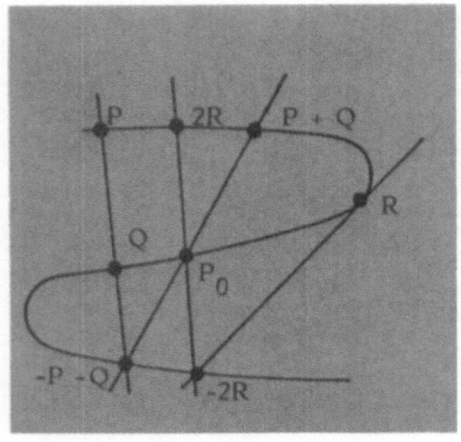

FIGURE 7-3

Fermat curve $X^3 + Y^3 + Z^3$ does not have good reduction at the prime 3. A more substantial example is the curve $Y^2Z = X^3 - 17Z^3$. It too is non-singular as a curve over **Q**, but it becomes singular when reduced modulo either 2, 3, or 17.

Shafarevich conjectured [13] that for a fixed genus g and a fixed finite set S of primes in a number field K, there were at most a finite number of isomorphism classes of curves of genus g defined over K with good reduction outside S. For example, as applied to the collection of elliptic curves $y^2 = X(X - 1)(X - \lambda)(\lambda \in \mathbf{Z})$, the conjecture follows from finiteness of the set of λ such that λ and $\lambda - 1$ are divisible only by primes of S.

Suppose C is such a curve with $g \geq 2$ and P is a K-point of C. Parshin showed that after a finite extension K' of K one could find another curve C' with genus g' and good reduction outside S' together with a map $C' \to C$ which is ramified only over P. (A map of curves is ramified at a point P if the number of points in the inverse image of P is strictly less than the degree of the mapping.) Further, K', g', and S' depend only on g and S, not on P. Shafarevich's conjecture then says there are only a finite number of such C'. Suppose now C has an infinite number of K-points P_r, $r = 1, 2, 3, \ldots$ An infinite number of the corresponding C'_r would necessarily be isomorphic to a common curve C'' which would admit an infinite number of distinct maps to C. By classical Riemann surface theory, the set of maps from one surface of genus ≥ 2 to another is finite, however, so the set of K-points of C is finite.

The Passage to the Jacobian

Subtle questions about curves are often profitably treated by "passing to the Jacobian," which number-theoretically means generalizing from elliptic curves to abelian varieties. Indeed, in the present case, Shafarevich formulated and Faltings proves a stronger version of Shafarevich's conjecture, dealing with abelian varieties.

To explain, I need to recall a bit more about Riemann surfaces (viz. compact complex manifolds of dimension 1). The curves we have been discussing give rise to Riemann surfaces by choosing an embedding of the coefficient field K in the complex numbers and then looking at all complex points. The genus g of the curve equals the number of "holes" in the surface (Figure 7-4). A holomorphic differential 1-form on an open set U in the complex plane is an expression $f(z)dz$, where z is the complex parameter and f is a holomorphic function. If p is a path in U, we may compute the integral $\int_p f(z)\, dz$. A Riemann surface is covered by such open sets U_i, and we may define a holomorphic 1-form on the surface to be a collection of these on the U_i which agree on $U_i \cap U_j$. The space V of holomorphic 1-forms has dimension equal to the genus of the surface.

A path p on the surface defines an element $\int_p$ in the dual space V^* to V by integration, and the collection of such functionals for p running over all *closed* paths forms a lattice L of maximal rank in V^*. The quotient $J = V^*/L$ is a complex torus called the Jacobian of the surface. There is a natural divisor θ on J, so called because its inverse image in V^* is defined by the vanishing of the Riemann theta function. It turns out that θ determines a *polarization*, i.e., there is an embedding of J in projective space P^n and a hyperplane section H such that $H \cap J =$ a multiple of θ. The Torelli theorem says that the pair J, θ actually determines the curve.

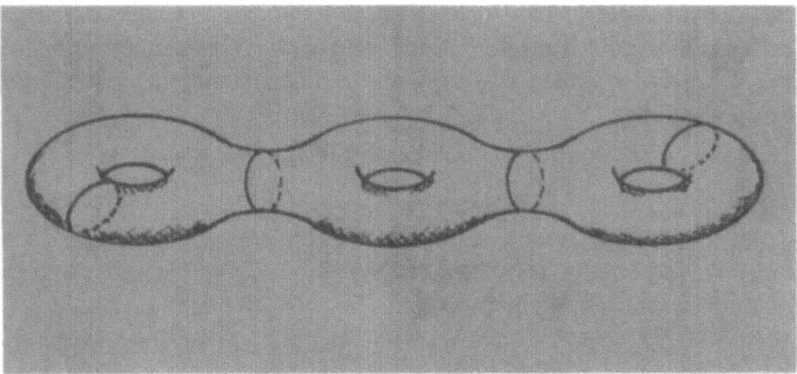

FIGURE 7-4 *Curve of genus 3.*

The construction of the Jacobian can be carried out algebraically and yields an *abelian variety J* defined over the same field K as the original curve. An abelian variety is a closed, connected subvariety of projective space which has a group law. To say it is defined over K means that it is the zero set of a collection of polynomials all of whose coefficients are in K. Over the complex numbers such a thing is necessarily a complex torus, but not every complex torus can be embedded in projective space as the zero set of polynomials. The simplest abelian varieties are those of dimension one; they are the elliptic curves already mentioned above. The Shafarevich conjecture for abelian varieties says that the set of principally polarized abelian varieties of dimension g defined over K with good reduction outside a given finite set of places S is finite. It is this result Faltings actually proves. One sees from the Torelli theorem that it implies the Shafarevich conjecture for curves, which by the Parshin argument implies the Mordell conjecture.

Height

There is a classical method, due to Fermat, for proving that some numerical problems have no solutions, and Faltings' work is in part a generalization of that method. Fermat showed that there are no solutions in non-zero integers to $x^4 + y^4 = z^2$, by associating a height to any triple (p, q, r) satisfying this equation—for example, $|r|$. The method of infinite descent then allowed him to show that if a triple (p, q, r) with positive height lies on the curve, then so does a new triple $(\tilde{p}, \tilde{q}, \tilde{r})$ which has smaller height, $0 < |\tilde{r}| < |r|$. It follows that there are no integer points on $x^4 + y^4 = z^2$ other than $(0, 0, 0)$.

How does Faltings generalize this? As before, suppose V is the vector space of holomorphic 1-forms on a projective curve. If $v_1, v_2, \ldots, v_g$ are a basis for the coordinate functions on V^*, the differential $dv_1 \wedge dv_2 \wedge \ldots \wedge dv_g$ is invariant under translations and so descends to a g-form w on J. In fact w can be defined algebraically for any g-dimensional abelian variety and is canonical up to multiplication by a scalar in K. Faltings associates to an abelian variety over K a numerical invariant $h(A)$ called the height. This is a sum of local terms associated to prime ideals of K (non-archimedean places) as well as real or complex completions (archimedean places). For a prime ideal p of K, the theory of the Neron model gives a preferred way to reduce the variety A modulo p. If the differential w vanishes (respectively, has a pole) when reduced mod p, the local term is an appropriate multiple of $-\log (Np)$ (resp. $+\log (Np)$), where Np is the number of elements in the finite field at p. (When $K = \mathbf{Q}$, $Np = p$.) For each embedding of K into $\mathbf{R}$ or $\mathbf{C}$, Faltings takes a suitable multiple of the log volume of A, $\frac{1}{2}\log (\int_A w \wedge \bar{w})$. He adds these local invariants to get the height. By the product formula in number theory, $h(A)$ is independent of the choice of w.

The notion of height in diophantine geometry is more commonly applied to points in projective space. Suppose, for example, x is a point in P^n defined over the rationals, corresponding to a line in $\mathbf{Q}^{n+1}$. Let $x_0, x_1, \ldots, x_n$ be coordinates of a point on the line such that the x_r are all integers with no common factors, and define

$$h(x) = \max\{\log |x_r|\}.$$

There is an analogous construction when $\mathbf{Q}$ is replaced by any number field K. If the number field K and a constant c are fixed, there are at most a finite number of points of height $\leq c$.

Faltings relates the two notions of height by means of the Siegel moduli space $\mathcal{N}_g$ parametrizing principally polarized abelian varieties of dimension g. He notes that $\mathcal{N}_g$ can be embedded as a locally closed subvariety of a projective space in such a way that if x corresponds to an abelian variety A, then $h(A)$ and $h(x)$ coincide up to a fixed non-zero multiple plus an error term having at worst logarithmic growth near infinity on $\mathcal{N}_g$. (Infinity makes sense here because $\mathcal{N}_g$ is not complete. Abelian varieties can degenerate.) It follows that there exist at most a finite number of g-dimensional principally polarized semi-stable abelian varieties of height $\leq c$. I will refer to this as the principle of bounded height. It is the main idea of the proof.

Isogeny

The next step is to study how heights behave under *isogeny*. By definition, *an isogeny of abelian varieties is a surjective homomorphism of algebraic groups with finite kernel*. The classical case is a map between elliptic curves $f: A \to B$. This lifts to a map on the covering spaces $f: \mathbf{C} \to \mathbf{C}$ which satisfies $f(z + \lambda) = f(z)$ whenever λ lies in the lattice of periods Λ of A ($A \cong \mathbf{C}/\Lambda$). Accordingly the derivative of f is periodic and holomorphic, and is therefore a constant, say α, by Liouville's theorem. Consequently Λ is mapped into Λ', the lattice of periods for B, as a subgroup of finite index, so it has a finite kernel.

Isogeny is often the "right" notion of equivalence for abelian varieties. For example, Poincaré's complete reducibility theorem allows one, given an isogeny P from A to B with image A', to replace A and B by isogenous varieties $A' \times A$ and $A' \times B$, so that f becomes the composite $A' \times A \to A' \to A' \times B$ of a projection and an injection. Consequently many problems concerning abelian varieties are profitably tackled by first considering the variety and its isogeneous images as equivalent, and Faltings' paper proceeds in just this way. So we must look at how height varies under isogeny.

The phenomenon of interest is possible ramification of an isogeny on reduction mod p. Consider as an example the one-dimensional abelian variety defined by $X^3 + Y^3 - 1 = 0$. If we take $(1, 0)$ as origin for the group law, then multiplication by -2 can be described geometrically as associating to a point P on the curve the third point of intersection of the tangent to P with the curve (Figure 7-3). The tangent line to the curve at a point (x, y) is defined by $x^2 X + y^2 Y - 1 = 0$, so one deduces for multiplication by -2 the coordinate transformation

$$(x, y) \to \left(\frac{x^4 - 2x}{1 - 2x^3}, \frac{y^4 - 2y}{1 - 2y^3} \right).$$

This mapping has a vertical tangent over the point $(1, 0)$ when we reduce mod 2 (i.e., set $2 = 0$ in the formula). The change of height under isogeny is given by one half the log of the degree of the isogeny minus a correction term which measures this sort of ramification.

Tate's Conjecture

To explain the application of this formula, which will lead us to Tate's conjecture, I must talk a bit about the arithmetic of abelian varieties. Consider the kernel $_nA$ of multiplication by an integer n on

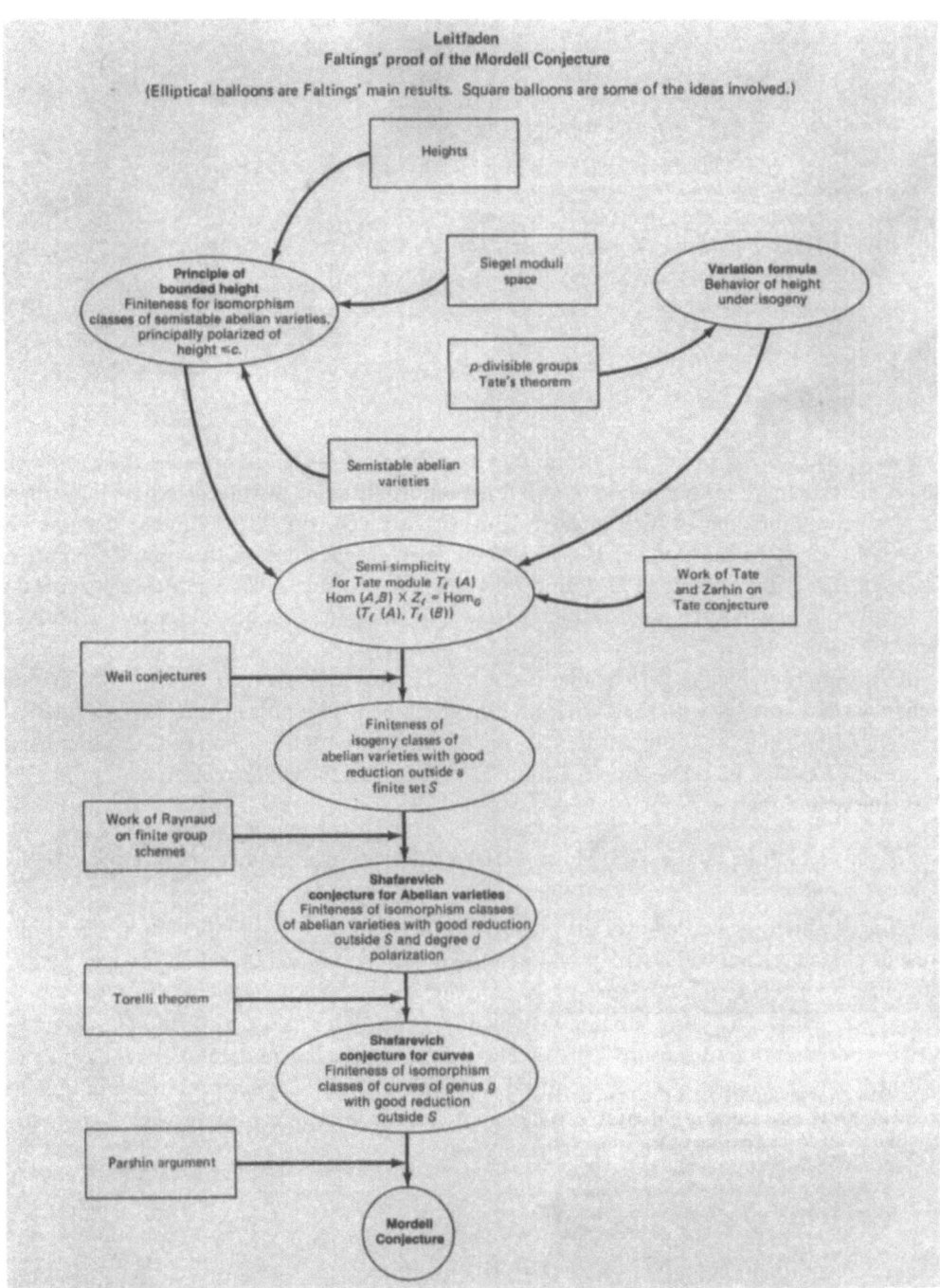

Leitfaden
Faltings' proof of the Mordell Conjecture

(Elliptical balloons are Faltings' main results. Square balloons are some of the ideas involved.)

FIGURE 7-5

an abelian variety A defined over a field K. Recalling that the complex points of A look like $\mathbf{C}^g/L$ for a lattice L, we see that as an abstract group $_nA$ looks like $(\mathbf{Z}/n\mathbf{Z})^{2g}$. But $_nA$ has additional structure. The coordinates of the finite set $_nA$ of points are algebraic over K but need not themselves lie in K. Thus the galois group G of the algebraic closure of K acts on $_nA$, giving a representation of G into $GL_{2g}(\mathbf{Z}/n\mathbf{Z})$. It is convenient to fix a prime ℓ and consider simultaneously all $_nA$ as n runs through

FIGURE 7-6 *Gerd Faltings was born on July 28, 1954 in Gelsenkirchen-Buer, West Germany. He spent his school years there and then moved on to study mathematics with Professor H-J. Nastold in Münster. He received his Ph.D. in 1978.*

Since then he has been a Research Fellow at Harvard (1978–79), an assistant at Münster (1979–82), and is presently a Professor at Wuppertal. His mathematical interests began in commutative algebra but have since moved to algebraic geometry.

all powers of ℓ. Using multiplication by ℓ to join them, $_nA \to {}_nA$, we can take the inverse limit to obtain the Tate module, denoted by $T_\ell(A)$. It is isomorphic as an abstract group to $\mathbf{Z}_\ell^{2g}$, where $\mathbf{Z}_\ell$ is the ℓ-adic completion of the integers. But G also acts on $T_\ell(A)$, giving a representation $G \to GL_{2g}(\mathbf{Z}_\ell)$. To gain further perspective on $T_\ell(A)$, again think of the complex points of A as $\mathbf{C}^g/L$. Then $_nA$ is isomorphic to L/nL, and $T_\ell(A)$ to $L \otimes \mathbf{Z}_\ell$. Since L corresponds in a natural way to the first homology of $\mathbf{C}^g/L$, we can think of $T_\ell(A)$ as being the first homology of A with $\mathbf{Z}_\ell$-coefficients.

If B is another abelian variety, one can ask to compare the group $\mathrm{Hom}(A, B)$ of mappings of abelian varieties over K with the corresponding group of G-mappings $\mathrm{Hom}_G(T_\ell(A), T_\ell(B))$, where the latter is the group of homomorphisms which commute with the action of G. An important conjecture of Tate (that he proved for K a finite field) was that the galois representation on $T_\ell(A)$ was semi-simple and that

$$\mathrm{Hom}(A, B) \otimes Z_\ell \cong \mathrm{Hom}_G\big(T_\ell(A), T_\ell(B)\big) \quad \text{(isomorphism of groups)},$$

assuming K finitely generated over the prime field. This important result fits into the general framework of characterizing when a map between the homology of two objects is induced by an actual geometric map between the objects.

Suppose $W \subset T_\ell(A)$ is a submodule which is stable under the galois group G. We can write W as an inverse limit of finite groups $W_n \subset A$ and define A_n to be the quotient A/W_n. A key step in Tate's argument was that under a certain technical condition concerning polarizations an infinite number of the A_n were isomorphic. Then if, say, A_n and A_{mn} are isomorphic, he constructs a geometric map from A to itself by taking a chain of maps

$$A \to A_{mn} \cong A_n \to A.$$

But it turns out that Faltings' formula of the behavior of heights under isogeny, together with a theorem of Tate on p-divisible groups, applied to the A_n as above (but now defined over a number field) gives a uniform bound for $h(A_n)$. The principle of bounded height enables him to deduce that an infinite number of the A_n are isomorphic, and now Tate's method and arguments due to Zarhin [18] lead to a proof of the Tate conjecture.

Back to the Shafarevich Conjecture

Faltings now had to prove the Shafarevich conjecture that there exist at most a finite number of principally polarized abelian varieties of dimension g defined over a number field K with good reduction outside a given finite set of primes S. Consider first the problem of showing there exist at most a finite number of isogeny classes of g-dimensional abelian varieties with good reduction outside S (i.e., we identify two abelian varieties if there is an isogeny from one to the other). Since we have the Tate conjecture, this amounts to showing there are only a finite number of isomorphism classes among the $T_\ell(A) \otimes Q$; since the representations are semi-simple it suffices to show they give rise to only a finite number of different trace functions on the galois group.

To do this, Faltings shows that there is a finite subset of g consisting of *Frobenius elements* such that the image of this finite subset in the endomorphism ring of $T_\ell(A)$ generates the same subring as all of G. Further, the finite set depends only on the prime ℓ, the dimension of A, and the set S; it is otherwise independent of A. This reduces one to showing that for each given Frobenius element, the trace of its action on $T_\ell(A)$ can take on only finitely many values. But now one can apply the well-known Weil conjectures, which say that the trace is an integer (i.e., in $\mathbf{Z}$) with a precisely bounded absolute value.

The proof of the Shafarevich conjecture again invokes the principle of bounded height. Given a set of g-dimensional principally polarized abelian varieties over K, with good reduction outside S, it will suffice to show the heights are bounded. By the result of the previous paragraph, we may assume the varieties are all isogenous. The argument is completed by applying the variation of height formula, together with a theorem of Raynaud on galois actions on points of group schemes and another application of the Weil conjectures for abelian varieties.

References

(I have included those papers which appear in Falting's bibliography, together with a few others of historical interest.)

1. Arakelov, S. (1971) Families of curves with fixed degeneracies, *Math. USSR Izvestija 5:* 1277–1302.
2. Baily, W. L. and Borel, A. (1966) Compactification of arithmetic quotients of bounded symmetric domains, *Ann. of Math. 84:* 442–528.
3. Deligne, P. and Mumford, D. (1969) The irreducibility of the space of curves of a given genus, *Publ. Math. 36:* 75–110.
4. Faltings, G. (1984) Calculus on arithmetic surfaces, *Ann. of Math. 119:* 387–424.
5. Faltings, G. (1983) Arakelov's theorem for abelian varieties, *Inv. Math. 73:* 337–347.
6. Manin, Ju (1963) *Dokl. Akad. Nauk SSSR 152* (1963) English translation, *Soviet Math. 4:* 1505–1507.
7. Mordell, L. J. (1922) *Proc. Cambridge Philos. Soc.,* 179–192.
8. Moret-Bailly, L. (1983) Variétés abeliennes polarisées sur les corps de fonctions, *C. R. Acad. Sc. Paris 296:* 267–270.
9. Namikawa, Y. (1980) *Toroidal compactification of Siegel spaces,* Springer Lecture Notes Vol. 812. New York: Springer-Verlag.
10. Parshin, A. N. (1968) Algebraic curves over function fields I, *Math. USSR Izvestija 2:* 1145–1170.
11. Raynaud, M. (1974) Schemas en groupes de type $(p, \ldots, p)$, *Bull. Soc. Math. France 102:* 241–280.
12. Grauert, H. (1965) Mordell's Vermutung über rationale Punkte aut algebraischen Kurven und Funktionenkörper, *Publ. Math. I.H.E.S.,* Vol. 25.
13. Shafarevich, I. R. (1962) I.C.M.

14. Szpiro, L. (1979) Sur le théorème de rigidité de Parshin et Arakelov, *Astérisque 64:* 169–202.
15. Tate, J. (1967) *p*-divisible groups, in *Proceedings of a conference on local fields* (Driebergen, 1966), 158–183. Berlin: Springer-Verlag.
16. Tate, J. (1966) Endomorphisms of abelian varieties over finite fields, *Inv. Math 2:* 134–144.
17. Weil, A. (1980) *Collected Works,* Vol. III, p. 454. Berlin: Springer-Verlag.
18. Zarhin, J. G. (1974) Isogenies of abelian varieties over fields of finite characterics, *Math. USSR Sbornik 24:* 451–461.
19. Zarhin, J. G. (1974) A remark on endomorphisms of abelian varieties over function fields of finite characteristics, *Math. USSR Izvestija 8:* 477–480.

8

Adventures In Arithmetick, or: How to Make Good Use of a Fourier Transform

R. C. Vaughan

1. Introduction

It is, I believe, normal in an inaugural lecture at Imperial College for a professor to choose a topic which is close to his research work yet which is accessible to a wider audience than that which normally attends his seminars. A pure mathematician is at a particular disadvantage in this context. His work is often difficult for another mathematician to follow, and many highly educated people are even quite happy to exaggerate their lack of understanding of mathematics.

At least number theory possesses a large number of remarkable problems and beautiful theorems which are not beyond popular comprehension, although there is nothing in the least popular about its methods.

In this lecture I want to tell you about some problems that have interested me and to explain to you how various analytical techniques can be useful in their elucidation.

Some of the problems I will discuss are of a great age, with a considerable history. But pure mathematicians are used to working on a vast time scale. For example, let me remind you of the classical theorem on perfect numbers.

A perfect number is a natural number that is the sum of its aliquot parts

$$s(n) = \sum_{\substack{d|n \\ d<n}} d, \, n \text{ perfect} \Leftrightarrow n = s(n).$$

For example, 6, 28, 496 are perfect.

The theorem alluded to gives a necessary and sufficient condition for all even numbers to be perfect.

THEOREM. An even number n is perfect if and only if it is of the form $n = 2^r(2^{r+1} - 1)$ and $2^{r+1} - 1$ is a prime number.

Volume 9, No. 2 (Spring 1987), 53–60.

That the condition is sufficient occurs in Book IX, Proposition 36 of Euclid. Necessity was established by Euler in work published posthumously. Thus we have an example of a theorem that took 2,000 years to prove. I, therefore, make no apology for working on problems that are only a few hundred years old, such as the

2. Goldbach Conjectures

These arose as the result of an exchange of letters between Euler and Goldbach in 1742, and I state them in their modern forms as follows:

Goldbach conjectures (1742).

1. Every even number greater than 2 is a sum of two primes.
2. Every odd number greater than 5 is a sum of three primes.

There has been a great deal of work on these problems in the last sixty-five years. Of one of the two major lines of attack that have been developed, namely that of sieve methods initiated by Brun in 1920, I shall say *nothing* today. Instead I will concentrate on the method introduced by Hardy and Littlewood in the early 1920s. Their work was dependent on an unproved hypothesis, the so-called generalized Riemann hypothesis, which has not been established to this day. However in 1937 Vinogradov introduced an important refinement which avoided the necessity for the assumption of any unproved hypothesis. Thus Vinogradov was able to establish that every *large* odd natural number is the sum of three primes.

$$\text{Vinogradov (1937)} \; n > n_o, n \text{ odd} \Rightarrow n = p + p' + p''.$$

The binary problem has been more resistant but by using a variant of the Hardy-Littlewood-Vinogradov method, Estermann, Chudakov, and van der Corput (independently) were able to show that at any rate *almost every even number* is the sum of two primes. More precisely they established that if

$$E(x) = \text{card}\{n \le x : 2 \mid n, n \ne p + p'\}$$

then

$$\text{Estermann, Chudakov, van der Corput (1937)} \quad \forall c > 0, E(x) = O\big(x(\log x)^{-c}\big).$$

More recently Montgomery and I were able to show that

$$\text{Montgomery, Vaughan (1975)} \quad \exists \delta > 0: E(x) = O(x^{1-\delta})$$

and J. R. Chen (1980) has calculated the constants in our argument and has shown that

$$\delta = \frac{1}{100}$$

is permissible.

The Hardy-Littlewood-Vinogradov method can be described as follows. Consider the trigonometrical polynomial

$$S(\alpha) = \sum_{p \leq N} e^{2\pi i p \alpha} (\alpha \in \mathbb{R}),$$

where α is a real number and $i = \sqrt{-1}$. Then on taking the sth power of S and collecting together those terms for which the sum of the frequencies is a given natural number n, we obtain

$$(S(\alpha))^s = \sum_n R(n; N)e^{2\pi i n \alpha},$$

where $R(n; N)$ is the number of ordered s-tuples $(p_1, \ldots, p_s)$ of primes not exceeding N whose sum is n,

$$R(n; N) = \text{card}\{(p_1, \ldots, p_s): p_j \leq N, p_1 + \ldots + p_s = n\}.$$

Now $R(n; N)$ is the fourier coefficient of $(S(\alpha))^s$. Thus it is the fourier transform of $(S(\alpha))^s$ on the torus $\mathbb{T} = \mathbb{R}/\mathbb{Z}$, and in particular, if we take $n = N$, then

$$\text{card}\{(p_1, \ldots, p_s): p_1 + \cdots + p_s = N\} = \int_0^1 (S(\alpha))^s e^{-2\pi i N \alpha} d\alpha.$$

The essence of the Hardy-Littlewood method is an alternative means of evaluating the integral, at least asymptotically. In the case of the Goldbach *ternary* problem, this can be done as follows. When α is rather close to a rational number with a denominator that is small compared with N, e.g.

$$\text{if } \left| \alpha - \frac{a}{q} \right| \leq \frac{(\log N)^{100}}{N}, (a, q) = 1, q \leq (\log N)^{100},$$

information about the distribution of primes in arithmetic progressions enables one to establish that for the union $\mathcal{M}$ of all such α, modulo one, the contribution from $\mathcal{M}$ satisfies

$$\int_{\mathcal{M}} (S(\alpha))^3 e^{-2\pi i N \alpha} d\alpha \sim \mathcal{G}(N) \frac{N^2}{(\log N)^3},$$

where $\mathcal{G}(N)$ is an oscillatory factor that satisfies

$$\frac{1}{2} < \mathcal{G}(N) < 2 \quad \text{when } N \text{ is odd.}$$

The intervals making up $\mathcal{M}$ are called the major arcs. If it can be shown that the contribution from the so-called minor arcs

$$\mathpzc{m} = [0, 1] \setminus \mathcal{M}$$

is of a smaller order of magnitude, then it follows, of course, that

$$\text{card}\{(p, p', p''): p + p' + p'' = N\} \sim \mathcal{G}(N) \frac{N^2}{(\log N)^3},$$

and hence that every large odd natural number is the sum of three primes.

To estimate the contribution from the minor arcs one observes that

$$\int_{\mathcal{M}} |S(\alpha)|^3 d\alpha \leq \left(\sup_{\alpha \in \mathcal{M}} |S(\alpha)|\right) \int_0^1 |S(\alpha)|^2 d\alpha$$

and by Parseval's identity and the prime number theorem applied to the last integral we have

$$\int_{\mathcal{M}} |S(\alpha)|^3 d\alpha = O\left(\frac{N}{\log N} \sup_{\alpha \in \mathcal{M}} |S(\alpha)|\right).$$

Thus one requires a non-trivial estimate for the trigonometrical polynomial

$$S(\alpha) = \sum_{p \leq N} e^{2\pi i p \alpha}$$

when α is *not* well approximated by a rational number with a relatively *small* denominator.

Thus the Goldbach ternary problem is reduced to that of considering a sum over prime numbers

$$\sum_{p \leq N} f(p),$$

where f is oscillatory, unimodular, and, we hope, of small mean value. Note that in this case there is an obvious extension of f to all of the natural numbers, namely,

$$f(n) = e^{2\pi i n \alpha},$$

and non-trivial estimates in terms of α are available for such sums as

$$\sum_{n \leq N} f(mn) \text{ and } \sum_{n \leq N} f(m_1 n) \bar{f}(m_2 n)$$

since in each case they are simply the sum of a geometric progression. Now I want to outline a general method of getting to grips with such sums.

As a preliminary let me take you on a short detour.

3. The Riemann Zeta Function

The Riemann zeta function $\zeta(s)$ is defined for Re $s > 1$ by

$$\zeta(s) = \sum_{n=1}^{\infty} \frac{1}{n^s}$$

and it can be analytically continued to the whole complex plane and is regular at every point except $s = 1$ where it has a simple pole with residue 1. When Re $s > 1$, it also has the Euler product

$$\prod_p (1 - p^{-s})^{-1}.$$

Now logarithmic differentiation of the Euler product, expansion of $(1 - p^{-s})^{-1}$ as an infinite series, and rearranging the terms in their natural order gives a Dirichlet series representation for $-\zeta'/\zeta$,

$$-\frac{\zeta'}{\zeta}(s) = \sum_{n=1}^{\infty} \frac{\Lambda(n)}{n^s}, \quad \Lambda(n) = \begin{cases} \log p & (n = p^k), \\ 0 & (\text{otherwise}), \end{cases}$$

where Λ is von Mangoldt's function, namely Λ is 0 unless n is a power of a prime in which case it is the logarithm of the prime.

Since the squares of primes, the cubes of primes, and so on, are relatively infrequent, von Mangoldt's function essentially just counts the primes with logarithmic weight.

It is easily seen by inverting the order of summation and integration that the Mellin transform of the sum

$$\sum_{n \leq x} \Lambda(n) \text{ is } -\frac{1}{s} \frac{\zeta'}{\zeta}(s), \quad \int_1^{\infty} \left(\sum_{n \leq x} \Lambda(n) \right) x^{-s-1} dx = -\frac{1}{s} \frac{\zeta'}{\zeta}(s).$$

If one makes the obvious change of variable $x = e^y$, this is a Laplace transform. Now the inverse Laplace transform is a fourier transform. Thus one expects that

$$\sideset{}{'}\sum_{n \leq x} \Lambda(n) = \frac{1}{2\pi i} \int_{c-i\infty}^{c+i\infty} \left(-\frac{\zeta'}{\zeta}(s) \right) \frac{x^s}{s} ds \qquad (c > 1),$$

where the $'$ indicates that if x is an integer, we have to take the average of the upper and lower limits.

The Riemann zeta function has so-called trivial zeros at the negative even integers and all its other zeros ρ, say, have real part between 0 and 1. Thus formally moving the line of integration to $-\infty$ and invoking Cauchy's residue theorem gives the formula

$$\sum_{n \leq x} \Lambda(n) = x - \sum_{\rho} \frac{x^{\rho}}{\rho} + \sum_{n=1}^{\infty} \frac{x^{-2n}}{2n} - \frac{\zeta'}{\zeta}(0).$$

All this was known essentially to Riemann, but there are various problems of convergence, and the first rigorous proof is due to von Mangoldt in 1895.

The prime number theorem is equivalent to the assertion that the contribution from the zeros is small compared with x, and this was also established first in the 1890s, by Hadamard and de la Vallée Poussin.

It is, I think, quite well known that the modern theory of entire functions was largely founded in order to establish these results.

The famous Riemann hypothesis is the assertion that all the non-trivial zeros ρ have their real parts equal to $1/2$. This is unproved to this day, although it is known that the first 200 million non-trivial zeros are all on the $1/2$-line. There are many generalizations of the Riemann hypothesis and it is undoubtedly the most important unsolved problem in mathematics today.

Let me now return to the problem of estimating

4. Sums over Prime Numbers

$$\sum_{p \leq N} f(p).$$

By partial summation it would suffice to obtain an estimate for

$$S = \sum_{n \leq N} \Lambda(n) f(n).$$

Now consider the identity

$$-\frac{\zeta'}{\zeta} = G + (-\zeta'F) - \zeta FG - (\zeta F - 1)\left(-\frac{\zeta'}{\zeta} - G\right). \tag{1}$$

As Littlewood has said, all identities are trivial, and none more so than this one, which can easily be verified by multiplying out the last term on the right. Note that F and G are quite arbitrary.

The way to try and make use of this is to take F and G to be reasonable approximations to ζ^{-1} and $-\zeta'/\zeta$ respectively. There is a very simple Dirichlet series for ζ^{-1} obtained via the Euler product for ζ. Thus

$$\zeta(s)^{-1} = \prod_p (1 - p^{-s}) = \sum_{n=1}^{\infty} \frac{\mu(n)}{n^s}$$

where μ, the Möbius function, is ± 1 or 0. Indeed it is evident by multiplying out the Euler product and appealing to uniqueness of factorization that $\mu(n) = 0$ when n is divisible by the square of a prime and otherwise is ± 1 according as n has an even or odd number of prime factors. Thus good choices for F and G are

$$F(s) = \sum_{n \leq u} \mu(n) n^{-s}, \; G(s) = \sum_{n \leq v} \Lambda(n) n^{-s},$$

FIGURE 8-1

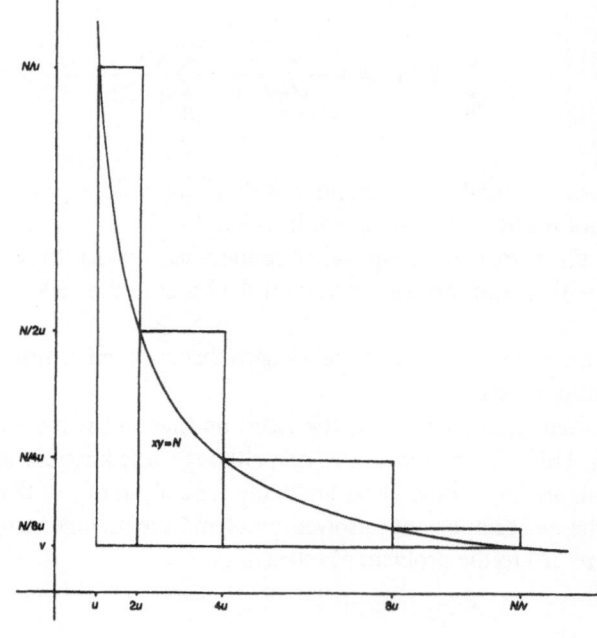

where u and v are parameters at our disposal.

Each of the terms on the right of (1) then has a Dirichlet series expansion so we may write

$$-\frac{\zeta'}{\zeta}(s) = D_0(s) + D_1(s) - D_2(s) - D_3(s).$$

Now the uniqueness theorem for Dirichlet series tells us that we have a partition of von Mangoldt's function

$$\Lambda(n) = c_0(n) + c_1(n) - c_2(n) - c_3(n),$$

where the c_j are the coefficients of the respective Dirichlet series on the right-hand side. Multiplying by an arbitrary arithmetical function f and summing gives us an abstraction of the original identity (1).

$$S = \sum_{n \leq N} \Lambda(n) f(n) = S_0 + S_1 - S_2 - S_3.$$

By examining the coefficients that arise from (1) with the given choice of F and G, we obtain the following expressions for $S_0, \ldots, S_3$

$$S_0 = \sum_{n \leq v} \Lambda(n) f(n), \qquad S_1 = \sum_{m \leq u} \mu(m) \sum_{n \leq N/m} (\log n) f(mn),$$

$$S_2 = \sum_{m \leq uv} c_m \sum_{n \leq N/m} f(mn), \qquad S_3 = \sum_{m > u} \sum_{\substack{n > v \\ mn \leq N}} d_m \Lambda(n) f(mn),$$

where c_m and d_m have somewhat complicated definitions but can be shown at any rate to satisfy

$$|c_m| \leq \log m, \quad \sum_{m \leq X} |d_m|^2 = O\big(X (\log X)^3\big).$$

Let me explain the philosophy underlying the use of this identity. In any one particular application there are often many details to attend to but I can at any rate outline the principal ideas.

We think of u and v as tending to infinity rather slowly as $N \to \infty$. Thus the sum S_0 will only make a comparatively small contribution and so can be ignored. The sums S_1, S_2 are essentially of the kind

$$\sum_m a_m \sum_n f(mn) \tag{I}$$

with

$$|a_m| \leq \log m, m \leq u \text{ or } m \leq uv, n \leq N/m.$$

The log in S_1, by the way, being smooth, can always be removed by partial summation.

The sum S_3 is of the form

$$\sum_m \sum_n a_m b_n f(mn) \tag{II}$$

with

$$\sum_{m \le X} |a_m|^2 = O\big(X (\log X)^3\big), \; |b_n| \le \log n.$$

The type (I) sums should be easy to estimate by performing the summation over n. For example when

$$f(n) = e^{2\pi i n \alpha},$$

the inner sum is the sum of a g.p. Since the sum over m is restricted by u and v, which are assumed to be not very large relative to N, the cancellation occurring in the inner sum should give a non-trivial estimate in terms of α.

In the type (II) sum we observe that we have summation conditions of the form

$$m > u, n > v, mn \le N.$$

Thus

$$m < N/v, n < N/u.$$

In fact we are summing over the ordered pairs (m, n) in a subregion of the area under the rectangular hyperbola $xy = N$. Moreover we are able to avoid the parts under the hyperbola near infinity. Also we wish to approximate the region under consideration by rectangles. This can be done by splitting up the type (II) sum into sub-sums

$$\sum_{u2^k < m \le u2^{k+1}} \; \sum_{v < n \le N/m} a_m b_n f(mn). \tag{IIk}$$

Then by taking $f_{m,n} = f(mn)$ for m, n with $u2^k < m < u2^{k+1}$ and $v < n \le N/m$, and $f_{m,n} = 0$ for $u2^k < m \le u2^{k+1}$ and $N/m < n \le N/(u2^k)$, the type (II)$_k$ sum can be written, in matrix notation, as

$$\mathbf{a} \mathcal{F} \mathbf{b}^T$$

Now this can be viewed, of course, as a bilinear form in variables $\mathbf{a}$ and $\mathbf{b}$. By Cauchy's inequality

$$|\mathbf{a}\mathcal{F}\mathbf{b}^T|^2 \le \mathbf{a}\mathcal{F}\mathcal{F}^*\mathbf{a}^* |\mathbf{b}|^2$$

where * denotes the complex conjugate transpose. Moreover

$$H = \mathcal{F}\mathcal{F}^*$$

is a Hermitian matrix. It will suffice, therefore, to bound the largest eigenvalue λ of H,

$$\lambda = \text{largest eigenvalue of } H,$$

for then we would have

$$|\mathbf{a}\mathcal{F}\mathbf{b}^T|^2 \le \lambda |\mathbf{a}|^2 |\mathbf{b}|^2.$$

An elementary classical theorem tells us that if h_{rs} is the general entry in H,

$$H = (h_{rs}),$$

then λ is bounded by the maximum over the rows of the sum of the moduli of the entries in the row.

$$\lambda \leq \max_r \sum_S |h_{rs}|.$$

Now we can interpret this in terms of f and obtain an upper bound for λ

$$\lambda \leq \max_{u2^k < m_1 \leq u2^{k+1}} \sum_{u2^k < m_2 \leq u2^{k+1}} \left| \sum_{v < n \leq \min\left(\frac{N}{m_1}, \frac{N}{m_2}\right)} f(m_1 n) \bar{f}(m_2 n) \right|.$$

Once more we have an innermost sum which we can hope to perform.

This technique works very well in Goldbach's problem and enables us to establish Vinogradov's theorem. There are many other problems to which this method may also be applied.

For example, Heath-Brown and Patterson have used it in resolving an old problem of Kummer. Gauss had shown in connection with the law of quadratic reciprocity that the trigonometrical sum

$$\sum_{n=1}^{p} e^{2\pi i n^2/p} = \begin{cases} \sqrt{p} & p \equiv 1 \pmod 4 \\ i\sqrt{p} & p \equiv 3 \pmod 4 \end{cases}.$$

Here p is a prime number.

In 1847 Kummer raised the question of the distribution of the values of the cubic sum

$$\sum_{n=1}^{p} e^{2\pi i n^3/p}, \quad \text{when } p \equiv 1 \pmod 3$$

as p ranges over the primes leaving the remainder 1 on division by 3. By the way, the condition $p \equiv 1 \pmod 3$ is necessary because otherwise the sum is 0 for trivial reasons. It corresponds to the implicit condition that p be odd in Gauss' sum. In particular one can ask if there is a simple formula as in Gauss' sum.

It was known that

$$-2\sqrt{p} \leq \sum_{n=1}^{p} e^{2\pi i n^3/p} \leq 2\sqrt{p}$$

so that

$$\sum_{n=1}^{p} e^{2\pi i n^3/p} = 2p^{\frac{1}{2}} \cos \vartheta_p, 0 \leq \vartheta_p \leq \pi.$$

Calculations by von Neumann and Goldstine, and by Emma Lehmer, seemed to indicate that the ϑ_p are rather randomly distributed. This was confirmed by Heath-Brown and Patterson who showed that the ϑ_p are uniformly distributed in the interval $[0,\pi]$.

Heath-Brown, Patterson (1979) ϑ_p u.d. in $[0, \pi]$.

In particular ϑ_p must take on infinitely many different values, so there is no simple formula for the sum such as that which occurs for Gauss' sum.

There are applications of the method to give information on the distribution of primes in arithmetic progressions, for example

$$p \equiv a(\text{mod } q), p \leq X, q \leq X^{17/32} \tag{*}$$

when the modulus q is permitted to be quite large by comparison with the primes under consideration. Some of this work is based on an identity of Heath-Brown rather than the one I described earlier, and this can be considered an abstraction of the analytic identity

$$\text{Heath-Brown} \quad -\frac{\zeta'}{\zeta} = \sum_{r=1}^{k} (-1)^r \binom{k}{r} \zeta' \zeta^{r-1} F^r - \frac{\zeta'}{\zeta} (1 - \zeta F)^k.$$

Again this identity is trivial, this time because of the binomial theorem. Such estimates have played a role in the very recent work of Adleman, Fouvry, and Heath-Brown in which they show that there are infinitely many prime exponents for which the first case of Fermat's last theorem is true

$$\text{Adleman, Fouvry, Heath-Brown (1984)} \quad \exists \infty \text{ many } p \text{ s.t. } x^p + y^p = z^p \Rightarrow p \mid xyz.$$

6. Coefficients of Cyclotomic Polynomials

Let me divert you now with a rather different adventure. The n-th cyclotomic polynomial Φ_n is defined as the monic polynomial whose roots are the primitive n-th roots of unity

$$\Phi_n(z) = \prod_{\substack{m=1 \\ (m,n)=1}}^{n} (z - e^{2\pi im/n}).$$

The degree of the n-th polynomial is, of course, Euler's function $\varphi(n)$, so we may write

$$\Phi_n(z) = \sum_{m=0}^{\varphi(n)} a(m, n) z^m.$$

These polynomials have many interesting properties. For instance, Φ_n is the maximal irreducible factor, over the rationals, of

$$z^n - 1.$$

Also Φ_n can be written as the product

$$\Phi_n(z) = \prod_{d|n} (z^{n/d} - 1)^{\mu(d)}, \tag{2}$$

where μ is our old friend Möbius' function. This shows, by the way, that Φ_n has integer coefficients. Many special cases are well known; for example, when p is prime,

$$p \text{ prime:} \quad \Phi_p(z) = 1 + z + \ldots + z^{p-1}.$$

Also it follows very easily from the formula (2) above that if N is the product of the distinct prime divisors of n,

$$N = \prod_{p|n} p,$$

then

$$\Phi_n(z) = \Phi_n(z^{n/N})$$

and if n is odd, then

$$n \text{ odd} \implies \Phi_{2n}(z) = \Phi_n(-z).$$

Thus in any investigation of these polynomials, we can reduce our enquiry to the case when n is odd, square-free, and composite.

Here is a table of the coefficients of Φ_n for all such n less than 100 (Figure 8-2). A cursory glance indicates at once that the coefficients are all ± 1 or 0. This, combined with our earlier observations, proves that the coefficients of the first 104 cyclotomic polynomials are all ± 1 or 0:

$$a(m, n) = \pm 1 \text{ or } 0 \text{ whenever } n \leq 104.$$

Any engineer in the audience should now be utterly convinced that all cyclotomic polynomials have their coefficients ± 1 or 0.

However it was shown by Migotti in 1883 that the seventh coefficient of the 105th polynomial is equal to -2.

$$\text{Migotti (1883)} \quad a(7,105) = -2.$$

More recently Schur showed that there are cyclotomic polynomials whose coefficients are arbitrarily large,

$$A(n) = \max_m |a(m, n)|$$

$$\text{Schur (c. 1935)} \quad \limsup_{n \to \infty} A(n) = \infty$$

and Erdős has shown that occasionally the coefficients can get very large indeed

$$\text{Erdős (1949)} \quad \exists \, c > 0 \text{ s.t. i.o. } A(n) > \exp\left(\exp\left(\frac{c \log n}{\log \log n}\right)\right). \tag{3}$$

At about the same time Bateman showed that they are never substantially larger than this

$$\text{Bateman (1940)} \quad A(n) < \exp\left(\exp\left((\log 2 + o(1))\frac{\log n}{\log \log n}\right)\right),$$

and this led Erdős to conjecture that his lower bound (3) should hold for any $c < \log 2$.

```
n = 15    phi(n) = 8
 1 -1   0   1  -1   1   0  -1   1
n = 21    phi(n) = 12
 1 -1   0   1  -1   0   1   0  -1   1   0  -1   1
n = 33    phi(n) = 20
 1 -1   0   1  -1   0   1  -1   0   1  -1   1   0  -1   1   0  -1   1   0  -1   1
n = 35    phi(n) = 24
 1 -1   0   0   0   1  -1   1  -1   0   1  -1   1  -1   1   0  -1   1  -1   1   0   0   0  -1   1
n = 39    phi(n) = 24
 1 -1   0   1  -1   0   1  -1   0   1  -1   0   1   0  -1   1   0  -1   1   0  -1   1   0  -1   1
n = 51    phi(n) = 32
 1 -1   0   1  -1   0   1  -1   0   1  -1   0   1  -1   0   1  -1   1   0  -1   1   0  -1   1   0  -1
 1  0  -1   1   0  -1   1
n = 55    phi(n) = 40
 1 -1   0   0   0   1  -1   0   0   0   1   0  -1   0   0   1   0  -1   0   0   1   0   0  -1   0   1
 0  0  -1   0   1   0   0   0  -1   1   0   0   0  -1   1
n = 57    phi(n) = 36
 1 -1   0   1  -1   0   1  -1   0   1  -1   0   1  -1   0   1  -1   0   1   0  -1   1   0  -1   1   0
-1  1   0  -1   1   0  -1   1   0  -1   1
n = 65    phi(n) = 48
 1 -1   0   0   0   1  -1   0   0   0   1  -1   0   1  -1   1  -1   0   1  -1   1  -1   0   1  -1   1
 0 -1   1  -1   1   0  -1   1  -1   1   0  -1   1   0   0   0  -1   1   0   0   0  -1   1
n = 69    phi(n) = 44
 1 -1   0   1  -1   0   1  -1   0   1  -1   0   1  -1   0   1  -1   0   1  -1   0   1  -1   1   0  -1
 1  0  -1   1   0  -1   1   0  -1   1   0  -1   1   0  -1   1
n = 77    phi(n) = 60
 1 -1   0   0   0   0   0   1  -1   0   0   1  -1   0   1  -1   0   0   1  -1   0   1   0  -1   0   1
-1  0   1   0  -1   0   1   0  -1   1   0  -1   0   1   0  -1   1   0   0  -1   1   0  -1   1   0   0
-1  1   0   0   0   0   0  -1   1
n = 85    phi(n) = 64
 1 -1   0   0   0   1  -1   0   0   0   1  -1   0   0   0   1  -1   1  -1   0   1  -1   1  -1   0   1
-1  1  -1   0   1  -1   1  -1   1   0  -1   1  -1   1   0  -1   1  -1   1   0  -1   1  -1   1   0   0
 0 -1   1   0   0   0  -1   1   0   0   0  -1   1
n = 87    phi(n) = 56
 1 -1   0   1  -1   0   1  -1   0   1  -1   0   1  -1   0   1  -1   0   1  -1   0   1  -1   0   1  -1
 0  1  -1   1   0  -1   1   0  -1   1   0  -1   1   0  -1   1   0  -1   1   0  -1   1   0  -1   1   0
-1  1   0  -1   1
n = 91    phi(n) = 72
 1 -1   0   0   0   0   0   1  -1   0   0   0   0   1   0  -1   0   0   0   0   1   0  -1   0   0   0
 1  0   0  -1   0   0   0   1   0   0  -1   0   0   1   0   0   0  -1   0   0   1   0   0   1   0   0   0  -1   0
 1  0   0   0   0  -1   0   1   0   0   0   0  -1   1   0   0   0   0   0  -1   1
n = 93    phi(n) = 60
 1 -1   0   1  -1   0   1  -1   0   1  -1   0   1  -1   0   1  -1   0   1  -1   0   1  -1   0   1  -1
 0  1  -1   0   1   0  -1   1   0  -1   1   0  -1   1   0  -1   1   0  -1   1   0  -1   1   0  -1   1
 0 -1   1   0  -1   1   0  -1   1
n = 95    phi(n) = 72
 1 -1   0   0   0   1  -1   0   0   0   1  -1   0   0   0   1  -1   0   0   1   0  -1   0   0   0   1   0
-1  0   0   1   0  -1   0   0   1   0  -1   0   1   0   0  -1   0   1   0   0  -1   0   1   0   0  -1
 0  1   0   0  -1   1   0   0   0  -1   1   0   0   0  -1   1   0   0   0  -1   1
```

FIGURE 8-2

Erdős (1957) conjectured

$$c < \log 2 \text{ permissible.}$$

Well, let me sketch for you a proof of this conjecture and even that

$$c = \log 2 \text{ permissible.}$$

Consider the formula

$$\Phi_n(z) = \prod_{d|n} (z^{n/d} - 1)^{\mu(d)} \tag{2}$$

that I mentioned earlier, and consider the real part of the logarithm along a ray from the origin through a primitive fifth root of unity and then consider the Mellin transform. One obtains

$$\frac{4n^{-\sigma}}{\Gamma(\sigma)} \int_0^\infty \left(\log \left| \Phi_n \left(e^{-\frac{1}{x} + 2\pi i a/5} \right) \right| \right) x^{-\sigma - 1} dx$$

$$= \zeta(\sigma + 1) \prod_{p | 5n} \left(1 - \frac{1}{p^\sigma} \right) - 5^{\frac{1}{2}} L(\sigma + 1) \chi(an) \prod_{p | n} \left(1 - \frac{\chi(p)}{p^\sigma} \right)$$

where L is the Dirichlet series formed from the quadratic character χ modulo 5:

$$L(s) = \sum_{m=1}^\infty \frac{\chi(m)}{m^s}, \quad \chi(m) = \begin{cases} 1, & m \equiv \pm 1 (\text{mod } 5) \\ -1, & m \equiv \pm 2 (\text{mod } 5) \\ 0, & m \equiv 0 (\text{mod } 5) \end{cases}.$$

Now if we assume that n is chosen so that for each prime divisor p of n we have $\chi(p) = -1$ and then a is chosen so that $\chi(an) = -1$, and we let $\sigma \to 0+$, we obtain

$$\log \left| \Phi_n (e^{2\pi i a/5}) \right| \geq \frac{d(n)}{2} \log \frac{1 + \sqrt{5}}{2},$$

where d is the divisor function.

By invoking standard theorems about the distribution of primes one can construct such n so that

$$d(n) > 2 \frac{\log n}{\log \log n} + c_1 \frac{\log n}{(\log \log n)^2},$$

where c_1 is a positive constant. It follows that Φ_n is large and therefore has large coefficients of the required size. More precisely we have shown that

$$z^n - 1$$

can have irreducible factors with very large coefficients, greater than

$$\exp \left(2 \frac{\log n}{\log \log n} \right).$$

This theorem is often quoted in the literature on factorization of polynomials as a warning that the size of the coefficients of the polynomial to be factored is *not* a good measure of the ease with which the polynomial can be factored (into its irreducible components).

Remarkably, as far as I have been able to ascertain, this is the only piece of research undertaken at Imperial College to be quoted in Knuth's encyclopaedic work on "The Art of Computer Programming."

I hope that with this problem and the other problems that I have discussed today that I have been able to demonstrate to you that one of the oldest areas of mathematical research is still live and well and acting, at the center of the mathematical stage, as a steady source of mathematical ideas.

Bibliography

Adleman, L. M., and Heath-Brown, D. R. (1985). The first case of Fermat's Last Theorem, *Invent. Math.* 79, 409–416.

Bateman, P. T. (1949). Note on the coefficients of the cyclotomic polynomial, *Bull. Amer. Math. Soc.* 55, 1180–1181.

Brent, R. P., van de June, J., te Riele, H. J. T., and Winter, D. T. (1982). The first 200,000,001 zeros of Riemann's zeta function. *Computational methods in number theory, Part II*, 384–403. Math. Centre Tracts, 155, Math Centrum, Amsterdam.

Brun, V. (1920). Le crible d'Eratosthène et le théorème de Goldbach, *Skrifter utgit av Videnskapsselskapet i Kristiania, mat.-naturv.* Kl. 1:3.

Chen Jing Run and Pan Cheng Dong (1980). On the exceptional set of Goldbach numbers, *Sci. Sinica* 23, 219–232.

Chen Jing Run (1983). On the exceptional set of Goldbach numbers (II), *Sci. Sinica* A, 327–342.

Chudakov, N. G. (1938). On the density of the set of even integers which are not representable as a sum of two odd primes, *Izv. Akad. Nauk SSSR, Ser. Mat.* 1, 25–40.

Corput, J. G. van der (1937). Sur l'hypothèse de Goldbach pour presque tous les nombres pairs, *Acta Arith.* 2, 266–290.

Erdős, P. (1949). On the coefficients of the cyclotomic polynomial, *Portugal Math.* 8, 63–71.

Erdős, P. (1957). On the growth of the cyclotomic polynomial in the interval (0,1), *Proc. Glasgow Math. Assoc.* 3, 102–104.

Estermann, T. (1938). Proof that almost all even positive integers are sums of two primes, *Proc. London Math. Soc.* 44, 307–314.

Euler, L. (1742). Letter to Goldbach, 30 June 1742. *Corresp. Math. Phys.* (P. H. Fuss, ed.) 1 (1843), 135.

Euler, L. (1849). De numeris amicabilibus, *Comm. Arith.* 2, 630. See also *Opera postuma* 1 (1862), 88.

Fouvry, E. (1985). Théorème de Brun-Titchmarsh. Application au Théorème de Fermat, *Invent. Math.* 79, 383–407.

Gauss, C. F. (1808). Summatio quarumdam serierum singularium, *Werke* II, 11–45.

Goldbach, C. (1742). Letter to Euler, 7 June 1742. *Corresp. Math. Phys.* (P. H. Fuss, ed.) 1 (1843), 127.

Goldstine, H. H., and von Neumann, J. (1953). A numerical study of a conjecture of Kummer. *Math. Comp.* 7, 133–143.

Hadamard, J. (1896). Sur la distribution des zéros de la fonction $\zeta(s)$ et ses conséquences arithmétiques, *Bull. Soc. Math. de France* 24, 199–220.

Hardy, G. H., and Littlewood, J. E. (1922). Some problems of "Partitio numerorum" (III): On the expression of a number as a sum of primes, *Acta Math.* 44, 1–70.

Hardy, G. H., and Littlewood, J. E. (1923). Some problems of "Partitio numerorum" (V): A further contribution to the study of Goldbach's problem, *Proc. London Math. Soc.* (2) 22, 46–56.

Heath-Brown, D. R. (1982). Prime numbers in short intervals and a generalized Vaughan identity, *Can. J. Math.* 34, 1365–1377.

Heath-Brown, D. R., and Patterson, S. J. (1979). The distribution of Kummer sums at prime arguments, *J. reine ang. Math.* 310, 111–130.

Ingham, A. E. (1932). *The distribution of prime numbers,* Cambridge University Press.

Knuth, D. E. (1981). *The art of computer programming. Vol. 2: Semi-numerical algorithms.* Second edition. Addison-Wesley, Reading, Mass.

Kummer, E. E. (1846). De residuis cubicis disquisitiones nonnullae analyticae, *J. reine angew. Math.* 32, 341–359 (= *Coll. Papers,* Vol. 1, 145–163, Springer-Verlag, Berlin, 1975).

Lehmer, E. (1936). On the magnitude of the coefficients of the cyclotomic polynomial, *Bull. Amer. Math. Soc.* 42, 389–392.

Lehmer, E. (1956). On the location of Gauss sums, *Math. Comp.* 10, 194–202.

Mangoldt, H. von (1895). Zu Riemann's Abhandlung "Ueber die Anzahl der Primzahlen unter einer gegebenen Grösse," *J. reine ang. Math.* 114, 255–305.

Migotti, A. (1883). Zur Theorie der Kreisteilungsgleichung, Z. B. der Math.-Naturwiss. *Classe der Kaiserlichen Akademie der Wissenschaften,* Wien, 87, 7–14.

Montgomery, H. L., and Vaughan, R. C. (1975). The exceptional set in Goldbach's problem, *Acta Arith.* 27, 353–370.

Schur, I. (c. 1935). Letter to Landau; see E. Lehmer (1936).

Vallée Poussin, C.-J. de la (1896). Recherches analytiques sur la théorie des nombres; Première partie: La fonction $\zeta(s)$ de Riemann et les nombres premiers en général, *Ann. Soc. Sci. Bruxelles,* 20, 183–256.

Vaughan, R. C. (1977). Sommes trigonométriques sur les nombres premiers, *C.R. Acad. Sci., Paris,* A, 285, 981–983.

Vaughan, R. C. (1981). *The Hardy-Littlewood method,* Cambridge University Press.

Vinogradov, I. M. (1937). Some theorems concerning the theory of primes, *Rec. Math.* 2 (44), 179–195.

9

Solving Polynomial Systems

Tien-Yien Li

1. Introduction

A high school student should be able to solve the following problem:

$$x^2 + y^2 = 5 \tag{1}$$
$$x - y = 1 \tag{2}$$

That is, from (2), we get $x = y + 1$ and then substitute into (1). The solutions $x = 2, y = 1$, and $x = -1, y = -2$ are obtained.

QUESTION: Can this method be used to solve a general polynomial system consisting of n equations with n unknowns?

The general version of this method is known as "elimination theory" in modern algebra. Due to the instability of solving high degree polynomials, however, this method is extremely difficult to implement on computers.

Even junior high school students should be able to solve the following system:

$$x^2 = 1 \tag{3}$$
$$y = 1 \tag{4}$$

Since the unknowns are not mixed, the solutions $x = 1, y = 1$ and $x = -1, y = 1$ are obvious. Now, in order to solve the system (1), (2), we build in one more parameter t. That is, we look at the system

$$(1 - t) \begin{bmatrix} x^2 - 1 \\ y - 1 \end{bmatrix} + t \begin{bmatrix} x^2 + y^2 - 5 \\ x - y - 1 \end{bmatrix} = \begin{bmatrix} 0 \\ 0 \end{bmatrix} \tag{5}$$

for $t \in [0, 1]$. When $t = 0$, we are facing the system (3), (4), for which we know the solutions. When $t = 1$, we are facing the problem for which we want the answers. For each individual t in $[0,1]$, (5) is

Volume 9, No. 3 (Summer 1987), 33–39

FIGURE 9-1

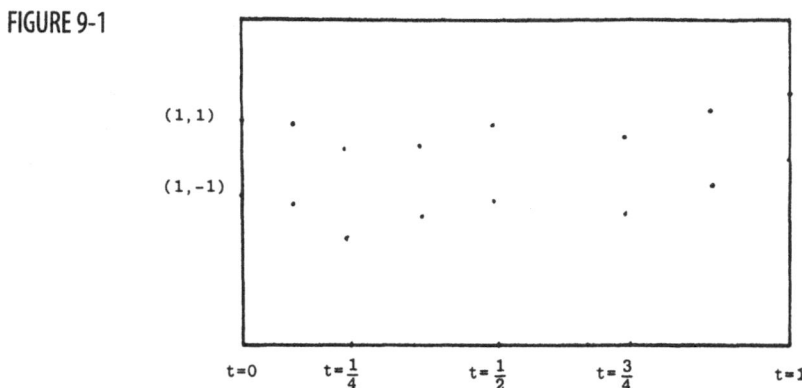

$t=0$ $t=\frac{1}{4}$ $t=\frac{1}{2}$ $t=\frac{3}{4}$ $t=1$

a system of two polynomial equations that, in general, has two solutions, as in Figure 9-1. For example, for $t = 1/4$, the system (5) becomes

$$x^2 + \frac{1}{4}y^2 - 2 = 0, \quad \frac{1}{4}x + \frac{1}{2}y - 1 = 0$$

which has two solutions $x = -9.656$, $y = 6.828$, and $x = 1.65$, $y = 1.172$.

Intuitively, everything is continuous and differentiable, and there does not seem to be any reason to suspect the whole picture should not be as in Figure 9-2. Suppose it is. Denote $z = (x, y)$ and write system (5) as

$$H(z, t) = \begin{bmatrix} H_1(z, t) \\ H_2(z, t) \end{bmatrix} = \begin{bmatrix} 0 \\ 0 \end{bmatrix} = 0. \tag{6}$$

As in Figure 9-2, the solution z is a function of t; thus we may write (6) as

$$H\big(z(t), t\big) = 0. \tag{7}$$

Differentiating (7) with respect to t, we have

$$H_z \frac{dz}{dt} + H_t = 0 \quad \text{and} \quad \frac{dz}{dt} = -H_z^{-1} H_t, \tag{8}$$

where H_z, H_t are partial derivatives of H with respect to z and t, respectively.

FIGURE 9-2

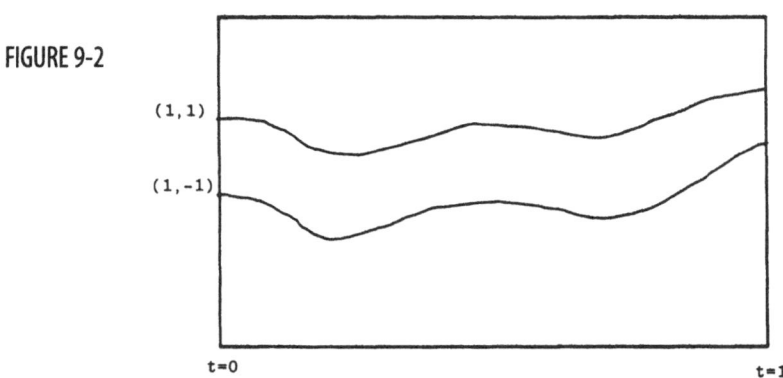

$t=0$ $t=1$

FIGURE 9-3

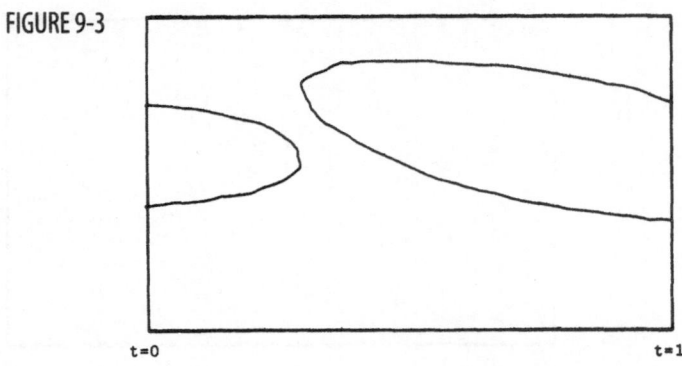

t=0　　　　　　　　　　　　　　　　　　t=1

FIGURE 9-4

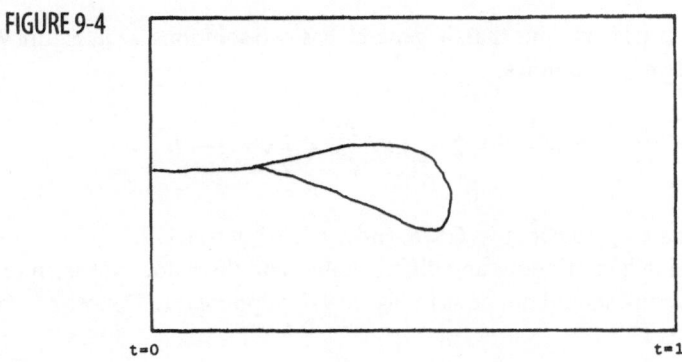

t=0　　　　　　　　　　　　　　　　　　t=1

FIGURE 9-5

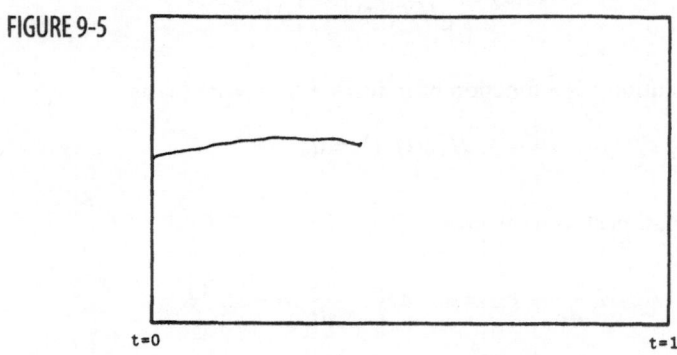

t=0　　　　　　　　　　　　　　　　　　t=1

FIGURE 9-6

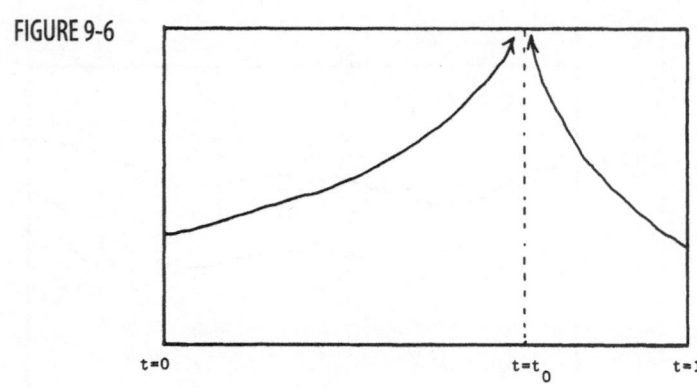

t=0　　　　　　　　　　　　$t=t_0$　　　　t=1

Notice that if the curves $z(t)$ are as in Figure 9-2, then H_z is invertible. The two curves are the integral solutions of the initial value problems $z(0) = (1, -1)$ and $z(0) = (1, 1)$ of the ordinary differential equation (8). There is a well-established theory of approximating the solutions of initial value problems of general ordinary differential equations. The solutions $z(t)$ of (8) at $t = 1$ are the answers we seek for the polynomial system (1), (2) that we want to solve.

This method can be easily generalized to cover the general case of n equations in n unknowns. However, what might go wrong? For example, the curves in Figure 9-2 suggest the following questions:

1. Theoretically speaking, are they really curves?
2. Even if they are nice curves, can they turn around as in Figure 9-3?
3. Can they bifurcate as in Figure 9-4?
4. Will they stop as in Figure 9-5?
5. Will they "blow up" as in Figure 9-6? (That is, for some $t_0 < 1$, the corresponding polynomial system has a root at infinity.)

The main theme of this method, which is known as the "continuation method" (see [1]), is the following. In order to approximate the solutions of a polynomial system, we start off, at $t = 0$, with some system for which we know the solutions. Then some curves, as integral solutions of some ordinary differential equation, are followed. We reach the solutions of the unknown system at $t = 1$. As one can see easily, if any of the above questions (1–5) has a positive answer, this method fails.

2. Homotopy Continuation Method

Let us formulate the problem formally. Let

$$P_1(x_1, \ldots, x_n) = 0, \quad \ldots, \quad P_n(x_1, \ldots, x_n) = 0$$

be n polynomials in n unknowns and write $x = (x_1, \ldots, x_n)$ and $P(x) = (p_1(x), \ldots, p_n(x))$. In order to approximate all isolated zeros of $P(x) = 0$, we are looking for a *homotopy* $H : [0, 1] \times \mathbf{C}^n \to \mathbf{C}^n$ starting with a trivial (i.e., easily solved) set of polynomial equations

$$q_1(x_1, \ldots, x_n) = 0, \quad \ldots, \quad q_n(x_1, \ldots, x_n) = 0$$

or, $Q(x) = 0$, where $Q(x) = (q_1(x), \ldots, q_n(x))$, and $H(0, x) = Q(x), H(1, x) = P(x)$, and such that for each isolated zero $\bar{x}$ of P, there is a smooth curve Γ of zeros of H in $[0, 1] \times \mathbf{C}^n$ leading from some zero of Q at $t = 0$ to $(1, \bar{x})$.

Drexler [3] and Garcia and Zangwill [4] independently and almost simultaneously started looking at solving polynomial systems by this method. A few years later, their homotopies were superseded by the following [6, 7, 11]:

THEOREM A. For a polynomial system $P(x) = (p_1(x), \ldots, p_n(x))$, where $x \in \mathbf{C}^n$ and degree $p_i = d_i$, define $H:[0, 1] \times \mathbf{C}^n \to \mathbf{C}^n$ by

$$H(t, x) = (1 - t)Q(x) + tP(x)$$

where $Q(x) = (q_1(x), \ldots, q_n(x))$ with

$$q_1(x) = a_1 x_1^{d_1} - b_1, \quad \ldots, \quad q_n(x) = a_n x_n^{d_n} - b_n.$$

FIGURE 9-7

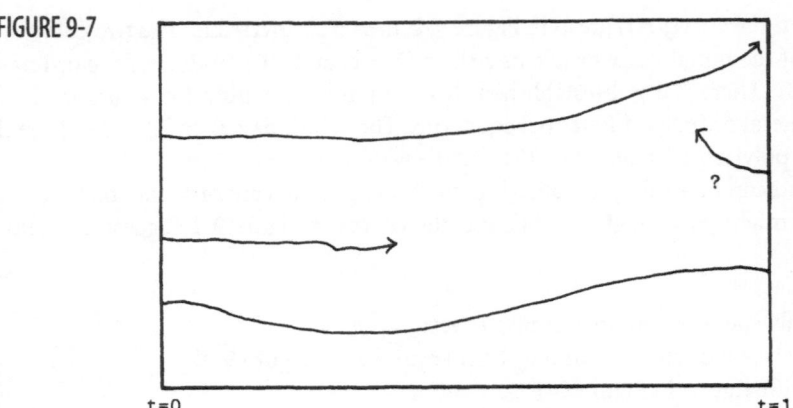

t=0   t=1

Then for "almost all" $(a, b) \in \mathbf{C}^n \times \mathbf{C}^n$ with $a = (a_1, \ldots, a_n)$, $b = (b_1, \ldots, b_n)$,

(a) the solution set of $H(t, x) = 0$ consists of smooth one-manifolds parametrized by $t \in [0, 1)$ and
(b) each isolated solution of $P(x) = 0$ is reached by a finite path emanating from a solution of $Q(x) = 0$.

Theorem A mainly says that for randomly chosen $a = (a_1, \ldots, a_n)$, $b = (b_1, \ldots, b_n)$, the patho-logical problems (1–5) mentioned in the last section disappear. First of all, the solution set of $H(t, x) = 0$ consists of curves! By simple arguments using Sard's Theorem and the implicit function theorem, the curves never stop, turn around, or bifurcate. Therefore, they can be parametrized by t and can be considered as solutions of ordinary differential equations (as in (8)). Theorem A also says that these curves $r(t)$ stay finite for t in $[0,1)$. Therefore, as in Figure 9-7, the curves emanating from $t = 1$ must merge with one of the curves starting from $t = 0$. Thus, by following all the curves of $H^{-1}(0)$, we obtain all the isolated zeros of the polynomial system $P(x)$ we want to solve.

3. Deficient Polynomial Systems

The classical Bezóut Theorem in algebraic geometry says that the upper bound for the number of isolated zeros of a polynomial system $P(x) = (p_1(x), \ldots, p_n(x))$, counting multiplicities, is $d \equiv d_1 \times d_2 \times \cdots \times d_n$ where d_i equals the degree of p_i. The number d is usually called the *Bezóut number*. For polynomial systems $P(x)$ whose number of total isolated zeros, counting multiplicities, reaches this upper bound, Theorem A seems to give a "perfect" method for approximating all the iso-lated solutions. In this case, there are d curves emanating from $t = 0$, and each one reaches one of the isolated zeros of $P(x)$. Unfortunately, almost all systems of polynomials we encounter in applica-tions have fewer than, and in some cases only a small fraction of, the Bezóut number of solutions. We will call such a system *deficient*. For a deficient system $P(x) = 0$, some of the d curves in Theo-rem A reach the isolated solutions of $P(x) = 0$ and the rest of the curves diverge to infinity as t approaches 1. In this situation, following curves that eventually diverge to infinity as $t \rightarrow 1$ is a waste. Let us demonstrate the significance of this waste by a concrete example. Consider an eigen-value problem

$$Ax = \lambda x \tag{9}$$

where $A = (a_{ij}) \in \mathbf{C}^{n \times n}$, $x = (x_1, \ldots, x_n) \in \mathbf{C}^n$. We may consider (9) as the following $n + 1$ poly-nomial equations in $n + 1$ unknowns $(\lambda, x_1, \ldots, x_n)$

$$P_1 = \lambda x_1 - (a_{11}x_1 + \cdots + a_{1n}x_n) = 0, \quad \cdots, \quad P_n = \lambda x_n - (a_{n1}x_1 + \cdots + a_{nn}x_n) = 0, \tag{10}$$
$$P_{n+1} = b_1 x_1 + b_2 x_2 + \cdots + b_n x_n - 1 = 0$$

where $b = (b_1, \ldots, b_n) \in \mathbf{C}^n$ is chosen at random. Notice that if (λ, x) is an eigenpair of (9), then (λ, kx) also satisfies (9) for any $k \in \mathbf{C}$. To make the solutions isolated, we add the equation P_{n+1} to *normalize* the eigenvectors.

In real applications, $n = 100$ is just a moderate problem. In this situation, the Bezóut number is 2^{100} and the number of isolated zeros of (10) is 100, in general. That is, there are 2^{100} curves to follow in Theorem A, and all but 100 of them will diverge to infinity as $t \rightarrow 1$. But which 100 curves? Without any information, one has to follow all 2^{100} curves in order to find the 100 curves that converge. To follow $2^{100} \cong 10^{60}$ curves is extremely difficult, if not impossible.

4. To Deal with Deficient Polynomial Systems

For a polynomial system $P(x) = (p_1(x), \ldots, p_n(x))$ with $x = (x_1, \ldots, x_n)$ and $d_i = $ degree p_i, write

$$p_1(x) = p_1^1(x) + p_1^2(x), \quad \ldots, \quad p_n(x) = p_n^1(x) + p_n^2(x), \tag{11}$$

where $p_i^1(x)$ consists of all the terms of $p_i(x)$ with degree d_i and $p_i^2(x) = p_i(x) - p_i^1(x)$. We now consider the homogeneous polynomial system

$$p_1^1(x_1, \ldots, x_n) = 0, \quad \ldots, \quad p_n^1(x_1, \ldots, x_n) = 0. \tag{12}$$

It is obvious that $(0, \ldots, 0)$ is a solution of $p^1(x) = (p_1^1(x), \ldots, p_n^1(x)) = 0$.

THEOREM B.   If 0 is the only solution of (12), then the system (11) has exactly $d = d_1 \times d_2 \times \cdots \times d_n$ isolated solutions, counting multiplicities.

This theorem was proved by van der Waerden [13] and rediscovered by Garcia and Li [5]. An interesting aspect of Theorem B is that it generalizes two fundamental facts of undergraduate mathematics:

1.   For an $n \times n$ matrix $A = (a_{ij}) \in \mathbf{C}^{n \times n}$, $x = (x_1, \ldots, x_n) \in \mathbf{C}^n$ and $b = (b_1, \ldots, b_n) \in \mathbf{C}^n$, consider the matrix equation

$$Ax = b \tag{13}$$

as a system of polynomial equations:

$$p_1(x_1, \ldots, x_n) = a_{11}x_1 + a_{12}x_2 + \cdots + a_{1n}x_n - b_1 = 0$$

$$\vdots \tag{14}$$

$$p_n(x_1, \ldots, x_n) = a_{n1}x_1 + a_{n2}x_2 + \cdots + a_{nn}x_n - b_n = 0.$$

Each p_i is of degree 1, i.e., $d_i = 1$ for all i. The corresponding polynomial system (12) is

$$p_1^1(x_1, \ldots, x_n) = a_{11}x_1 + \cdots + a_{1n}x_n = 0$$

$$\vdots$$

$$p_n^1(x_1, \ldots, x_n) = a_{n1}x_1 + \cdots + a_{nn}x_n = 0,$$

or

$$Ax = 0 \tag{15}$$

in matrix form. Theorem B says that if 0 is the only solution of (15) (A is *non-singular* in linear algebra terminology), then the system (14) has $d = d_1 \times d_2 \times \cdots \times d_n = 1 \times \cdots \times 1 = 1$ isolated zero. In other words, the matrix equation (13) has a unique solution.

2.   Consider an n-degree polynomial of one unknown

$$p(x) = x^n + a_1 x^{n-1} + \cdots + a_n. \tag{16}$$

The n-degree term in this case is

$$p^1(x) = x^n.$$

Since 0 is the only solution of $p^1(x) = 0$, Theorem B implies that (16) has n roots. This result is the well-known "Fundamental Theorem of Algebra."

The non-zero solutions of (12) are normally called the "zeros at infinity" of the polynomial system $P(x) = (p_1(x), \ldots, p_n(x))$ in (11). The deficient system mentioned in the previous section is a polynomial system $P(x) = (p_1(x), \ldots, p_n(x))$ having non-zero solutions for $P^1(x) = (p_1^1(x), \ldots, p_n^1(x)) = 0$ as in (12). That is, a deficient polynomial system has "zeros at infinity." As we discussed in the last section, almost all the polynomial systems we come across in applications have "zeros at infinity"—and in some cases lots! For those problems, the homotopy $H(t,x)$ introduced in Theorem A may not be appropriate, since some (perhaps many) of the curves diverge to infinity as $t \to 1$.

One can easily see, however, that for the "trivial" system $Q(x) = (q_1(x), \ldots, q_n(x))$, where

$$q_1(x) = a_1 x_1^{d_1} - b_1, \quad \ldots, \quad q_n(x) = a_n x_n^{d_n} - b_n$$

as in Theorem A, 0 is the only solution of its highest degree homogeneous part

$$q_1^1 = a_1 x_1^{d_1}, \quad \ldots, \quad q_n^1 = a_n x_n^{d_n}.$$

FIGURE 9-8

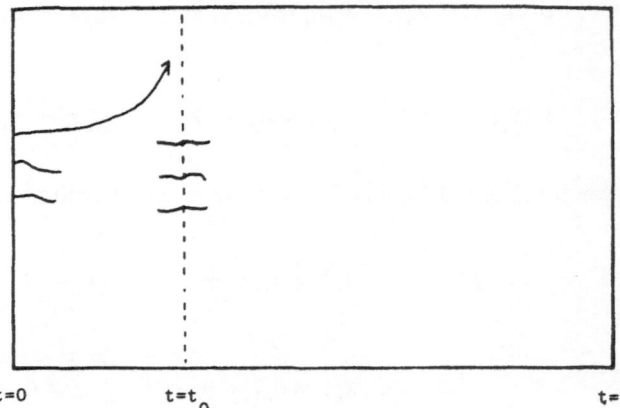

t=0   t=t$_0$ t=1

That is, the polynomial system $Q(x)$ is non-deficient. It thus has $d = d_1 \times d_2 \times \cdots \times d_n$ isolated zeros, as one can easily show. Theorem A mainly says that for randomly chosen $a = (a_1, \ldots, a_n)$, $b = (b_1, \ldots, b_n)$ in $\mathbf{C}^n$, for the polynomial system corresponding to each fixed t in $[0,1)$, 0 is the only solution of its highest degree homogeneous part. Therefore, the solution curves of $H(t, x) = 0$ stay finite for $t \in [0, 1)$, since for each fixed t in $[0,1)$, the corresponding polynomial system has no zeros at infinity.

From this observation, we may deal with a deficient polynomial system $P(x) = (p_1(x), \ldots, p_n(x))$ with the following adjustment:

1. First of all, we still must choose a system $Q(x) = (q_1(x), \ldots, q_n(x))$ that can be easily solved to begin.
2. For $i = 1, \ldots, n$, $q_i(x)$ must have the same degree as $p_i(x)$.
3. The system $Q(x)$ is also deficient and has the same "zeros at infinity" as $P(x)$ does.
4. Make sure that for each t in $[0,1)$, the corresponding polynomial system in the homotopy

$$H(t, x) = (1 - t)Q(x) + tP(x)$$

has the same "zeros at infinity" as $Q(x)$ does.

Roughly speaking, (2) ensures that the sum of the number of finite isolated zeros plus the number of "zeros at infinity" of $P(x)$ and $Q(x)$ stay the same. Then, (4) guarantees that for each t in $[0,1)$, the number of finite isolated zeros of the corresponding polynomial system $H(t, x)$ stays constant. Consequently, the curves emanating from finite isolated zeros of $Q(x)$ at $t = 0$ stay finite for $t \in [0, 1)$. For, if the number of finite isolated zeros of Q is k, then the corresponding polynomial systems $H(t, x)$ also have k finite isolated zeros for all $t \in [0, 1)$. As in Figure 9-8, if any of the curves $r(t) \to \infty$ as $t \to t_0 < 1$, then for t in the neighborhood $(t_0 - \epsilon, t_0)$ of t_0 the corresponding polynomial system $H(t, x)$ has more than k finite isolated zeros, which is a contradiction.

By far the most general result obtained in this direction is Theorem C below. By a *random product homotopy* we mean a homotopy of the form

$$H(t, x) = (1 - t)cQ(x) + tP(x)$$

where $Q(x)$ is a "random product"

$$q_1 = (L_{11} + b_{11}) \cdots (L_{1d_1} + b_{1d_1}), \quad \cdots, \quad q_n = (L_{n1} + b_{n1}) \cdots (L_{nd_n} + b_{nd_n}). \qquad (17)$$

Here $L_{ij}(x_1, \ldots, x_n)$ are linear forms in $x_1, \ldots, x_n$ and $b_{ij}, c \in \mathbf{C}$ are randomly chosen constants. The solutions of $H(0, x) = Q(x)$ are simple to find because of the product structure of $Q(x)$, so it passes the test of being trivial.

THEOREM C [10].   Let $p_1, \ldots, p_n$ be polynomials in the variables $x_1, \ldots, x_n$ and let $q_1, \ldots, q_n$ be of random product form (17) with $\deg p_i = \deg q_i$. Suppose that

(a) the set S of "zeros at infinity" of $Q = (q_1, \ldots, q_n)$ is "smooth,"
(b) every point of S is also a "zero of infinity" of $P = (p_1, \ldots, p_n)$. Then, for $b = (b_{ij})$ belonging to an open dense subset of $\mathbf{C}^{d_1 \times \cdots \times d_n}$ with full measure,
(c) the solution paths of $H(t, x) = 0$ beginning at $t = 0$ are smooth one-manifolds parameterized by $t \in [0, 1)$, and
(d) each isolated solution of $P(x) = 0$ is reached by a finite path emanating from a solution of $Q(x) = 0$.

To illustrate the application of the above theorem, we consider the following example. The equilibrium points of the four-dimensional Lorentz Attractor [11] are given by the zeros of the polynomial system

$$p_1(x_1, x_2, x_3, x_4) = x_1(x_2 - x_3) - x_4 + h$$
$$p_2(x_1, x_2, x_3, x_4) = x_2(x_3 - x_4) - x_1 + h$$
$$p_3(x_1, x_2, x_3, x_4) = x_3(x_4 - x_1) - x_2 + h$$
$$p_4(x_1, x_2, x_3, x_4) = x_4(x_1 - x_2) - x_3 + h,$$

where h is a given constant. Here, the "zeros at infinity" of $P = (p_1, p_2, p_3, p_4)$ are the zeros of the polynomial system

$$x_1(x_2 - x_3) = 0, x_2(x_3 - x_4) = 0, x_3(x_4 - x_1) = 0, x_4(x_1 - x_2) = 0.$$

It can be easily calculated by hand that the non-zero solutions are $(a, 0, 0, 0)$, $(0, b, 0, 0)$, $(0, 0, c, 0)$, $(0, 0, 0, d)$ and (r, r, r, r) where a, b, c, d, r are arbitrary non-zero constants. In order to keep the same zeros at infinity, let $Q = (q_1, q_2, q_3, q_4)$ be

$$q_1 = (x_1 - r_1 x_4 + \alpha_1)(x_2 - x_3 + \beta_1)$$
$$q_2 = (x_2 - r_2 x_1 + \alpha_2)(x_3 - x_4 + \beta_2)$$
$$q_3 = (x_3 - r_3 x_2 + \alpha_3)(x_4 - x_1 + \beta_3)$$
$$q_4 = (x_4 - r_4 x_3 + \alpha_4)(x_1 - x_2 + \beta_4).$$

For randomly chosen $\alpha = (\alpha_1, \alpha_2, \alpha_3, \alpha_4)$, $\beta = (\beta_1, \beta_2, \beta_3, \beta_4)$, $r = (r_1, r_2, r_3, r_4)$, the system $Q = (q_1, q_2, q_3, q_4)$ has eleven finite isolated zeros, and they can be found by solving eleven combinations of 4×4 non-singular matrix equations. The homotopy

$$H(t, x) = (1 - t)cQ(x) + tP(x) = 0$$

with randomly chosen c has eleven solution curves, none of which goes to infinity for t in $[0,1)$. All the isolated solutions of $P = (p_1, p_2, p_3, p_4)$ are obtained by following the eleven curves, starting from the isolated zeros of $Q = (q_1, q_2, q_3, q_4)$ at $t = 0$.

As another example, consider the eigenvalue problem (10) as we mentioned in Section 3.

$$p_1 = \lambda x_1 - (a_{11}x_1 + \cdots + a_{1n}x_n), \quad \cdots, \quad p_n = \lambda x_n - (a_{n1}x_1 + \cdots + a_{nn}x_n)$$
$$p_{n+1} = b_1 x_1 + b_2 x_2 + \cdots + b_n x_n - 1.$$

We may get "zeros at infinity" of $P = (p_1, \ldots p_{n+1})$ by solving

$$p_1^1(x_1, \ldots, x_n, \lambda) = \lambda x_1, \quad \cdots, \quad p_n^1(x_1, \ldots, x_n, \lambda) = \lambda x_n,$$
$$p_{n+1}^1(x_1, \ldots, x_n, \lambda) = b_1 x_1 + \cdots + b_n x_n.$$

It can be easily seen that they are

$$z_1 = \{(x_1, \ldots, x_n, \lambda) \in \mathbf{C}^{n+1} \mid \lambda = 0, b_1 x_1 + \cdots + b_n x_n = 0\},$$
$$z_2 = \{(x_1, \ldots, x_n, \lambda) \in \mathbf{C}^{n+1} \mid x_1 = x_2 = \cdots = x_n = 0\}.$$

To keep zeros at infinity the same, we choose

$$q_1(x_1, \ldots, x_n, \lambda) = (\lambda + \alpha_1)(x_1 + \beta_1), \quad \ldots, \quad q_n(x_1, \ldots, x_n, \lambda) = (\lambda + \alpha_n)(x_n + \beta_n),$$
$$q_{n+1}(x_1, \ldots, x_n, \lambda) = b_1 x_1 + b_2 x_2 + \cdots + b_n x_n + \beta_{n+1}.$$

For $\alpha_i \neq \alpha_j$ and $\beta_i \neq \beta_j$ ($1 \leq i, j \leq n + 1$), we can see that $Q = (q_1, \ldots, q_{n+1})$ has n isolated zeros which are $(-\beta_1, -\beta_2, \ldots, -\beta_{i-1}, h_i, -\beta_{i+1}, \ldots, -\beta_n, -\alpha_i)$, where $h_i = -\beta_{n+1} + \sum_{j \neq i} b_j \beta_j$ for $i = 1, \ldots, n$. The homotopy

$$H(t, x) = (1 - t)cQ + tP$$

has n curves, none of which goes to ∞ and we obtain all the eigenpairs.

5. Conclusion

The research in this area is in its preliminary stage. Using Theorem A to compute all isolated solutions of a polynomial system requires an amount of computational effort proportional to the product of the degrees—roughly, proportional to the size of the system. Homotopies for solving deficient systems for which the computational effort is instead proportional to the actual number of solutions are being developed. In general, homotopies are constructed to respect the special structure of the deficient system. Theorem C helped to solve problems in dynamical systems [11], the load-flow equations in power systems [9], and various eigenvalue problems [2,8,12]. Many polynomial systems in engineering models are not covered by Theorem C, however; see, for example [14,15,17].

An interesting feature of using the homotopy continuation method to solve polynomial systems is that the curves followed by the scheme are computed independently of one another. Therefore, the algorithm is an excellent candidate for exploiting the advantages of parallel processing—the most important special structure of fifth-generation computers. For example, the homotopy continuation method for calculating eigenvalues and eigenvectors of a matrix is a serious alternative to the currently most popular approach—EISPACK [6]. The preliminary computation results are extremely promising.

References

1. E. Allgower and K. Georg, "Simplicial and continuation methods for approximating fixed points and solutions to systems of equations," *SIAM Review 22* (1980), 28–85.
2. M. Chu, T. Y. Li, T. Sauer, "A homotopy method for general λ-matrix problems," *Siam J. Matrix Anal. Appl. 9* (1988), 528–536.
3. F. J. Drexler, "A homotopy method for the calculation of all zero-dimensional polynomial ideals," *Continuation Methods*, 69–93, H. Wacker (ed.), Academic Press, New York, 1978.
4. C. B. Garcia and W. I. Zangwill, "Finding all solutions to polynomial systems and other systems of equations," *Math. Programming 16* (1979), 159–176.
5. C. B. Garcia and T. Y. Li, "On number of solutions to polynomial systems of equations," *SIAM J. of Numer. Anal. 17* (1980), 540–546.
6. T. Y. Li, "On Chow, Mallet-Paret and Yorke homotopy for solving systems of polynomials," *Bull. Institute of Mathematics, Academica Sinica 11* (1983), 433–437.
7. T. Y. Li and T. Sauer, "Regularity results for solving systems of polynomials by homotopy method," *Num. Math 50* (1987), 283–289.

8. T. Y. Li and T. Sauer, "Homotopy methods for generalized eigenvalue problems," *Lin. Alg. Appl.* *91* (1987), 65–74.

9. T. Y. Li, T. Sauer, J. Yorke, "Numerical solution of a class of deficient polynomial systems," *SIAM J. Num. Anal. 24* (1987), 435–451.

10. T. Y. Li, T. Sauer, J. Yorke, "The random product homotopy and deficient polynomial systems," *Num. Math. 51* (1987), 481–500.

11. E. Lorenz, "The local structure of a chaotic attractor in four dimensions," *Physica 13D* (1984), 90–104.

12. A. Morgan, "A homotopy for solving polynomial systems," *Applied Math. and Comp. 18* (1986), 87–92.

13. B. L. van der Waerden, "Die Alternative bei nichtlinearen Gleichungen," *Nachrichten der Gesellschaft der Wissenschaften zu Göttingen, Math. Phys. Klasse* (1928), 77–87.

14. S. Richter and R. De Carlo, "A homotopy method for eigenvalue assignment using decentralized state feedback," *IEEE Trans. Auto. Control, AC-29* (1984), 148–158.

15. M. G. Safonov, "Exact calculation of the multivariable structured-singular-value stability margin," *IEEE Control and Decision Conference*, Las Vegas, Dec. 12–14, 1984.

16. B. T. Smith, J. M. Boyle, J. J. Dongarra, B. S. Garbow, Y. Ikebe, V. C. Klema, C. B. Moler, *Matrix Eigensystem Routines—EISPACK Guide*, Springer-Verlag, 1976.

17. L.-W. Tsai and A. Morgan, "Solving the kinematics of the most general six- and five-degree-of-freedom manipulators by continuation methods," *ASME J. Mechanisms, Transmissions and Automation in Design 107* (1985), 48–57.

10

Artin's Conjecture for Primitive Roots

M. Ram Murty

Introduction

In his preface to *Diophantische Approximationen,* Hermann Minkowski expressed the conviction that the "deepest interrelationships in analysis are of an arithmetical nature." Gauss described one such remarkable interrelationship in articles 315–317 of his *Disquisitiones Arithmeticae.* There, he asked why the decimal fraction of 1/7 has period length 6:

$$\frac{1}{7} = 0.142857\ 142857\ 142857\ldots$$

whereas 1/11 has period length of only 2:

$$\frac{1}{11} = 0.09\ 09\ 09\ldots.$$

Why does 1/99007599, when written as a binary fraction (that is, expanded in base 2), have a period of nearly 50 million 0s and 1s? To answer these questions, Gauss introduced the concept of a primitive root.

To motivate our discussion, let p be a prime ($\neq 2, 5$), and let

$$\frac{1}{p} = 0.a_1 a_2 \ldots a_k \ldots$$

be its decimal expansion with period k. Then, it is easily seen that

$$\frac{1}{p} = \left(\frac{a_1}{10} + \frac{a_2}{10^2} + \cdots + \frac{a_k}{10^k}\right)\left(1 + \frac{1}{10^k} + \frac{1}{10^{2k}} + \cdots\right) = \frac{M}{10^k - 1},$$

Volume 10, No. 4 (Fall 1988), 59–67

where M is some integer. Therefore, $10^k - 1 = Mp$. That is,

$$10^k \equiv 1 (\text{mod } p). \tag{1}$$

The period k must satisfy the above congruence, and k is characterized as the smallest exponent for which (1) is satisfied.

If k is the smallest integer satisfying (1), we say that 10 has *order k* (mod p). By Fermat's little theorem,

$$10^{p-1} \equiv 1 (\text{mod } p)$$

and therefore

$$0 < k \leq p - 1.$$

Thus the largest period of the decimal expansion of $1/p$ can occur if and only if 10 has order $p - 1$ (mod p). In such a case, 10 is called a *primitive root* (mod p). More generally, if $p \nmid a$ and the smallest k such that

$$a^k \equiv 1 (\text{mod } p)$$

is $p - 1$, then a is called a *primitive root* (mod p). If n is the product of distinct primes p_i, the period of $1/n$ in base a is the least common multiple of the orders of a (mod p_i), provided a and n are relatively prime. In the case $n = 99007599 = (9851)(9949)$, 9851 and 9949 are both prime, 2 is a primitive root for both primes, and the period of $1/99007599$ in base 2 is

$$\text{lcm}[9850, 9948] = 48,993,900.$$

Therefore, $1/99007599$ has a binary expansion of period 48,993,900. Such facts are used by the computer scientists to generate pseudorandom binary sequences.

This remarkable interrelationship does not end here. Gauss raised the question of how often 10 is a primitive root modulo p, as p varies over the primes, but made no specific conjecture. A precise conjecture was formulated by E. Artin [1], pages viii–x, in 1927 during a conversation with H. Hasse. He hypothesized that for any given non-zero integer a other than 1, -1, or a perfect square, there exist infinitely many primes p for which a is a primitive root (mod p). Moreover if $N_a(x)$ denotes the number of such primes up to x, he conjectured an asymptotic formula of the form

$$N_a(x) \sim A(a) \frac{x}{\log x}$$

as $x \to \infty$, where $A(a)$ is a certain constant depending on a. This is now known as Artin's conjecture. Artin wrote [1], page 534:

> We all believe that mathematics is an art. The author of a book, the lecturer in a classroom tries to convey the structural beauty of mathematics to his readers, to his listeners. In this attempt, he must always fail. Mathematics is logical to be sure, each conclusion is drawn from previously derived statements. Yet the whole of it, the real piece of art, is not linear; worse than that, its perception should be instantaneous. We all have experienced on some rare occasions the feeling of

elation in realizing that we have enabled our listeners to see at a moment's glance the whole architecture and all its ramifications.

Artin had a profound effect in the shaping of the mathematics of our time. His deep insights into class field theory reciprocity laws, non-abelian L series, and the arithmetic of function fields have guided the development of modern number theory. Artin's conjecture is one celebrated instance of his mathematical intuition and creativity. It is the focal point of diverse areas of mathematics such as group theory, algebraic and analytic number theory, and algebraic geometry. In fact, Artin's motivation originated in algebraic number theory. His intuition was as follows.

For a to be a primitive root (mod p), it is necessary and sufficient that

$$a^{(p-1)/q} \not\equiv 1 (\text{mod } p)$$

for every prime divisor q of $p - 1$. For if k is the order of a mod p, then $k \mid (p - 1)$, and if $k \neq p - 1$, then $k \mid (p - 1)/q$ for some prime divisor q of $p - 1$. Heuristically, a is a primitive root (mod p) if the "events"

$$p \equiv 1 (\text{mod } q), \quad a^{(p-1)/q} \equiv 1 (\text{mod } p) \tag{2}$$

do not occur. To invert the problem, fix q and find the probability that a prime p satisfies the above conditions. By Dirichlet's theorem, $p \equiv 1$ (mod q) is true for primes p with frequency

$$\frac{1}{q - 1}.$$

One would expect that $a^{(p-1)/q} \equiv 1 (\text{mod } p)$ occurs with probability $1/q$. The probability that both events occur is $1/q(q - 1)$, since these events can be assumed to be independent. To ensure that a is a primitive root (mod p), the above events must not occur for any q. This suggests a probability of

$$\prod_q \left(1 - \frac{1}{q(q-1)}\right)$$

for such primes.

In 1967, Hooley [10] proved both Artin's conjecture and an asymptotic formula for $N_a(x)$ subject to the assumption of the generalized Riemann hypothesis. This hypothesis, which is still unproved, is the natural extension of the classical Riemann hypothesis to the Dedekind zeta function of a number field (see next section). The implication of Hooley's theorem is that if Artin's conjecture is false, then the generalized Riemann hypothesis is false.

In 1983, Rajiv Gupta and the author [6] proved, *without any hypothesis*, that there is a specific set of thirteen numbers such that, for at least one of these thirteen numbers, Artin's conjecture is true. This established, for the first time, the existence of some a for which Artin's conjecture is true. Moreover, this proof demonstrated that the conjecture was also true for almost all a. Gupta, Kumar Murty, and the author [8] subsequently reduced the size of this set to 7. In 1985, Heath-Brown [9] refined this result to obtain a set of three numbers, by an application of the "Chen-Iwaniec switching." (Switching first occurs in Lemma 4.4 of Iwaniec [13] in the study of primes of the form $\phi(x,y) + A$, where ϕ is a quadratic form. It was also discovered independently by Chen in his quasi-resolution of the twin prime problem and the Goldbach conjecture.) More precise results will be stated below. One consequence of the Heath-Brown refinement is the following theorem.

THEOREM 1.   One of 2, 3, 5 is a primitive root (mod p) for infinitely many primes p.

In order that a be a primitive root for a prime p not dividing a, it is clearly necessary and sufficient that for each prime q,

$$p \equiv 1 \pmod{q} \Rightarrow a^{(p-1)/q} \not\equiv 1 \pmod{q}. \tag{3}$$

Using this criterion, several nineteenth-century mathematicians observed that 2 is a primitive root (mod p) whenever p is of the form $4q + 1$, where q is prime. In such a case, $p - 1$ has only two prime divisors, namely 2 and q. Since q is odd,

$$p = 4q + 1 \equiv 5 \pmod{8}$$

and

$$2^{(p-1)/2} \equiv -1 \pmod{p}$$

by a special case of quadratic reciprocity. Also,

$$2^{(p-1)/q} = 2^4 \equiv 1 \pmod{p}$$

implies that $p = 3$ or 5 and neither of these primes is of the form $4q + 1$. Therefore, (3) is satisfied and 2 is a primitive root (mod p). It is a classic unsolved problem to determine whether there are infinitely many primes of the form $(p - 1)/4$. It is known by sieve methods that $(p - 1)/4$ is infinitely often a product of at most two primes and both these prime factors are greater than p^θ, with $\theta > 1/4$. Thus, there are not many conditions specified by (3) to ensure that a is a primitive root (mod p) for such primes. This is the essential fact that enables us to prove Theorem 1.

1. Intuition of Artin and Hooley's Theorem

An *algebraic number* α is a complex number that satisfies an equation of the form

$$c_n\alpha^n + c_{n-1}\alpha^{n-1} + \cdots + c_1\alpha + c_0 = 0, \tag{4}$$

where c_i, $0 \le i \le n$, are rational numbers. If n is the smallest natural number for which an equation of the form (4) holds, α is said to be of *degree n*. The set of all numbers of the form

$$b_0 + b_1\alpha + \cdots + b_{n-1}\alpha^{n-1}$$

with $b_i \in \mathbf{Q}$, forms a field K, which we denote by $\mathbf{Q}(\alpha)$, and we say that K has degree n. The set of all $\beta \in \mathbf{Q}(\alpha)$ which satisfy a relation of the form

$$\beta^n + a_{n-1}\beta^{n-1} + \cdots + a_1\beta + a_0 = 0$$

with $a_i \in \mathbf{Z}$ forms a *ring* called the *ring of integers* of K, which we denote by O_K.

O_K enjoys some remarkable properties that were first discovered and systematically used by Dedekind. For instance, every ideal of O_K can be factored uniquely into a product of prime ideals. This is the number field analog of the ancient theorem that every natural number is a unique product of prime numbers.

If p is a prime number, the ideal generated by p in O, namely pO_K, factorizes as a product of distinct prime ideals P_i:

$$pO_K = P_1^{e_1} \ldots P_g^{e_g}.$$

Dedekind proved the important relation

$$n = \sum_{i=1}^{g} e_i f_i,$$

where f_i, defined by the cardinality of the quotient ring $O_K/P_i = p^{f_i}$, is called the *degree* of the prime ideal P_i. A prime p is said to *split completely* in K if

$$pO_K = P_1 \ldots P_n$$

with distinct prime ideals P_i of degree one. Let $\pi_K(x)$ denote the number of primes $p \le x$ that split completely in K. If K is a normal extension of $\mathbf{Q}$, then a theorem of Chebotarev states that the density of primes p that split completely in K is $1/n$. That is,

$$\lim_{x \to \infty} \frac{\pi_K(x)}{\pi_{\mathbf{Q}}(x)} = \frac{1}{n}.$$

Artin realized that the two conditions of (2) are satisfied if and only if p splits completely in the (normal) extension $L_q = \mathbf{Q}(\zeta_q, a^{1/q})$, where ζ_q denotes a primitive qth root of unity. If a is squarefree, then the degree of $L_q/\mathbf{Q}$ is $q(q-1)$, and by the theorem of Chebotarev, the density of primes that split completely in L_q is

$$\frac{1}{q(q-1)}.$$

Now, a is a primitive root (mod p) if and only if the above two conditions (2) do not hold for all q. That is, a is a primitive root (mod p) if and only if p does not split completely in any L_q. We would therefore expect the density of such primes to be

$$\prod_q \left(1 - \frac{1}{q(q-1)}\right).$$

This was the heuristic reasoning that led Artin to formulate his conjecture. Computations by Lehmer revealed that some adjustment is needed in the above conjectured density for a general a, in order to take into account the possible dependence in the fields L_q. It is clear that if we let (for each squarefree integer k),

$$L_k = \prod_{q|k} L_q$$

be the composition of the fields L_q for primes q dividing k and let n_k be the degree of L_k over $\mathbf{Q}$, then the set of primes that do not split completely in any L_k has density

$$A(a) = \sum_{k=1}^{\infty} \frac{\mu(k)}{n_k}.$$

by the inclusion-exclusion principle. The Möbius function μ is defined by

$$\mu(k) = \begin{cases} (-1)^r, & \text{if } k \text{ is the product of } r \text{ distinct primes} \\ 0, & \text{otherwise.} \end{cases}$$

It can be shown that if k is odd, then the fields L_q for $q \mid k$ are completely linearly disjoint. Therefore, for odd k,

$$n_k = \prod_{q \mid k} n_q,$$

and for even subscripts, n_{2k} is equal to n_k or $2n_k$ according to whether $\sqrt{a}$ is or is not contained in the field of kth roots of unity. If $a = bc^2$ with b squarefree, then the criterion for $\sqrt{a}$ to be contained in the field of kth roots of unity (for odd squarefree k) is that $b \mid k$ and $b \equiv 1 \pmod 4$. The formula

$$A(a) = \delta \prod_q \left(1 - \frac{1}{n_q}\right),$$

where

$$\delta = \begin{cases} 1 & \text{if } b \not\equiv 1 \pmod 4 \\ 1 - \mu(b) \prod_{q \mid b} 1/(n_q - 1) & \text{if } b \equiv 1 \pmod 4, \end{cases}$$

is the "obvious" modification for the formulation of the precise conjecture. Subject to the generalized Riemann hypothesis, Hooley proved that this modified density is the correct density of primes for which a is a primitive root.

To make this precise, we need an effective version of the Chebotarev density theorem. In retrospect, it seems rather surprising that such a theorem was not proved until half a century later. One reason for this might be inadequate communication between analytic and algebraic number theorists.

The *Dedekind zeta function* of a number field K is defined by

$$\zeta_K(s) = \sum_A N(A)^{-s},$$

where $N(A)$ denotes the cardinality of O_K/A, and the sum is over all ideals A of O_K. $\zeta_K(s)$ converges absolutely for $Re(s) > 1$. Hecke first showed that $\zeta_K(s)$ admits an analytic continuation to the entire complex plane, except for a simple pole at $s = 1$ and satisfies a functional equation analogous to that of the classic Riemann zeta function. In the same spirit, there is the *generalized Riemann hypothesis*: for $Re(s) > 0$,

$$\zeta_K(s) = 0 \Rightarrow Re(s) = \frac{1}{2}.$$

Under this hypothesis, Hooley proceeded as follows. If π_q denotes the set of primes p that split completely in L_q, and $\pi_k(x)$ denotes the number of primes p up to x contained in $\cap_{q \mid k} \pi_q$, then by the inclusion-exclusion principle

$$N_a(x) = \sum_{k=1}^{\infty} \mu(k) \pi_k(x).$$

Because $\pi_k(x) = 0$ for $k > x$, this is a finite sum. If the analog of the Riemann hypothesis for the Dedekind zeta function corresponding to the field L_d is assumed, then one can prove along classical lines the following prime number theorem:

$$\pi_d(x) = \frac{\mathrm{li}\, x}{n_d} + O(x^{1/2} \log dx), \tag{5}$$

where

$$\mathrm{li}\, x = \int_2^x \frac{dt}{\log t}$$

and the constant implied by the O symbol is absolute. (Here and elsewhere, the O notation means the following: we write that $f(x) = O(g(x))$ if $|f(x)| \le Cg(x)$ for some constant C.) If we insert this in the above formula for $N_a(x)$, then the contribution from the error term is clearly too large. For this reason, Hooley decomposed the sum in the following way, for $k = \Pi_{p < z} p$:

$$N_a(x) \le \sum_{d|k} \mu(d) \pi_d(x),$$

because the right-hand side enumerates primes that satisfy a proper subset of the conditions specified by (3). Moreover, if $M(x; z, w)$ denotes the number of primes $p \le x$ that satisfy (2) for some $z < q < w$, then

$$N_a(x) \ge \sum_{d|k} \mu(d) \pi_d(x) - M(x; z, x).$$

If $z = \frac{1}{6} \log x$, then from (5),

$$N_a(x) = A(a)\mathrm{li}\, x + O\left(\frac{x}{\log^2 x}\right) + O\big(M(x; z, x)\big).$$

To treat the term $M(x; z, x)$, write

$$M(x; z, x) \le M\left(x; z, \frac{x^{\frac{1}{2}}}{\log^2 x}\right) + M\left(x; \frac{x^{\frac{1}{2}}}{\log^2 x}, x^{\frac{1}{2}} \log x\right) + M(x; x^{\frac{1}{2}} \log x, x).$$

The first two terms on the right-hand side are shown to be

$$O\left(\frac{x \log \log x}{\log^2 x}\right)$$

by using (5) and estimates derived from sieve methods. The generalized Riemann hypothesis is unable to estimate the third term in a satisfactory way. The treatment of $M(x; x^{1/2} \log x, x)$ is ingenious and simple. A variation of this method appears in our quasi-resolution of Artin's conjecture [6]. The term in question enumerates primes p such that

$$a^{(p-1)/q} \equiv 1 \,(\mathrm{mod}\, p)$$

for some prime $q > x^{1/2} \log x$. Thus, such a prime p divides

$$\prod_{m < x^{1/2}/\log x} (a^m - 1)$$

because

$$\frac{p-1}{q} < \frac{x^{1/2}}{\log x}.$$

But the number of prime divisors of $a^m - 1$ is at most $m \log a$ (using the fact that a natural number n has at most $\log n$ prime factors). Therefore, the total number of prime factors in question cannot exceed

$$\sum_{m < x^{1/2}/\log x} m \log a = O\left(\frac{x}{\log^2 x}\right).$$

Putting these three estimates together, we conclude that

$$M(x; z, x) = O\left(\frac{x \log \log x}{\log^2 x}\right).$$

This proves that, subject to the generalized Riemann hypothesis,

$$N_a(x) = A(a)\frac{x}{\log x} + O\left(\frac{x \log \log x}{\log^2 x}\right),$$

which completes the account of Hooley's theorem.

2. Elliptic Analogs

In the fall of 1983, at the Institute for Advanced Study in Princeton, the author and Rajiv Gupta were considering some unresolved conjectures of Lang and Trotter concerning elliptic curves. In 1976, Lang and Trotter [14] formulated the elliptic analog of the Artin conjecture. More precisely, let E be an elliptic curve over $\mathbf{Q}$. That is, E can be viewed as the solutions of the equation

$$y^2 = x^3 + g_2 x + g_3, \quad g_2, g_3 \in \mathbf{Q}.$$

If K is a field, then we define

$$E(K) = \left\{(x, y) \mid x, y \in K, y^2 = x^3 + g_2 x + g_3\right\}.$$

Jacobi turned this set into an additive abelian group by defining the addition of two points $P = (x_1, y_1), Q = (x_2, y_2)$ as $R = (x_3, y_3)$, where for $x_1 \neq x_2$,

$$x_3 = -x_1 - x_2 + \frac{(y_2 - y_1)^2}{(x_2 - x_1)^2}$$

and

$$y_3 = -\left\{ x_3 \left(\frac{y_2 - y_1}{x_2 - x_1} \right) + \frac{x_2 y_1 - x_1 y_2}{x_2 - x_1} \right\},$$

and if $x_1 = x_2$, then

$$x_3 = -2x_1 + \left(\frac{3x_1^2 + g_2}{2y_1} \right)^2, \quad y_3 = (x_1 - x_3) \left(\frac{3x_1^2 + g_2}{2y_1} \right) + y_1.$$

(These formulas are valid for any field of characteristic $\neq 2, 3$.) A classic theorem of Mordell and Weil states that $E(K)$ thus defined is a finitely generated abelian group. The number of independent generators is called the *rank* of $E(K)$. Geometrically, the addition of two points of the cubic is the reflection in the x-axis of the third point determined by the secant (or tangent) joining the two given points (see Figure 10-1).

The Mordell-Weil theorem says therefore that all of the points with rational coordinates can be obtained by the tangent-secant process from a finite number of points. The map ϕ_n defined by $\phi_n(P) = nP$ for $P \in E(K)$ defines an endomorphism of $E(K)$ for every integer n. Therefore, the endomorphism ring of $E(K)$ contains a natural copy of $\mathbf{Z}$. Deuring proved the remarkable fact that if the endomorphism ring is strictly larger than the copy of $\mathbf{Z}$, then it can be naturally identified to a subring of the ring of integers in an imaginary quadratic field. These are called elliptic curves *with complex multiplication* whenever these extra endomorphisms exist. The corresponding imaginary quadratic field is called the *CM* field. If a is a rational point of infinite order, the elliptic analog of Artin's conjecture is to determine the density of primes p for which $E(\mathbf{F}_p)$ (the rational points on the curve E viewed over the finite field $\mathbf{F}_p$) is generated by $\bar{a}$, the reduction of a (mod p). Such a point is called a primitive point. Lang and Trotter conjectured that the density of primes p for which a is a primitive point always exists. Attempts to prove this conjecture by the method of Hooley, outlined in section 1, proved futile. The error terms in the Chebotarev density theorem were too large. Nevertheless, Gupta and the author were successful in carrying out a variation of the method of Hooley for curves with complex multiplication and primes p that split completely in the corresponding *CM* field. The problem still remains open in the non-*CM* case.

Lang and Trotter formulated higher rank analogs of their conjecture. A natural question arises: Does the problem get simpler if the rank of the curve goes up? The elliptic analog of Lemma 3 in the next section was the key. Thanks to Lemma 3 the problem does become simpler when the rank increases and the full strength of the Riemann hypothesis need not be invoked. In fact, Gupta and the author proved:

THEOREM 2. Let E be an elliptic curve with *CM* by an order in an imaginary quadratic field. If the rank of $E(Q) \geq 6$, then there is a specific set S of 2^{18} rational points such that at least one of these points is a primitive point (mod p) for infinitely many primes p.

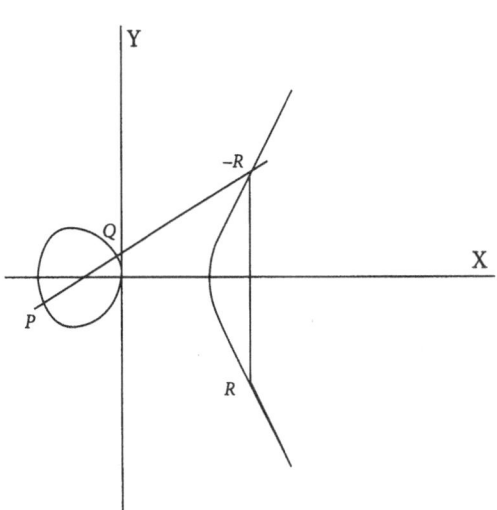

FIGURE 10-1 *Addition on a cubic.*

What about the classic case? Apparently no one had previously formulated the higher rank analog of the classic conjecture of Artin. Going back and formulating the higher rank analog, the authors of [7] were led to the beginnings of the results described in the previous section. Together with Heath-Brown's refinement, the theorem stated at the outset is the most that we can say on Artin's conjecture at present.

3. Quasi-Resolution of Artin's Conjecture

Suppose now that r integers $a_1, a_2, \ldots, a_r$ are given that are multiplicatively independent. That is, if there are integers $n_1, n_2, \ldots, n_r$ such that

$$a_1^{n_1} \ldots a_r^{n_r} = 1,$$

then $n_1 = n_2 = \cdots = n_r = 0$. Let Γ be the subgroup of $\mathbf{Q}^\times$ generated by $a_1, \ldots, a_r$. Denote by Γ_p the image of Γ mod p). Thus, if $r = 1$ and $a = a_1$, then a is a primitive root (mod p) if and only if Γ_p is the full group of coprime residue classes (mod p). In the general case of arbitrary r, consider the size of Γ_p as p varies. The key lemma is:

LEMMA 3.   The number of primes p such that $|\Gamma_p| \leq y$ is $O(y^{1+1/r})$.

PROOF.   Consider the set $S = \{a_1^{n_1} \ldots a_r^{n_r} : 0 \leq n_i \leq y^{1/r}, 1 \leq i \leq r\}$. As $a_1, a_2, \ldots, a_r$ are multiplicatively independent, the number of elements of S exceeds

$$\left([y^{1/r}] + 1\right)^r > y.$$

If p is a prime such that $|\Gamma_p| \leq y$, then two distinct elements of S are congruent (mod p). Hence, p divides the numerator N of

$$a_1^{m_1} \ldots a_r^{m_r} - 1$$

for some $m_1, m_2, \ldots, m_r$ satisfying

$$|m_i| \leq y^{1/r}, \quad 1 \leq i \leq r.$$

For a fixed choice of $m_1, \ldots, m_r$, the number of such primes is bounded by

$$\log N \leq y^{1/r} \sum_{i=1}^{r} \log a_i = O(y^{1/r}).$$

Taking into account the number of possibilities for $m_1, \ldots, m_r$, the total number of primes p cannot exceed

$$O(y^{1+(1/r)}).$$

This completes the proof of the lemma.

REMARK. This is the higher dimensional version of the argument used to treat

$$M(x; x^{1/2} \log x, x)$$

of the previous section. In a different context, this lemma was first proved by Matthews [15].

A higher rank analog of Artin's conjecture can be formulated. Let Γ be the subgroup of $\mathbf{Q}^\times$ generated by r multiplicatively independent natural numbers $a_1, \ldots a_r$.

QUESTION. When is Γ_p the set of coprime residue classes (mod p) for infinitely many primes p?

This question is considered for $r \geq 3$ in a leisurely fashion, ignoring technicalities. Suppose that $(p - 1)/2$ is a product of two primes each greater than p^θ, with $\theta > 1/4$. Let B be the set of such primes and denote by $B(x)$ the number of such primes up to x. By sieve methods, it can be shown that

$$B(x) \geq \frac{cx}{\log^2 x}$$

for some positive constant c. The interested reader should consult the excellent exposition of Bombieri [2], pages 65–75, for the technical details. Then, since Γ_p is a subgroup of the coprime residue classes (mod p), the size of Γ_p does not have many possibilities. This is because $|\Gamma_p|$ divides $p - 1$. Consequently, if either of the two prime factors divides the index of Γ_p in $\mathbf{F}_p^\times$, then

$$|\Gamma_p| \leq p^{1-\theta} \leq x^{1-\theta},$$

for $p \leq x$. Because $r \geq 3$, we find by the lemma that the number of such primes does not exceed

$$O(x^{4(1-\theta)/3})$$

and because $\theta > 1/4$, this bound is certainly

$$O\left(\frac{x}{\log^2 x} \right).$$

Therefore, if the primes for which $|\Gamma_p| \leq p^{1-\theta}$ are thrown away, then for almost all primes contained in B,

$$|\Gamma_p| > p^{1-\theta}.$$

For $p \in B$, the only divisors of $p - 1$ that satisfy the above inequality are $(p - 1)/2$ and $p - 1$. Therefore

$$|\Gamma_p| = \frac{p - 1}{2} \text{ or } p - 1.$$

If the first possibility can be eliminated, then Γ_p is the group of coprime residue classes (mod p). If $|\Gamma_p| = (p-1)/2$, then $\Gamma_p = (\mathbf{F}_p^\times)^2$. Impose the restriction that at least one of $a_1, \ldots, a_r$ is not a perfect square. The sieve methods alluded to earlier produce a set of primes B' such that $|\Gamma_p| \neq (p-1)/2$ and $(p-1)/2$ is either prime or a product of two primes, each greater than p^θ, $\theta > 1/4$. The method indicated above forces $\Gamma_p = \mathbf{F}_p^\times$.

Thus, if $r \geq 3$, and if at least one of $a_1, a_2, \ldots, a_r$ is not a perfect square, then there are infinitely many primes p such that Γ_p is the set of coprime residue classes $(\bmod\, p)$.

To take a specific case, we deduce that 2, 3, and 5 together generate the coprime residue classes $(\bmod\, p)$ for at least

$$\frac{cx}{\log^2 x}$$

primes $p \leq x$. To produce an a that generates $\mathbf{F}_p^\times$, consider a variation of Hooley's argument in section 2. Suppose that none of 2, 3, 5 is a primitive root $(\bmod\, p)$ for $p \in B$. If $(p-1)/2$ is a prime, then one of 2, 3, 5 is a primitive root $(\bmod\, p)$; otherwise, 2, 3, 5 would generate a subgroup strictly smaller than $\mathbf{F}_p^\times$, which is a contradiction. Therefore $(p-1)/2$ is not prime. In this case,

$$\frac{p-1}{2} = q_1 q_2, \quad q_1 < q_2,$$

with $q_1 > p^\theta$, $\theta > 1/4$, the order $(\bmod\, p)$ of each of 2, 3, 5 must be one of $q_1, q_2, 2q_1$, or $2q_2$. Clearly $q_1 < p^{1/2}$; otherwise, $(p-1)/2 > p$, a contradiction.

Now let $\eta > 0$ be a parameter to be chosen. For a fixed q_1, the number of solutions of

$$\frac{p-1}{2} = q_1 q_2$$

where p and q_2 are primes can be shown by elementary sieve methods to satisfy the bound

$$\frac{Cx}{q_1 \left(\log \frac{x}{q_1}\right)^2}$$

for some positive constant C and for $q_1 < x$. From this, it can be deduced that the number of $p \in B'$ such that $q_1 > p^{(1/2)-\eta}$ cannot exceed

$$\sum_{x^{(1/2)-\eta} < q_1 < x^{1/2}} \frac{Cx}{4q_1 \log^2 x} \leq C_1 \eta \frac{x}{\log^2 x}$$

for some absolute constant C_1. If η is chosen sufficiently small, then it may be assumed without loss that for a certain positive constant c_1, at least

$$\frac{c_1 x}{\log^2 x}$$

primes $p \in B'$, $p \leq x$ are such that $(p-1)/2$ is either prime or

$$\frac{p-1}{2} = q_1 q_2, \quad p^\theta < q_1 < p^{1/2-\eta} < q_2$$

with $\theta > 1/4$. This facilitates matters considerably because if a is any natural number such that the order of a (mod p) is $< p^{(1/2)-\eta}$, then applying the lemma with $r = 1$, and $\Gamma = \langle a \rangle$, the subgroup of $\mathbf{Q}^\times$ generated by a, shows that the number of such primes cannot exceed $O(x^{1-2\eta})$. Thus the number of primes p such that 2, 3, or 5 has order (mod p) equal to q_1 or $2q_1$ is at most $O(x^{1-2\eta})$. Therefore, if these primes are eliminated from the set B', then for x sufficiently large, there remain at least $c_1 x/\log^2 x$ primes such that the order (mod p) of 2, 3, and 5 is q_2 or $2q_2$. But then, 2, 3, 5 together generate a subgroup strictly smaller than $\mathbf{F}_p^\times$, which contradicts the fact that they generate $\mathbf{F}_p^\times$. Therefore, one of 2, 3, 5 is a primitive root modulo p for infinitely many primes p.

By a variation of the method observed earlier, if $(p-1)/2$ is prime, then Artin's conjecture can be proved for some specific a. Sieve methods are unable to separate those primes p such that $(p-1)/2$ has *exactly* two prime factors, and this constitutes the famous parity problem of sieve theory.

4. Refinements and Concluding Remarks

In 1977, Iwaniec [11, 12] discovered an improved form of the remainder term in the linear sieve, which leads to the result that there are at least $cx/\log^2 x$ primes $p \leq x$ such that $c > 0$ and $(p-1)/2$ has all prime factors $> p^\theta$ with $\theta = (1/4) - \epsilon$. In 1982, Fouvry and Iwaniec [4] proved a theorem of the form

$$\sum_{m < x^{(9/17)-\epsilon}} \lambda(m) \left(\pi(x, m, 1) - \frac{\mathrm{li}\, x}{\phi(m)} \right) = O\left(\frac{x}{\log^A x} \right), \tag{6}$$

where $\lambda(m)$ is a certain convolution of arithmetical functions, $\pi(x, m, 1)$ denotes the number of primes $p \leq x$, $p \equiv 1 \pmod{m}$, and ϕ denotes Euler's function. With the improved form of the error term in the linear sieve referred to above, this result produces the desired proportion of primes with $\theta = (9/34) - \epsilon > 1/4$. Such a result gives an affirmative answer to the question of the previous section for the rank $r \geq 3$. Indeed, utilizing (6), Rajiv Gupta and the author [5] proved:

THEOREM 4. For any distinct primes q, r, s, at least one element in the set

$$\{qs^2, q^3r^2, q^2r, r^3s^2, r^2s, q^2s^3, qr^3, q^3rs^2, rs^3, q^2r^3s, q^3s, qr^2s^3, qrs\}$$

is a primitive root (mod p) for infinitely many primes p.

In his doctoral thesis, Fouvry [5] discussed various results and techniques to extend the range in (6). In particular, he proved that the exponent $9/17$ can be improved to $17/32$. Finally, in 1983, Bombieri, Friedlander, and Iwaniec [3] proved that the exponent can be improved to $4/7$. This result enables us to obtain $cx/\log^2 x$ primes $p \geq x$ and $c > 0$, such that $(p-1)/2$ is a product of three prime factors, each greater than $p^{(2/7)-\epsilon}$. Heath-Brown observed that those primes p such that $(p-1)/2$ is a product of precisely three prime factors can be removed from this set, so that many primes p for which $(p-1)/2$ is a product of *two* large prime factors are obtained. (As was mentioned earlier, Iwaniec and Chen previously used such an idea, in different contexts.) This yields at least $cx/\log^2 x$ primes $p \leq x$ such that $(p-1)/2$ is either prime or

$$\frac{p-1}{2} = q_1 q_2, \quad p^\theta < q_1 < p^{(1/2)-\eta} < q_2, \quad \theta > \frac{1}{4}.$$

By the method described in section 3, it follows that one of 2, 3, 5 is a primitive root modulo p for infinitely many primes p.

It is conjectured by Halberstam and Elliott that

$$\sum_{m<x^{1-\epsilon}} \left| \pi(x, m, 1) - \frac{\text{li } x}{\phi(m)} \right| = O\left(\frac{x}{\log^A x}\right). \tag{7}$$

In fact, if we had (6) with an exponent of $2/3 + \epsilon$, instead of $(9/17) - \epsilon$, then this would give the result that one of 2 or 3 is a primitive root (mod p) for infinitely many primes p. Any further improvement in (6) does not seem to give any better result. Within a decade (6) may be proved with the exponent $2/3 + \epsilon$. (Recently, it was announced [3] that (7) is true with $x^{1-\epsilon}$ replaced by $x^{1/2}(\log x)^{1987}$ and $A < 3$ —certainly a significant development.) This still would not resolve Artin's conjecture. Perhaps some new simple idea is still lurking in the background that would settle the whole conjecture.

The methods under discussion actually give better results than stated. A special case version of the results was adopted for the sake of clarity. Clearly one can prove along the same lines that if there are three distinct prime numbers q, r, s, then at least one of them is a primitive root (mod p) for infinitely many primes p. It follows that there can be at most two exceptional primes for which Artin's conjecture is false.

Let E be the set of integers, which are not perfect squares, for which Artin's conjecture is false. Let $E(x)$ denote the number of elements of E which are $\leq x$. Srinivasan and the author [16] proved that

$$E(x) = O(\log^6 x)$$

utilizing (6). Heath-Brown [9] independently obtained the slightly finer result

$$E(x) = O(\log^2 x)$$

by incorporating the results of [3]. In a similar vein, he proved that there are at most three square-free integers for which Artin's conjecture is false.

The result (6) with an exponent of $2/3 + \epsilon$ would prove $E(x) = O(\log x)$ and one exceptional a for which Artin's conjecture is false. Of course, if there is an exceptional a, then the generalized Riemann hypothesis would be false in view of Hooley's result.

Problems in mathematics, and number theory in particular, largely serve as motivating forces for the discovery and understanding of new concepts. They provide the background for the play of ideas. We may not see a resolution of Artin's conjecture in the near future. Nevertheless, it has provided us with a rich interplay of algebraic and analytic number theory and demonstrated a profound relationship between analysis and arithmetic.

References

1. E. Artin, *Collected Papers*, Reading, MA: Addison-Wesley (1965).
2. E. Bombieri, Le grand crible dans la théorie analytique des nombres, *Astérique* 18 (1974).
3. E. Bombieri, J. B. Friedlander, and H. Iwaniec, Primes in arithmetic progressions to large moduli, *Acta Math.* 156 (1986), 203–251.
4. E. Fouvry and H. Iwaniec, Primes in arithmetic progressions, *Acta Arith.* 42 (1983), 197–218.
5. E. Fouvry, Autour du théorème de Bombieri-Vinogradov, *Acta Math.* 152 (1984), 219–244.
6. R. Gupta and M. Ram Murty, A remark on Artin's conjecture, *Inventiones Math.* 78 (1984), 127–130.

7. R. Gupta and M. Ram Murty, Primitive points on elliptic curves, *Compositio Math.* 58 (1986), 13–44.

8. R. Gupta, V. Kumar Murty, and M. Ram Murty, The Euclidean algorithm for *S* integers, *CNS Conference Proceedings*, Vol. 7 (1985), 189–202.

9. D. R. Heath-Brown, Artin's conjecture for primitive roots, *Quart. J. Math. Oxford* (2) 37 (1986), 27–38.

10. C. Hooley, On Artin's conjecture, *J. reine angew. Math.* 226 (1967), 209–220.

11. H. Iwaniec, Rosser's sieve, *Acta Arith.* 36 (1980), 171–202.

12. H. Iwaniec, A new form of the error term in the linear sieve, *Acta Arith.* 37 (1980), 307–320.

13. H. Iwaniec, Primes of the type $\phi(x, y) + A$, where ϕ is a quadratic form, *Acta Arith.* 21 (1972), 203–224.

14. S. Lang and H. Trotter, Primitive points on elliptic curves, *Bulletin Amer. Math. Soc.* 83 (1977), 289–292.

15. C. R. Matthews, Counting points modulo p for some finitely generated subgroups of algebraic groups, *Bulletin London Math. Soc.* 14 (1982), 149–154.

16. M. Ram Murty and S. Srinivasan, Some remarks on Artin's conjecture, *Canadian Math. Bull.* 30 (1987), 80–85.

11

Representation Theory of Finite Groups: from Frobenius to Brauer

Charles W. Curtis

This article is dedicated to the memory of my friend and collaborator, Irving Reiner.

The representation theory of finite groups began with the pioneering research of Frobenius, Burnside, and Schur at the turn of the century. Their work was inspired in part by two largely unrelated developments which occurred earlier in the nineteenth century. The first was the awareness of characters of finite abelian groups and their application by some of the great nineteenth-century number theorists. The second was the emergence of the structure theory of finite groups, beginning with Galois' brief outline of the main ideas in the famous letter written on the eve of his death and continuing with the work of Sylow and others, including Frobenius himself.

My aim is to give an account of some of the early work, the problems considered, the conjectures made, and then to trace a few threads in the development of the mathematical ideas from their origins to their place in Brauer's theory of modular representations.

Characters of Finite Abelian Groups and Nineteenth-Century Number Theory

A *character* of a finite abelian group A is a homomorphism from A into the multiplicative group of the field $\mathbf{C}$ of complex numbers, in other words, a function $\chi: A \to \mathbf{C}^* = \mathbf{C} - \{0\}$, which satisfies the condition:

$$\chi(ab) = \chi(a)\chi(b) \text{ for all } a, b \text{ in } A.$$

The simplest examples, which occur in elementary number theory, involve the additive and multiplicative groups of the finite field $\mathbf{Z}_p = \mathbf{Z}/p\mathbf{Z}$ of residue classes $\bar{a} = a + p\mathbf{Z}$, for a prime p. Additive characters of $\mathbf{Z}_p$ are characters of the additive group of $\mathbf{Z}_p$, with the defining property that $\chi(\bar{a} + \bar{b}) = \chi(\bar{a})\chi(\bar{b})$, for all residue classes $\bar{a}$ and $\bar{b}$. These are obtained by taking powers of a pth root of unity, so $\chi(\bar{a}) = \omega^a$, where $\omega^p = 1$ in $\mathbf{C}$. Multiplicative characters of $\mathbf{Z}_p$ are characters of the multiplicative group of $\mathbf{Z}_p$ and include Legendre's quadratic residue symbol $(a/p) = \pm 1$, for a nonzero residue class $\bar{a}$ with $(a/p) = 1$ if $x^2 \equiv a \pmod{p}$ has a solution and -1 if not.

Volume 14, No. 4 (Fall 1992), 48–57

Gauss combined additive and multiplicative characters χ and π, respectively, to form certain sums of roots of unity (today called *Gauss sums*), which have the form

$$g(\chi, \pi) = \sum \chi(\bar{t})\pi(\bar{t}), \quad \bar{t} \neq 0 \text{ in } \mathbf{Z}_p.$$

In §358 of the *Disquisitiones Arithmeticae* [23], he derived the polynomial equations satisfied by the expressions $g(\chi,\pi)$ in some special cases, using information about the number of solutions of congruences, such as $x^n + y^n \equiv 1 \pmod{p}$. In terms of what we know now, this appears to have been a case of putting the cart before the horse. In fact, Gauss sums have proved to be fundamental for obtaining formulas for the number of solutions of a wide class of polynomial congruences and for the more general problem of counting the number of solutions of polynomial equations over finite fields. A nice account of these matters, with historical comments, can be found in the first part of Weil's paper on the number of solutions of equations over finite fields [43] (see also [27], §8.3).

Multiplicative characters were used by Dirichlet in his reinterpretation (see [11]) of some of Gauss' work on genera of binary quadratic forms, where the character-theoretic nature of the quadratic residue symbol (a/p) was applied. He also used them in his definition of L-series, and in the proof, using L-series, that certain arithmetic progressions contain infinitely many primes [10].

Dedekind edited Dirichlet's lectures on number theory for publication and added supplements containing material of his own. In view of the different ways characters had been applied in Dirichlet's work, he called attention to the general notion of characters of abelian groups in one of the supplements (see [11], page 345, footnote, and pages 611, 612). Weber had also become interested in abelian group characters, had published a paper on them, and gave a full account of them in his *Lehrbuch der Algebra* [41], including their construction using the factorization of abelian groups as direct products of cyclic groups.

The starting point of the representation theory of finite groups was Dedekind's work, apparently unpublished, on the factorization of the group determinant of a finite abelian group, and his suggestion, in a letter to Frobenius in 1896, that perhaps Frobenius might be interested in the same problem for general (not necessarily abelian) groups. Here is a statement of the problem.

Let $\{x_g\} = \{x_{g_1}, \ldots, x_{g_n}\}$ be a set of n indeterminates over the field $\mathbf{C}$ of complex numbers, indexed by the elements $\{g_1, \ldots, g_n\}$ of a finite group G of order n. Form the $n \times n$ matrix whose entry in the ith row and jth column is the indeterminate $x_{g_i g_j^{-1}}$. The group determinant of G is the determinant $\Theta = |x_{gh^{-1}}|$ of this matrix, and is a polynomial in the indeterminates x_{g_i}, with integer coefficients. Dedekind had proved the elegant result that, for a finite abelian group, the group determinant Θ factors over the complex numbers as a product of linear factors, whose coefficients are given by the different characters of χ of the group:

$$\Theta = \prod_\chi \left(\chi(g)x_g + \chi(g')x_{g'} + \cdots\right).$$

As he communicated to Frobenius, he had also investigated the factorization of Θ for nonabelian groups in some special cases and had observed that, in the cases he had examined, Θ had irreducible factors of degree greater than one.

The factorization of Θ is not as special a problem as it appears. It is related to the problem of factoring the characteristic polynomial, in the regular representation, of an element of the group algebra $\Sigma x_g g$ with indeterminate coefficients $x_{g'}$, into its irreducible factors. Exactly the same idea was used, with great success, by Killing and Cartan and by Cartan and Molien to obtain the structure of semi-simple Lie algebras and associative algebras, over the field of complex numbers [26].

Frobenius' First Papers on Character Theory

With Dedekind's letter as a spur, Ferdinand Georg Frobenius (1849–1917) burst onto the scene with three papers, published in 1896, in which he created the theory of characters of finite groups, factored the group determinant for non-abelian groups, and established many of the results that have become standard in the subject. At this point in his career, he had assumed Kronecker's chair in Berlin and was already widely known for his research on theta functions, determinants and bilinear forms, and the structure of finite groups, all of which contributed ideas he was able to use in his new venture.

His first task in "Über Gruppencharaktere" [16] was to define characters of non-abelian finite groups. The key to his approach was the study of the multiplicative relations satisfied by the conjugacy classes $\{C_1, \ldots, C_s\}$ in a finite group G. From his previous work in finite group theory, he was well aware of the importance of counting the numbers of solutions of equations in a group G. His starting

FIGURE 11-1 *Ferdinand Georg Frobenius.*

point was the consideration of the integers $\{h_{ijk}\}$, denoting the numbers of solutions of the equations $abc = 1$, with $a \in C_i$, and $c \in C_k$. From them, he defined a new set of integers, $a_{ijk} = h_{i'jk}/h_i$, where $C_{i'} = C_i^{-1}$, and h_i is the number of elements in the class C_i. He then made the crucial observation that the a_{ijk} satisfy identities which imply that the bilinear multiplication defined on a vector space E over $\mathbf{C}$ with basis elements $\{e_1, \ldots, e_s\}$ by the formulas

$$e_j e_k = \sum a_{ijk} e_i \tag{1}$$

is associative and commutative; that is, we have

$$e_i(e_j e_k) = (e_i e_j)e_k \quad \text{and} \quad e_i e_j = e_j e_i$$

for all i, j, k.

This was a situation familiar to him, in view of "Über vertauschbare Matrizen" [15]. He summoned into play a theorem on what we now call the irreducible representations of commutative, semi-simple algebras. The theorem asserts that, under a condition equivalent to semi-simplicity of the algebra, there exist $s = \dim E$ linearly independent numerical solutions $(\rho_1, \ldots, \rho_s)$ of the equations (1), so that $\rho_j \rho_k = \sum a_{ijk} \rho_i$. The condition is that $\det(p_{k\ell}) \neq 0$, where $(p_{k\ell})$ is the matrix with entries

$$p_{k\ell} = \sum_{i,j} a_{ijk} a_{ji\ell}.$$

He proved it, in this case, by an ingenious direct argument based on properties of the class intersection numbers $\{h_{ijk}\}$. Special cases of the result had been obtained by Dedekind, Weierstrass, and Study ([6], [42], [39]), and the definitive theorem, with a new proof, was given by Frobenius himself in "Über vertauschbare Matrizen" [15], the first paper in the 1896 series.

The characters $\chi = (\chi_1, \ldots, \chi_s)$ of the finite group G were defined in terms of the solutions ρ_j of the equations (1), by the formulas

$$h_j \chi_j / f = \rho_j,$$

where f is a proportionality factor, and $h_j = |C_j|$ as above. This is hardly an intuitively satisfying definition. Things become a little clearer if we realize, as Frobenius did, that the characters can be viewed as complex-valued class functions $\chi : G \to \mathbf{C}$, constant on the conjugacy classes (this is what it means to be a class function), satisfying the relations

$$\chi_j \chi_k = f \sum a_{ijk} \chi_i, \tag{2}$$

where $\chi_j = \chi(x)$ for $x \in C_j$ and the constant f, called the degree of the character χ, is $\chi(1)$, the value of χ at the identity element 1 of G. The algebra E, as Frobenius realized somewhat later ([18], §6), is isomorphic to the center of the group algebra of G, so that for abelian groups, the constants a_{ijk} describe the multiplication in the group algebra and the equations (2) are clear generalizations of the definition, given previously, of characters of abelian groups.

The first main results about characters were what are now called the *orthogonality relations*, for two characters χ and ψ, which assert that

$$\frac{1}{|G|} \sum_{g \in G} \chi(g) \psi(g^{-1}) = \begin{cases} 1 \text{ if } \chi = \psi \\ 0 \text{ if } \chi \neq \psi \end{cases}.$$

(These involve the choice of the constant f taken above.) By the theorem used to obtain the characters, the number of different characters and the number s of conjugacy classes are the same, so the characters define an $s \times s$ matrix, called the *character table* of G, whose (i,j)th entry is the value of the ith character at an element in the jth conjugacy class. The orthogonality relations express the fact that, in a certain sense, the rows and columns of the character table are orthogonal.

The question arises, what information about a finite group G is contained in its character table? Frobenius took up the problem himself, and it has fascinated group-theorists ever since. His first contribution to it followed easily from his approach to characters ([15], §4). Using the orthogonality relations, he deduced a formula for the class intersection numbers h_{ijk} in terms of the character table, a result which later proved to be fundamental for applications of character theory to finite groups.

Another interpretation of the orthogonality formulas is that the characters form an orthonormal basis for the vector space of class functions on G, with respect to the hermitian inner product defined by

$$(\zeta, \eta) = |G|^{-1} \sum_{g \in G} \zeta(g) \overline{\eta(g)}, \text{ for class functions } \zeta, \eta. \tag{3}$$

This makes it possible to do a kind of Fourier analysis in the vector space of class functions, in which the "Fourier coefficients" a_χ in the expansion of a class function $\zeta = \Sigma a_\chi \chi$ in terms of the characters, are given by the inner products $a_\chi = (\zeta, \chi)$, for each character χ.

After establishing the foundations of character theory in the second 1896 paper, he turned, in the third, to the solution of the problem raised by Dedekind, about the factorization of the group determinant $\Theta = |x_{gh^{-1}}|$ of a finite group G. He settled the problem with a flourish, proving that $\Theta = \Pi \Phi^f$, with s irreducible factors Φ, whose coefficients are given in terms of the s different characters of G, and the really difficult result, which he called the fundamental theorem in the theory of the group determinant, that the degree of each irreducible factor Φ and the multiplicity with which it occurs in the factorization of Θ coincide and are both equal to the degree f of the corresponding character. He pointed out the consequence that if n is the order of the finite group G, then n is the sum of the squares of the degrees of the characters:

$$n = \sum f_\chi^2, \text{ where } f_\chi = \deg \chi.$$

The Group Determinant of the Symmetric Group S_3

Elements of the Group:

$$g_1 = 1, g_2 = (12), g_3 = (23), g_4 = (13), g_5 = (123), g_6 = (132)$$

Indeterminates: $x_1, x_2, \ldots, x_6$ with $x_i = x_{g_i}$

The group determinant:

$$\Theta = |x_{g_i g_j^{-1}}| = \begin{vmatrix} x_1 & x_2 & x_3 & x_4 & x_6 & x_5 \\ x_2 & x_1 & x_5 & x_6 & x_4 & x_3 \\ x_3 & x_6 & x_1 & x_5 & x_2 & x_4 \\ x_4 & x_5 & x_6 & x_1 & x_3 & x_2 \\ x_5 & x_4 & x_2 & x_3 & x_1 & x_6 \\ x_6 & x_3 & x_4 & x_2 & x_5 & x_1 \end{vmatrix}$$

The factorization of Θ as predicted by Frobenius [17] (see also [29], Section 4):

$$\Theta = F_1 F_2 (F_3)^2,$$

with

$$F_1 = x_1 + x_2 + x_3 + x_4 + x_5 + x_6$$
$$F_2 = x_1 - x_2 - x_3 - x_4 + x_5 + x_6$$
$$F_3 = x_1^2 - x_2^2 + x_2 x_3 - x_3^2 + x_2 x_4 + x_3 x_4 - x_4^2 - x_1 x_5 + x_5^2 - x_1 x_6 - x_5 x_6 + x_6^2.$$

As we noted earlier, the definition (2) of characters of a finite group G does not have the same immediate relation to the structure of G enjoyed by the concept of characters of abelian groups. In the following year, 1897, he clarified the situation by introducing, for the first time, the concept of representation of a finite group. This he defined, as we do today, to be a homomorphism $T : G \to GL_d(\mathbf{C})$, where $GL_d(\mathbf{C})$ is the group of invertible $d \times d$ matrices over $\mathbf{C}$ and d is called the *degree of the representation*, so we have

$$T(gh) = T(g)T(h), \text{ for all } g, h \in G.$$

For an abelian group, the characters, defined previously, are representations of degree one. In the general case, he defined two representations T and $T' : G \to GL_{d'}(\mathbf{C})$ to be *equivalent* if they have the same degree, $d = d'$, and the representations T and T' are intertwined by a fixed invertible matrix X so that $T(g)X = XT'(g)$, or $X^{-1}T(g)X = T'(g)$, for all $g \in G$; in other words, the representation T' is obtained from T by a change of basis in the underlying vector space. In particular, the matrices $T(g)$ and $T'(g)$ are similar, for $g \in G$ and therefore have the same numerical invariants associated with similarity: the same set of eigenvalues, the same characteristic polynomial, trace, and determinant. The important invariant for representation theory is the trace function,

$$\chi(g) = \text{Trace } T(g), \quad g \in G,$$

which Frobenius called the *character of the representation*. The characters defined earlier by the formulas (2) turned out to be the trace functions of certain representations characterized by the irreducibility of polynomials analogous to the group determinant associated with them.

False modesty was not a weakness of Frobenius. From the beginning of his research on the theory of characters, he was keenly aware of its potential importance for algebra and group theory. He was on to a good thing, and he knew it. Altogether, he published more than twenty papers between 1896 and 1907, extending the theory of characters and representations in various directions, and applying the results to finite group theory.

One of the highlights among the papers published after 1896 was a deep analysis of the relation between characters of a group G and the characters of a subgroup H of G [19]. As he stated in the introduction, an understanding of this relationship is crucial for the practical computation of representations and characters—a statement as true now as it was then! One of the main ideas in the paper was the definition of the *induced class function* ψ^G, for a class function ψ on a subgroup H of G, by the formula

$$\psi^G(g) = |H|^{-1} \sum_{x \in G} \dot{\psi}(xgx^{-1}), \quad g \in G,$$

where $\dot{\psi}$ is the function on G defined by

$$\dot{\psi}(g) = \begin{cases} \psi(g) & \text{if } g \in H \\ 0 & \text{if } g \notin H. \end{cases}$$

He proved the fundamental result, now called the Frobenius Reciprocity Law, which states that

$$(\psi^G, \zeta)_G = (\psi, \zeta|_H)_H,$$

for class functions ψ on H and ζ on G, respectively, where $(\ ,\ )_G$ and $(\ ,\ )_H$ are the inner products (3) on the vector spaces of class functions on G and H and $\zeta|_H$ denotes the restriction of the class function ζ to H. Using the Fourier analysis for expansions of class functions in terms of characters, the Reciprocity Law implies that ψ^G is the character of a representation of G if ψ is the character of a representation of H, and gives the desired information about the relationship between characters of G and H.

Frobenius relished computations, the more challenging the better, and rounded out this great series of papers with computations of the character tables of all the groups in the infinite families consisting of the projective unimodular groups $PSL_2(p)$, for odd primes p (in [16], §10); the symmetric groups S_n (in [20]); and the alternating groups A_n (in [21]). The methods he developed of carrying out these computations involved the full range of his ideas on characters, combined with new techniques from combinatorics and algebra, far ahead of their time, which continue to have a strong influence on research in these areas.

A comprehensive historical analysis of Frobenius' first papers on character theory, his correspondence with Dedekind, and other contemporary work in algebra and representation theory, was provided by T. Hawkins [24], [25], [26].

Character Theory and the Structure of Finite Groups: William Burnside (1852–1927)

At about the same time that Frobenius' first papers on character theory appeared, Burnside published his treatise, *Theory of Groups of Finite Order* (1897). After graduating from Cambridge in 1875, Burnside had followed the Cambridge tradition in applied mathematics, with his research in hydrodynamics, until his appointment as Professor of Mathematics at Greenwich, in 1885. His work in group theory began with a paper on automorphic functions in 1892 and continued with research on dis-

FIGURE 11-2 *William Burnside.*

continuous groups and then finite groups, leading to his book [5].

At first he was not optimistic about the possible applications of representations to finite group theory. In the preface to the first edition of his book, in reply to the question of why he devoted considerable space to permutation groups while groups of linear transformations were not referred to, he explained, "My answer to this question is that while, in the present state of our knowledge, many results in the pure theory are arrived at most readily by dealing with properties of substitution groups [i.e., groups of permutations], it would be difficult to find a result that could be most directly obtained by the consideration of groups of linear transformations."

He was aware of Frobenius' work, however, and developed independently his own approach to representations and characters. It is interesting to speculate on how the work of each one influenced the other. They frequently referred to each other's work in their publications, but as far as I know, they never met, or corresponded extensively with each other.

In the preface to the second edition (1911), he stated, ". . . the reason given in the original preface for omitting any account of it no longer holds good. In fact, it is more true to say that for further advances in the abstract theory one must look largely to the representation of a group by linear substitutions." Later (on p. 269, footnote), he described his indebtedness to the work of Frobenius: "The theory of the representation of a group of finite order as a group of linear substitutions was largely, and the allied theory of group characteristics was entirely, originated by Prof. Frobenius." He then listed the papers of Frobenius discussed in the preceding section, and continues, "In this series of memoirs Prof. Frobenius' methods are, to a considerable extent, indirect; and the same is true of two memoirs, 'On the continuous group that is defined by any given group of finite order,' I and II, Proc. L.M.S. Vol. XXIX (1898) in which the author obtained independently the chief results of Prof. Frobenius' earlier memoirs."

Frobenius expressed himself on the matter, in one of his letters to Dedekind, as follows ([25], page 242; see also [29]): "This is the same Herr Burnside who annoyed me several years ago by quickly rediscovering all the theorems I had published on the theory of groups, in the same order and without exception. . . ."

One of Burnside's best-known achievements in group theory is the theorem, proved using character theory, that every finite group G whose order is divisible only by two primes is solvable: $|G| = p^\alpha q^\beta$, for primes p, q, implies that G is solvable. The $p^\alpha q^\beta$-theorem implies, among other things, that the order of a finite, simple, non-abelian group is divisible by at least three different prime numbers. *Simple* means having no non-trivial normal subgroups. Every finite group has a composition series whose factors are simple groups, so that, in a sense, simple groups are the building blocks of all finite groups. Burnside took a great interest in the classification of finite simple groups, a problem that dominated research in finite group theory until its solution in the 1980s.

It was in this connection that he remarked, in note M of the second edition of his book, "There is in some respects a marked difference between groups of even and those of odd order." He went on to discuss the possible existence of non-abelian simple groups of odd order, remarking that he had shown that the number of possible prime factors of a simple group of composite odd order is at least 7. He continued with the statement, "The contrast that these results shew between groups of odd and even order suggests inevitably that non-abelian simple groups of odd order do not exist."

Further progress on this problem was a long time coming. A breakthrough came with M. Suzuki's proof [40] in 1957 that there are no simple groups of composite odd order having the property that the centralizers of all non-identity elements are abelian. In his proof, he made heavy use of a subtle extension of Frobenius' work on induced characters, called the theory of exceptional characters. The next step was the theorem of Feit, Hall, and Thompson [13], that the same result held for groups with the property that centralizers of non-identity elements are nilpotent.

The culmination of this line of research came in 1963, with the publication of Walter Feit and John Thompson of what has become known as the odd-order paper [14], containing one theorem: All finite groups of odd order are solvable. Although a purely group-theoretic proof (not using characters) has been found for Burnside's $p^\alpha q^\beta$-theorem, the proof of the odd-order theorem contains an apparently essential component based on character theory. Feit and Thompson's proof of it takes about 250 pages of close reasoning which to this day resists significant simplification, so perhaps the fifty-year wait following Burnside's statement of the problem is not so surprising.

New Foundations of Character Theory: Issai Schur (1875–1941)

Issai Schur entered the University of Berlin in 1894 to study mathematics and physics. Among his instructors, he expressed special thanks, in a brief autobiographical note at the end of his dissertation, to Professors Frobenius, Fuchs, Hensel, and Schwarz. The dissertation itself, on the classification of the polynomial representations of the general linear group, was a work of such distinction as to place him at once on an equal footing with his illustrious predecessors in representation theory.

The difficulty of the proofs of the main theorems in Frobenius' approach to character theory has already been mentioned. If all persons wishing to enter the field had to master the intricacies of group determinants, the representation theory of finite groups might well have remained a closed book to all but a few.

Burnside's account of the foundations of the theory made important strides towards greater accessibility. In particular, he was apparently the first to take irreducible representations and complete reducibility as concepts of central importance. A representation T of a finite group G is called *reducible* if it is equivalent to a representation T' of the form

$$T'(g) = \begin{pmatrix} T_1(g) & A(g) \\ 0 & T_2(g) \end{pmatrix} \text{ for all } g \in G,$$

for representations T_1 and T_2 of lower degree. If this does not occur, the representation is said to be irreducible. Using results of Loewy [31] and E. H. Moore [33] on the existence of G-invariant hermitian forms, Maschke [32] had proved that every representation T is completely reducible, that is, T is either irreducible or equivalent to a direct sum of irreducible representations.

It remained to Schur, however, to give a wholly elementary and self-contained exposition of the main facts about representations and characters [36]. His starting point was the result, now called Schur's Lemma, which as he pointed out had also played an important role in Burnside's account of the theory. He stated the result, in two parts, as follows:

I.   Let T and T' be irreducible representations of a finite group G, of degrees d and d', respectively. Let P be a constant $d \times d'$ matrix, such that

$$T(g)P = PT'(g), \text{ for all } g \text{ in } G.$$

Then either $P = 0$, or T and T' are equivalent, and P is an invertible $d \times d$ matrix.

II. The only matrices P which commute with all the matrices $T(g)$, $g \in G$, for an irreducible representation T, are scalar multiples of the identity matrix.

As a consequence, he gave short, understandable proofs of the orthogonality properties of the matrix coefficient functions $\{a_{ij}(g)\}$, and for the characters, of irreducible representations $T(g) = (a_{ij}(g))$, $g \in G$. He also gave a new proof of Maschke's theorem on complete reducibility, replacing an appeal to the existence of invariant bilinear forms by a simple, direct argument, in much the same spirit as the standard proof used today. This work, along with what were by all accounts clear and beautifully presented courses of lectures, put the subject within reach, for students, and professional mathematicians, without requiring a specialized background.

One of his students, Walter Ledermann, remarking on the popularity of his lectures, recalls attending his algebra course in a lecture theater filled with about four hundred students, and sometimes having to use opera glasses to follow the speaker when he was unlucky enough to get a seat in the back [30].

Schur's research opened up two more important lines of investigation. In the first [38], he introduced what are called *projective representations* of a finite group G, that is, homomorphisms τ from G into the projective general linear group $PGL_n(\mathbf{C}) = GL_n(\mathbf{C})/\{\text{scalars}\}$. He analyzed precisely when such a representation τ could be lifted to an ordinary representation T of a suitably defined covering group $\tilde{G}$, so that the diagram

$$\begin{array}{ccc} \tilde{G} & \xrightarrow{T} & GL_d(\mathbf{C}) \\ \downarrow & & \downarrow \\ G & \xrightarrow{\tau} & PGL_d(\mathbf{C}) \end{array}$$

is commutative, and the kernel of the homomorphism from $\tilde{G}$ to G is contained in the center of $\tilde{G}$. He constructed a universal covering group $\tilde{G}$, which can be put in the diagram above (for some choice of T) for all projective representations τ. The methods used to construct $\tilde{G}$ and the kernel of the homomorphism from $\tilde{G}$ to G are the beginnings of a major chapter in group theory known as the cohomology of groups.

Another theme was Schur's search for arithmetical properties of representations, which brought out connections with algebraic number theory. The central idea is the concept of a splitting field K of a finite group G. This is a subfield K of the complex field $\mathbf{C}$ with the property that each irreducible representation $T : G \to GL_d(\mathbf{C})$ is equivalent to a K-representation $T' : G \to GL_d(K)$. A splitting field is minimal if no proper subfield is a splitting field. From the work of Frobenius, it was known that a given finite group G has a splitting field K, which is an algebraic number field, that is, a finite extension of the rational field. The splitting field problem was to determine, for a finite group G, the algebraic number fields K which are minimal splitting fields. Splitting fields reflect, in some mysterious way, the structure of the group. For example, splitting fields for cyclic groups require the addition of roots of unity to the rational field, while the field of rational numbers is a splitting field for the symmetric groups S_n.

Both Burnside and Schur were interested in the splitting field problem and had evidence to support the conjecture that the cyclotomic field of mth roots of unity, where m is the least common multiple of the orders of the elements of G, is always a splitting field. Using a subtle device, known as the Schur index, Schur was able to prove the conjecture for all solvable groups [37].

The Dawn of the Modern Age of Representation Theory: Emmy Noether (1882–1935)

By finding simple algebraic ideas to express the essential structure of a mathematical theory, Emmy Noether reshaped many different parts of twentieth-century mathematics. Representation theory of finite groups was no exception: It has never been the same since the publication of her article

FIGURE 11-3 *Emmy Noether.*

"Hyperkomplexe Grössen und Darstellungstheorie" (1929) [34]. She presented a basic set of ideas underlying the representation theory of a finite-dimensional algebra over an arbitrary field. In the case of representations of finite groups, the algebra involved was the group algebra KG of the group G over a field K. This is the associative ring whose additive group is the vector space over K with a distinguished basis indexed by the elements of the group G. In order to define multiplication in KG, it is enough to define it for a pair of basis elements, corresponding to elements g and h in G; their product is defined to be the basis element corresponding to the product $g \cdot h$ in G.

Representations of G over the field K may be viewed as homomorphisms $T : G \to GL(V)$, where V is a finite-dimensional vector space over K and $GL(V)$ is the group of invertible linear transformations on V. The definition given earlier amounts to choosing a basis in V and taking note of the resulting isomorphism: $GL(V) \cong GL_d(K)$, where d is the dimension of V over K.

Noether's critical observation was that each representation $T : G \to GL(V)$ defines the structure of a left KG-module on V, with the module operation $a \cdot v$ defined by setting

$$a \cdot v = \sum_{g \in G} \alpha_g T(g)v, \text{ for } v \in V \quad \text{and} \quad a = \sum_{g \in G} \alpha_g g \in KG.$$

(Here we have identified the element $g \in G$ with the basis element of KG corresponding to it.) Conversely, each finitely-generated left KG-module V defines a representation $T : G \to GL(V)$ by reversing the procedure given above.

It is easily checked that two representations are equivalent if and only if the KG-modules corresponding to them are isomorphic. Thus the main problem of representation theory, which is the classification of the representations of a finite group G up to equivalence, becomes the problem of construction and classification of modules over the group algebra. The problem makes sense for arbitrary fields K. For a field K of characteristic zero, or of prime characteristic p not dividing the order of the group, the left KG-modules are semi-simple, that is, direct sums of simple modules, by a version of Maschke's Theorem. This implies that the group algebra KG is semi-simple, and the main facts about representations, such as that the number of equivalence classes of irreducible representations in a splitting field is the same as the number of conjugacy classes, become straightforward applications of the Wedderburn structure theorems for semi-simple algebras.

For more detailed analysis of the contents of her paper [37], see the article by Jacobson [28].

Richard Brauer (1901–1977) and Modular Representation Theory

The next great surge of activity in the representation theory of finite groups, and one that ties up some of the threads started earlier in this article, centered around the work of Richard Brauer, and its continuation by his students and successors, on modular representation theory. Brauer had been a student in Berlin, and completed his dissertation under Schur's supervision in 1926. His early work on representation theory and the theory of simple algebras, including the invention of what has become known, over his objections, as the Brauer group, firmly established his position in the European mathematical community. When Hitler came to power in 1933, Brauer, Emmy Noether, and many other Jewish university teachers in Germany were dismissed from their positions and came to the United States. Brauer's major publications on modular representations began to appear soon after his arrival in the United States, and the subject remained a focus of his research throughout the rest of his life.

My interest in representation theory was kindled by lectures given by Brauer that I heard as a graduate student, including all four of his 1948 AMS Colloquium Lectures at the summer meetings in Madison and a lecture on his solution of Artin's conjecture on L-series with general group characters at another meeting in New York, when he was awarded the Cole Prize of the American Mathematical Society for this work. I can't say I understood the lectures very well at the time, but they made a strong impression.

A few years later, in 1954, Irving Reiner and I spent a year at the Institute for Advanced Study in Princeton. Neither of us knew much about representation theory, but it seemed to us to be a subject in which exciting things were happening, especially those connected with Brauer's work. We organized an informal seminar devoted to Brauer's work on modular representations and other topics in character theory. This led to our book-writing projects, as a way of learning the subject.

Modular representation theory is the classification of kG-modules, where kG is the group algebra of a finite group G over a field k of characteristic $p > 0$. In case p divides the order of the group G, the group algebra kG is not semi-simple, and the kG-modules are not necessarily direct sums of simple modules, so their classification is much more difficult. Modular representations were first considered by Leonard Eugene Dickson [7], [8], [9]; among other things, he was the first to point out the different nature of the theory in case the order of the group is divisible by the characteristic of the field.

One of Brauer's first results in this subject was the theorem ([1], 1935), that the number of equivalence classes of irreducible representations of a finite group G, in a splitting field of characteristic $p > 0$, is equal to the number of conjugacy classes in G containing elements of order prime to p. If p does not divide the order of G, every conjugacy class has this property, and the result agrees with known properties of representations in the complex field $\mathbf{C}$.

Brauer maintained a steady interest in the relation between properties of the irreducible complex-valued characters and the structure of finite groups. One of his objectives was to use modular representation theory to obtain new information about the values of the irreducible characters in $\mathbf{C}$ and to apply it to problems on the structure of groups. Many of his lectures at meetings and research conferences contained lists of unsolved problems, often involving finite simple groups and properties of their characters.

In order to develop the connection between modular representations and complex-valued characters, he introduced what is now called a p-modular system, consisting of an algebraic number field K, which is a splitting field for G; a discrete valuation ring R with quotient field K, maximal ideal P, and residue field $k = R/P$ of characteristic p, for a fixed prime number p. As an application of the character theory he had developed in connection with his prize-winning proof [3] of Artin's conjecture, he had also succeeded in proving the splitting field conjecture of Burnside and Schur [2]. For the cyclotomic field containing the nth roots of unity, where n is the order of G, it follows that K, and the residue field $k = R/P$, are both splitting fields.

The next step explains how modular representations are related to representations in the field K. Each KG-module V defines a representation $T : G \to GL_d(K)$. Since R is a principal ideal domain, it follows that there exists a representation $T' : G \to GL_d(R)$ which is equivalent to $T : T'(g) = XT(g)X^{-1}$, for all $g \in G$. The homomorphism $R \to R/P = k$ can be applied to the entries of the matrices $T'(g)$; this procedure yields a representation $\bar{T}' : G \to GL_d(k)$ and a kG-module $M = \bar{V}$. The representation $\bar{T}'$ and the module $M = \bar{V}$ are obtained from T and V by what is called *reduction mod P*, so $\bar{T}'$ is a modular representation of G. But there is a difficulty connected with this process. The kG-module $M = \bar{V}$ is not determined up to isomorphism by the isomorphism class of the KG-module V. Nevertheless, in a fundamental joint paper [4], Brauer and his Toronto Ph.D. student Cecil Nesbitt proved that the composition factors of M are uniquely determined.

Using the process of reduction mod P, they defined the *decomposition matrix D* as follows. The rows of D are indexed by the isomorphism classes of simple KG-modules (or, what amounts to the same thing, by the equivalence classes of irreducible representations $T : G \to GL_d(K)$), the columns by the isomorphism classes of simple kG-modules, and an entry d_{ij} of D gives the number of times the jth simple kG-module appears as a composition factor in the module obtained by reduction mod P from the ith simple KG-module. They also introduced the Cartan matrix C, whose entry c_{ij} counts the number of times the jth simple kG-module occurs as composition factor in the ith indecomposable left ideal occurring in a suitably indexed list of indecomposable direct summands of kG. In 1937 they proved the remarkable fact that the Cartan matrix and the decomposition matrix satisfy the relation $C = {}^tDD$, where ${}^tD$ is the transpose of D. This establishes a deep connection between the representation theory of G in the field k of characteristic zero and the representation theory of G in the field k of characteristic p. Its proof used a result of Frobenius [22], which was a refinement of his previous work on the factorization of the group determinant.

The preceding result is only the beginning of Brauer's theory. The refinements of character theory he was seeking came from his theory of p-blocks. These describe a partition of the set of irreducible characters in subsets, called p-blocks, corresponding to the decomposition of the group algebra RG as a direct sum of indecomposable two-sided ideals. To each p-block of irreducible characters, he associated a certain p-subgroup of G, called the defect group of the block. He obtained precise information about the values of the irreducible characters in a given p-block, using the modular representation theory of the defect group and its normalizer. This work, in turn, led to applications of the theory of p-blocks of characters by Brauer, Suzuki, and others to important early steps in the classification of finite simple groups (see [12], Chapters VIII and XII).

The belief that representation theory of finite groups had a bright future was shared by Frobenius, Burnside, Schur, Noether, and Brauer. The high level of current research activity in the subject and its connections with other parts of mathematics seem to support their judgment.

References

1. R. Brauer, *Über die Darstellungen von Gruppen in Galoischen Feldern*, Actualités Scientifiques et Industrielles 195, Hermann, Paris, 1935.

2. R. Brauer, "On the representation of a group of order g in the field of gth roots of unity," *Amer. J. Math.* 67 (1945), 461–471.

3. R. Brauer, "On Artin's L-series with general group characters," *Ann. of Math.* (2) 48 (1947), 502–514.

4. R. Brauer and C. J. Nesbitt, "On the modular representations of groups of finite order I," *Univ. of Toronto Studies*, Math. Ser. 4, 1937.

5. W. Burnside, *Theory of Groups of Finite Order*, Cambridge, 1897; Second Edition, Cambridge, 1911.

6. R. Dedekind, "Zur Theorie der aus n Haupteinheiten gebildeten complexen Grössen," *Göttingen Nachr.* (1885), 141–159.

7. L. E. Dickson, "On the group defined for any given field by the multiplication table of any given finite group," *Trans. A.M.S.* 3 (1902), 285–301.

8. L. E. Dickson, "Modular theory of group matrices," *Trans. A.M.S.* 8 (1907), 389–398.

9. L. E. Dickson, "Modular theory of group characters," *Bull. A.M.S.* 13 (1907), 477–488.

10. P. G. Lejeune Dirichlet, "Beweis des Satzes, dass jede unbegrenzte arithmetische Progression, deren erstes Glied und Differenz ganze Zahlen ohne gemeinschaftlichen Factor sind, unendlich viele Primzahlen enthält," *Abh. Akad. d. Wiss. Berlin* (1837), 45–81. *Werke I*, 313–342.

11. P. G. Lejeune Dirichlet, *Vorlesungen über Zahlentheorie*, 4th ed. Published and supplemented by R. Dedekind, Vieweg, Braunschweig, 1894.

12. W. Feit, *The Representation Theory of Finite Groups*, North-Holland, Amsterdam, 1982.

13. W. Feit, M. Hall, and J. G. Thompson, "Finite groups in which the centralizer of any non-identity element is nilpotent," *Math. Z.* 74 (1960), 1–17.

14. W. Feit and J. G. Thompson, "Solvability of groups of odd order," *Pacific J. Math.* 13 (1963), 775–1029.

15. F. G. Frobenius, "Über vertauschbare Matrizen," *S'ber. Akad. Wiss. Berlin* (1896), 601–614; *Ges. Abh. II*, 705–718.

16. F. G. Frobenius, "Über Gruppencharaktere," *S'ber. Akad. Wiss. Berlin* (1896), 985–1021; *Ges. Abh. III*, 1–37.

17. F. G. Frobenius, "Über die Primfactoren der Gruppendeterminante," *S'ber. Akad. Wiss. Berlin* (1896), 1343–1382; *Ges. Abh. III*, 38–77.

18. F. G. Frobenius, "Über die Darstellung der endlichen Gruppen durch lineare Substitutionen," *S'ber. Akad. Wiss. Berlin* (1897), 994–1015; *Ges. Abh. III*, 82–103.

19. F. G. Frobenius, "Über Relationen zwischen den Charakteren einer Gruppe und denen ihrer Untergruppen," *S'ber. Akad. Wiss. Berlin* (1898), 501–515; *Ges. Abh. III*, 104–118.

20. F. G. Frobenius, "Über den Charaktere der symmetrischen Gruppe," *S'ber. Akad. Wiss. Berlin* (1900), 516–534; *Ges. Abh. III*, 148–166.

21. F. G. Frobenius, "Über die Charaktere der alternirenden Gruppe," *S'ber. Akad. Wiss. Berlin* (1901), 303–315; *Ges. Abh. III*, 167–179.

22. F. G. Frobenius, "Theorie der hyperkomplexen Grössen," *S'ber. Akad. Wiss. Berlin* (1903), 504–537; *Ges. Abh. III*, 284–317.

23. C. F. Gauss, *Disquisitiones Arithmeticae*, Leipzig, 1801; English translation by A. A. Clarke, Yale University Press, New Haven, 1966.

24. T. Hawkins, "The origins of the theory of group characters," *Archive Hist. Exact Sc.* 7 (1971), 142–170.

25. T. Hawkins, "New light on Frobenius' creation of the theory of group characters," *Archive Hist. Exact Sc.* 12 (1974), 217–243.

26. T. Hawkins, "Hypercomplex numbers, Lie groups, and the creation of group representation theory," *Archive Hist. Exact Sc.* 8 (1971), 243–287.

27. K. Ireland and M. Rosen, *A Classical Introduction to Modern Number Theory*, Springer-Verlag, New York, 1980.

28. N. Jacobson, Introduction, in *Emmy Noether, Ges. Abh.*, Springer-Verlag, Berlin, 1983; 12–26.

29. W. Ledermann, "The origin of group characters," *J. Bangladesh Math. Soc.* 1 (1981), 35–43.

30. W. Ledermann, "Issai Schur and his school in Berlin," *Bull. London Math. Soc.* 15 (1983), 97–106.

31. A. Loewy, "Sur les formes quadratiques définies à indéterminées conjuguées de M. Hermite," *Comptes Rendus Acad. Sci. Paris* 123 (1896), 168–171.

32. H. Maschke, "Beweis des Satzes, dass diejenigen endlichen linearen Substitutionsgruppen, in welchen einige durchgehends verschwindende Coefficienten auftreten, intransitiv sind," *Math. Ann.* 52 (1899), 363–368.

33. E. H. Moore, "A universal invariant for finite groups of linear substitutions: with applications in the theory of the canonical form of a linear substitution of finite period," *Math. Ann.* 50 (1898), 213–219.

34. E. Noether, "Hyperkomplexe Grössen und Darstellungstheorie," *Math. Z.* 30 (1929), 641–692; *Ges. Abh.* 563–992.

35. I. Schur, *Über eine Klasse von Matrizen, die sich einer gegebenen Matrix zuordnen lassen*, Dissertation, Berlin, 1901; *Ges. Abh. I*, 1–72.

36. I. Schur, "Neue Begründung der Theorie der Gruppencharaktere," *S'ber. Akad. Wiss. Berlin* (1905), 406–432; *Ges. Abh. I*, 143–169.

37. I. Schur, "Arithmetische Untersuchungen über endliche Gruppen linearer Substitutionen" *S'ber. Akad. Wiss. Berlin* (1906), 164–184; *Ges. Abh. I*, 177–197.

38. I. Schur, "Untersuchungen über die Darstellung der endlichen Gruppen durch gebrochene lineare Substitutionen," *J. reine u. angew. Math.* 132 (1907), 85–137; *Ges. Abh. I*, 198–205.

39. E. Study, "Über Systeme von complexen Zahlen," *Göttingen Nach.* (1889), 237–268.

40. M. Suzuki, "The non-existence of a certain type of simple group of odd order," *Proc. A.M.S.* 8 (1957), 686–695.

41. H. Weber, *Lehrbuch der Algebra*, vol. 2, Vieweg, Braunschweig, 1896.

42. K. Weierstrass, "Zur Theorie der aus n Haupteinheiten gebildeten complexen Grössen," *Göttingen Nach.* (1884), 395–414.

43. A. Weil, "Numbers of solutions of equations in finite fields," *Bull. A.M.S.* 55 (1949), 497–508; *Collected Papers, I*, 399–410.

12

Quaternionic Determinants

Helmer Aslaksen

Introduction

The classical matrix groups are of fundamental importance in many parts of geometry and algebra. Some of them, like $Sp(n)$, are most conceptually defined as groups of quaternionic matrices. But, the quaternions not being commutative, we must reconsider some aspects of linear algebra. In particular, it is not clear how to define the determinant of a quaternionic matrix. Over the years, many people have given different definitions. In this article I will discuss some of these.

Let us first briefly recall some basic facts about quaternions. The quaternions were discovered on October 16, 1843 by Sir William Rowan Hamilton. (For more on the history, I recommend [19], [31], [47], and [48].) They form a non-commutative, associative algebra over $\mathbb{R}$:

$$\mathbb{H} = \{a + ib + jc + kd \mid a, b, c, d \in \mathbb{R}\},$$

where

$$i^2 = j^2 = k^2 = -1, \quad ij = k = -ji, \quad jk = i = -kj, \quad ki = j = -ik.$$

We can also express $z \in \mathbb{H}$ in the form $z = x + jy$, where $x, y \in \mathbb{C}$, but then we have to remember that $yj = j\bar{y}$ for $y \in \mathbb{C}$. Notice that $\mathbb{H}$ is not an algebra over $\mathbb{C}$, since the center of $\mathbb{H}$ is only $\mathbb{R}$. Conjugation in $\mathbb{H}$ is defined by $\overline{a + ib + jc + kd} = a - ib - jc - kd$ and satisfies $\overline{uv} = \bar{v}\bar{u}$. We will call the quaternions of the form $ib + jc + kd$ with $b, c, d \in \mathbb{R}$ the *pure quaternions*.

For any ring R, we let R^* denote the set of units in R, i.e., the invertible elements of R. If R is a skewfield, then $R^* = R - \{0\}$. Let $M(n, R)$ be the ring of $n \times n$ matrices with entries in R. We will denote the set of invertible $n \times n$ matrices over R by $GL(n, R)$. (Some readers might worry about our definition of invertible in $M(n, \mathbb{H})$: Is there a distinction between left and right inverses? We will see later that there is no such problem. See also [15] and [32].)

Cayley

The most simple-minded approach when trying to define the determinant of a quaternionic matrix would be to use the usual formula. But then the question is: Which usual formula? For a 2×2 determinant we could use $a_{11}a_{22} - a_{12}a_{21}$ (expanding along the first row) or $a_{11}a_{22} - a_{21}a_{12}$ (expanding

Volume 18, No. 3 (Summer 1996), 57–65

along the first column), or some other ordering of the factors in the usual formula. To a modern mathematician, this lack of a canonical definition is an indication that this is *not* the correct approach. But we might still ask ourselves: What exactly would happen if we tried one of these formulas?

In 1845, just two years after Hamilton's discovery of the quaternions, Arthur Cayley [10, 35] did precisely this. He chose to expand both the original matrix and all the minors along the first column (or vertical row as he called it). If we denote the Cayley determinant by Cdet, we get

$$\text{Cdet} \begin{pmatrix} a_1 & b_1 \\ a_2 & b_2 \end{pmatrix} = a_1 b_2 - a_2 b_1$$

and

$$\text{Cdet} \begin{pmatrix} a_1 & b_1 & c_1 \\ a_2 & b_2 & c_2 \\ a_3 & b_3 & c_3 \end{pmatrix} = a_1(b_2 c_3 - b_3 c_2) - a_2(b_1 c_3 - b_3 c_1) + a_3(b_1 c_2 - b_2 c_1).$$

Is this a good definition? Cayley himself points out that if two rows are the same in a 2×2 matrix, then

$$\text{Cdet} \begin{pmatrix} a & b \\ a & b \end{pmatrix} = ab - ab = 0,$$

whereas if two columns are the same in a 2×2 matrix, then

$$\text{Cdet} \begin{pmatrix} a & a \\ b & b \end{pmatrix} = ab - ba,$$

which in general is non-zero. For some reason, this didn't seem to bother Cayley much, and he happily proceeded to write a couple more pages about his new function. But it should bother us.

Let us try to clarify the situation by first deciding on which properties we want the determinant to satisfy. Based on our experience with complex matrices, we will call $d : M(n, \mathbb{H}) \mapsto \mathbb{H}$ a determinant if it satisfies the following three axioms.

AXIOM 1. $d(A) = 0$ if and only if A is singular.

AXIOM 2. $d(AB) = d(A)d(B)$ for all $A, B \in M(n, \mathbb{H})$.

AXIOM 3. If A' is obtained from A by adding a left-multiple of a row to another row or a right-multiple of a column to another column, then $\text{d}(A') = \text{d}(A)$.

Let me make some comments about these axioms. It can be shown [7] that if d is not constantly equal to 0 or 1, then Axiom 2 implies that $d(A) = 0$ for all singular matrices. Thus, we need only to define the determinant of invertible matrices.

Notice that in Axiom 3 there is a distinction between left and right scalar multiplication. Consider the mapping $T(v) = cv$. Then, for $f \in \mathbb{H}$,

$$T(fv) = c(fv)$$

is in general different from

$$fT(v) = f(cv),$$

whereas

$$T(vf) = c(vf) = cvf = T(v)f.$$

We see that we must write the coefficients of a linear transformation on the opposite side of what we use for the vector space structure. I will identify vectors with columns and identify linear transformations with matrices on the left, but consider all vector spaces to be right vector spaces.

Axiom 3 can be expressed in terms of matrix multiplication. Let e_{ij} be the matrix with a 1 in the (i, j) entry, and 0 otherwise. Define

$$B_{ij}(b) = I_n + be_{ij} \text{ for } i \neq j.$$

Multiplying a matrix A by $B_{ij}(b)$ on the left adds the jth row multiplied by b on the left to the ith row, whereas multiplying A by $B_{ij}(b)$ on the right adds the ith column multiplied by b on the right to the jth column. So Axiom 3 can be restated (using Axiom 2) as saying that $d(B_{ij}(b)) = 1$.

It is easy to see that

$$B_{ij}(b)^{-1} = B_{ij}(-b),$$

so it follows that products of $B_{ij}(b)$'s generate a subgroup of $GL(n, \mathbb{H})$, which we will denote by $SL(n, \mathbb{H})$. Notice that when K is a field, we define $SL(n, K)$ to be the set of matrices with determinant equal to 1. But because we don't have a determinant yet, we must define $SL(n, \mathbb{H})$ in some other way and then hope that once we have our determinant, it will have $SL(n, \mathbb{H})$ as its kernel. That Axiom 3 can be restated as saying that matrices in $SL(n, \mathbb{H})$ have determinant equal to 1 is therefore promising.

An obvious question is now whether such determinants exist. Let me first state a simple obstruction.

THEOREM 1.   Assume that d is a determinant, i.e., d satisfies our three axioms. Then the image $d(M(n, \mathbb{H}))$ is a commutative subset of $\mathbb{H}$.

This theorem essentially says that when trying to define a quaternionic determinant, we must keep it complex-valued. This rules out Cayley's definition, since Cdet is onto $\mathbb{H}$.

The proof of Theorem 1 depends on the next two lemmas. We first observe that the definition of $B_{ij}(b)$ only involves two indices. We can, therefore, often assume without loss of generality that $n = 2$. A simple calculation proves the following lemma.

LEMMA 2.   Let $a \neq 0$ and d be a determinant. Then

$$\begin{pmatrix} a & 0 \\ 0 & a^{-1} \end{pmatrix} = \begin{pmatrix} 1 & 0 \\ -a^{-1} & 1 \end{pmatrix} \begin{pmatrix} 1 & a-1 \\ 0 & 1 \end{pmatrix} \begin{pmatrix} 1 & 0 \\ 1 & 1 \end{pmatrix} \begin{pmatrix} 1 & a^{-1}-1 \\ 0 & 1 \end{pmatrix}$$

and

$$d\left(\begin{pmatrix} a & 0 \\ 0 & a^{-1} \end{pmatrix} \right) = 1.$$

The next lemma is crucial.

LEMMA 3.   Every $A \in GL(n, \mathbb{H})$ can be written in the form

$$A = D(x)B,$$

where

$$D(x) = \begin{pmatrix} 1 & & & \\ & \ddots & & \\ & & 1 & \\ & & & x \end{pmatrix}$$

and $B \in SL(n, \mathbb{H})$.

PROOF.   Because A is invertible, there must be at least one non-zero element in the first row, say $a_{1j} \neq 0$. By adding the jth column multiplied by $a_{1j}^{-1}(1 - a_{11})$ on the right to the first column, we get a matrix with $a_{11} = 1$. We can then make all the other entries in the first row equal to zero, and proceed by induction.

The observant reader may now be wondering about the uniqueness of the $A = D(x)B$ decomposition. But it is more urgent to prove Theorem 1.

PROOF OF THEOREM 1.   Define $f : \mathbb{H} \to \mathbb{H}$ by

$$f(x) = d\big(D(x)\big).$$

It follows from Lemma 3 that $f(\mathbb{H}) = d(M(n, \mathbb{H}))$ For simplicity of notation we will assume that $n = 2$. We have

$$d\begin{pmatrix} x & 0 \\ 0 & 1 \end{pmatrix} = d\left(\begin{pmatrix} x & 0 \\ 0 & x^{-1} \end{pmatrix} \begin{pmatrix} 1 & 0 \\ 0 & x \end{pmatrix} \right) = f(x)$$

by Axiom 2 and Lemma 2. But then

$$f(x)f(y) = d\left(\begin{pmatrix} x & 0 \\ 0 & 1 \end{pmatrix} \begin{pmatrix} 1 & 0 \\ 0 & y \end{pmatrix} \right) = d\begin{pmatrix} x & 0 \\ 0 & y \end{pmatrix} = d\left(\begin{pmatrix} 1 & 0 \\ 0 & y \end{pmatrix} \begin{pmatrix} x & 0 \\ 0 & 1 \end{pmatrix} \right) = f(y)f(x),$$

and we see that $f(\mathbb{H}) = d(M(n, \mathbb{H}))$ is commutative.

It is now time to ask how Cayley's definition fits into this. It clearly cannot satisfy all the three axioms. In fact, it doesn't satisfy any of them! Consider the matrix

$$M = \begin{pmatrix} k & j \\ i & 1 \end{pmatrix}.$$

It is easy to prove that if

$$M\begin{pmatrix} x \\ y \end{pmatrix} = \begin{pmatrix} 0 \\ 0 \end{pmatrix},$$

then $x = y = 0$, so M is invertible. But

$$M^t \begin{pmatrix} -1 \\ j \end{pmatrix} = \begin{pmatrix} 0 \\ 0 \end{pmatrix},$$

so M^t is singular. But Cdet $M = 0$ and Cdet $M^t = 2k$, so we see that Axiom 1 fails. This also shows that the transpose is not a very useful concept in quaternionic linear algebra. The reason is that it is neither an automorphism nor an antiautomorphism! (But notice that Hermitian involution, $M^* = \bar{M}^t$, is an antiautomorphism, i.e., $(MN)^* = N^*M^*$.) For similar reasons, the concept of rank is also more complicated. The right column-rank is the same as the left row-rank, but they might be distinct from the left column-rank, which is equal to the right row-rank [12]. Noting that

$$\text{Cdet}\left(\begin{pmatrix} 1 & i \\ j & k \end{pmatrix}\begin{pmatrix} k & j \\ i & 1 \end{pmatrix}\right) = 2 - 2k$$

whereas

$$\text{Cdet}\begin{pmatrix} 1 & i \\ j & k \end{pmatrix}\text{Cdet}\begin{pmatrix} k & j \\ i & 1 \end{pmatrix} = 0$$

we see that Axiom 2 also fails.

As for Axiom 3, we have

$$\text{Cdet}\begin{pmatrix} ab & b \\ a & 1 \end{pmatrix} = 0,$$

but after subtracting the second row multiplied by b on the left from the first row, we get

$$A' = \begin{pmatrix} ab - ba & 0 \\ 0 & 1 \end{pmatrix},$$

and $\text{Cdet}(A') = ab - ba$, which in general is non-zero.

This clearly shows that Cdet is not the way to go. A more promising lead is before us, in Lemma 3. It will be followed up later.

Let me finish this section with a remark about Theorem 1. It is inspired by a related theorem proved by the physicist and mathematician Freeman J. Dyson in 1972 [21]. He used a different third axiom:

AXIOM 3′.   Let $A = (a_{ij})$, $B = (b_{ij})$, and $C = (c_{ij})$. If for some row index r we have

$$a_{ij} = b_{ij} = c_{ij}, \quad i \neq r, \quad \text{and} \quad a_{ri} + b_{ri} = c_{ri},$$

then

$$d(A) + d(B) = d(C).$$

In other words, d should be additive in the rows. He then proved that if d satisfies axioms 1, 2, and 3′, the image of d is commutative. It is easy to see that Axioms 1, 2, and 3′ imply Axiom 3. We just have to prove that $d(B_{ij}(b)) = 1$. Let B' be the matrix obtained by replacing the ith entry along the diagonal in $B_{ij}(b)$ by a 0. Then B' is singular, and it follows from Axiom 3′ that $d(B_{ij}(b)) = 1$.

So his definition of determinant is more restrictive than ours. But it is, in fact, too restrictive. Determinants satisfying his three axioms simply don't exist over the quaternions! Why? It follows from Axiom 2 that $d(I_n) = 1$. Define

$$D(x) = \begin{pmatrix} 1 & & & \\ & \ddots & & \\ & & 1 & \\ & & & x \end{pmatrix}.$$

Since $I_n + D(-1) = 2D(0)$ is singular, it follows from Axioms 1 and 3' that $d(D(-1)) = -1$. Because $-1 = iji^{-1}j^{-1}$, we get $D(-1) = D(i)D(j)D(i)^{-1}D(j)^{-1}$, so $D(-1)$ is a commutator in $GL(n, \mathbb{H})$. But Axiom 2 and Theorem 1 then imply that $d(D(-1)) = 1$, which is a contradiction.

Study

Concerning quaternionic determinants, nothing much happened during the 75 years after Cayley. In the second (posthumous) edition of W. R. Hamilton's book *Elements of Quaternions* [24] from 1889, the editor added an appendix, which was just a restatement of Cayley's paper. Also, a paper by J. M. Peirce [38] from 1899 is just a laborious elaboration on the Cayley determinant. But in 1920 a very interesting paper by Eduard Study appeared [44]. (For more details, see also [16], [23], and [46].) His idea was to transform a quaternionic matrix into a complex $2n \times 2n$ matrix and then take the determinant.

I will start by discussing some important homomorphisms between quaternionic, complex, and real matrices. Recall that any complex $n \times n$ matrix can be written uniquely as $N = C + iD$, where C and D are real $n \times n$ matrices. We can then define an injective algebra homomorphism $\phi : M(n, \mathbb{C}) \rightarrow M(2n, \mathbb{R})$ by

$$\phi(C + iD) = \begin{pmatrix} C & -D \\ D & C \end{pmatrix}.$$

Set

$$J = \begin{pmatrix} 0 & -I_n \\ I_n & 0 \end{pmatrix}.$$

Let R_i be right-multiplication by i on $\mathbb{C}^n$. The corresponding matrix is iI, and $J = \phi(iI) = \phi(R_i)$. (I will sometimes identify a linear transformation and its standard matrix.) This gives a complex structure on $\mathbb{R}^{2n}$, and we know that $P \in M(2n, \mathbb{R})$ corresponds to a complex linear transformation if and only if P commutes with the complex structure. Hence,

$$\phi(M(n, \mathbb{C})) = \{P \in M(2n, \mathbb{R}) \mid JP = PJ\}.$$

In a similar way, any quaternionic $n \times n$ matrix can be expressed uniquely in the form $M = A + jB$, where A and B are complex $n \times n$ matrices. (We write j on the left since we work with right vector spaces.) We can, therefore, define $\psi : M(n, \mathbb{H}) \rightarrow M(2n, \mathbb{C})$ by

$$\psi(A + jB) = \begin{pmatrix} A & -\overline{B} \\ B & \overline{A} \end{pmatrix}.$$

It is straightforward to show that this map is an injective algebra homomorphism. [This implies in particular that there is no distinction between left- and right-inverses in $GL(n, \mathbb{H})$.]

Let R_j be right-multiplication by j on $\mathbb{H}^n$. Notice that any $\mathbb{H}$-linear transformation commutes with R_j, but that R_j is *not* $\mathbb{H}$-linear. Thus, there is no matrix associated to R_j and it doesn't make sense to talk about $\psi(R_j)$, but we can still consider the corresponding map of $\mathbb{C}^{2n}$ given by $\tilde{R}_j(x, y) = (-\bar{y}, \bar{x})$. We see that $\tilde{R}_j$ corresponds to first multiplying by J and then conjugating. This gives a quaternionic structure on $\mathbb{C}^{2n}$, and we know that $N \in M(2n, \mathbb{C})$ corresponds to a quaternionic linear transformation if and only N commutes with the quaternionic structure. Since $N\overline{Jv} = \overline{NJv}$, we have $N\overline{Jv} = \overline{JNv}$ if and only if $\bar{N}J = JN$, so

$$\psi\left(M(n, \mathbb{H})\right) = \left\{N \in M(2n, \mathbb{C}) \mid JN = \bar{N}J\right\}. \tag{1}$$

Notice that this is simply a generalization of the formula $jz = \bar{z}j$ for $z \in \mathbb{C}$.

It follows immediately from (1) that $\det_{\mathbb{C}} \psi(M) \in \mathbb{R}$, but we will soon see that, in fact, we have $\det_{\mathbb{C}} \psi(M) \geq 0$. (I will sometimes write $\det_{\mathbb{R}}$ or $\det_{\mathbb{C}}$ to stress that I'm taking the determinant of a real or complex matrix.)

By applying the homomorphism $\phi_1 : \mathbb{C} \cong M(1, \mathbb{C}) \to M(2, \mathbb{R})$ to each element of $M \in M(n, \mathbb{C})$, we get a map $\tilde{\phi} : M(n, \mathbb{C}) \to M(2n, \mathbb{R})$. [$\phi(N)$ consists of four n-blocks, whereas $\tilde{\phi}(N)$ consists of n^2 2-blocks.] The important thing here is that the 2-blocks in $\tilde{\phi}(N)$ are easier to manage than the n-blocks in $\phi(N)$. Since $\mathbb{C}$ is commutative and ϕ_1 is a homomorphism, the 2-blocks in $\tilde{\phi}(N)$ commute. This allows us to use the following folklore theorem. [It has been rediscovered numerous times, but to the best of my knowledge it is originally due to M. H. Ingraham [26].]

THEOREM 4.   If $A = (A_{ij})$ is a square block matrix, where the A_{ij} are mutually commutative $m \times m$ matrices, and B is the m $\times$ m matrix obtained by taking the determinant of A with the A_{ij} as elements, then $\det A = \det B$.

For example, if A_{11}, A_{12}, and A_{22} are mutually commutative, then

$$\det \begin{pmatrix} A_{11} & A_{12} \\ A_{21} & A_{22} \end{pmatrix} = \det(A_{11}A_{22} - A_{12}A_{21}). \tag{2}$$

In other words, you evaluate by "taking the determinant twice."

By shuffling some rows and columns, we see that $\det_{\mathbb{R}} \phi(N) = \det_{\mathbb{R}} \tilde{\phi}(N)$, and we can now apply Theorem 4 to get [6]

$$\det_{\mathbb{R}} \phi(N) = \det_{\mathbb{R}} \tilde{\phi}(N) = \det_{\mathbb{R}}\left(\phi_1(\det_{\mathbb{C}} N)\right)$$

$$= \det_{\mathbb{R}} \begin{pmatrix} \text{Re } \det_{\mathbb{C}} N & -\text{Im } \det_{\mathbb{C}} N \\ \text{Im } \det_{\mathbb{C}} N & \text{Re } \det_{\mathbb{C}} N \end{pmatrix} = |\det_{\mathbb{C}} N|^2,$$

for $N \in M(n, \mathbb{C})$. This discussion leads to the following important theorem.

THEOREM 5.   For any complex matrix N, we have

$$\det_{\mathbb{R}} \phi(N) = |\det_{\mathbb{C}} N|^2 \geq 0. \tag{3}$$

For any quaternionic matrix M, we have

$$\det_{\mathbb{C}} \psi(M) = \sqrt{\det_{\mathbb{R}} \phi\big(\psi(M)\big)} \geq 0. \tag{4}$$

PROOF. The first part follows from (2). It follows from (1) that $\det_{\mathbb{C}} \psi(M) \in \mathbb{R}$, and since $\det \phi(GL(n, \mathbb{H}))$ is a connected subset of $\mathbb{R}$, we get that Sdet $M \geq 0$ for quaternionic matrices. We then deduce (4) from (2).

We are finally ready to define the Study determinant Sdet by

$$\text{Sdet } M = \det_{\mathbb{C}} \psi(M).$$

The obvious question is now which axioms the Study determinant satisfies. The Study determinant satisfies Axiom 2 because ψ is a homomorphism. Let us show that Axiom 1 holds. (Notice that the proof of this statement is wrong in both editions of the otherwise excellent book by Morton L. Curtis [16].) We know that if Sdet $M = \det_{\mathbb{C}} \psi(M) \neq 0$, then $\psi(M)$ is invertible in $M(2n, \mathbb{C})$, but we need to know that the inverse actually lies in $\psi(M(n, \mathbb{H}))$. By conjugating and inverting the formula $J\psi(M) = \overline{\psi(M)}J$, we see that $J\psi(M)^{-1} = \overline{\psi(M)^{-1}}J$. But then it follows from (1) that $\psi(M)^{-1}$ lies in $\psi(M(n, \mathbb{H}))$.

To show that Axiom 3 holds, it suffices to prove that Sdet $B_{ij}(b) = 1$. If $b = b_1 + jb_2$, then

$$\psi\big(B_{ij}(b)\big) = \begin{pmatrix} I_n + b_1 e_{ij} & -\overline{b}_2 e_{ij} \\ b_2 e_{ij} & I_n + \overline{b}_1 e_{ij} \end{pmatrix}.$$

But $e_{ij}e_{ij} = 0$, so we can apply Theorem 4 to get $\det(\psi(B_{ij}(b))) = \det(I_n) = 1$.

Thus, the Study determinant satisfies all our axioms, and it is used frequently in differential geometry and Lie theory [23]. Bear in mind that it is a *quadratic* function of the entries, not multi-linear in the rows and the columns like the usual determinant.

Let me finish this section with a couple of additional comments. The Study determinant was defined above by identifying $\mathbb{H}$ with $\mathbb{C}^2$. What would happen if we instead identified $\mathbb{H}$ with $\mathbb{R}^4$? After all, the center of $\mathbb{H}$ is $\mathbb{R}$, not $\mathbb{C}$, so the quaternions form an $\mathbb{R}$-algebra. We can write $M \in M(n, \mathbb{H})$ uniquely as $M = A_0 + iA_1 + jA_2 + kA_3$ where $A_0, A_1, A_2,$ and A_3 are real $n \times n$ matrices, and apply the homomorphism $\mu : M(n, \mathbb{H}) \to M(4n, \mathbb{R})$ given by

$$\mu(A_0 + iA_1 + jA_2 + kA_3) = \begin{pmatrix} A_0 & -A_1 & -A_2 & -A_3 \\ A_1 & A_0 & -A_3 & A_2 \\ A_2 & A_3 & A_0 & -A_1 \\ A_3 & -A_2 & A_1 & A_0 \end{pmatrix}.$$

Notice that

$$\phi\psi(A_0 + iA_1 + jA_2 + kA_3) = \begin{pmatrix} A_0 & -A_2 & -A_1 & A_3 \\ A_2 & A_0 & A_3 & A_1 \\ A_1 & -A_3 & A_0 & -A_2 \\ -A_3 & -A_1 & A_2 & A_0 \end{pmatrix} \neq \mu(A_0 + iA_1 + jA_2 + kA_3),$$

but it is easy to see that by shuffling some rows, columns, and signs, we get (see also [4] and [30])

$$\det_{\mathbb{R}} \mu(M) = \det_{\mathbb{R}} \phi\big(\psi(M)\big) = \mathrm{Sdet}(M)^2.$$

I also note that in general

$$\psi(M^t) = \psi(A^t + jB^t) = \begin{pmatrix} A^t & -\bar{B}^t \\ B^t & \bar{A}^t \end{pmatrix} \neq \begin{pmatrix} A^t & B^t \\ -\bar{B}^t & \bar{A}^t \end{pmatrix} = \psi(M)^t,$$

but

$$\psi(M^*) = \psi(\bar{A}^t + \overline{jB}^t) = \psi(\bar{A}^t - \bar{B}^t j)$$
$$= \psi(\bar{A}^t - jB^t) = \begin{pmatrix} \bar{A}^t & \bar{B}^t \\ -B^t & A^t \end{pmatrix} = \psi(M)^*.$$

Hence, $\mathrm{Sdet}M^* = \overline{\mathrm{Sdet}M} = \mathrm{Sdet}M$; but in general $\mathrm{Sdet}M^t \neq \mathrm{Sdet}M$, for, as we saw earlier, M can be invertible while M^t is singular.

Dieudonné

Study was not the only one studying quaternionic determinants in his time. In the next ten years, A. Heyting, E. H. Moore, O. Ore, and A. R. Richardson all wrote about this topic [25, 34, 36, 42, 43]. The paper by Oystein Ore [36] is important because it introduces the concept of the ring of fractions for a non-commutative ring. But from the point of view of determinants, the most interesting are the papers by A. R. Richardson [42, 43] (this is the Richardson in the Littlewood–Richardson rule, but Littlewood is not the one in Hardy-Littlewood). His main contribution was to make it apparent that commutators play a key role. His papers are filled with formulas involving commutators.

Let us go back to studying $SL(n, \mathbb{H})$ and take a closer look at Lemma 3. It is easy to see that $SL(n, \mathbb{H})$ is a normal subgroup of $GL(n, \mathbb{H})$, and it can be shown [1, 15, 17, 40] that $SL(n, \mathbb{H})$ is the commutator subgroup of $GL(n, \mathbb{H})$.

LEMMA 6. $SL(n, \mathbb{H}) = [GL(n, \mathbb{H}), GL(n, \mathbb{H})].$

Let me mention in passing that for any field k, the commutator of $GL(n, k)$ is $SL(n, k)$, except when $n = 2$ and k is $\mathbb{Z}_2$ or $\mathbb{Z}_3$ [15].

The main reason why Lemma 3 is so crucial is that it shows that we only need to define our determinant on the matrices $D(x)$. But you may be impatient for me to get back to the issue of uniqueness. Since $SL(n, \mathbb{H})$ is normal in $GL(n, \mathbb{H})$, the question becomes: for which $x \in \mathbb{H}$ does $D(x)$ lie in $SL(n, \mathbb{H})$? The answer is given by the following lemma.

LEMMA 7.

$$D(x) = \begin{pmatrix} 1 & & & \\ & \ddots & & \\ & & 1 & \\ & & & x \end{pmatrix}$$

is a commutator in $GL(n, \mathbb{H})$ [i.e., it lies in $SL(n, \mathbb{H})$] if and only if x is a commutator in $\mathbb{H}^*$.

PROOF. One direction is trivial:

$$\begin{pmatrix} 1 & 0 \\ 0 & aba^{-1}b^{-1} \end{pmatrix} = \begin{pmatrix} 1 & 0 \\ 0 & a \end{pmatrix} \begin{pmatrix} 1 & 0 \\ 0 & b \end{pmatrix} \begin{pmatrix} 1 & 0 \\ 0 & a^{-1} \end{pmatrix} \begin{pmatrix} 1 & 0 \\ 0 & b^{-1} \end{pmatrix}.$$

The other direction, however, is not so easy. It is essentially equivalent to showing that the Dieudonné determinant is well defined, so it is an easy consequence of results in [1], [17], and [40], and I refer the reader to those excellent sources for the details.

It follows that in the decomposition $A = D(x)B$, neither x nor B is unique but the coset $x[\mathbb{H}^*, \mathbb{H}^*] \in \mathbb{H}^*/[\mathbb{H}^*, \mathbb{H}^*]$ *is* unique. This is exactly what Jean Dieudonné used in his 1943 paper [17]. His goal was to show how the determinant could be expressed in terms of group theory. We would expect

$$\det \begin{pmatrix} a & 0 \\ 0 & b \end{pmatrix} = \det \begin{pmatrix} b & 0 \\ 0 & a \end{pmatrix},$$

but then we probably need the determinant to take values in a commutative ring, and we get that by considering $\mathbb{H}^*/[\mathbb{H}^*, \mathbb{H}^*]$. His main theorem states that for any skewfield K, there is an isomorphism

$$GL(n, K)/[GL(n, K), GL(n, K)] \rightarrow K^*/[K^*, K^*].$$

For $K = \mathbb{H}$, this is immediate from Lemmas 3 and 7. We therefore define the Dieudonné determinant by

$$\det A = \det(D(x)B) = x[\mathbb{H}^*, \mathbb{H}^*].$$

Thanks to Lemma 7, we see that this is well defined and that the kernel is precisely $SL(n, \mathbb{H})$, i.e., our definition of $SL(n, \mathbb{H})$ agrees with the usual one, once we have the determinant.

If we now extend to $M(n, \mathbb{H})$, we get a determinant that takes values in $\mathbb{H}^*/[\mathbb{H}^*, \mathbb{H}^*] \cup \{0\}$. But what does this set look like? We need the following lemma.

LEMMA 8. $[\mathbb{H}^*, \mathbb{H}^*]$ is isomorphic to the set of quaternions of length 1.

PROOF. It is clear that every commutator has length 1. The set of quaternions of length 1 can be identified with S^3, and $\psi(S^3) = SU(2)$. But every element of $SU(2)$ is conjugate to a diagonal element, so it follows that every element in S^3 is conjugate to an element of S^1, the unit circle of $\mathbb{C} \subset \mathbb{H}$. (This also follows from the Noether-Skolem Theorem.) So, given $z \in S^3$, we can write $z = xyx^{-1}$ with $y \in S^1$.

We can identify the pure quaternions with $\mathbb{R}^3$, and for $p, q \in \mathbb{R}^3$ we have $p^{-1} = \bar{p}/|p|^2 = -p/|p|^2$ and

$$pq = -\langle p, q \rangle + p \times q,$$

where $\langle , \rangle$ is the usual inner product on $\mathbb{R}^3$ and $\times$ is the vector product in $\mathbb{R}^3$. From this, we can easily deduce that every quaternion can be written as the product of two pure quaternions.

Since y is complex, we can find $w \in \mathbb{C}$ with $y = w^2$, and it follows from the above that we can write $w = pq$, where $p, q \in \mathbb{R}^3$. Since $|w| = |y| = 1$, we can also assume that $|p| = |q| = 1$, so $p^{-1} = -p$ and $q^{-1} = -q$. But then

$$z = xpqpqx^{-1} = xpq(-p)(-q)x^{-1} = xpqp^{-1}q^{-1}x^{-1}$$
$$= (xpx^{-1})(xqx^{-1})(xpx^{-1})^{-1}(xqx^{-1})^{-1}.$$

For other proofs, see [9], [17], and [50]. It follows that $\mathbb{H}^*/[\mathbb{H}^*, \mathbb{H}^*]$ is isomorphic to the positive real numbers. Define

$$\omega : \mathbb{H}^*/[\mathbb{H}^*, \mathbb{H}^*] \to \mathbb{R}^+ \quad \text{by} \quad \omega\big(x[\mathbb{H}^*, \mathbb{H}^*]\big) = |x|,$$

and define the normalized Dieudonné determinant by

$$\mathrm{Ddet}(M) = \omega\big(\det(M)\big).$$

Dieudonné showed [17] that any determinant function satisfying our three axioms will be of the form

$$d(M) = \mathrm{Ddet}^r(M) \tag{5}$$

for some $r \in \mathbb{R}$. In particular, we can easily check the following theorem.

THEOREM 9. We have

$$\mathrm{Sdet}\, M = \det_{\mathbb{C}}\big(\psi(M)\big) = \mathrm{Ddet}^2(M), \tag{6}$$

$$\det_{\mathbb{R}} \mu(M) = \det_{\mathbb{R}} \phi\big(\psi(M)\big) = \mathrm{Ddet}^4(M). \tag{7}$$

Let me also point out that it follows from (6) that the Study determinant corresponds to the reduced norm [15].

Equation (5) has been generalized by L. E. Zagorin [52]. If v is a homomorphism of $\mathbb{H}$ into $M(s, \mathbb{C})$, and $\bar{v}$ is the corresponding homomorphism of $M(n, \mathbb{H})$ into $M(ns, \mathbb{C})$ then $\det_{\mathbb{C}} \bar{v}(M) = \mathrm{Ddet}^s(M)$.

In addition to satisfying our thee axioms, the Dieudonné determinant has several other properties [1, 17, 40]. Interchanging rows i and j corresponds to left-multiplying by the matrix $P_{ij} = B_{ij}(1)B_{ji}(-1)B_{ij}(1)$. But $-1 \in [\mathbb{H}^*, \mathbb{H}^*]$, so $\det P_{ij} = 1[\mathbb{H}^*, \mathbb{H}^*]$: interchanging two rows doesn't change the determinant.

When $n = 2$,

$$\det \begin{pmatrix} a & b \\ c & d \end{pmatrix} = \det \begin{pmatrix} a & b \\ 0 & d - ca^{-1}b \end{pmatrix} = (ad - aca^{-1}b)[\mathbb{H}^*, \mathbb{H}^*] \quad \text{if } a \neq 0,$$

and

$$\det \begin{pmatrix} 0 & b \\ c & d \end{pmatrix} = \det \begin{pmatrix} c & d \\ 0 & b \end{pmatrix} = cb[\mathbb{H}^*, \mathbb{H}^*] = -bc[\mathbb{H}^*, \mathbb{H}^*].$$

We can also show that multiplying a row on the left by m or multiplying a column on the right by m multiplies the determinant by $m[\mathbb{H}^*, \mathbb{H}^*]$. (This last product can be either on the left or on the right, since $\mathbb{H}^*/[\mathbb{H}^*, \mathbb{H}^*]$ is commutative.) On the other hand,

$$\det \begin{pmatrix} 1 & a \\ b & ab \end{pmatrix} = (ab - ba)[\mathbb{H}^*, \mathbb{H}^*],$$

but

$$b[\mathbb{H}^*, \mathbb{H}^*] \det \begin{pmatrix} 1 & a \\ 1 & a \end{pmatrix} = 0,$$

and we see that we cannot factor out a right multiple of a row.

Moreover, it doesn't behave well with respect to addition. Consider the determinant as a function of the first row, keeping the other rows fixed. Denote this function by $m(v)$. Define addition in $\mathbb{H}^*/[\mathbb{H}^*, \mathbb{H}^*]$ by setting

$$a[\mathbb{H}^*, \mathbb{H}^*] + b[\mathbb{H}^*, \mathbb{H}^*] = \{ak_1 + bk_2 \mid k_1, k_2 \in [\mathbb{H}^*, \mathbb{H}^*]\}.$$

It can then be shown [1] that

$$m(v_1 + v_2) \subset m(v_1) + m(v_2).$$

If we use Ddet instead of det and denote the corresponding function by $M(v)$, we get a sort of triangle inequality:

$$M(v_1 + v_2) \leq M(v_1) + M(v_2).$$

Moore

We started out by showing what was wrong with the Cayley determinant. But sometimes it does work. Granted that his formula doesn't make sense in general, does it still make sense for certain matrices? The answer is that if we restrict to Hermitian quaternionic matrices ($M^* = M$), then we get a useful function by specifying a certain ordering of the factors in the $n!$ terms in the sum. This was first studied by Eliakim Hastings Moore (for biographical information about Moore, see [37]), and I will denote it by Mdet.

Let σ be a permutation of n. Write it as a product of disjoint cycles. Permute each cycle cyclically until the smallest number in the cycle is in front. Then sort the cycles in decreasing order according to the first number of each cycle. In other words, write

$$\sigma = (n_{11} \cdots n_{1l_1})(n_{21} \cdots n_{2l_2}) \cdots (n_{r1} \cdots n_{rl_r}),$$

where for each i, we have $n_{i1} < n_{ij}$ for all $j > 1$, and $n_{11} > n_{21} > \cdots > n_{r1}$. Then we define

$$\text{Mdet } M = \sum_{\sigma \in S_n} |\sigma| m_{n_{11}n_{12}} \cdots m_{n_{1l_1}n_{11}} m_{n_{21}n_{22}} \cdots m_{n_r l_r n_{r1}}.$$

If H is Hermitian, then Mdet H is a real number. I will not go into details, but refer to the work of Moore, Jacobson, Dyson, Mehta, Chen, Van Praag, and Piccinni [5, 11, 12, 20, 21, 27, 28, 32, 33, 34, 39, 49, 50]. But I would again like to make some comments.

In general, it is difficult to talk about eigenvalues of a quaternionic matrix [13, 29]. As we work with right vector spaces, we must consider right eigenvalues. If

$$Mx = x\lambda,$$

then for $q \neq 0$, we get

$$M(xq) = x\lambda q = (xq)q^{-1}\lambda q.$$

Hence, all the conjugates of λ are also eigenvalues.

Let us study the conjugacy classes more closely. For $q \in \mathbb{H}$, we define $\rho(q)$ by $\rho(q)(x) = qxq^{-1}$. Since $\rho(q)$ leaves the real axis invariant and is orthogonal, we can restrict to $\mathbb{R}^3$. It is easy to see [18] that if we write $q = q_0 + q'$ with $q_0 \in \mathbb{R}$ and $q \in \mathbb{R}^3$, then $\rho(q)$ represents the rotation of $\mathbb{R}^3$ with the axis q' and angle $2\arctan(|q'|/q_0)$. From this we get that if x is real, then the conjugacy class of x is just $\{x\}$, whereas for $x \in S^3 \setminus \{\pm 1\}$, we get a copy of S^2 containing x and orthogonal to the real axis. Suppose that $\lambda = \lambda_0 + \lambda'$ with $\lambda_0 \in \mathbb{R}$ and $\lambda' \in \mathbb{R}^3$. Then $q\lambda q^{-1} = \lambda_0 + q\lambda'q^{-1}$, and the conjugacy class of λ' intersects the i axis at $\pm|\lambda'|i$. It follows that the conjugacy class of a non-real eigenvalue contains exactly two complex numbers and that they are conjugate.

If p is complex and $v = u + jw$, then $Mv = vp$ if and only if $\psi(M)\,(uw)^t = (uw)^tp$, and it can be proved by induction [29] that the eigenvalues of $\psi(M)$ occur in conjugate pairs. It follows that the eigenvalues of $\psi(M)$ are precisely the $2n$ numbers $\lambda_1, \ldots, \lambda_n$ and $\overline{\lambda_1}, \ldots, \overline{\lambda_n}$, whereas the eigenvalues of M are the elements of the conjugacy classes of $\lambda_1, \ldots, \lambda_n$, where we can replace λ_i by $\overline{\lambda_i}$.

It is now easy to show [29] that M is symplectically similar to a triangular matrix with diagonal elements d_i, where d_i equals λ_i or $\overline{\lambda_i}$. For more about normal forms of quaternionic matrices, see [27], [29], [41], [45], and [51].

If we restrict to a Hermitian matrix, H, then it turns out that all its eigenvalues are real (and there are, therefore, precisely n of them, since each conjugacy class only contains one element) and that the matrix can be symplectically diagonalized; that is, we can find $P \in GL(n, \mathbb{H})$ such that

$$PH\bar{P}^t = D,$$

where $\bar{P}^t = P^{-1}$ and D is diagonal and real.

We can now prove the following theorem that relates the Moore determinant to the other determinants.

THEOREM 10.   Let H be a Hermitian quaternionic matrix. Then

$$|\text{Mdet}\, H| = \text{Ddet}\, H \quad and \quad \text{Mdet}\, H[\mathbb{H}^*, \mathbb{H}^*] = \det H. \tag{8}$$

For any quaternionic matrix M, we have

$$\text{Sdet}\, M = \text{Mdet}(M M^*). \tag{9}$$

PROOF.   It can be shown that for a Hermitian matrix, the Moore determinant is equal to the product of the eigenvalues, so Mdet H is real-valued. But the normalized Dieudonné determinant of a diagonal matrix is the norm of the product of the diagonal elements, so (8) follows. To prove (9), we just have to observe that the eigenvalues of AA^* are positive, and use the product rule and (6).

Finally, if H is Hermitian, then

$$\left(J\psi(H)\right)^t = -\psi(H)^tJ = -J\overline{\psi(H)^t} = -J\psi(H),$$

so $J\psi(H)$ is skew-symmetric, and we can take its Pfaffian [14]. But

$$\text{pf}\left(-J\psi(H)\right)^2 = \det_{\mathbb{C}}\left(-J\psi(H)\right) = \text{Ddet}^2 H = \text{Mdet}^2 H,$$

so

$$\mathrm{Mdet}(H) = \mathrm{pf}\big(-J\,\psi(H)\big).$$

For other applications of the Pfaffian, see [2] and [3].

SP(n)

I would like to finish with a simple application of these ideas. As mentioned in the introduction, the group $SP(n)$ can be defined as the group preserving the norm on $\mathbb{H}^n$. But the usual description of this group is by considering its image under ψ in $M(\mathbb{C}, 2n)$. It is easy to see that all such matrices have determinants ± 1. There are different ways of proving that in fact the determinant is equal to 1, but this also follows from the results above, since all matrices in $\psi\,(GL(\mathbb{H}, n))$ have positive determinants.

In conclusion, I would also like to mention the recent work of Gelfand and Retakh [22]. Unfortunately, it is beyond the scope of this article to report on it.

References

1. E. Artin, *Geometric Algebra,* New York: Interscience, 1957; reprinted by Wiley, New York, 1988.
2. H. Aslaksen, *SO(2)* invariants of a set of 2 × 2 matrices. *Math. Scand.* 65 (1989), 59–66.
3. H. Aslaksen, E.-C. Tan, and C. Zhu, Invariant theory of special orthogonal groups, *Pacific J. Math.* 168 (1995), 207–215.
4. A. Bagazgoitia, A determinantal identity for quaternions, in *Proceedings of the 1983 Conference on Algebra Lineal y Aplicaciones, Vitoria-Gasteiz, Spain, 1984,* pp. 127–132.
5. R. W. Barnard and E. Hastings Moore, *General analysis. Part 1,* Memoirs of the American Philosophical Society, 1935.
6. J. Brenner, Expanded matrices from matrices with complex elements, *SIAM Rev.* 3 (1961), 165–166.
7. J. Brenner, Applications of the Dieudonné determinant, *Linear Algebra Appl.* 1 (1968), 511–536.
8. J. Brenner, Corrections to "Applications of the Dieudonné determinant," *Linear Algebra Appl.* 13 (1976), 289.
9. J. Brenner and J. De Pillis, Generalized elementary symmetric functions and quaternion matrices, *Linear Algebra Appl.* 4 (1971), 55–69.
10. A. Cayley, On certain results relating to quaternions, *Philos. Mag.* 26 (1845), 141–145; reprinted in *The Collected Mathematical Papers Vol. 1,* Cambridge: Cambridge University Press, 1989, pp. 123–126.
11. L. Chen, Definition of determinant and Cramer solution over the quaternion field, *Acta Math. Sinica (N.S.)* 7 (1991), 171–180.
12. L. Chen, Inverse matrix and properties of double determinant over quaternion field, *Sci. China Ser. A* 34 (1991), 528–540.
13. P. M. Cohn, The similarity reduction of matrices over a skew field, *Math. Z.* 132 (1973), 151–163.
14. P. M. Cohn, *Algebra, vol. 1,* 2nd ed., New York: Wiley, 1991.
15. P. M. Cohn, *Algebra, vol. 3,* 2nd ed. New York: Wiley, 1991.
16. M. L. Curtis, *Matrix Groups,* New York: Springer-Verlag, 1979; 1984.
17. J. Dieudonné, Les déterminants sur un corps non-commutatif, *Bull. Soc. Math. France* 71 (1943), 27–45.
18. J. Dieudonné. *Special Functions and Linear Representations of Lie Groups,* CBMS 42, Providence, RI, American Mathematical Society, 1980.

19. R. Dimitrić and B. Goldsmith, Sir William Rowan Hamilton, *Math. Intelligencer* 11 (1989), no. 2, 29–30.
20. F. J. Dyson, Correlations between eigenvalues of a random matrix, *Commun. Math. Phys.* 19 (1970), 235–250.
21. F. J. Dyson, Quaternion determinants, *Helv. Phys. Acta* 45 (1972), 289–302.
22. I. M. Gelfand and V. S. Retakh, Determinants of matrices over non-commutative rings, *Functional Anal. Appl.* 25 (1991), 91–102.
23. F. Reese Harvey, *Spinors and Calibrations*, New York: Academic Press, 1990.
24. W. R. Hamilton, *Elements of Quaternions*, 2nd ed., London: Longman, 1889.
25. A. Heyting, Die Theorie der linearen Gleichungen in einer Zahlenspezies mit nichtkommutativer Multiplikation, *Math. Ann.* 98 (1927), 465–490.
26. M. H. Ingraham, A note on determinants, *Bull. Amer. Math. Soc.* 43 (1937), 579–580.
27. N. Jacobson, Normal semi-linear transformations, *Amer. J. Math.* 61 (1939), 45–58.
28. N. Jacobson, An application of E. H. Moore's determinant of a Hermitian matrix, *Bull. Amer. Math. Soc.* 45 (1939), 745–748.
29. H. C. Lee, Eigenvalues and canonical forms of matrices with quaternionic entries, *Proc. Roy. Irish Acad. Sect. A*, 52 (1949), 253–260.
30. D. W. Lewis, A determinantal identity for skewfields, *Linear Algebra Appl.* 7 (1985), 213–217.
31. K. O. May, The impossibility of a division algebra of vectors in three dimensional space, *Amer. Math. Monthly* 73 (1966), 289–291.
32. Madan Lal Mehta, Determinants of quaternion matrices, *J. Math. Phys. Sci.* 8 (1974), 559–570.
33. Madan Lal Mehta, *Elements of Matrix Theory*, Delhi Hindustan Pub. Corp., 1977.
34. E. H. Moore, On the determinant of an hermitian matrix of quaternionic elements, *Bull. Amer. Math. Soc.* 28 (1922), 161–162.
35. Thomas Muir, *The Theory of Determinants, Vol. 2*, London: MacMillan, 1911.
36. O. Ore, Linear equations in non-commutative fields, *Ann. Math.* 32 (1931), 463–477.
37. K. Hunger Parshall and D. E. Rowe, *The Emergence of the American Mathematical Research Community, 1876–1900: J. J. Sylvester, Felix Klein and E. H. Moore*, Providence, RI: American Mathematical Society, 1994.
38. J. M. Peirce, Determinants of quaternions, *Bull. Amer. Math. Soc.* 5 (1899), 335–337.
39. P. Piccinni, Dieudonné determinant and invariant real polynomials on $\mathfrak{gl}\,(n, \mathbb{H})$, *Rend. Mat. (7)2* (1982), 31–45.
40. R. S. Pierce, *Associative Algebras*, New York: Springer-Verlag, 1982.
41. J. Radon, Lineare Scharen orthogonaler Matrizen, *Abh. Math. Sem. Univ. Hamburg* 1 (1922), 2–14.
42. A. R. Richardson, Hypercomplex determinants, *Messenger of Math.* 55 (1926), 145–152.
43. A. R. Richardson, Simultaneous linear equations over a division algebra, *Proc. London Math. Soc.* 28 (1928), 395–420.
44. E. Study, Zur Theorie der linearen Gleichungen, *Acta Math.* 42 (1920), 1–61.
45. O. Teichmüller, Operatoren im Wachsschen Raum, *J. Reine Angew. Math.* 174 (1935), 73–124.
46. C. L. Tong, Symplectic Groups, honours thesis, National Univ. of Singapore, 1991.
47. B. Leednert van der Waerden, Hamilton's discovery of quaternions, *Math. Mag.* 49 (1976), 227–234.
48. B. Leednert van der Waerden, *A History of Algebra*, New York: Springer-Verlag, 1985.
49. P. Van Praag, Sur les déterminants des matrices quaterniennes, *Helv. Phys. Acta* 62 (1989), 42–46.
50. P. Van Praag, Sur la norme réduite du déterminant de Dieudonné des matrices quaterniennes, *J. Algebra* 136 (1991), 265–274.
51. L. A. Wolf, Similarity of matrices in which the elements are real quaternions, *Bull. Amer. Math. Soc.* 42 (1936), 737–743.
52. L. E. Zagorin, The determinants of matrices over a field (Russian), *Proc. First Republican Conf. Math. Byelorussia, Izdat*, Minsk: "Vyssaja Skola", 1965, pp. 151–152

Part Three

ANALYSIS

13

The Surfaces of Delaunay

James Eells

1. Background

In 1841 the astronomer/mathematician C. Delaunay isolated a certain class of surfaces in Euclidean space, representations of which he described explicitly [1]. In an appendix to that paper, M. Sturm characterized Delaunay's surfaces variationally; indeed, as the solutions to an isoperimetric problem in the calculus of variations. That in turn revealed how those surfaces make their appearance in gas dynamics; soap bubbles and stems of plants provide simple examples. See Chapter V of the marvellous book [8] by D'Arcy Thompson for an essay on the occurrence and properties of such surfaces in nature.

More than 130 years later E. Calabi pointed out to me that the solutions to a certain pendulum problem of R. T. Smith [7] could be interpreted via the Gauss maps of Delaunay's surfaces [2]. And Eells and Lemaire [4] found that the Gauss map of one of those surfaces produces a solution to an existence problem in algebraic/differential topology.

The purpose of this article is to retrace those steps in an expository manner—as a revised version of [2].

2. Roulettes of a Conic

The first step is to derive the equations describing the trace of a focus F of a non-degenerate conic ℓ as K rolls along a straight line in a plane. (Perhaps these derivations were better known a century ago!) We examine various cases separately.

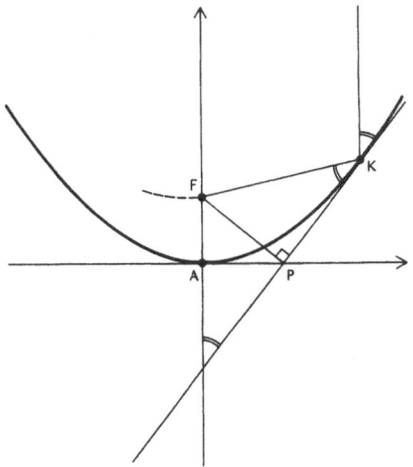

FIGURE 13-1 *ℓ is a parabola*

ℓ IS A PARABOLA (SEE FIGURE 13-1)
Here A is the vertex of ℓ. The line PK is tangent to ℓ at the point K. The following properties are elementary:

1. Correspondingly marked angles are equal;
2. FP is orthogonal to PK.

Thus we obtain

$$\overline{FA} = \overline{FP}\cos\angle AFP = \overline{FP}\cos\angle PFK.$$

Volume 9, No. 1 (Winter 1987), 53–57

Now we change our viewpoint and think of the tangent line PK as the axis—the x-axis—along which the parabola ℓ rolls. We denote the ordinate of F by y and observe that

$$\cos \angle PFK = \frac{dx}{ds}$$

describes the rate of change of abscissa of F with respect to arc length s; i.e.,

$$\frac{dx}{ds} = \alpha,$$

where α denotes the angle made by the tangent with the x-axis. Thus setting $c = \overline{FA}$, we obtain the differential equation

$$c = y\frac{dx}{ds} = \frac{y}{\sqrt{1+y'^2}}, \quad \text{or} \quad y = \sqrt{\frac{y^2 - c^2}{c^2}}.$$

Its solution is the *catenary*

$$y = \frac{c}{2}(e^{x/c} + e^{-x/c}) = c\cosh x/c. \tag{2.1}$$

That equation describes the shape of a flexible inextensible free-hanging cable—thereby explaining its name. In that context we can obtain the equation of the catenary as the Euler-Lagrange equation of the potential energy integral

$$P(y) = \int_{x_0}^{x_1} y\sqrt{1+y'^2}dx,$$

subject to variations holding fixed the length integral

$$\int_{x_0}^{x_1} \sqrt{1+y'^2}dx = L.$$

Indeed, from general principles we are asked to find a real number a and an extremal of the integral

$$J(y) = \int_{x_0}^{x_1} \left\{\sqrt{1+y'^2} + ay\sqrt{1+y'^2}\right\} dx.$$

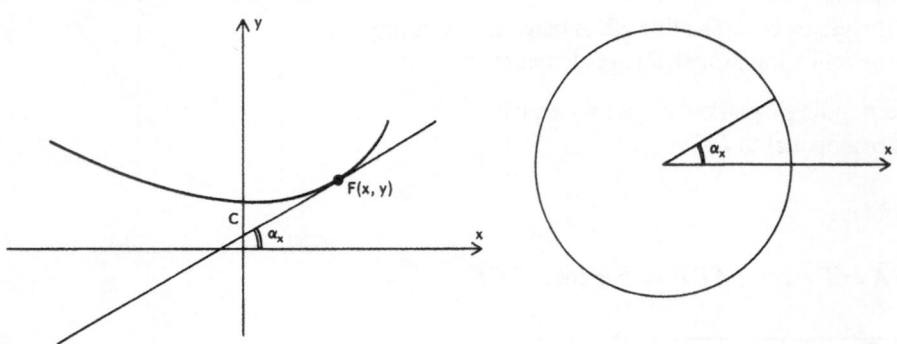

FIGURE 13-2

Its Euler-Lagrange equation has first integral

$$y' = \sqrt{\frac{(1+ay)^2 - b^2}{b^2}} \quad \text{for } b \in \mathbf{R}.$$

The equation of the catenary is derived from this, choosing suitable normalizations.

The curvature of ℓ is measured by the amount of turning of its tangent. That is expressed by the *Gauss map* of ℓ into the unit circle, given by $x \to \alpha_x$ (see Figure 13-2), where

$$\cos \alpha_x = \frac{dx}{ds} = \frac{c}{y}.$$

The Gauss map of the roulette of the parabola is injective onto an open semi-circle.

ℓ IS AN ELLIPSE (SEE FIGURE 13-3)

Here F and F' are the foci of ℓ; the O is its center. The line PKP' is tangent to ℓ at K. Letting a and b denote the lengths of the semi-axes of ℓ, we obtain the following properties:

1. $\overline{FK} + \overline{F'K} = 2a > 0$;
2. the pedal equation $\overline{PF} \cdot \overline{P'F'} = b^2$ (see [9, Ch. VIII 6]);
3. the normal to the locus of F passes through K.

Again using PK as x-axis,

$$\frac{y}{\overline{FK}} = \sin \angle FKP = \cos \angle FTP = \frac{dx}{ds}$$
$$\frac{y'}{\overline{F'K}} = \sin \angle F'KP' = \cos \angle F'TP' = \frac{dx}{ds}.$$

From these we derive

$$y + y' = 2a\frac{dx}{ds}, \quad yy' = b^2,$$

so that

$$y^2 - 2ay\frac{dx}{ds} + b^2 = 0.$$

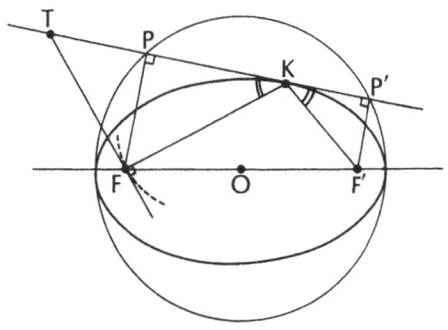

FIGURE 13-3 ℓ *is an ellipse*

By analyzing all cases and taking $a \leq b$, we obtain

$$y^2 \pm 2ay\frac{dx}{ds} + b^2 = 0. \tag{2.2}$$

The solutions to that differential equation can be given explicitly in terms of elliptic functions; see [1], [5, pp. 416–418].

The locus of either focus will be called the *undulary* (See Figure 13-4). Its Gauss map is given by $x \to \alpha_x$, where

$$\cos \alpha_x = \mp \frac{y^2 + b^2}{2ay}.$$

It maps ℓ onto a closed arc of the unit circle.

There are two limiting cases, which are perhaps best handled separately. When $b \to a$ the undulary degenerates to a straight line, the locus of the centre of a circle rolling on a line. And where $b \to 0$ the undulary becomes a semi-circle centered on the x-axis.

ℓ IS AN HYPERBOLA (SEE FIGURE 13-5)
In analogy with the case of the ellipse, we have

1. $\overline{FK} - \overline{F'K} = 2a > 0$;
2. $\overline{PF} \cdot \overline{P'F'} = b^2$.

Thus we obtain the following differential equation for the locus of F, given as a first integral of an Euler-Lagrange equation:

$$y^2 \pm 2ay\frac{dx}{ds} - b^2 = 0. \tag{2.3}$$

The loci of the two foci fit together to form the curve which we shall call the *nodary*. Its Gauss map $x \to \alpha_x$ is governed by

$$\cos \alpha_x = \mp \frac{y^2 - b^2}{2ay}.$$

The Gauss map has no extreme points, and direct verification shows that it is surjective.

A *roulette of a conic* is a catenary, undulary, nodary, a straight line parallel to the x-axis, or a semi-circle centered on the x-axis.

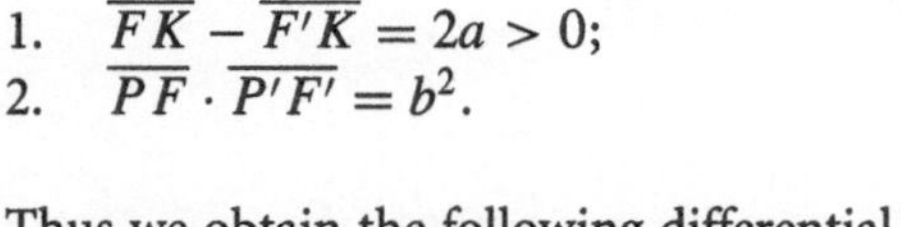

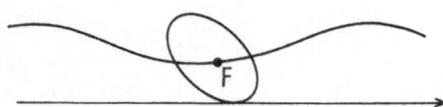

FIGURE 13-4 *The undulary.*

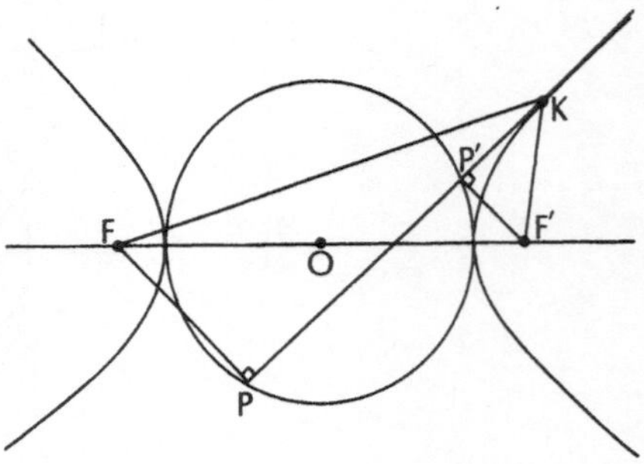

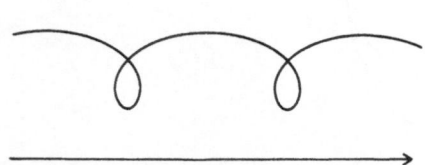

FIGURE 13-6 *The nodary.*

FIGURE 13-5 ℓ *is an hyperbola.*

3. Surfaces of Revolution with Constant Mean Curvature

Rotating each of the roulettes about its axis of rolling produces five types of surfaces in Euclidean 3-space $\mathbf{R}^3$, called *the surfaces of Delaunay*: the *catenoids*, *unduloids*, *nodoids*, the *right circular cylinders*, and the *spheres*.

VARIATIONAL CHARACTERIZATION: We formulate the following isoperimetric principle, for the unduloid and nodoid (only minor technical changes being required for the other cases).
   Consider graphs in $\mathbf{R}^2$ of non-negative functions

$$y: [x_0, x_1] \to \mathbf{R}(\geq 0)$$

with fixed volume of revolution

$$V(y) = \pi \int_{x_0}^{x_1} y^2 dx;$$

and extremize their lateral area

$$A(y) = 2\pi \int_{x_0}^{x_1} y^2 ds$$

holding the endpoints fixed. By general principles of constraint (under the heading of Lagrange's method of multipliers for isoperimetric problems [5]), we are led to the Euler-Lagrange equation associated with the integral

$$F(y) = \pi \int_{x_0}^{x_1} (y^2 dx + 2ay\, ds) = \pi \int_{x_0}^{x_1} (y^2 + 2ay\sqrt{1 + y'^2})dx.$$

Here a is a convenient real parameter. Its integrand f does not involve x explicitly, so we obtain a first integral from

$$0 = y'\left(f_y - \frac{d}{dx}f_{y'}\right) = \frac{d}{dx}(f - y'f_{y'}).$$

Thus $f - y'f_{y'} = \pm b^2$, where b is another real parameter. Consequently,

$$y^2 + \frac{2ay}{\sqrt{1 + y'^2}} \mp b^2 = 0.$$

But

$$\frac{1}{\sqrt{1 + y'^2}} = \frac{dx}{ds}$$

so the extremal equation for our variational problem coincides with that of the roulette of the ellipse or hyperbola ((2.2) and (2.3)).

GAUSS MAPS: In an analogy with the case of oriented curves in the plane (§2), we associate to any oriented surface M immersed in $\mathbf{R}^3$ its Gauss map $\gamma : M \to S$ (the unit 2-sphere centered at the origin in $\mathbf{R}^3$), defined by assigning to each point $x \in M$ the positive unit vector orthogonal to the oriented tangent plane to M at x. Its differential $d\gamma(x)$ can be interpreted as a symmetric bilinear form on the tangent space $T_x M$. Its eigenvalues λ_1, λ_2 are well determined up to order. The symmetric functions $K_x = \lambda_1 \lambda_2$ and $H_x = (\lambda_1 + \lambda_2)/2$ are called the *curvature* of M and the *mean curvature of the immersion* at x, respectively. For instance,

1. the cylinder has $K \equiv 0$ and constant mean curvature $H \neq 0$;
2. the sphere of radius R has a constant curvature $K = 1/R^2$ and constant mean curvature $H = 1/R$;
3. the catenoid has variable curvature K and mean curvature $H \equiv 0$;
4–5. the unduloid and nodoid have variable curvature K and constant mean curvature $H \neq 0$.

These five surfaces were recognized by Plateau, using soap film experiments.

Say that a surface of constant mean curvature in $\mathbf{R}^3$ is *complete* if it is not part of a larger such surface. From Sturm's variational characterization, we obtain

DELAUNAY'S THEOREM: The complete immersed surfaces of revolution in $\mathbf{R}^3$ with constant mean curvature are precisely those obtained by rotating about their axes the roulettes of the conics.

Thus Delaunay's surfaces are those surfaces of revolution M in $\mathbf{R}^3$ which are maintained in equilibrium by the pressure of a field of force which acts everywhere orthogonally to M.

4. Harmonic Gauss Maps

An easy yet vitally important theorem of Ruh-Vilms [6] states that:

A surface M immersed in R^3 has constant mean curvature if and only if its Gauss map $\gamma \colon M \to S$ satisfies the equation

$$\Delta \gamma = \|d\gamma\|^2 \gamma,$$

where Δ denotes the Laplacian of M with conformal structure induced from that of $\mathbf{R}^3$, and vertical bars the Euclidean norm at each point. Indeed, (4.1) *is the condition for harmonicity of the map* γ [3]—and is the Euler-Lagrange equation associated to the *energy* (or *action*) *integral*

$$E(\gamma) = \frac{1}{2} \int_M \|d\gamma\|^2.$$

E is a conformal invariant of M.

SMITH'S MECHANICS: Motivated by certain mechanical analogies, R. T. Smith [7] found solutions to equation (4.1) as maps $\gamma \colon \mathbf{R}^2 \to S$, as follows:

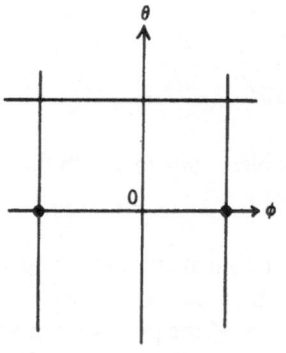
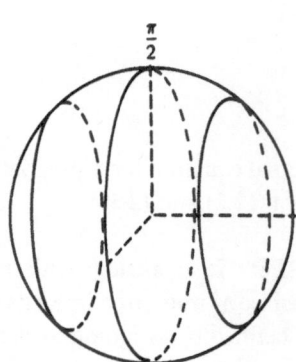

FIGURE 13-7

Think of points of $\mathbf{R}^2$ parametrized by angles (ϕ, θ), and use spherical coordinates on the sphere S, as shown in Figure 13-7.

If we restrict our attention to maps γ of the special form

$$(\phi, \theta) = \left(e^{i\theta} \sin \alpha(\phi), \cos \alpha(\phi)\right),$$

then the equation of harmonicity becomes the pendulum equation

$$\alpha'' = \frac{A}{2} \sin 2\alpha. \tag{4.3}$$

We assume that $\alpha(0) = \pi/2$, so that the solution oscillates symmetrically about $\pi/2$.

Now a first integral of (4.3) is given by

$$\alpha' = \sqrt{\frac{C - A \cos^2 \alpha}{2}}.$$

Again, that has an explicit solution in terms of elliptic functions. Furthermore, the *associated map* $\gamma \colon \mathbf{R}^2 \to S$ *is doubly periodic, factoring through the torus* $T = \mathbf{R}^2/\mathbf{Z}^2$ to produce a map $\gamma \colon T \to S$, as desired. Incidentally, the integrand of E is

$$\|d\gamma\|^2 = \alpha'^2 + \frac{A}{2} \sin^2 \alpha.$$

Calabi made the beautiful observation that *Smith's maps* $\gamma \colon T \to S$ *are the Gauss maps of certain surfaces of Delaunay* [2].

A HARMONIC REPRESENTATIVE IN A HOMOTOPY CLASS: If we represent the torus T in the form $T = \mathbf{R}/a\mathbf{Z} \times \mathbf{R}/2\pi\mathbf{Z}$ and use polar coordinates (r, θ) on the unit sphere S, then a map from the cylinder to S of the form

$$r = \Phi(x), \theta = y$$

subject to the conditions $\Phi(0) = 0$, $\Phi(a) = \pi$ is harmonic if and only if Φ satisfies the pendulum equation (4.3) with $A = 1$. There are such solutions. Indeed [4], *the Gauss map of the nodoid induces a harmonic map of a Klein bottle* $\gamma \colon K \to S$. *Furthermore, that map is not deformable to a constant map.*

Hopf's classification theorem insures that the maps $K \to S$ are partitioned by homotopy into just two classes. Thus the harmonic map γ represents the non-trivial class.

References

1. C. Delaunay, *Sur la surface de révolution dont la courbure moyenne est constante.* J. Math. pures et appl. Sér. 1 (6) (1841), 309–320. With a note appended by M. Sturm.
2. J. Eells, *On the surfaces of Delaunay and their Gauss maps.* Proc. IV Int. Colloq. Diff. Geo. Santiago de Compostela (1978), 97–116.
3. J. Eells and L. Lemaire, *A report on harmonic maps.* Bull. London Math. Soc. 10 (1978), 1–68.
4. J. Eells and L. Lemaire, *On the construction of harmonic and holomorphic maps between surfaces.* Math. Ann. 252 (1980), 27–52.
5. A. R. Forsyth, *Calculus of variations.* Cambridge (1927).
6. E. A. Ruh and J. Vilms, *The tension field of the Gauss map.* Trans. Amer. Math. Soc. 149 (1970), 569–573.
7. R. T. Smith, *Harmonic mappings of spheres.* University of Warwick Thesis (1972).
8. D'A. W. Thompson, *Growth and form.* Cambridge (1917).
9. C. Zwikker, *The advanced geometry of plane curves and their applications.* Dover (1963).

14

The Banach-Tarski Theorem

Robert M. French

It is theoretically possible, believe it or not, to cut an orange into a finite number of pieces that can then be reassembled to produce two oranges, each having exactly the same size and volume as the first one. That's right: With sufficient diligence and dexterity, from any three-dimensional solid we can produce two new objects exactly the same as the first one!

Mathematicians, upon first hearing of this result (otherwise known as the Banach-Tarski Theorem), are generally somewhat blasé; they know that funny counter-intuitive things crop up all the time whenever infinity is involved. Most mathematicians encounter the result for the first time in graduate school and file it away in their strange results category (along with space-filling curves, Cantor functions, and non-measurable sets). But in spite of the relative simplicity of the proof, discovered by Stefan Banach and Alfred Tarski in 1924 and hinging on the Axiom of Choice, many mathematicians go no further than the lay scientist who comes across the result.

The mathematics of infinity is almost always counter-intuitive and has been so ever since its inception at the end of the nineteenth century when Georg Cantor proved the completely astounding result that infinity came in different sizes. This result initially so upset the mathematical community that Henri Poincaré once maligned it as a disease from which mathematics would have to recover [1].

The purpose of this article is neither to explain the subtleties of infinity nor to give a rigorous proof of the Banach-Tarski Theorem. Instead, a few simple notions about infinity will be explained that will serve as the basis for the subsequent explanation of the main ideas of the proof of this wonderful theorem.

Let's start with a bit of elementary geometry. Two subsets of the plane are said to be *congruent* when one can be made to coincide precisely with the other using only translations and rotations in the plane. The essence of congruence is that the distances between the points of the first set remain unchanged after it has been moved to coincide with the second set. Congruence, however, is not to be confused with one-to-one correspondence. The set of even numbers $\{2, 4, 6, \ldots\}$, for example, is not congruent to the set of natural numbers $\{1, 2, 3, \ldots\}$ because there is no way to *overlay* one set on the other even though the two sets can be put into one-to-one correspondence. Nothing prevents an infinite set from being congruent to a proper subset of itself, however. Consider, for example, the two infinite sets $\{1, 2, 3, \ldots\}$ and $\{5, 6, 7, \ldots\}$. Congruence is demonstrated by shifting all of the elements of the first set four units to the right.

Equivalence by Finite Decomposition

Let's return to our orange. Before we actually begin converting it into two oranges, we need the notion of "equivalence by finite decomposition." In spite of its complex name, the idea is simple.

Volume 10, No. 4 (Fall 1988), 21–28

Basically, we divide an object X into a finite number of disjoint parts and then rearrange them into a new object Y. (Note that "rearranging" a given set means that the set, in its initial position, is congruent to the set in its final position.) Under these circumstances, we say that X is equivalent by finite decomposition to Y. This type of equivalence is transitive. In other words, if a set X is equivalent to Y, which in turn is equivalent to Z, then X and Z are also equivalent by finite decomposition.

Now, let's consider our first little "paradox": The set of positive integers, $\mathbf{N}$, is equivalent by finite decomposition to the integers with a one-element "hole" in them—for example, the set of integers with one of its members, say 5, removed. There are various demonstrations of this fact but the one given below will turn out to be the most instructive for what follows. First, create two subsets of $\mathbf{N}$: the set B consisting of all multiples of 5 (i.e., $\{5, 10, 15, \ldots\}$) and its complement A containing all non-multiples of 5 (i.e., $\{1, 2, 3, 4, 6, 7, \ldots\}$). By definition, these two sets are disjoint, and their union is equal to all of $\mathbf{N}$. We now are in a position to introduce the key technique of this proof and all of the others in this article, including the Banach-Tarski Theorem itself. We will call this technique "shifting toward infinity." We shift B toward infinity by 5 units, thus producing a new set B' equal to $\{10, 15, 20, \ldots\}$ which is, by definition, congruent to B. We now have, on the one hand, a disjoint union of sets $(A \cup B)$, which is equal to the positive integers, and a second disjoint union of sets $(A \cup B')$, which is equal to the positive integers with the elements 5 removed. But, as we have said, B and B' are obviously congruent, as is, even more obviously, A with itself. We can therefore conclude that the set of integers and the set of integers with 5 removed are equivalent by finite decomposition.

The next proof is slightly more complicated but is based on the same principle of shifting toward infinity that was used to show that $\mathbf{N}$ and $\mathbf{N} \setminus \{5\}$ were equivalent by finite decomposition. This time we will consider a circle and a circle with a one-point "hole" in it. The claim is that these two sets are equivalent by finite decomposition. Here is an outline of how the proof goes.

Let C be a circle with radius 1 unit, and let 0 be some point (in fact, the one we are going to "remove") on the circumference of C. From point 0, we move counter-clockwise along the circumference a distance of exactly one unit, the radius of the circle. Call the point at which we stopped 1, and then continue walking. Exactly one unit later, stop and mark the point where you stopped by 2, etc. Call B the set of all points $\{0, 1, 2, 3, \ldots\}$.

Just as in the previous demonstration, A will designate the points (of the circle this time) that are not in B. Now imagine that the set B is the channel selector dial of a television. Turn the dial one click to the left. This dial-turning superimposes the set $\{0, 1, 2, \ldots\}$ on the set $\{1, 2, 3, \ldots\}$. The latter set we call B; and obviously B and B' are congruent. Since the circle C is equal to $A \cup B$ and the circle without the point 0 is equal to $A \cup B'$, we conclude that the circle and the circle with a one-point hole are equivalent to each other by finite decomposition.

Next, we wish to demonstrate that a closed one-by-one square can be decomposed and reassembled to form a closed isosceles triangle whose altitude is equal to one of the sides of the original square.

The first thing we might try is to cut the square along one of its diagonals, thus obtaining two right triangles (see Figure 1). The desired isosceles triangle would be produced by reassembling the two right triangles in such a way that two of their legs would coincide and their hypotenuses would meet at a point. This method, however, does not work. When we cut the square, we do not produce two complete right triangles. The diagonal of the square can only be used to constitute *one* of the hypotenuses, not both. Furthermore, the definition of equivalence by finite decomposition requires that the constituent parts must be *disjoint* (this, at least, does correspond to our real-world notion of what is meant by the separate parts of something). When we cut an object in half we do not allow some points to belong to both halves. Figure 14-1 shows a way of cutting the square that satisfies this condition.

Unfortunately, there are two candidates for the altitude of the isosceles triangle and no points at all along one of its sides. Why not try to remove one of the extra altitudes and "paste" it along the

FIGURE 14-1 *How to transform a*
square into an isosceles triangle.

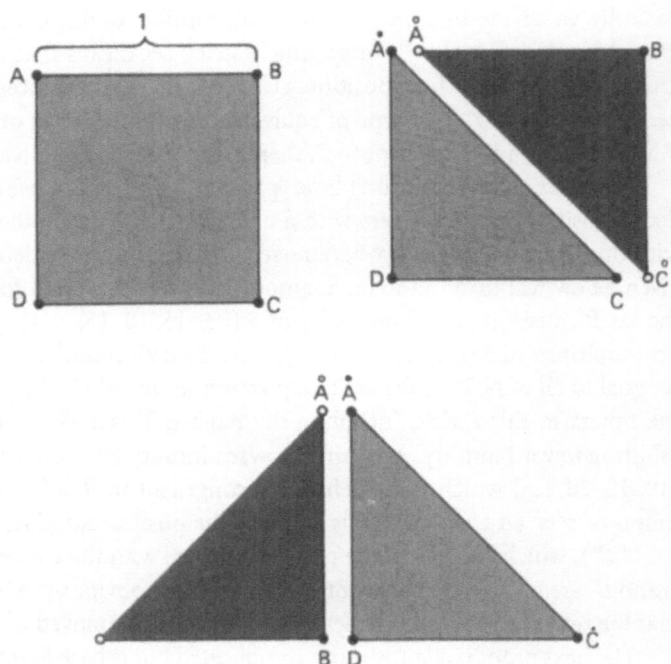

FIGURE 14-1 *How to transform a square into an isosceles triangle.*

edge of the triangle that is without points? It turns out that this trick almost works, but, as you can see in Figure 14-2, it falls slightly short of the mark. Even after pasting in this altitude of length 1, a "hole" remains. We still need a segment of length $\sqrt{2} - 1$. We will use the technique of shifting toward infinity to excise from the original square a line segment of the required length and then will show that this "theft" is of absolutely no deleterious mathematical consequence. Then, our minds at ease, we will finish our construction by plugging the hole in the side of the triangle with the purloined line segment.

How do we filch the required line segment from the original square? Basically, we show that the square and the square minus the desired line segment are equivalent by finite decomposition. The precise specifications of this line segment require it to have a length of $\sqrt{2} - 1$ with one end including its endpoint and the other end not. The excision technique is virtually identical to the one that

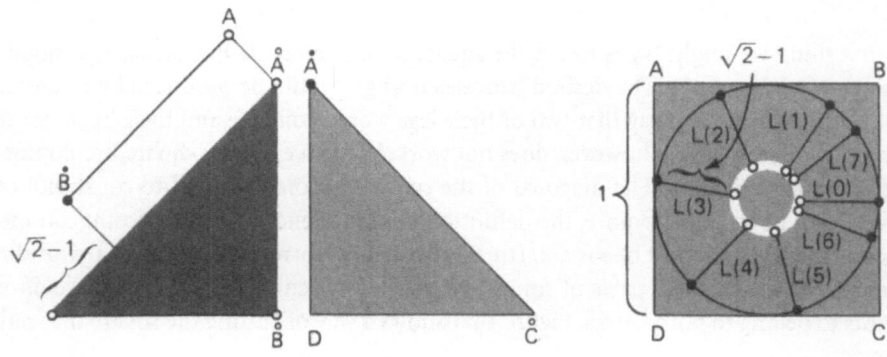

FIGURE 14-2 *The unit square is equivalent to the square minus*

allowed us to show a circle and a circle missing a point were equivalent by finite decomposition. Now, instead of removing a point from a circle, we will remove a line segment from a disk. We therefore begin by inscribing the circle C of the preceding demonstration in the square. We will be considering the closed disk D whose boundary is C. To each of the points $0, 1, 2, \ldots$ attach a segment of length $\sqrt{2} - 1$ (see Figure 14-2). Call these segments $L(0), L(1), \ldots$. The remainder of the proof is exactly the same as before except that C, the circle, is now replaced by D, the disk, and the point 0 by the line segment $L(0)$. We have therefore shown that the disk and the disk with a line segment missing are equivalent by finite decomposition. Further, because our theft of the line segment didn't affect any part of the square outside of the disk, we can safely assert that the square and the square without the missing line segment are also equivalent by finite decomposition. Finally, we insert $L(0)$ in the hole along the side of the triangle and obtain the desired result; the closed square is equivalent by finite decomposition to the closed isosceles triangle.

Until now we have only performed our shift-toward-infinity vanishing act on sets whose size was insignificantly small compared to the sets that contained them: a point taken from a circle and a line segment excised from a disk. While this may be mildly interesting, it's hardly spectacular. Let's now take a look and see how these techniques can also be applied to much larger sets, such as the entire volume of a solid ball.

Hausdorff's Paradox

We now have the tools necessary to produce two oranges from one. The heart of the proof of the Banach-Tarski Theorem is based on a result of Felix Hausdorff. Hausdorff's result concerns only the skin of the orange (a "skin" of thickness zero). The Hausdorff paradox, as it is called, shows that it is possible to divide this skin, once an insignificantly small (more precisely, countably infinite) set of points has been removed, into three disjoint sets of points A, B, and C such that A, B, C, and $B \cup C$ are all congruent to one another. Now, that is positively weird! The mutual congruence of these three sets means that A is congruent to the disjoint union of *two copies of itself*. This is referred to as a paradoxical decomposition of A. Essentially, by carefully reassembling these sets, A, B, and C, we obtain a set of two complete orange skins each of which is equivalent by finite decomposition to the original orange skin. In other words, one sphere can be cut up and reassembled into two spheres identical to the first.

Can we do the same thing to solid balls? The answer is yes; intuitively, we must imagine applying the Hausdorff technique to hollow balls whose skins get progressively thicker and thicker. Finally, we apply this construction to a "hollow" ball whose inside consists only of a single point to produce two equivalent balls each missing its center point. We have thus done the construction for the closest thing possible to a solid ball—namely, a solid ball without its center. Having gone that far, it is a relatively easy matter to show that the solid ball and the solid ball without its center are equivalent by finite decomposition. This completes the proof of the Banach-Tarski Theorem: A ball is equivalent by finite decomposition to two copies of itself.

Now let's take a look at the proof of Hausdorff's paradox, the mainstay of the Banach-Tarski Theorem. Recall that Hausdorff's construction is only concerned with the surface of the ball (i.e., the sphere) and not the ball.

Given a sphere S, we will select two axes of this sphere, F and G. The angle formed by these two axes at the center of the sphere is to be 45°. We will designate by f a clockwise rotation of the sphere by 180° about the F axis and by g a clockwise rotation of the sphere by 120° about the G axis. We call f and g *transformations of the sphere* (see Figure 14-3). We will use combinations of these transformations to describe different sequences of rotations of the sphere. For example, the composite transformation $g^2 f$ specifies the operation consisting of turning the sphere 180° about F, followed by two rotations of 120° about G. To avoid an unnecessary proliferation of exponents, we will write $\bar{g}$ to

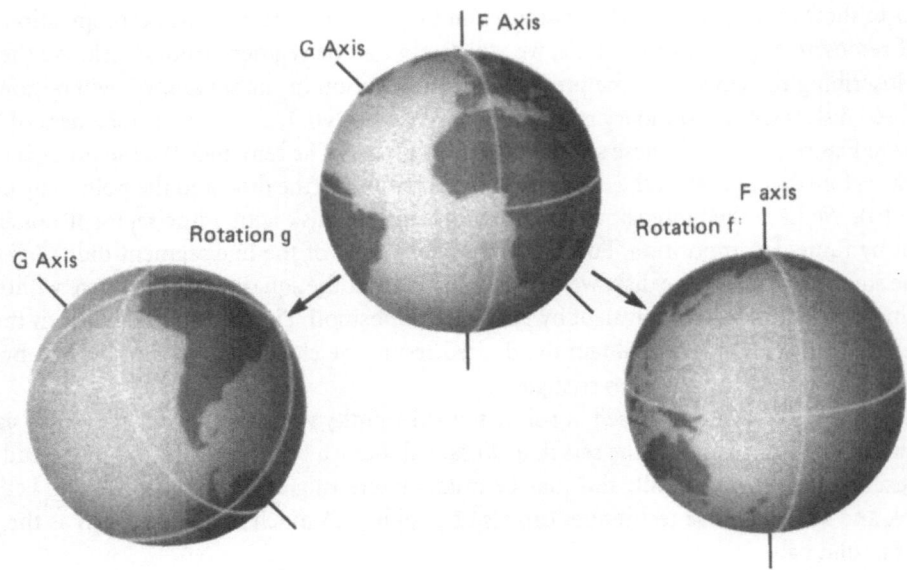

FIGURE 14-3 *The transformations of the sphere are composed of rotations about the F and G axes. The angle between the axes is chosen so that fg is not equal to gf.*

designate g^2. (Remember that $\bar{g}$ represents not only a clockwise rotation of 240° but also a rotation of 120° in the opposite direction.) From now on we will call f, g, and $\bar{g}$ *elementary transformations*. From a given position of the sphere, if we apply f twice in a row, the sphere will be returned to its initial position. We write $f^2 = 1$, where 1 is the identity transformation—in other words, the transformation that does not change the position of the sphere. Similarly, since g represents a rotation of 120°, $g^3 = 1$.

These two observations allow us to reduce complex transformations to a simpler form. For example, $g^5f^3 = (g^3)(g^2)(f^2)(f) = 1 \cdot (g^2) \cdot 1 \cdot f = g^2f$. On the other hand, there is no way to simplify the composite transformation $gfgfgf\bar{g}$ because the position of the two axes F and G was carefully selected in such a way that fg does not equal gf. In other words, starting with a given position of the sphere, when we perform the transformations in the order g, then f, the sphere will be in a different position than had we done first f, then g.

We are only interested in transformations reduced to their lowest form, and we will use an iterative machine (see Figure 14-4) to produce the set Q of all of these transformations. To start the machine running, we put the identity transformation 1 into the hopper. The machine executes the following three rules: (1) when 1 is the only transformation in the hopper, it produces the three elementary transformations f, g, and $\bar{g}$; (2) when a transformation whose leftmost element is f goes through the rule box, two new transformations are produced—the first by adding an additional rotation by g to the transformation in the box, and the second by adding an additional rotation by $\bar{g}$ (for example; if $f\bar{g}f$ goes into the rule box, $gf\bar{g}f$ and $\bar{g}f\bar{g}f$ will come out); (3) when a transformation whose leftmost element is g or $\bar{g}$ goes through the rule box, new transformation is produced by adding an additional rotation by f to it (for example, if $gf\bar{g}$ goes into the rule box, $fgf\bar{g}$ will come out).

The transformations produced in the rule box then drop into a transformation copier, which produces a copy of each transformation and sends it back up to the hopper; the original transformation then drops into the large collection bag below the machine. This is how Q, the set of transformations, 1, f, g, $\bar{g}$, fg, gf, $\bar{g}f$, ..., is produced. The angle between the axes F and G was chosen

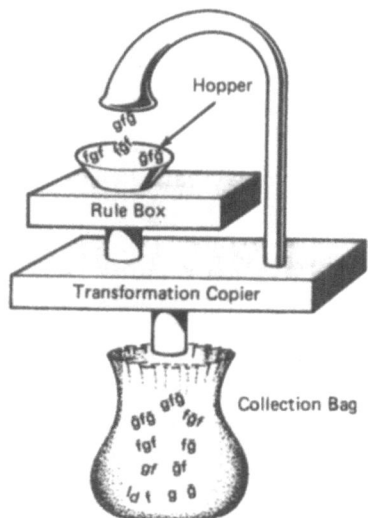

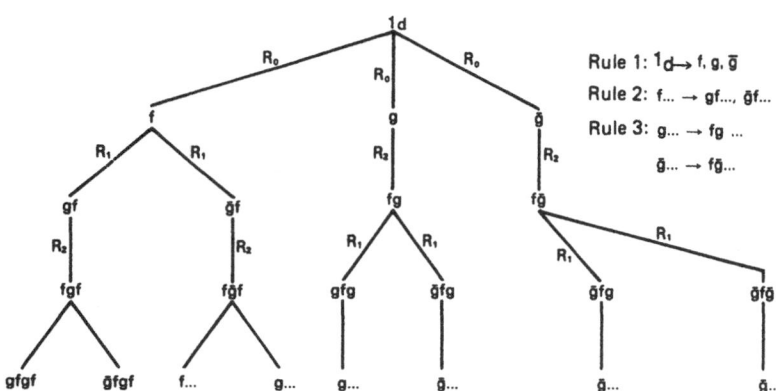

FIGURE 14-4 *The simple machine used to produce the set of all possible transformations.*

to ensure that each of the elements of Q represents a *unique* position of the sphere with respect to its initial position.

A Full Iterative Machine

This four-part machine with its hopper, rule box, transformation copier, and collection bag forms the basis of the more powerful iterative machine that we need for the Banach-Tarski theorem. The full iterative machine not only must be capable of producing all of the transformations in Q, but also must be able to *sort* them into three disjoint subsets, I, J, and K whose union is equal to all of Q and that have the following properties:

$$fI = J \cup K; \quad gI = J; \quad \bar{g}I = K.$$

What do these equalities mean? Consider the first one, $fI = J \cup K$, which means that if you apply f (a clockwise rotation of 180°) to all of the transformations in I, you obtain exactly the set of

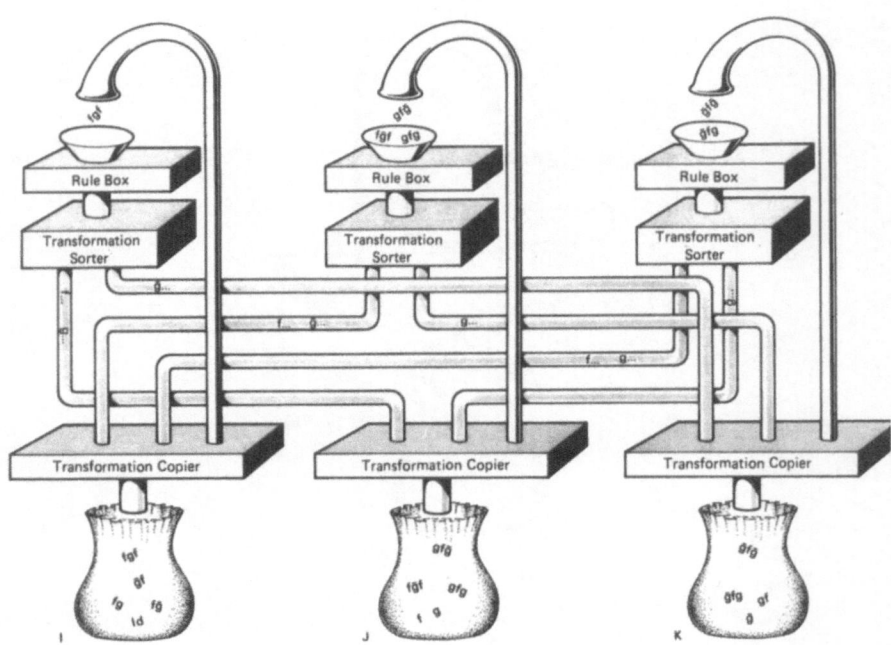

FIGURE 14-5 *The full-blown machine used to generate and sort all of the transformations.*

transformations $J \cup K$. In other words, I is congruent to $J \cup K$. Similarly, $gI = J$ means that upon applying g (a clockwise rotation of 120°) to all of the transformations in I, you obtain exactly all of the transformations in J. Thus, I is congruent to J. Similarly, we find that I is congruent to K.

Figure 14-5 shows the full-blown iterative machine, that will create these three sets of transformations I, J, and K. The major conceptual difference with the basic machine (Figure 14-4) is the addition of three transformation sorters. The role of these sorters is simple: Based solely on the leftmost elementary transformation of any transformation entering the sorter, they determine the tube down which it will be sent. The machine operates sequentially. First, it processes all of the transformations in its I hopper, then everything in its J hopper, and finally everything in its K hopper before returning to the I hopper. For this reason, we can talk of cycles of the machine. Were I to be asked to put my finger on the key technique of the proof of the Banach-Tarski Theorem, I wouldn't hesitate to single out this clever way of generating the three disjoint subsets, I, J, and K of the set Q of all transformations of the sphere. Figure 14-6 indicates several stages of production of this machine.

It should be clear, at least empirically, that we now have the desired relationships between the various subsets of Q, namely: $fI = J \cup K$; $gI = J$; $\bar{g}I = K$.

Are some of the pieces starting to fall together? The Hausdorff paradox states that we can divide the sphere (minus a countable set) into three disjoint subsets of *points* A, B, and C such that A, B, C, and $B \cup C$ are pairwise congruent. We produced with our iterative machine three disjoint subsets of *transformations* of the sphere I, J, and K such that I, J, K, *and* $J \cup K$ are pairwise congruent. If you

	CYCLE 0	CYCLE 1	CYCLE 2 –	CYCLE 3 –
Contents of Bag I	1	1	1, fg, gf, fg,	1, fg, $\bar{g}f$, fg, fgf^-
Contents of Bag J	empty	f, g	f, g	f, g, $f\bar{g}f$, $gf\bar{g}$, $\bar{g}fg$

FIGURE 14-6 *The results of the first iterations of the transformation-producing and sorting machine.*

think this is too much of a coincidence to be an accident, you are right. We are indeed closing in on the result.

Two Spheres from One

Let's return to our sphere. No matter how many times you rotate it in any imaginable way about a fixed center, when you are finally done, you can always find exactly one axis that would have allowed you to go from the initial position of the sphere to its final position in just one rotation. This is what we do for all of the transformations in Q. For each transformation, regardless of its length, we determine the axis of rotation that would have allowed us to go from the initial position directly to the final position of the sphere. This axis cuts the sphere at two points that we call, not surprisingly, *poles*. We then collect in a set D both poles associated with each transformation in Q. This set D represents the points on the sphere that, for at least one transformation of Q, *do not move*. (It turns out that D, being a countable set, is infinitesimally small compared to the entire sphere.) All of the other points on the sphere move for every transformation in Q. This set of points, which we will call D^* or, alternately, $S \setminus D$ (where S is the sphere), is the one that interests us and is virtually the entire sphere anyway.

How should we go about defining the three other sets of points A, B, and C whose disjoint union will be equal to D^*? To each point p in D^*, apply all of the transformations in Q, collecting the resulting points in a set called $Q(p) = \{p, f(p), g(p), \bar{g}(p), fg(p), \dots\}$. It is easy to show that for any two distinct points p and p', the sets $Q(p)$ and $Q(p')$ are either identical or disjoint. From each of the sets created in this way, pick a point. Collect all of these points together in a set M. (The possibility of creating this set M implies our tacit acceptance of the Axiom of Choice. Before devoting time to a discussion of this axiom, let's finish the proof.) A moment's reflection will convince you that the set D^* is equal to the set obtained by applying all of the transformations of Q to the points in M.

The last little step in the proof consists of dividing D^* into three disjoint subsets A, B, and C such that A, B, C, and $B \cup C$ are pairwise congruent. With the means now at our disposal, this will be easy. Recall the three subsets of transformations I, J, and K that we constructed so carefully. Define A as the set of points resulting from the application of all of the transformations of I to the set M. Similarly, B and C will be produced by the application of all of the transformations of J and K, respectively, to M. This construction gives the desired disjoint decomposition of D^* into A, B, and C. Because fI is equal to $J \cup K$, however, $f(A)$ is obviously equal to $B \cup C$. Since f is simply a rotation by 180°, we can conclude that A and $B \cup C$ are congruent. By similar reasoning, clearly $gI = J$ implies that A and B are congruent, and $\bar{g}I = K$ implies that A and C are congruent. The transitivity of congruence allows us to conclude that A, B, C, and $B \cup C$ are all pairwise congruent.

Now we are ready to use our trick of shifting to infinity. Recall the image of the television channel selector knob. Imagine instead that we now have a spherical knob with two axes of rotation. Suppose that the transformation f represents a click of the button by 180° about its first axis; to make the set A coincide with the set $B \cup C$, we need only turn the knob by one click about this axis. In a similar fashion, one 120° click about the second axis (i.e., the transformation g) brings the set A directly onto B, while two clicks make A coincide with C.

Finally, we are in possession of the result that will take us directly to the decomposition we need to finish the Banach-Tarski construction. Remember that our goal is to cut the ball into a finite number of pieces that will be reassembled into two balls of the same size and volume as the first one. As we have already said, our starting point will be the Hausdorff paradox. The idea is as follows. Given that the surface of the ball can be cut into four disjoint sets A, B, C, and D such that A, B, C, and $B \cup C$ are *all* mutually congruent, we can use the set $B \cup C$ as a "cutting template" to produce the pairs of sets that will eventually be reassembled into two separate spheres. Lay this template on top of A and cut out two sets A_1 and A_2 that are congruent to B and C, respectively. Since both B and C are each congruent to A, the decomposition of A into A_1 and A_2 is paradoxical. We then decompose

B and C in a similar fashion into B_1 and B_2, and C_1 and C_2. In other words, we can decompose S into disjoint subsets as follows:

$$S = A \cup B \cup C \cup D$$
$$= (A_1 \cup A_2) \cup (B_1 \cup B_2) \cup (C_1 \cup C_2) \cup D$$
$$= (A_1 \cup B_1 \cup C_1 \cup D) \cup (A_2 \cup B_2 \cup C_2).$$

From $(A_1 \cup B_1 \cup C_1 \cup D)$ we make one sphere S_1, which is equivalent by finite decomposition to the original sphere S (since A is congruent to A_1, B is congruent to B_1, etc.). Only one tiny detail remains to be shown. Can we construct a second sphere from $(A_2 \cup B_2 \cup C_2)$?

Rest assured, we haven't come this far for the answer to be no! First, notice that $A_2 \cup B_2 \cup C_2$ can *almost* be reassembled to make a second sphere S_2, identical to the original one; all that is missing is the set D whose size, as we have already pointed out is "insignificant" compared to that of S. The demonstration that a sphere and a sphere with the set D removed are equivalent by finite decomposition is essentially the same as the proof that a circle and the same circle with a point missing are equivalent by finite decomposition. Thus $S_2 \setminus D$ and S_2 are equivalent by finite decomposition, and that concludes the proof. We have shown that the sphere S, when properly dissected, can be decomposed and then reassembled into two spheres S_1 and S_2, each of which is equivalent by finite decomposition to S!

Applications

So we have now shown that one basketball, if it is cut up carefully enough, can spawn two. So much the better for the sports world, but what about the banking community? Can a bank note, even of the smallest denomination, produce two of its kind? Unfortunately not. The mathematician A. Lindenbaum proved that no bounded set in the plane can have a paradoxical decomposition [2], and a bank note, sad to say, is a bounded set in the plane.

We have already described the "thickening" technique by which we transform the spheres into balls. To produce these two copies by means of the Banach-Tarski Theorem, we need the Axiom of Choice. What could be more intuitively obvious than this axiom, which claims that it is possible to start with any collection of non-empty sets and create a new set by selecting one element from each of the sets in the collection.

The validity of the Axiom of Choice, like that of Euclid's Fifth Postulate some 200 years before, was a hotly debated subject within the mathematical community this century. The question was finally resolved around the beginning of the 1960s. The fate of this axiom resembled that of the Fifth Postulate. It turned out that the Axiom of Choice, like the Fifth Postulate, is neither true nor false, but independent of the other axioms of the system. If we accept it as true—and what could be more natural?—we are mathematically obliged to accept the strange result of Banach and Tarski that derives from it.

So much for theory. Now let's move on to some amusing practical applications. All you need is a sharp knife, a small loaf of bread, a few fish, and a large audience. Then if you go about carefully doing the cuts and reassemblies indicated in this article, who knows where it all might lead.

References

1. Martin Gardner, *Mathematical Carnival*, New York: Vintage Books (1965), Ch. 3, p. 27.
2. A. Lindenbaum, *Fund. Math.* (1926), 8, p. 218.

15

Painlevé's Conjecture

Florin N. Diacu

A-t-on tout à fait le droit d'établir une séparation entre les deux grands aspects de la vie de Painlevé, son côté scientifique et son côté humain? Ce n'est point certain et, devant nous, récemment, l'homme d'État qui a peut-être été le plus près de sa pensée et de son action, faisait ressortir l'unité secrète par laquelle toutes les manifestations de cette admirable nature sont solidaires les unes des autres.

Jacques Hadamard: *L'oeuvre scientifique de Paul Painlevé*
Revue de Métaphysique XLI (1934), 289-325

This is a story about celestial mechanics and mathematics and about a question older than Bieberbach's conjecture; a question that died close to its 100th birthday but which—like any good question—left behind it many other unanswered questions as well as a universe of intellectual achievements.

The *n*-Body Problem

The roots of the *n*-body problem get lost somewhere in the early history of humankind, but we can easily recognize its modern birth certificate signed by Isaac Newton in his fundamental *Philosophiae Naturalis Principia Mathematica,* published for the first time in 1687. The clear formulation of the problem in terms of differential equations is based on the inverse-square law of mutual attraction between particles and can be stated in the following way: Consider n particles in the ambient space whose positions are given by the vectors $\mathbf{q}_i$, $i = 1, \ldots, n$ (with respect to a fixed frame), and let $\mathbf{q} = (\mathbf{q}_1, \ldots, \mathbf{q}_n)$ be the *configuration* of the system. Determine the motion of the n particles by finding the general solution $(\mathbf{q}, \dot{\mathbf{q}})$ of the second-order system

$$\ddot{\mathbf{q}} = M^{-1}\nabla U(\mathbf{q}),$$

where $U: \mathbf{R}^{3n} \setminus \Delta \to \mathbf{R}_+$, $U(\mathbf{q}) = \sum m_i m_j |\mathbf{q}_i - \mathbf{q}_j|^{-1}$ is called the *potential function* (or *force function*) of the system of particles, $\Delta = \cup\{\mathbf{q} \mid \mathbf{q}_i = \mathbf{q}_j\}$ is the *collision set*, and $M = $ diag $(m_1, m_1, m_1, \ldots, m_n, m_n, m_n)$ is a 3n-dimensional diagonal matrix, $m_1, m_2, \ldots, m_n$ being the masses of the n particles. The usual formulation is that of an initial-value problem for a system of 6n differential equations: solve

$$\dot{\mathbf{q}} = M^{-1}\mathbf{p}, \quad \dot{\mathbf{p}} = \nabla U(\mathbf{q}) \tag{1}$$

Volume 15, No. 2 (Spring 1993), 6-12

subject to the initial conditions $(\mathbf{q}, \mathbf{p})(0) \in (\mathbf{R}^{3n} \setminus \Delta) \times \mathbf{R}^{3n}$, where $\mathbf{p} = M\dot{\mathbf{q}}$ denotes the *momentum* of the system.

For $n = 2$, the problem is not difficult, and its solution can be found in any celestial mechanics or astronomy textbook under the name of the *two-body problem* or the *Kepler problem* (in honor of the famous German astronomer Johannes Kepler who actually provided Newton the inspiration for the inverse-square attraction law). Depending on the initial conditions, the motion of one particle with respect to the other can be an ellipse (including possibly a circle), a parabola, a branch of a hyperbola, or a line. This last case, of *rectilinear* motion, is the only one when collisions between the two particles can take place. It is interesting that the complete solution as described above was not given by Newton as one would expect, but by Johann Bernoulli, and only in 1710 (see [24]).

For $n \geq 3$, the problem is still open even after three centuries of intense efforts to find its solution. Almost all important mathematicians up to the first quarter of this century attacked some aspect of the n-body problem, bringing important contributions to the understanding of the subject. In spite of this, the global image we have today is still far from complete.

There are several ways to approach the problem. A modern method for tackling systems of differential equations in nineteenth-century mathematics was to find *first integrals* and, consequently, to reduce the dimension of the system. More precisely, a function

$$F: (\mathbf{R}^{3n} \setminus \Delta) \times \mathbf{R}^{3n} \to \mathbf{R}$$

is said to be a *first integral* for Equations (1) if $F(\mathbf{q}, \mathbf{p}) = c$ (constant) along a solution $(\mathbf{q}, \mathbf{p})$ of it. A relation like this between the components of a solution reduces the dimension of the system by 1. It is known that systems of k equations have (locally) k linearly independent first integrals, and it was an important goal to find as many integrals as possible. For Equations (1), 10 of them were easy to obtain: three integrals of the momentum, three integrals of the center of mass, three of the angular momentum, and one energy integral, namely,

$$\sum \mathbf{p}_i = \mathbf{a}, \quad \sum m_i \mathbf{q}_i - \mathbf{a}t = \mathbf{b}, \quad \sum \mathbf{q}_i \times \mathbf{p}_i = \mathbf{c}, \quad T(\mathbf{p}) - U(\mathbf{q}) = h,$$

where $\mathbf{a}, \mathbf{b}, \mathbf{c}$ are constant vectors and h is a real constant, with T denoting the *kinetic energy*.

Any further attempt to find new ones was unsuccessful, and people started to look for other methods. The decisive result which stopped completely the search for first integrals was published in 1887 by Bruns. In a long paper [2] he proved the following negative statement:

THEOREM 1. The only linearly independent integrals of Equations (1), algebraic with respect to $\mathbf{q}, \mathbf{p}$, and t, are the 10 described above.

This was an important moment in the history of mathematics, which changed the way of thinking prevalent since Galilei. After a long period of quantitative methods, mathematicians understood that the class of problems solvable in this way is very small, and a large window on qualitative methods was opened. The new era was signaled by Liapunov stability criteria, obtained approximately at the same time, and also motivated by celestial mechanics.

Approximately one hundred years ago, interest in the problem reached a high level. Advised by Gustav Mittag-Leffler (at that time Editor-in-Chief of *Acta Mathematica*), King Oscar II of Sweden and Norway, a protector and supporter of science and especially of mathematics, established in 1887 an important prize for solving the 3-body problem. The formulation was very precise: *one must obtain, for any choice of the initial data, a solution expressing the coordinates as a power series, convergent for all real values of the time variable.* The idea of attacking the problem in this way is attributed to Dirichlet (see [19]). Bruns' result was at that time still too fresh to change the belief in quantitative methods. Unexpectedly, nobody could provide the desired solution. In spite of this, the prize was awarded to Henri Poincaré in 1889 for his memoir *Le problème des trois corps et les équations de la dynamique,* published in *Acta Mathematica* one year later [13]. This was in recognition of this paper's stimulating value for further research in mathematics and mechanics, and indeed this choice was a good one. Poincaré's interest was aroused by this success and he continued investigation into the mysterious *n*-body problem for many years. He also wrote the famous *Les nouvelles méthodes de la mécanique céleste,* in three volumes [14], where the idea of *chaos* appears for the first time.

Not only were many mathematical theories born from the study of the *n*-body problem but also the strength of new theories is checked today by trying to find applications of them to this old problem. It has been studied by classical analysis, differential equations, and sometimes function theory, but nowadays also by new fields like dynamical systems, differential topology, differential geometry, Morse theory, algebraic geometry, algebraic topology, symplectic manifolds, Lie groups and algebras, ergodic theory, numerical analysis and computers, operator theory, and C^*-algebras.

The Conjecture of Painlevé

In 1895, at 32 years of age, Paul Painlevé was already one of the most famous mathematicians of his time, and King Oscar II invited him to give a series of lectures at the University of Stockholm in September-November of that year. The event was considered of paramount importance, and even the King attended the introductory lecture. The notes were published in 1897 in handwritten form under the title *Leçons sur la théorie analytique des équations différentielles* [11] and can be found today also in Painlevé's Complete Works [12]. The last pages contain an application of the results to the 3-body problem and an opinion of the author concerning the *n*-body case, formulated as a statement which was known afterwards as the *Conjecture of Painlevé.* First, let us try to understand its natural birth.

Standard results of differential equations ensure, for any $(\mathbf{q}, \mathbf{p})(0) \in (\mathbf{R}^{3n} \setminus \Delta) \times \mathbf{R}^{3n}$, the existence and uniqueness of an analytic solution of Equations (1) defined locally on (let's say) (t^-, t^+), with 0 contained in this interval. Due to the symmetry of mechanical laws with respect to the past and future, one can study the problem on $(t^-, 0]$ or on $[0, t^+)$, without loss of generality. Because many scientists have a natural desire to predict future phenomena, let us choose the second interval. We can extend the solution analytically to a maximum interval $[0, t^*)$, with $0 < t^+ \leq t^* \leq \infty$. In case $t^* = \infty$, the motion is called regular, whereas if t^* is finite, we say that the solution experiences a *singularity.* What is the physical meaning of such a singularity and is it important? One obvious possible way for a solution to encounter a singularity is for a collision to occur. Indeed, the configuration vector $\mathbf{q}$ will then so tend to the set Δ that at least two position vectors have the same value, consequently ∇U tends to infinity and the equations of motion (1) become meaningless. The creation of the prize made the importance of such a study very clear. Because a series expansion of the coordinates convergent for every real value of t was asked, solutions leading to singularities were expected to be extended somehow beyond the singularity.

Although very young in 1887, Painlevé was working on his doctoral thesis and knew about the famous problem. He tried, therefore, to understand whether in the 3-body problem the only possible singularities are collisions. His worry about the occurrence of other singularities was motivated by the possible appearance of large oscillations (suspected already by Poincaré). For example, one particle could oscillate between the other two without colliding but coming closer and closer to a collision at each close encounter. Under such circumstances, one can find a subsequence t_n of times converging to a finite t^* such that $\nabla U(\mathbf{q}(t_n)) \to \infty$. This again makes Equations (1) meaningless, and such t^* is also a singularity. In modern terminology,

DEFINITION. Let $(\mathbf{q}, \mathbf{p})$ be a solution of Equations (1) defined on $[0, t^*)$ with t^* a singularity. Then t^* will be called a collision singularity if $\mathbf{q}(t)$ tends to a definite limit when $t \to t^*$, $t < t^*$. If the limit does not exist, then the singularity will be called a pseudocollision or non-collision singularity.

It is clear that these singularities (especially the non-collision ones) are an important obstacle to accomplishing King Oscar's goal. Indeed, one might try to extend a collision as an elastic bounce and possibly obtain a globally convergent power series, but how to do that with pseudocollisions? Painlevé doubted that pseudocollisions can actually appear and he proved for the 3-body case

THEOREM 3.   For $n = 3$, any solution of the Equations (1) defined on $[0, t^*]$ with t^* a finite singularity, experiences a collision when $t \to t^*$.

Attempts to extend this result to the n-body problem ($n > 3$) failed, and the intuition of Painlevé was that pseudocollisions may, indeed, arise for more than 4 bodies. Thus, his Stockholm lectures end with the following:

CONJECTURE: For $n \geq 4$, Equations (1) admit solutions with non-collision singularities.

Painlevé understood that this is a very hard problem; his subsequent mathematical work contains some papers dealing with singularities, none, however, attempting to prove the conjecture. After 1905, Painlevé's scientific activity becomes less intense because of his deep involvement in politics. Paul Painlevé was elected several times as deputy, holding the War, then Finance, and finally Air portfolios, and serving as President of the Chamber of Deputies of France. In 1918, he became Président de l'Académie des Sciences, and in 1927 the University of Cambridge offered him the title of Doctor Honoris Causa. Indeed a remarkable and successful life! His famous conjecture remained open, however, for more than half a century after his death.

It is interesting to note that collision orbits are very improbable. Donald Saari proved that in the n-body problem they are of *Lebesgue measure zero* and of the *first Baire category*. Moreover, this is true for all singularities in the 4-body problem (see [15,17]). Some of these results were generalized and are expressed in terms of lower-dimensional manifolds [18]. It is also expected that, for any n, singularities are improbable. However, these results did not diminish the interest in the study of singularities.

Singularity Criteria

Many of Painlevé's contemporaries tried to find examples of solutions with pseudocollisions, but no one succeeded. Their attention was, therefore, directed towards understanding theoretical aspects and especially towards criteria for obtaining non-collision singularities.

A way of finding singularities had already been found, but it is quite hard to discern when and by whom:

THEOREM 4.   Consider a solution $(\mathbf{q}, \mathbf{p})$ of Equations (1). Then t^* is a singularity of this solution iff

$$\liminf_{t \to t^*} \min_{i<j} q_{ij}(t) = 0, \qquad (2)$$

where $q_{ij} = |\mathbf{q}_i - \mathbf{q}_j|$.

Painlevé himself improved this result [11] in proving Theorem 3. He showed that condition (2) can actually be replaced by

$$\lim_{t \to t^*} \min_{i<j} q_{ij}(t) = 0.$$

The first important condition for the occurrence of non-collision singularities was found and published only in 1908 by a Swedish mathematician of German origin, Hugo von Zeipel [26]. His result has not only a nice formulation but also an unusual history and played a fundamental role in the story of Painlevé's conjecture.

THEOREM 5. If t^* is a collision singularity for a solution $(\mathbf{q}, \mathbf{p})$ of Equations (1), then $J(\mathbf{q}(t))$ tends to a definite limit when $t \to t^*$, where $J(\mathbf{q}) = \sum m_i |\mathbf{q}_i|^2$ is the moment of inertia.

This implies, of course, that a necessary condition for having a non-collision singularity is that the motion become unbounded in finite time, because the moment of inertia is a measure of the distribution of particles in space.

What is obvious is that at a singularity the whole $|(\mathbf{q}, \mathbf{p})|$ has to become unbounded. This always happens at a collision instant because the velocities are infinite. It is not clear what would happen in the configuration space (i.e., for the vector $\mathbf{q}$), and here lies von Zeipel's contribution. His paper appeared in a less famous journal (see [26]) and was, therefore, not well known. Personally I have tried to find it in several good university libraries in Eastern and Western Europe as well as in North America, but without success. An article of Dick McGehee [10], who spent a period in Stockholm and was interested in this subject, makes it less necessary to read the original.

The French astronomer Jean Chazy had announced Theorem 5 without making any reference to von Zeipel's paper [3]. Aurel Wintner wrote in 1941 that the proof of the Swedish mathematician has some gaps and there is no complete argument for the theorem [24]. Thirty years later, Hans Sperling gave a detailed proof [20], apparently ending the dispute. However, McGehee's paper cited above provides a translation in modern mathematical language of von Zeipel's proof, showing that it was actually correct from the beginning.

Today we know a beautiful generalization of this result which is due to Donald Saari from Northwestern University [16]. He proves that if $J \circ \mathbf{q}$ is a *slowly varying* function as $t \to t^*$ for a solution $(\mathbf{q}, \mathbf{p})$ of Equations (1), then the singularity t^* is necessarily a collision.

Theorem 5 is a fundamental contribution to the subject of singularities in the n-body problem, and the elucidation of Painlevé's conjecture would have been hard to imagine without it.

The Computer and the Idea

As has happened many times, the idea that was to solve Painlevé's conjecture came by looking for something else and depended on electronic computers.

In 1893, Meissel proposed the investigation of a so-called Pythagorean problem, in which three gravitationally attracting particles of masses 3, 4, and 5 are initially located at the vertices of a triangle with sides 3, 4, and 5 such that the corresponding point masses and sides are opposite. Releasing the particles with zero initial velocities from their positions, how will they move in the future? Burrau investigated the problem numerically in 1913 but without reaching important conclusions. Several computer investigations in 1966 and 1967 [21] helped to go much further by showing a surprising qualitative behavior. After passing close to a triple collision, two particles will form a binary while the other one is expelled with high velocity in the opposite direction, as in Figure 15-1 (see also [1]). The formation of the binary was an interesting point for astronomers, whereas the high-speed escape of the third particle attracted the attention of mathematicians. It provided the idea that it might be possible to construct an example of a non-collision singularity solution. The main reason for this qualitative feature is the triple approach of the particles, as was recognized in [8], [9], [22], [23].

We sketch crucial ideas from Dick McGehee's 1974 paper. He considered the case of the rectilinear 3-body problem, i.e., when the masses m_1, m_2, m_3 move all the time on a fixed line. He was inter-

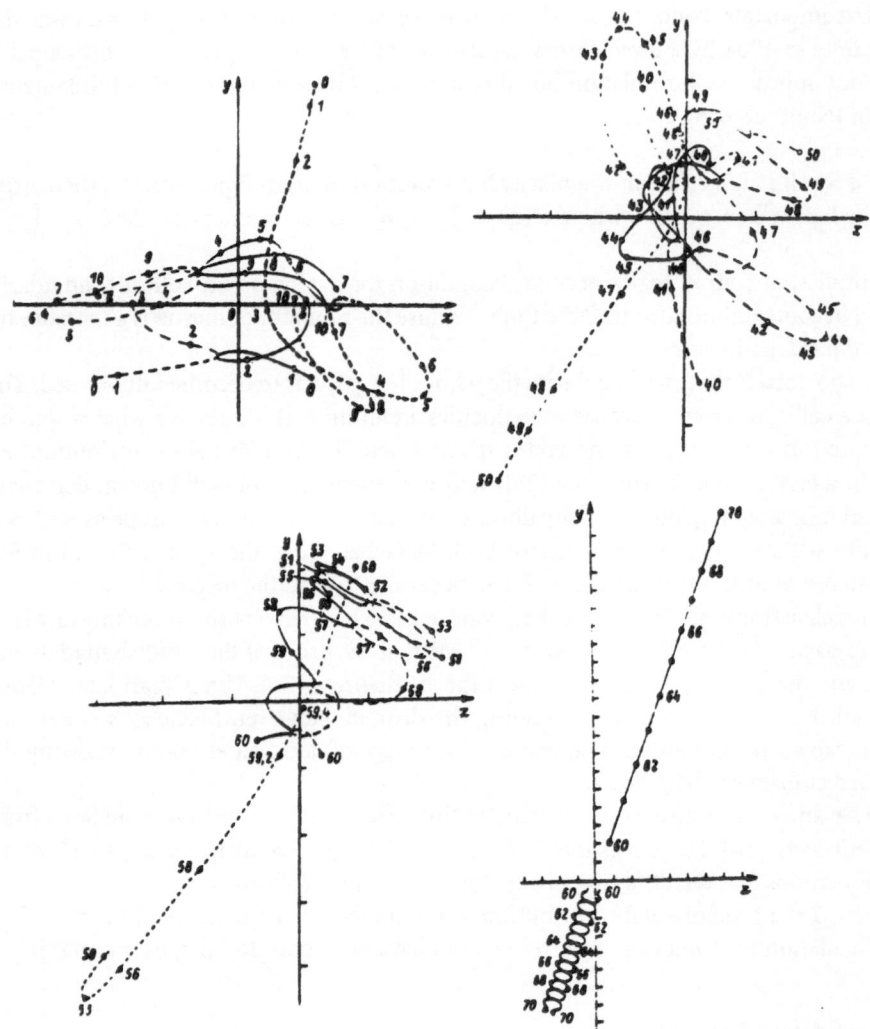

FIGURE 15-1 *Numerical results in the Pythagorean problem.*

ested in understanding the flow in a neighborhood of a triple collision solution. This was, indeed, a hard problem because previous numerical investigations suggested chaotic behavior near a *total collapse* (i.e., a simultaneous collision of all bodies). Only qualitatively speaking, the particles behave by forming a binary and an escape, numerical investigations showing a highly sensitive dependence with respect to initial data. For example, for some initial conditions the particles m_1 and m_2 form a binary and m_3 escapes, whereas, perturbing the data a little bit, it may be that m_1 and m_3 form a binary and m_2 escapes. Two such solutions look very different in a phase-space picture, in spite of being close to one another at some initial moment of time.

McGehee's idea was to restrict the equations of motion to an arbitrary energy level, then to blow up the singularity (by using certain transformations which today bear his name), and finally to introduce a so-called *collision manifold* which is proved to be independent of the chosen energy level. In the rectilinear 3-body problem, the collision manifold happens to be a sphere missing four points, as in Figure 15-2.

Roughly speaking, the McGehee coordinates are polar coordinates for the configuration vector and a decomposition of the velocity into a radial and a tangential component, rescaled by a suitable

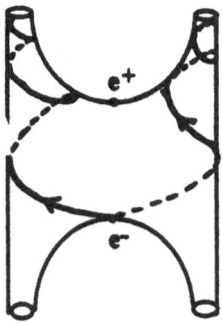

FIGURE 15-2 *The flow on the collision manifold.*

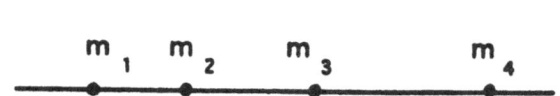

FIGURE 15-3 *The example of Mather and McGehee.*

transformation of time, which makes the collision manifold be approached asymptotically by the real flow when the new (fictitious) time variable goes to infinity.

The equations of motion restricted to the collision manifold do not describe a real physical situation. However, due to the continuity property of the solutions with respect to initial data, a study of the flow on this manifold provides valuable information on the behavior of solutions passing close to a triple collision. Many interesting theoretical results were proved in McGehee's paper using these powerful techniques, including a theorem on the occurrence of solutions with high-velocity escapes. Studies on collision singularities are hard to imagine today without McGehee's transformations.

The Example of Mather and McGehee

One of the results McGehee announced (without proof) in his 1974 paper is the construction of a solution with non-collision singularities in the rectilinear 5-body problem, using the idea of a high-speed escape. The trouble is not that collisions always appear in a rectilinear problem but that they always arise before an impending pseudocollision, as was shown by Saari [16]. I may now have confused the reader, for I said before that the solution was defined on a maximal interval $[0, t^*)$, t^* (finite) representing either a collision or a non-collision singularity. There is no inconsistency. Binary collision solutions can be analytically extended by a mathematical procedure called *regularization*. There is a vast literature on this subject (see, e.g., [4]). Physically, this means that an elastic bounce, without loss or gain of energy, takes place. I hope the sense of Saari's result is now clear.

Mather and McGehee [7] were later able to prove completely that a non-collision singularity can occur in the rectilinear 4-body problem, but only after an infinity of (regularized) binary collisions. Here is their scenario.

Four bodies of suitably chosen masses m_1, m_2, m_3, m_4 lie on a straight line at some initial moment (see Figure 15-3). The initial data (positions and velocities) are such that the particles m_1 and m_2 stay close together, so we say that they form a binary system. The particle m_3 oscillates between the binary system and the particle m_4. The motion is regularized beyond the binary collisions which take place at the instants $t_1, t_2, \ldots, t_k, \ldots$. This sequence converges as k goes to infinity. Meanwhile, the binary m_1, m_2 goes to $-\infty$, m_4 goes to $+\infty$, and m_3 bounces back and forth, with increased velocity after every close passage to a triple collision. This is possible because the distance between m_1 and m_2 tends to zero, the loss of potential energy of the binary being transferred into kinetic energy for the particle m_3. The proof of Mather and McGehee is not at all easy.

Whatever its mathematical beauty and interest for dynamical systems theory, the above example is not accepted as a proof of Painlevé's conjecture because the pseudocollision appears only after (infinitely many) collisions.

Gerver's First Example

In 1984, Joe Gerver from Rutgers University proposed a solution of a planar 5-body problem in which the particles escape to infinity in finite time [5]. Although he does not give a complex proof, he provides a lot of support for the existence of such a solution. We reproduce his scenario.

Consider the planar motion of five particles $m_1, \ldots, m_5$, with $m_3 = m_4$, m_2 somewhat greater but of the same order of magnitude as m_3, m_1 much smaller than m_2, and m_5 much smaller than m_1 (see Figure 15-4). Initially, m_1 is in a roughly circular orbit around m_2, whereas m_3 and m_4 are much further away. The bodies m_2, m_3, m_4 are approximately at the vertices of an obtuse triangle. Initially the triangle is slowly expanded while maintaining its shape. Meanwhile, m_5 moves rapidly around the triangle, coming close to each of the other four bodies, the velocity of m_5 being much greater than that of m_1. Each time m_5 passes close to m_1, it picks up a small amount of kinetic energy. This causes m_1 to fall into a lower orbit around m_2 such that the mean kinetic energy of m_1 in its orbit actually increases by about the same factor as for m_5. A small fraction of the kinetic energy of m_5 is transferred to m_2, m_3, and m_4, causing faster expansion of the triangle. The time required for one trip of m_5 around the triangle decreases each time (in spite of the expansion) by a factor slightly less than 1. After a finite time, the geometric progression of the time instants $t_1, t_2, \ldots, t_k, \ldots$ measuring a round trip will converge, and m_5 will have travelled an infinite number of times around the triangle. In the meantime, the triangle has become infinitely large.

Xia's Example

In his doctoral thesis written under the supervision of Donald Saari at Northwestern University, Jeff Xia proved in 1988 that a certain type of solution in the spatial 5-body problem leads to a non-collision singularity without involving an infinite number of binary collisions, as was the case in the example of Mather and McGehee. Painlevé's conjecture was finally proved.

The author considers two pairs of bodies, the particles in the same pair having equal masses, plus a fifth particle of small mass. The bodies in a pair move in highly eccentric orbits parallel with the (x, y)-plane (see Figure 15-5). The binaries are on opposite sides with respect to the (x, y)-plane and have an opposite rotation. The motion of the small particle is restricted to the z-axis, so that the total angular momentum is zero. The small particle will oscillate between the two binaries, determining an unbounded motion in finite time. More precisely, suppose the particle m_5 intersects the line connecting m_3 with m_4 from above, at a moment when these particles come near to their closest approach, the motion of m_3, m_4, and m_5 thus being close to a triple collision. The body m_5 goes a

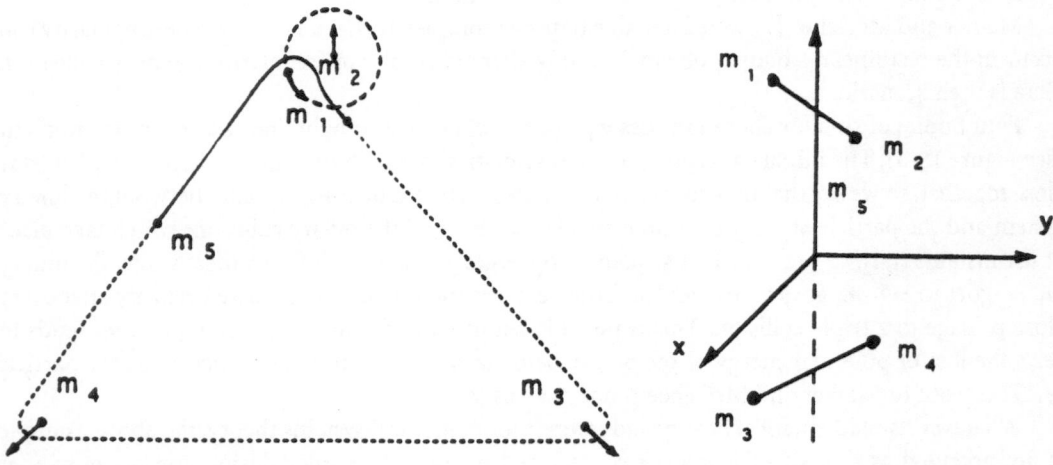

FIGURE 15-4 *Gerver's heuristic example.*   FIGURE 15-5 *The example of Xia.*

little under the line m_3m_4, whereas the particles m_3 and m_4 are at their closest approach. Thus, m_5 is strongly attracted backwards. It intersects the line m_3m_4 again when these point masses start to separate. This separation reduces the retaining force on the small particle which consequently moves very fast towards the other binary system. The action-reaction effect forces the binary m_3, m_4 to move further away from the plane (x, y). The same situation described above is now repeated (in mirror image) for the binary m_1, m_2. Iterating this procedure with higher and higher accelerations for m_5, the two binaries will be forced to tend to infinity in finite time. Simple though this scenario sounds, it is very hard to prove it is possible. For example, because the motion becomes unbounded in finite time, the acceleration effects on the small particle have to become infinitely large. The point masses in each binary must come closer and closer together, making it hard to guarantee non-occurrence of collisions.

There were mistakes in the first attempt of Xia, but he was able to correct them. The paper appeared in *Annals of Mathematics*.

His example can be extended to similar symmetric problems for any $N > 5$.

In spite of his youth (not even 30 years old in 1992), today associate professor at Georgia Tech, Xia has already brought a tremendous contribution to the field. He recently proved a new magnificent result, namely, that the very rare (and hard to detect) phenomenon called *Arnold diffusion* (a kind of chaos) takes place in a very natural system, the elliptical restricted 3-body problem. Arnold himself constructed in the 1960s a very sophisticated and artificial system to show for the first time that such a phenomenon exists. It is expected that Xia will make many other important contributions in years to come.

Gerver's Second Example

The idea of using radial symmetry, combined with the experience obtained by trying to prove his previous heuristic example, led Joe Gerver to the following solution for the planar case. Consider $3n$ bodies (n sufficiently large) in a plane as in Figure 15-5. $2n$ of the particles are arranged in n nearly circular orbiting pairs and all have the same mass. The center of mass of each binary lies at one of the vertices of a regular polygon. The other n bodies have small equal masses and move rapidly from one pair to the other as in Figure 15-6. When a small particle comes close to the binary, it takes some kinetic energy from the pair and transfers some momentum to it, forcing the binary to move into a tighter orbit and concomitantly to increase its distance from the center of the polygon. Iterating this process for a suitably chosen n, suitable values of the masses, and of the initial velocities, the size of the configuration will increase by each close encounter of a small particle with a binary. The sequence of times from one encounter to the next will converge to a finite value, whereas the system becomes unbounded in finite time. The complete proof contains very many computations and is, therefore, quite hard to follow (see [6]).

Gerver found out about Painlevé's conjecture nineteen years before he gave the solution. Xia succeeded in proving his example about six months before Gerver. However, Gerver's is the first confirmation of the conjecture for the case of planar solutions and is also very elementary, using mainly nineteenth-century mathematics. Seeing the proof, one sees that the conjecture would have been possible for Painlevé's contemporaries to prove, but nobody did it.

It was not the first time Gerver attacked a famous problem. As a graduate student at Columbia University

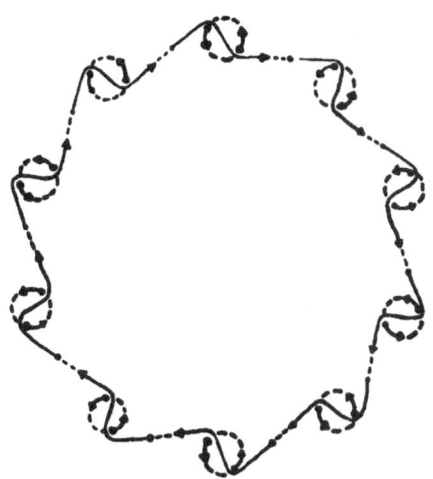

FIGURE 15-6 *Gerver's planar example.*

in 1969, he proved a conjecture of Riemann on the nowhere differentiability of the function $\sum_{n=1}^{\infty} \sin(n^2 x)/n^2$. But this was long before his work on Painlevé's conjecture started.

A comparison between the two solutions is hard to make. Each is interesting and valuable in its own way. Xia opened a new direction of work bringing fresh air into the field, whereas Gerver used the old methods showing that they can be successful too. Surely both achieved a most remarkable feat in an old and hard field where good new results are not at all easy to obtain.

References

1. V. I. Arnold, *Dynamical Systems III*, New York: Springer-Verlag, 1988.
2. H. Bruns, Über die Integrale des Vielkörper-Problems, *Acta Math.* 11 (1887), 25-96.
3. J. Chazy, Sur les singularités impossibles du problème des *n* corps, *C. R. Hebdomadaires Séances Acad. Sci. Paris* 170 (1920), 575-577.
4. F. N. Diacu, Regularization of partial collisions in the *N*-body problem, *Diff. Integral Eq.* 5 (1992), 103-136.
5. J. L. Gerver, A possible model for a singularity without collisions in the five-body problem, *J. Diff. Eq.* 52 (1984), 76-90.
6. J. L. Gerver, The existence of pseudocollisons in the plane, *J. Diff. Eq.* 89 (1991), 1-68.
7. J. Mather and R. McGehee, Solutions of the collinear four-body problem which become unbounded in finite time, *Dynamical Systems Theory and Applications* (J. Moser, ed.), Berlin: Springer-Verlag, 1975, 573-589.
8. R. McGehee, Triple collision in the collinear three-body problem, *Invent. Math.* 27 (1974), 191-227.
9. R. McGehee, Triple collision in Newtonian gravitational systems, *Dynamical Systems Theory and Applications* (J. Moser, ed.), Berlin: Springer-Verlag, 1975, 550-572.
10. R. McGehee, Von Zeipel's theorem on singularities in celestial mechanics, *Expo. Math.* 4 (1986), 335-345.
11. P. Painlevé, *Leçons sur la théorie analytique des équations différentielles*, Paris: Hermann, 1897.
12. *Oeuvres de Paul Painlevé*, Tome I, Paris Ed. Centr. Nat. Rech. Sci., 1972.
13. H. Poincaré, Sur le problème des trois corps et les équations de la dynamique, *Acta Math.* 13 (1890), 1-271.
14. H. Poincaré, *Les nouvelles méthodes de la mécanique céleste*, Paris: Gauthier-Villar et Fils, vol. I (1892), vol. II (1893), vol. III (1899).
15. D. G. Saari, Improbability of collisions in Newtonian gravitational systems, *Trans. Amer. Math. Soc.* 162 (1971), 267-271; 168 (1972), 521; 181 (1973), 351-368.
16. D. G. Saari, Singularities and collisions in Newtonian gravitational systems, *Arch. Rational Mech. Anal.* 49 (1973), 311-320.
17. D. G. Saari, Collisions are of first category, *Proc. Amer. Math. Soc.* 47 (1975), 442-445.
18. D. G. Saari, The manifold structure for collisions and for hyperbolic parabolic orbits in the *n*-body problem, *J. Diff. Eq.* 41 (1984), 27-43.
19. C. L. Siegel and J. K. Moser, *Lectures on Celestial Mechanics*, Berlin: Springer-Verlag, 1971.
20. H. J. Sperling, On the real singularities of the *N*-body problem. *J. Reine Angew. Math.* 245 (1970), 15-40.
21. V. Szebehely, Burrau's problem of the three bodies, *Proc. Nat. Acad. Sci. USA* 58 (1967), 60-65.
22. J. Waldvogel, The close triple approach, *Celestial Mech.* 11 (1975), 429-432.
23. J. Waldvogel, The three-body problem near triple collision, *Celestial Mech.* 14 (1976), 287-300.
24. A. Wintner, *The Analytical Foundations of Celestial Mechanics*, Princeton, NJ: Princeton University Press, 1941.
25. Z. Xia, The existence of non-collision singularities in newtonian systems, *Ann. of Math.* (2) 135 (1992), 411-468.
26. H. von Zeipel, Sur les singularités du problème des corps, *Arkiv för Mat. Astron. Fys.* 4, (1908), 1-4.

16

A Geometrization of Lebesgue's Space-Filling Curve

Hans Sagan

1. Introductory Remarks

G. Cantor demonstrated in 1878 that any two finite-dimensional smooth manifolds, no matter what their respective dimensions, have the same cardinality [1]. This is true, in particular, for the interval $\mathbf{I}$ = [0, 1] and the square $\mathbf{Q}$ = [0, 1] $\times$ [0, 1], meaning that there exists a bijective map from $\mathbf{I}$ onto $\mathbf{Q}$.

The question arose almost immediately whether or not such a mapping can possibly be continuous. E. Netto put an end to such speculation by showing in 1879 that such a bijective mapping is, by necessity, discontinuous [8]. Is it then, it was asked, still possible to obtain a continuous surjective mapping if the condition of bijectivity is dropped? G. Peano settled this question once and for all in 1890 by constructing the first "space-filling curve" [10]. (A "space-filling curve" is a continuous map from $\mathbf{I}$ to E^n ($n \geq 2$) whose image has positive n-dimensional Jordan content [11].) Other examples followed: [3, 6, 14].

However, this was not the end of it because the following related question arose: Whereas the interval $\mathbf{I}$ cannot be mapped continuously and bijectively onto an n-dimensional set of positive Jordan content (such as $\mathbf{Q}$ in E^2), can it be done with an image set of positive outer measure? In other words, are there Jordan curves (continuous injective maps from $\mathbf{I}$ into E^n) with positive n-dimensional outer measure? There are indeed, as W. F. Osgood showed in 1903 [9] when he constructed a one-parameter family of such curves. In fact, Osgood's curves are Lebesgue measurable. It is reasonable to assume that word of the Lebesgue measure had not reached Harvard by Thanksgiving of 1902 when Osgood submitted his paper for publication, because Lebesgue's pivotal thesis only appeared that year in the *Annali di Matematica pura et applicata* and there was no airmail. One does wonder, however, why Osgood did not make use of the Borel measure, which would have given him a stronger result. The limiting arc of Osgood's family is Peano's space-filling curve. This is not a coincidence. Peano's ingenious result undoubtedly inspired Osgood's construction.

Jordan curves with positive Lebesgue measure are now called Osgood curves.

In 1917 K. Knopp constructed another family of Osgood curves with Sierpiński's space-filling curve as its limit [4, 13, 14].

Apparently unaware of this earlier work, T. Lance and E. Thomas, in a recent note, developed the same idea, albeit by a different approach [5]. It is the purpose of this note to put Osgood's and Knopp's approach in juxtaposition to the one by Lance and Thomas and to point out how the latter

Volume 15, No. 4 (Fall 1993), 37–43

narrowly missed obtaining Lebesgue's space-filling curve as the limit of their family of Osgood curves and therewith the opportunity to put the first geometric generating process of Lebesgue's space-filling curve on record. Geometric generations of the space-filling curves by Peano, Hilbert, and Sierpiński have been known as long as the curves themselves in the cases of Hilbert's and Sierpiński's curves and almost as long in the case of Peano's curve [3, 7, 12, 13]. Besides being of interest by themselves, such geometric generations lead to uniformly convergent sequences of approximating polygons, which, in turn, lead to simple proofs of the continuity of the map. In the case of the Lebesgue curve, in particular, such a proof turns out to be much simpler than the conventional proof that is based on the structure of the Cantor set (such as the one in [2]).

2. The Curves by Osgood and Knopp

Osgood's construction of a family of Jordan curves with positive Lebesgue measure consists in the successive removal of grate-shaped regions from squares, starting out with the unit square and proceeding as indicated in Figure 16-1 for the first two steps. The shaded squares are what is left after each step. Starting with square S_1, they are then connected by "joins" as indicated by the bold line segments in Figure 16-1. The dimensions of the grate-shaped regions can be chosen so that the sum of their areas tends to some positive $\lambda < 1$. If A_n denotes the point set consisting of the 9^n squares and $9^n - 1$ joins that are obtained at the nth step and $J(A_n)$ its Jordan content, then the set $C = \bigcap_{n=1}^{\infty} A_n$ that is obtained after infinitely many steps has Lebesgue measure $\mu(C) = \lim_{n \to \infty} J(A_n) = 1 - \lambda > 0$. It represents a Jordan curve [9] that may be parametrized as follows: Dividing the interval $\mathbf{I}$ into seventeen congruent subintervals and excluding the even-numbered subintervals without beginning and endpoint, and repeating the process for each of the remaining nine closed subintervals, and then again for each of the remaining eighty-one closed subintervals, and then again and again, ad infinitum, generates a Cantor-type discontinuum Γ_{17}. At the first step, the remaining nine closed intervals are mapped into the squares $S_1, S_2, \ldots, S_9$ and the excluded eight open intervals are mapped linearly onto the joins (without beginning point and endpoints) from S_1 to S_2, S_2 to S_3, $\ldots$, S_8 to S_9, and the process is repeated ad infinitum. The complement of Γ_{17} is mapped bijectively onto the set of all joins, and Γ_{17} is mapped onto $\bigcap_{n=1}^{\infty} Q_n$, where Q_n represents the 9^n squares of the nth iteration. (Osgood mapped the excluded even-numbered closed subintervals onto the joins with beginning points and endpoints and the remaining odd-numbered open intervals into the squares. We have deviated from his construction with a view toward what we are going to do in Section 3.)

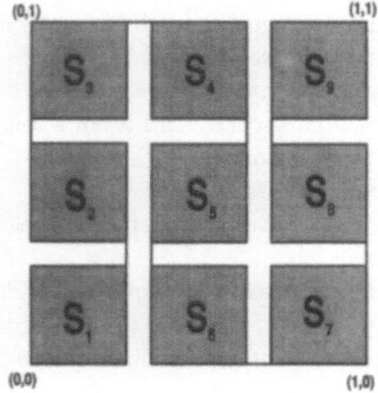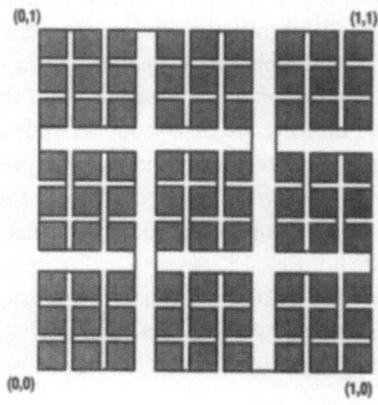

FIGURE 16-1 *Osgood's construction.*

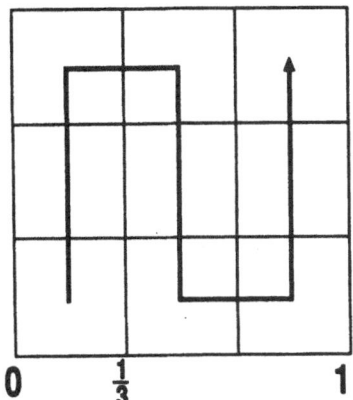 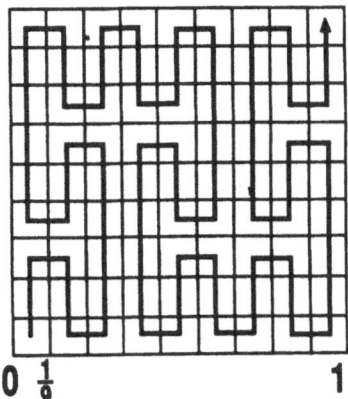

FIGURE 16-2 *Construction of Peano's space-filling curve.*

In the limit with $\lambda \to 0$, Peano's space-filling curve is obtained. The first two steps in the construction of Peano's curve are illustrated in Figure 16-2 where the bold polygonal lines indicate the order in which the squares have to be lined up. Joining the beginning points of these polygonal lines to $(0, 0)$ and their endpoints to $(1, 1)$ by straight line segments yields the approximating polygons we mentioned in Section 1. (Another notion of approximating polygons may be found in [7, 12, 13].) Compare Figure 16-2 with Figure 16-1. Since consecutive squares now have an edge in common, the injectivity of the mapping is lost. This is to be expected because manifolds of different dimensions cannot be homeomorphic.

Knopp's construction of a family of Osgood curves consists in the successive removal of triangular regions from an initial triangle as indicated for the first four steps in Figure 16-3, where the shaded triangles are the ones that are left after each step. (The initial triangle $\mathbf{T}$ need not be a right isosceles triangle.)

At the first step, we remove a triangle of area $r_1 m(\mathbf{T})$, where $m(\mathbf{T})$ is the area of the initial triangle $\mathbf{T}$ and where $r_1 \in (0, 1)$, to be left with two triangles $\mathbf{T}_0$, $\mathbf{T}_1$ of a combined area $m(\mathbf{T})(1 - r_1)$.

FIGURE 16-3 *Knopp's construction of Osgood curves.*

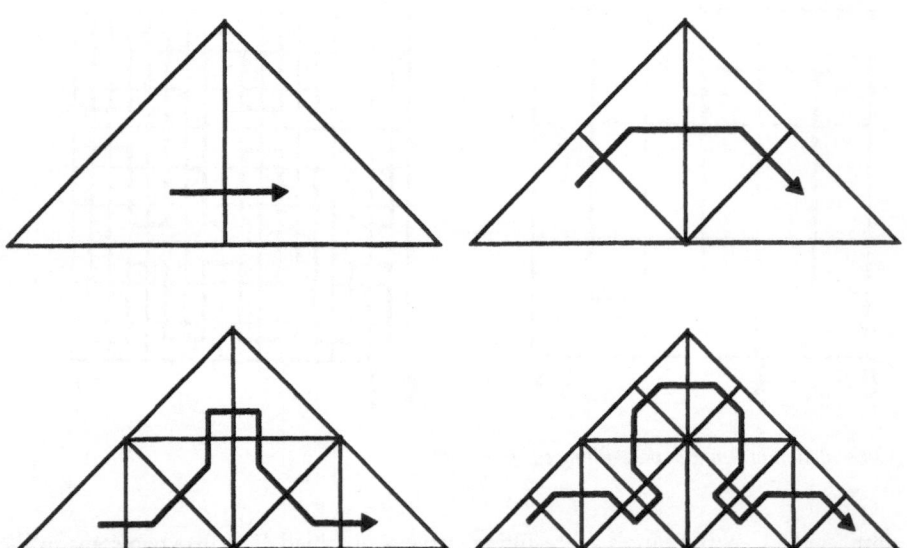

FIGURE 16-4 *Generating the Sierpiński curve.*

From $\mathbf{T}_0$ and $\mathbf{T}_1$, we remove triangles of area $r_2m(\mathbf{T}_0)$ and $r_2m(\mathbf{T}_1)$ for some $r_2 \in (0, 1)$ to be left with four triangles $\mathbf{T}_{00}, \mathbf{T}_{01}, \mathbf{T}_{10}$ and $\mathbf{T}_{11}$ of a combined area $m(\mathbf{T})(1 - r_1)(1 - r_2)$, etc. In the limit, we obtain a point set $\mathbf{C}$ of Lebesgue measure $\mu(\mathbf{C}) = m(\mathbf{T}) \prod_{k=1}^{\infty}(1 - r_k)$. If we choose the r_k such that $\sum_{k=1}^{\infty} r_k$ converges, then $\mu(\mathbf{C})$ is positive. If, at each step, the triangles that are to be removed are placed judiciously, then all dimensions of the remaining triangles tend to zero and the remaining triangles shrink into points. If the interval $[0, {}^1\!/_2]$ (all numbers $0_2 0a_2a_3\ldots$) is mapped into $\mathbf{T}_0$ and $[{}^1\!/_2, 1]$ (all numbers $0_2 1a_2a_3\ldots$) into $\mathbf{T}_1$ so that $[0, {}^1\!/_4]$ (all numbers $0_2 00a_3a_4\ldots$) is mapped into $\mathbf{T}_{00}, [{}^1\!/_4, {}^1\!/_2]$ (all numbers $0_2 01a_3a_4\ldots$) into $\mathbf{T}_{01}, [{}^1\!/_2, {}^3\!/_4]$ (all numbers $0_2 10a_3a_4\ldots$) into $\mathbf{T}_{10}$, and $[{}^3\!/_4, 1]$ (all numbers $0_2 11a_3a_4\ldots$) into $\mathbf{T}_{11}$, etc., with the points common to two adjacent intervals being mapped into the vertices common to the corresponding adjacent triangles, we see that the mapping from $[0, 1]$ to $\mathbf{C}$ is bijective and continuous and $\mathbf{C}$ is an Osgood curve of Lebesgue measure $\mu(\mathbf{C}) > 0$.

If we choose, for example, as initial triangle a right isosceles triangle of base 2 [and hence, $m(\mathbf{T})$ $= 1$] and $r_k = 1/4k^2, k = 1, 2, 3, \ldots$, we obtain from Weierstrass' factorization theorem that the corresponding Osgood curve has Lebesgue measure $2/\pi$. Setting $r_k = r/k^2$ instead and taking the limit as $r \to 0$, a space-filling curve, namely, the Sierpiński curve, is obtained because the combined area of the removed triangles shrinks to zero (see also [13]). Since adjoining triangles are then not only joined at vertices but also along edges, the injectivity of the mapping is lost. (See also Figure 16-4, where the bold polygons indicate the order in which the triangles have to be taken, and compare Figures 16-3 and 16-4. These polygonal lines, when extended by line segments to entry point and exit point, represent approximating polygons that converge uniformly to the Sierpiński curve.)

3. The Lance–Thomas Curve and the Lebesgue Curve

Instead of removing grates from squares or triangles, Lance and Thomas, by contrast, remove cross-shaped regions from squares as we have indicated for the first two steps in Figure 16-5 [5]. They connect the remaining squares by joins as indicated in Figure 16-5 by bold line segments. As in the preceding cases, the dimensions of the regions that are to be removed may be chosen so that the sum of their areas tends to some positive $\lambda < 1$. An Osgood curve of Lebesgue measure $1 - \lambda$ is obtained. Its parametrization may be accomplished as in the case of Osgood's original example, using instead

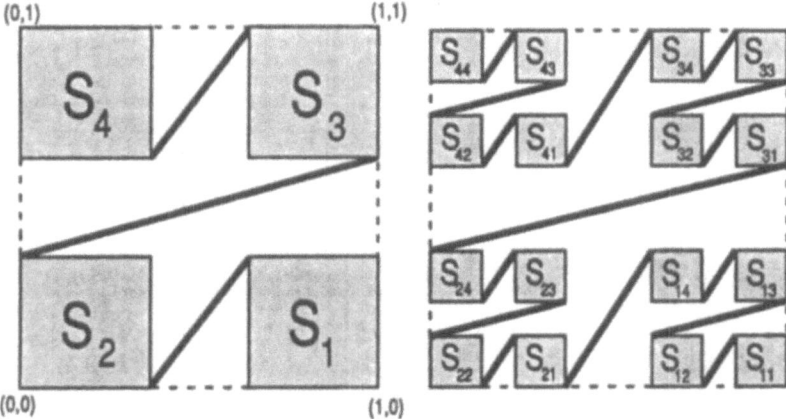

FIGURE 16-5 *Generating the Lance–Thomas curve.*

of a Cantor-type discontinuum the Cantor set itself, mapping the excluded (open) middle thirds linearly onto the joins and the remaining closed intervals into the squares. Specifically, at the first step, the closed intervals $[0, 1/9]$, $[2/9, 1/3]$, $[2/3, 7/9]$, $[8/9, 1]$ are mapped into the squares S_1 to S_4 with the endpoints going into the appropriate corners and the open intervals $(1/9, 2/9)$, $(1/3, 2/3)$, $(7/9, 8/9)$ linearly onto the joins (without beginning points and endpoints) from S_1 to S_2, S_2 to S_3, and S_3 to S_4. The process is continued *ad infinitum* to obtain a continuous bijective map from $\mathbf{I}$ onto $\bigcap_{n=1}^{\infty} A_n$, where A_n denotes the set consisting of the 4^n squares and $4^n - 1$ joins that are obtained at the nth iteration. As in the case of Osgood's example, $\bigcap_{n=1}^{\infty} A_n$ is Lebesgue measurable with measure $\lim_{n \to \infty} J(A_n) = 1 - \lambda > 0$.

While every part of Knopp's Osgood curve is again an Osgood curve, this is not the case for the examples by Osgood and Lance and Thomas because of the presence of joins. Because of this, Knopp leveled in [4], p. 109, footnote 2, some justifiable criticism at Osgood's construction. (In the same footnote, he dismisses an attempt by Sierpiński in [15] to construct a curve without this shortcoming as too complicated.) This criticism applies to the construction by Lance and Thomas as well, and it could be viewed as a throwback to Osgood's original attempt, were it not for the fact that a slight modification of their construction yields Lebesgue's space-filling curve as the limit.

Lebesgue's space-filling curve is defined on the Cantor set

$$\Gamma = \{0_3 2a_1 2a_2 2a_3 \ldots : a_j = 0 \text{ or } 1\} \tag{1}$$

by

$$x = 0_2 a_1 a_3 a_5 \ldots, \qquad y = 0_2 a_2 a_4 a_6 \ldots,$$

and on the complement $\Gamma^c = [0, 1] \setminus \Gamma$ of the Cantor set by linear interpolation (see [6] or [11]). If one replaces the joins in Figure 16-5 by the ones in Figure 16-6 and starts with the lower left corner, one obtains Lebesgue's space-filling curve as the limiting arc as $\lambda \to 0$. This may be seen as follows: By our construction, the interval $[0, 1/9]$ (all numbers of the form $0_3 00a_3 a_4 a_5 \ldots$) is mapped into S_1 in Figure 16-7 and the points of S_1 have coordinates $(0_2 0b_2 b_3 b_4 \ldots, 0_2 0c_2 c_3 c_4 \ldots)$. The interval $[2/9, 1/3]$ (all numbers $0_3 02a_3 a_4 \ldots$) is mapped into S_2 with points $(0_2 0b_2 b_3 b_4 \ldots, 0_2 1c_2 c_3 c_4 \ldots)$, the interval $[2/3, 7/9]$ (all numbers $0_3 20a_3 a_4 a_5 \ldots$) into S_3 with $(0_2 1b_2 b_3 b_4 \ldots, 0_2 0c_2 c_3 c_4 \ldots)$ and $[8/9, 1]$ (all numbers $0_3 22a_3 a_4 a_5 \ldots$) into S_4 with $(0_2 1b_2 b_3 b_4 \ldots, 0_2 1c_2 c_3 c_4 \ldots)$. The process is to be repeated with the intervals $[0, 1/81]$,

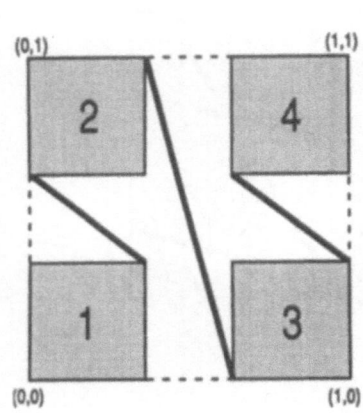

FIGURE 16-6 *Recursion operator leading to Lebesgue's curve.*

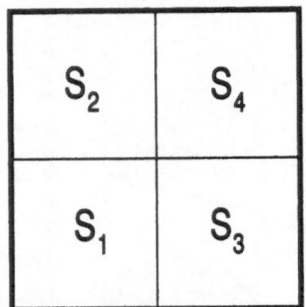

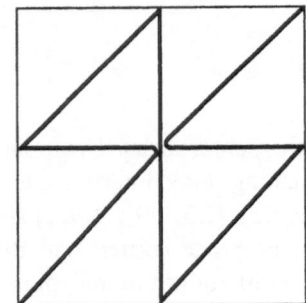

FIGURE 16-7 *Image of the Cantor set.*

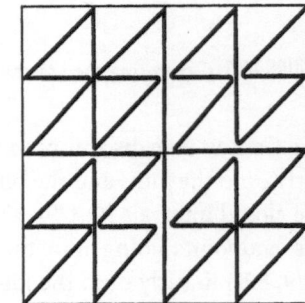

FIGURE 16-8 *Approximating Polygons for Lebesgue's space-filling*

[2/81, 1/27], ..., [80/81, 1] (all numbers of the type $0_3 0000a_5a_6a_7\ldots, 0_3 0002a_5a_6a_7\ldots, \ldots,$ $0_3 2222a_5a_6a_7\ldots$) and the squares $\mathbf{S}_{ij}$ within each of the squares $\mathbf{S}_i$ ($i, j = 1, 2, 3, 4$) with points $(0_2 00b_3b_4b_5\ldots, 0_2 00c_3c_4c_5\ldots)$, $(0_2 00b_3b_4b_5\ldots, 0_2 01c_3c_4c_5\ldots), \ldots,$ $(0_2 11b_3b_4b_5\ldots,$ $0_2 11c_3c_4c_5\ldots)$, etc. This process, continued *ad infinitum*, demonstrates that the mapping satisfies (1). Because $t = 1/3 = 0_2 0\bar{2}$ is mapped by (1) into the point $(1/2, 1)$ and $t = 2/3 = 0_3 2$ into $(1/2, 0)$, the limiting position of the join from the exit point of the second square in Figure 16-6 to the third square, as $\lambda \to 0$, represents the linear interpolation on $(1/3, 2/3) \subset \Gamma^c$; because $t = 1/9 = 0_3 00\bar{2}$ is mapped into $(1/2, 1/2)$ and $t = 2/9 = 0_3 02$ into $(0, 1/2)$, the limiting position of the join from the first square to the second square represents the linear interpolation on $(1/9, 2/9) \subset \Gamma^c$; because $t = 7/9 = 0_3 20\bar{2}$ has the image $(1, 1/2)$ and $t = 8/9 = 0_3 22$ the image $(1/2, 1/2)$, the limiting position of the join from the third square to the fourth square represents the linear interpolation on $(7/9, 8/9) \subset \Gamma^c$, etc. (See also Figure 16-6.) Repeating this argument *ad infinitum* reveals the limiting positions of the joins to represent the linear interpolation on Γ^c as called for by Lebesgue's definition.

We can now utilize this geometric generation of the Lebesgue curve to construct approximating polygons as follows: In each square, we join entry point and exit point by a straight line (diagonal) and leave the joins as they are, as we have indicated in Figure 16-8 for the first two steps. (In our illustration, we have rounded off some corners to prevent the polygon from bumping into itself and obscuring its progression.) These polygons are approximating polygons in the conventional sense because within each square, the distance from the polygon to the Lebesgue curve is bounded above by the length of the diagonal of the square, namely, by $2^{1/2-n}$, and the polygons coincide with the Lebesgue curve along the joins. Hence, they form a sequence that converges uniformly to the Lebesgue curve, the continuity of which is thus established.

In conclusion, let us note that Osgood's limiting curve, namely, the Peano curve, and Knopp's limiting curve, namely, the Sierpiński curve, are nowhere differentiable, whereas the limiting curve of the Lance–Thomas family is, just as the Lebesgue curve, differentiable a.e.

References

1. G. Cantor, Ein Beitrag zur Mannigfaltigkeitslehre, *Crelle J.* 84 (1878), 242–258.
2. A. Devinatz, *Advanced Calculus,* New York: Holt, Rinehart, Winston (1968), 253.
3. D. Hilbert, Ueber die stetige Abbildung einer Linie auf ein Flaechenstueck, *Math. Ann.* 38 (1891), 459–460.
4. K. Knopp, Einheitliche Erzeugung und Darstellung der Kurven von Peano, Osgood und von Koch, *Arch. Math. Phys.* 26 (1917), 103–115.
5. T. Lance and E. Thomas, Arcs with positive measure and a space-filling curve, *Amer. Math. Monthly* 98 (1991), 124–127.
6. H. Lebesgue, *Lecons sur l'Intégration et la Recherche des Fonctions Primitives,* Paris: Gauthier-Villars (1904), 44–45.
7. E. H. Moore, On certain crinkly curves, *Trans. Amer. Math. Soc.* 1 (1900), 72–90.
8. E. Netto, Beitrag zur Mannigfaltigkeitslehre, *Crelle J.* 86 (1879), 263–268.
9. W. F. Osgood, A Jordan curve of positive area, *Trans. Amer. Math. Soc.* 4 (1903), 107–112.
10. G. Peano, Sur une courbe qui remplit toute une aire plane, *Math. Ann.* 36 (1890), 157–160.
11. H. Sagan, Some reflections on the emergence of space-filling curves, *Franklin J.* 328 (1991), 419–430.
12. H. Sagan, On the geometrization of the Peano curve and the arithmetization of the Hilbert curve, *Int. J. Math. Educ. Sci. Technol.* 23 (1992), 403–411.
13. H. Sagan, Approximating polygons for the Sierpiński–Knopp space-filling curve, *Bull. Acad. Sci. Polon.* 40 (1992), 19–29.
14. W. Sierpiński, Sur une nouvelle courbe continue qui remplit tout une aire plane, *Bull. Acad. Cracovie (Sci. Mat. Nat. Serie A)* (1912), 462–478.
15. W. Sierpiński, Sur une courbe non quarrable, *Bull. Acad. Cracovie (Sci. Mat. Nat. Serie A)* (1913), 254–263.

17

Sophus Lie and Harmony in Mathematical Physics, on the 150th Anniversary of His Birth

Nail H. Ibragimov

"The extraordinary significance of Lie's work for the general development of geometry can not be overstated; I am convinced that in years to come it will grow still greater"—so wrote Felix Klein [13] in his nomination of the results of Sophus Lie on the group-theoretic foundations of geometry to receive the N. I. Lobachevskii prize. This prize was established by the Physical-Mathematical Society of the Imperial University of Kazan in 1895 and was to recognize works on geometry, especially non-Euclidean geometry, chosen by leading specialists. The first three prizes awarded were to the following:

1897: S. Lie   (Nominator: F. Klein)
1900: W. Killing   (Nominator: F. Engel)
1904: D. Hilbert   (Nominator: H. Poincaré).

There can be no doubt that the work of Lie in differential equations merits equally high evaluation. One of Lie's striking achievements in this domain was the discovery that the majority of the known methods of integration, which until then had seemed artificial and not intrinsically related to one another, could be introduced all together by means of group theory. Further, Lie gave a classification of ordinary differential equations of arbitrary order in terms of the admitted group, thereby identifying the full set of equations which could be integrated or reduced to lower-order equations by group-theoretic considerations. But these and a rich store of other results of his did not lend themselves to popular expositions and remained for a long time the special preserve of a few. Today we find that this is the case with methods of solution of the problems of

Friedrich Engel

Volume 16, No. 1 (Winter 1994), 20-28

mathematical physics. Many of them have a group-theoretic nature yet are taught as though they were the result of a lucky guess.

It was my good fortune to get interested in applications of groups to differential equations at the very beginning of my university work, and to write my first paper under the direction of Professor L. V. Ovsiannikov, who has done so much to awaken interest in this discipline and establish it as a contemporary scientific field. In my later work I saw over and over how effective a tool Lie theory is for solving complicated problems. It significantly widens and sharpens the intuitive notion of symmetry, supplies concrete methods to apply it, guides one to the proper formulation of problems, and often discloses possible approaches to solving them.

This article presents my view of the role of Lie group theory in mathematical physics, drawing on parts of some of my lectures over the years at Moscow University and Moscow Institute of Physics and Technology.

His Life Story

Marius Sophus Lie was born 17 December 1842 in the town of Nordfjordeid, Norway, the sixth and youngest child of the Lutheran pastor Johann Herman Lie. He studied in Christiania (now Oslo) from 1857, first in gymnasium and then (1859-1865) at the University. Among the events of Lie's life which set his creative course, these stand out: his independent study in 1868 of the geometric works of Chasles, Poncelet, and Plücker; his travels in Germany and France in 1869-1870; his contacts there with Felix Klein, Chasles, Jordan, and Darboux; and his close friendship with Klein, leading to a long collaboration. Lie worked at the University of Christiania from 1872 to 1886, then from 1886 to 1898 at Leipzig. He died 18 February 1899 in Christiania.

The life and intellectual development and works of the greatest Norwegian mathematician are described in reminiscences of his colleagues and later biographies (see, for example, [7, 22, 27, 29], and references therein). I call special attention to the painstaking introduction of F. Engel to Lie's Collected Works [21]. These give detailed insight into the essence of Lie's ideas and a picture of him as a person.

Symmetry of Differential Equations

The notion of differential equations really has two components. For an ordinary first-order differential equation, for example, it is necessary

1. to specify a surface $F(x, y, y') = 0$ in the space of the three variables x, y, y'; we will call this surface the *skeleton* of the differential equation;
2. to define the class of solutions; for example, a smooth solution is a continuously differentiable function $\varphi(x)$ such that the curve

$$y = \varphi(x), \quad y' = \frac{\partial \varphi(x)}{\partial x}$$

lies on the surface, i.e.,

$$F\left(x, \varphi(x), \frac{\partial \varphi(x)}{\partial x}\right) = 0$$

identically in x; going over to discontinuous or generalized solutions (keeping the same skeleton) changes the situation altogether.

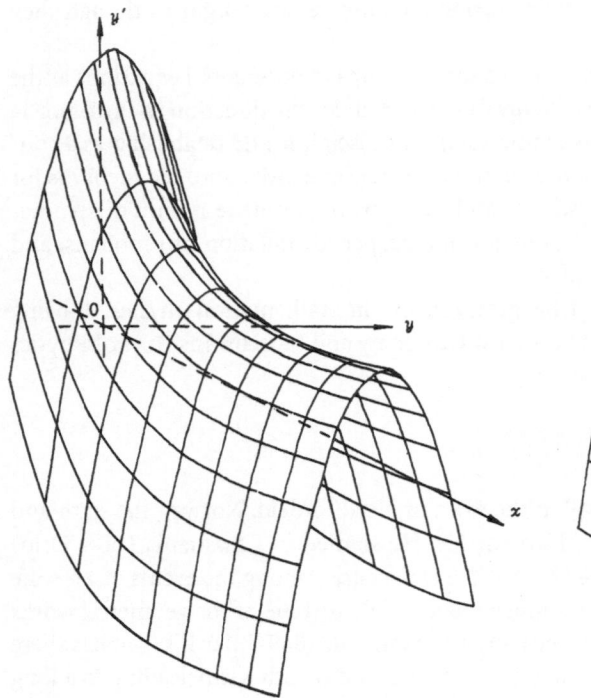

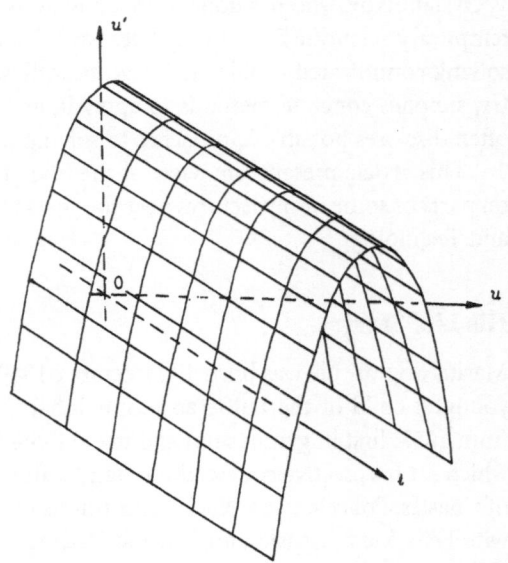

FIGURE 17-1 *The skeleton of the Riccati equation* $y' + y^2 - 2/x^2 = 0$ *is a surface invariant under the group of inhomogeneous deformations* $\bar{x} = xe^a$, $\bar{y} = ye^{-a}$, $\bar{y}' = y'e^{-2a}$.

FIGURE 17-2 *The skeleton of the equation* $u' + u^2 - u - 2 = 0$, *obtained from the Riccati equation* $y' + y^2 - 2/x^2 = 0$ *by the change of variables* $t = \ln x$, $u = xy$.

A decisive move in integrating differential equations is simplifying the skeleton by means of a suitable change of variables. For this purpose, one uses the *symmetry group* of the differential equation (or its *admissible group*), defined as the group of transformations of the (x, y)-plane whose extensions to the derivatives of $y', \ldots$ leave the equation's skeleton invariant.

EXAMPLE. The Riccati equation $y' + y^2 - 2/x^2 = 0$ admits the group of transformations $\bar{x} = xe^a$, $\bar{y} = ye^{-a}$, for the equation's skeleton (Figure 17-1) is invariant under the inhomogeneous stretching $\bar{x} = xe^a$, $\bar{y} = ye^{-a}$, $\bar{y}' = y'e^{-2a}$ which is obtained by extending the transformations of the group to the first derivative y'. The substitution $t = \ln x$, $u = xy$ leads to the differential equation $u' + u^2 - u - 2 = 0$. Thus, it straightens out the skeleton of the original Riccati equation, taking it to a parabolic cylinder (Fig. 17-2); concomitantly, the stretchings are replaced by a group of translations $\bar{t} = t + a$, $\bar{u} = u$, and $\bar{u}' = u'$.

Group Classification

In a short communication to the Scientific Society of Göttingen (3 December 1874), I gave, among other things, a listing of all continuous transformation groups in two variables x, y, and specially emphasized that this might be made the basis of a classification and rational integration theory of all differential equations $f(x, y, y', \ldots, y^{(m)}) = 0$ admitting a continuous group of transformations. The great program sketched there I have subsequently carried out in detail. (S. Lie [16], p. 187)

Table 17-1 **Lie's group classification of second-order equations.**

GROUP	BASIS OF THE LIE ALGEBRA	EQUATION
G_1	$X_1 = \frac{\partial}{\partial x}$	$y'' = f(y, y')$
G_2	$X_1 = \frac{\partial}{\partial x}, \quad X_2 = \frac{\partial}{\partial y}$	$y'' = f(y')$
	$X_1 = \frac{\partial}{\partial y}, \quad X_2 = x\frac{\partial}{\partial x} + y\frac{\partial}{\partial y}$	$y'' = \frac{1}{x} f(y')$
	$X_1 = \frac{\partial}{\partial x} + \frac{\partial}{\partial y}, \quad X_2 = x\frac{\partial}{\partial x} + y\frac{\partial}{\partial y}, \quad X_3 = x^2\frac{\partial}{\partial x} + y^2\frac{\partial}{\partial y}$	$y'' + 2\left(\frac{y' + Cy'^{3/2} + y'^2}{x-y}\right) = 0$
G_3	$X_1 = \frac{\partial}{\partial x}, \quad X_2 = 2x\frac{\partial}{\partial x} + y\frac{\partial}{\partial y}, \quad X_3 = x^2\frac{\partial}{\partial x} + xy\frac{\partial}{\partial y}$	$y'' = Cy^{-3}$
	$X_1 = \frac{\partial}{\partial x}, \quad X_2 = \frac{\partial}{\partial y}, \quad X_3 = x\frac{\partial}{\partial x} + (x + y)\frac{\partial}{\partial y}$	$y'' = Ce^{-y'}$
	$X_1 = \frac{\partial}{\partial x}, \quad X_2 = \frac{\partial}{\partial y}, \quad X_3 = x\frac{\partial}{\partial x} + ky\frac{\partial}{\partial y}$	$y'' = Cy'^{(k-2)(k-1)},$ $k \neq 0, \frac{1}{2}, 1, 2$
G_8	$X_1 = \frac{\partial}{\partial x}, \quad X_2 = \frac{\partial}{\partial y}, \quad X_3 = x\frac{\partial}{\partial y}, \quad X_4 = x\frac{\partial}{\partial x}, \quad X_5 = y\frac{\partial}{\partial x}$	$y'' = 0$
	$X_6 = y\frac{\partial}{\partial y}, \quad X_7 = x^2\frac{\partial}{\partial x} + xy\frac{\partial}{\partial y}, \quad X_8 = xy\frac{\partial}{\partial x} + y^2\frac{\partial}{\partial y}$	

This and the next two sections give some of the main results of implementing the program, as it applies to ordinary second-order differential equations. The restriction to second order is motivated not by anything essential about the method but by a desire to concentrate on concrete cases and give brief but definitive statements.

For second-order equations, the group classification [16] looks especially simple. The second-order classification result is stated briefly and explicitly in [18], §3, and appears here as Table 17-1. Remember that Lie carried out his classification in the complex domain, using complex substitutions and complex bases of algebras as needed. For example, the equation

$$y'' = C(1 + y'^2)^{3/2} e^{q \arctan y'}, \quad C, q = \text{const},$$

admits a 3-dimensional Lie algebra with basis

$$X_1 = \frac{\partial}{\partial x}, \quad X_2 = \frac{\partial}{\partial y}, \quad X_3 = (qx + y)\frac{\partial}{\partial x} + (qy - x)\frac{\partial}{\partial y}.$$

No real substitution takes this to any of the equations of Table 17-1. But it is transformed to the equation

$$\bar{y}'' = C\bar{y} \cdot \frac{k - 2}{k - 1}, \quad k = \frac{q + 1}{q - 1}$$

by the complex substitution $\bar{x} = \frac{1}{2}(y - ix), \bar{y} = \frac{1}{2}(y + ix)$.

Algorithm of Integration

I noticed that the majority of ordinary differential equations which were integrable by the old methods were left invariant under certain transformations, and that these integration methods consisted in using that property. Once I had thus represented many old integration methods from a common viewpoint, I set myself the natural problem: to develop a general theory of integration for all ordinary differential equations admitting finite or infinitesimal transformations. (S. Lie [17], p. iv)

Table 17-2 **Algorithm for integrating a second-order equation using a 2-dimensional Lie algebra.**

STEP	OPERATION	RESULT
1	Compute the admitted Lie algebra L_r.	A basis of L_r: $X_1, \ldots, X_r$
2	If $r = 2$, go to the next step; if $r > 2$, then distinguish any 2-dimensional subalgebra L_2 of L_r. (If $r = 1$, the order of the equation may be lowered; if $r = 0$, the group method is not applicable.)	A basis of L_2: X_1, X_2
3	Determine the type of the algebra L_2 obtained by Table 17.3. For this one computes the commutator $[X_1, X_2]$ of X_1 and X_2 and their pseudoscalar product $X_1 \vee X_2$; if $[X_1, X_2]$ is neither 0 nor X_1, then choose a new basis X_1', X_2', such that $[X_1', X_2'] = X_1'$.	Reduction of structure to a canonical form from Table 17.3
4	Bring the basis of L_2 into agreement with Table 17.3 by going over to canonical variables x, y. Rewrite the equation in canonical variables and integrate it.	Finding the integrating change of variables
5	Rewrite the solution in terms of the original variables.	Solution of the equation

If a second-order equation admits a Lie algebra of dimensionality $r \geq 2$, then it can be integrated by a group-theoretic quadrature method. This can be done in various ways, one of which is given in Table 17-2. It is based on the simple fact that in the complex case any Lie algebra of dimensionality $r > 2$ has a distinguished 2-dimensional subalgebra. But the structure of a 2-dimensional Lie algebra with basis

$$X_\alpha = \xi_\alpha(x, y)\frac{\partial}{\partial x} + \eta_\alpha(x, y)\frac{\partial}{\partial y}, \quad \alpha = 1, 2,$$

can be described simply in terms of the commutator

$$[X_1, X_2] = X_1 X_2 - X_2 X_1$$

and the pseudo-scalar product

$$X_1 \vee X_2 = \xi_1 \eta_2 - \eta_1 \xi_2.$$

This description is given in Table 17-3; for details, see [4, 6, 11, 12, 19, 23, 32].

Table 17-3 **Canonical form of 2-dimensional Lie algebras and invariant second-order equations.**

TYPE	L_2 STRUCTURE	BASIS OF L_2 IN CANONICAL VARIABLES		EQUATION
I	$[X_1, X_2] = 0$, $X_1 \vee X_2 \neq 0$	$X_1 = \frac{\partial}{\partial x}$,	$X_2 = \frac{\partial}{\partial y}$	$y'' = f(y')$
II	$[X_1, X_2] = 0$, $X_1 \vee X_2 = 0$	$X_1 = \frac{\partial}{\partial y}$,	$X_2 = x\frac{\partial}{\partial y}$	$y'' = f(x)$
III	$[X_1, X_2] = X_1$, $X_1 \vee X_2 \neq 0$	$X_1 = \frac{\partial}{\partial y}$,	$X_2 = x\frac{\partial}{\partial x} + y\frac{\partial}{\partial y}$	$y'' = \frac{1}{2}f(y')$
IV	$[X_1, X_2] = X_1$, $X_1 \vee X_2 = 0$	$X_1 = \frac{\partial}{\partial y}$,	$X_2 = y\frac{\partial}{\partial y}$	$y'' = f(x)y'$

EXAMPLE.   Let us apply the group algorithm to the equation

$$y'' = \frac{y'}{y^2} - \frac{1}{xy}.$$

First step: finding the admissible algebra. This is done by use of the so-called *determining equation.* It turns out, as a consequence of standard and straightforward computations, that our equation admits the Lie algebra L_2 with basis

$$X_1 = x^2 \frac{\partial}{\partial x} + xy \frac{\partial}{\partial y}, \quad X_2 = -x \frac{\partial}{\partial x} - \frac{y}{2} \frac{\partial}{\partial y}.$$

From Table 17-2 we see that we can pass at once to the third stage.
   Third step: finding the type of the algebra L_2. We have

$$[X_1, X_2] = X_1, \quad X_1 \vee X_2 = \tfrac{1}{2} x^2 y \neq 0.$$

Consequently, the algebra L_2 belongs to type III of Table 17-3.
   Fourth step: finding the integrating change of variables. From the equations

$$X_1(t) = 0, \quad X_1(u) = 1,$$

we find the substitution

$$t = \frac{y}{x}, \quad u = -\frac{1}{x},$$

taking the 1-parameter group generated by the operator X_1 (group of projective transformations) to the group of translations in u. After this substitution, the basis of L_2 takes the form

$$X_1 = \frac{\partial}{\partial u}, \quad X_2 = \frac{t}{2} \frac{\partial}{\partial t} + u \frac{\partial}{\partial u}$$

and coincides (up to the inessential coefficient $1/2$ in X_2) with the canonical basis for type III in Table 17-3. Here we exclude solutions of the form $y = Cx$. This substitution has put the equation into the integrable form

$$\frac{u''}{u'^2} + \frac{1}{t^2} = 0.$$

Solving it gives

$$u = -\frac{t^2}{2} + C \quad \text{and} \quad u = \frac{t}{C_1} + \frac{1}{C_1^2} \ln |C_1 t - 1| + C_2.$$

Fifth step: finding the solution in the given variables. Now replace t, u in the foregoing formulas by their expressions, and recall the excluded special solutions $y = Cx$. We obtain the general solution of the given second-order differential equation in the form

$$y = Cx, \quad y = \pm\sqrt{2x + Cx^2}, \quad C_1 y + C_2 x + x \ln \left| C_1 \frac{y}{x} - 1 \right| + C_1^2 = 0.$$

Linearization

In the study of ordinary differential equations it is useful to have simple tests for linearizability. Summing up Lie's results on this question, we can state the following theorem ([16], Part III, §1; see also [11, 12]).

THEOREM 1.   The following are equivalent:
(i)   the second-order ordinary differential equation

$$y'' = f(x, y, y')$$ (1)

can be linearized by change of variables;
(ii)   Equation (1) has the form (2):

$$y'' + F_3(x, y)y'^3 + F_2(x, y)y'^2 + F_1(x, y)y'^1 + F(x, y) = 0$$ (2)

with coefficients F_3, F_2, F_1, F satisfying the compatibility conditions of the auxiliary system

$$\frac{\partial z}{\partial x} = z^2 - Fw - F_1 z + \frac{\partial F}{\partial y} + FF_2,$$

$$\frac{\partial z}{\partial y} = -zw + FF_3 - \frac{1}{3}\frac{\partial F_1}{\partial x} + \frac{2}{3}\frac{\partial F_1}{\partial y},$$ (3)

$$\frac{\partial w}{\partial x} = zw - FF_3 - \frac{1}{3}\frac{\partial F_1}{\partial y} + \frac{2}{3}\frac{\partial F_1}{\partial x},$$

$$\frac{\partial w}{\partial y} = -w^2 + F_2 w + F_3 z + \frac{\partial F_3}{\partial x} - F_1 F_3;$$

(iii) Equation (1) admits an 8-dimensional Lie algebra;
(iv) Equation (1) admits a 2-dimensional Lie algebra with basis X_1, X_2 such that

$$X_1 \vee X_2 = 0.$$ (4)

EXAMPLE 1.   The equation $y'' = e^{-y'}$ from Table 17-1 is not linearizable, for it does not have the form (2) in (ii).

EXAMPLE 2.   Suppose in Eq. (2) that $F_1 = F_2 = F_3 = 0$. Then Equations (3) take the form

$$z_x = z^2 - Fw + F_y, \quad z_y = -zw, \quad w_x = zw, \quad w_y = -w^2$$

and the compatibility condition $z_{xy} = z_{yx}$ gives $F_{yy} = 0$. Consequently, the equation $y'' + F(x, y) = 0$ having an $F(x, y)$ not already linear in y cannot be linearized.

EXAMPLE 3.   Let us see when the equation $y'' = f(y')$ in Table 17-1 can be linearized. According to (ii) of Theorem 1, it is required for linearizability that $f(y')$ be a polynomial of at most third degree, i.e., that the equation be of the form

$$y'' + A_3 y'^3 + A_2 y'^2 + A_1 y' + A_0 = 0$$ (5)

with constant coefficients A_i. When one writes out the auxiliary system (3) for Eq. (5), one easily sees it is compatible. Consequently, Eq. (5) is linearizable for arbitrary coefficients A_i.

EXAMPLE 4.   Now consider this equation from Table 17-1:

$$y'' = \frac{1}{x} f(y').$$

Linearizability requires it to be of the form (2), i.e.,

$$y'' + \frac{1}{x}(A_3 y'^3 + A_2 y'^2 + A_1 y' + A_0) = 0 \tag{6}$$

with constant coefficients A_i. The compatibility conditions of its auxiliary system (3) come out to

$$A_2(2 - A_1) + 9 A_0 A_3 = 0, \quad 3 A_3(1 + A_1) - A_2^2 = 0.$$

Setting $A_3 = -a, A_2 = -b$, this gives

$$A_1 = -\left(1 + \frac{b^2}{3a}\right), \quad A_0 = -\left(\frac{b}{3a} + \frac{b^3}{27a^2}\right).$$

Hence, Eq. (6) can be linearized if and only if it is of the form

$$y'' = \frac{1}{x}\left[ay'^3 + by'^2 + \left(1 + \frac{b^2}{3a}\right)y' + \frac{b}{3a} + \frac{b^3}{27a^2}\right]. \tag{7}$$

It is convenient to find the linearizing substitution from assertion (*iv*). Let us do this for Eq. (7) in the case $a = 1, b = 0$:

$$y'' = \frac{1}{x}(y' + y'^3). \tag{8}$$

This equation admits the algebra L_2 with basis

$$X_1 = \frac{1}{x}\frac{\partial}{\partial x}, \quad X_2 = \frac{y}{x}\frac{\partial}{\partial x}, \tag{9}$$

satisfying condition (4). This algebra L_2 belongs to type II of Table 17-3, and the linearizing substitution is obtained by going over to the canonical variables $\bar{x} = y, \bar{y} = x^2/2$, relative to which Eq. (9) take the form $X_1 = \partial/\partial\bar{y}, X_2 = \bar{x}(\partial/\partial\bar{y})$. Aside from the particular solutions $y = $ const, this transforms Eq. (8) into $\bar{y}'' + 1 = 0$.

Invariant Solutions

Special types of exact solutions, now widely known as *invariant solutions,* have long been used to advantage on concrete problems. They have grown familiar in mathematics, mechanics, and physics even before there was any group theory, acquiring the status of folklore. Lie [20] elucidated their group-theoretical meaning and studied the possibility of integrating partial differential equations when the group is sufficiently rich (see [20], Chaps. III and IV).

Subsequently, group theory made it possible to clarify, sharpen, and extend many intuitive ideas, and incorporate the method of invariant solutions as an essential component of modern group analysis. It was exactly by the notion of invariant solution that group theory was able to transfer its area of application from ordinary differential equations to the problems of mathematical physics, especially thanks to the works [1, 5, 24, 26, 31].

EXAMPLE.   Consider the equation

$$y'' = y^{-3}$$

from Table 17-1, admitting a 3-parameter group. Its solution,

$$y = \sqrt{1 + x^2},$$

is invariant under a 1-parameter group, whose generator is

$$X_1 + X_3 = (1 + x^2)\frac{\partial}{\partial x} + xy\frac{\partial}{\partial y}.$$

Subjecting this invariant solution to the transformations of the 3-parameter admissible group gives

$$y = \left[C_1 x^2 + 2\sqrt{C_1 C_2 - 1}\,x + C_2\right]^{1/2},$$

which is the general solution. This means every solution of this equation is invariant under some 1-parameter subgroup of the admissible 3-parameter group (details in [12]).

The Invariance Principle in the Problems of Mathematical Physics

When we pass from ordinary to partial differential equations, it becomes impossible (with rare exceptions) and anyway not particularly useful to write out general solutions. But mathematical physics in any case seeks only those solutions that satisfy given side conditions—initial conditions, boundary conditions, etc. In solving many problems of mathematical physics it is advantageous to use the following semi-empirical rule, which is rigorously based only in certain cases (see [5, §89; 25, §29; 28]).

THE INVARIANCE PRINCIPLE.   If a boundary-value problem is invariant under a group, then we should seek a solution among functions invariant under this group.

Invariance of a boundary-value problem means invariance of the differential equation, also of the manifold where the data are given, and also of the data themselves.

When invariance of the boundary conditions is lost (as often happens), the principle stated can be put to use in other ways. This is what happens, for example, in the method of the majorant in the proof of the Cauchy-Kovalevskaya theorem (on the method of the invariant majorant, see [9]).

Another example is the Riemann method, which reduces the Cauchy problem with arbitrary (hence not invariant) data to the special Goursat problem, which is invariant and can be solved by the invariance principle. Here, we only indicate briefly the essence of this approach, referring for details to [12].

The Group Approach to Riemann's Method

This section is an attempt at synthesis, at combining Riemann's method [30] of integrating linear hyperbolic second-order equations with Lie's group classification [16] of such equations. Also important here is the invariant formulation (in terms of Laplace invariants) of Lie's results, as given by Ovsiannikov [25].

Riemann's method reduces the problem of integrating the equation

$$L[u] \equiv u_{xy} + a(x, y)u_x + b(x, y)u_y + c(x, y)u = f(x, y) \tag{10}$$

to the construction of an auxiliary function v satisfying the adjoint equation with given conditions on the characteristics:

$$L^*[v] = 0, \quad v\,|_{x=x_0} = \exp \int_{y_0}^{y} a(x_0, \eta)d\eta, \quad v\,|_{y=y_0} = \exp \int_{x_0}^{x} b(\xi, y_0)\,d\xi. \tag{11}$$

Once v is found, the solution of the Cauchy problem for Eq. (10) with data on an arbitrary non-characteristic curve is obtained by the known integral formula. The function v is called the *Riemann function*, and the boundary-value problem (11) which determines it is called the *characteristic Cauchy problem* or *Goursat problem*.

The quantities

$$h = a_x + ab - c, \quad k = b_y + ab - c$$

are called the *Laplace invariants* for Eq. (10). They remain unaltered by any linear transformation of u with variable coefficients, with x, y not being transformed. In contrast, the quantities

$$p = \frac{k}{h}, \quad q = \frac{1}{h}(\ln h)_{xy} \tag{12}$$

are invariant under the general equivalence transformation $\bar{x} = \alpha(x)$, $\bar{y} = \beta(y)$, $\bar{u} = \lambda(x, y)u$ of the homogeneous equation (10). These invariants are useful for the classification of Eq. (10) according to the dimensionality of the admissible group. Namely, *the homogeneous equation* (10) $(f = 0)$ *admits a 4-dimensional Lie algebra* [more precisely, the quotient algebra with respect to the ideal generated by $X = \phi(x, y)$ with $\phi(x, y)$ an arbitrary solution of Eq. (10)] *if the quantities* (12) *are constant; whereas if at least one of them fails to be constant, then Eq.* (10) *can admit at most a 2-dimensional algebra.* For proof, see [25], §9.6. Using this result, one proves the following theorem (see [11] or [12]):

THEOREM 2. Assume that Eq. (10) has constant invariants (12). Then the Goursat problem (11) admits a 1-parameter group. Therefore, the invariance principle is applicable, and the Riemann function can be found from a second-order ordinary differential equation.

EXAMPLE 1.   For the telegrapher's equation $u_{xy} + u = 0$ we have $p = 1, q = 0$. Hence, Theorem 2 applies. The Goursat problem (11), namely,

$$v_{xy} + v = 0, \quad v\mid_{x=x_0} = 1, \quad v\mid_{y=y_0} = 1 \qquad (13)$$

must by Theorem 2 admit a 1-parameter group with generator

$$X = (x - x_0)\frac{\partial}{\partial x} - (y - y_0)\frac{\partial}{\partial y}.$$

Functionally independent invariants of this group are v and $z = (x - x_0)(y - y_0)$. Therefore, the invariant solution has the form $v = V(z)$, and after substitution in Eqs. (13), we get a form of Bessel's equation: $zV'' + V' + V = 0$ with condition $V(0) = 1$. Consequently, for the telegrapher's equation, a Riemann function is the Bessel function J_0.

EXAMPLE 2.   Riemann ([30], §9) applied the technique he introduced to the equation

$$u_{xy} + \frac{\ell}{(x+y)^2}u = 0, \quad \ell = \text{const} \neq 0. \qquad (14)$$

In the corresponding problem (11), the condition on the characteristics has the form

$$v\mid_{x=x_0} = 1, \quad v\mid_{y=y_0} = 1. \qquad (14')$$

Riemann reduced the problem (14), (14′) to an ordinary differential equation (leading to the special hypergeometric function of Gauss), considering v as a function of the variable

$$z = \frac{(x - x_0)(y - y_0)}{(x_0 + y_0)(x + y)}. \qquad (15)$$

Here is how this looks from the group point of view. The invariants (12) of Eq. (14) are $p = 1$, $q = 2/\ell$. Hence, Theorem 2 applies here. Solving the determining equation, we find the operator

$$X = (x - x_0)(x + y_0)\frac{\partial}{\partial x} - (y - y_0)(y + x_0)\frac{\partial}{\partial y},$$

admissible for the Goursat problem (14), (14′). Invariants for this operator are v and the quantity z given by Eq. (15). Therefore, the invariant solution has the form $v = V(z)$. This is just the invariant solution found by Riemann!

EXAMPLE 3.   Next take the equation

$$u_{xy} + \frac{\ell}{x+y}u = 0, \quad \ell = \text{const} \neq 0,$$

an "intermediate case" between the telegrapher's equation (13) and Eq. (14). The invariants (12) are $p = 1, q = 1/\ell(x + y)$. But q not being constant, Theorem 2 is not applicable.

A full catalog of equations to which Theorem 2 applies is in [11].

Fundamental Solutions

Keeping the same orientation as in the preceding section—the application of the invariance principle to boundary-value problems with arbitrary data by reduction to an invariant problem of a special form—let us see what Lie group theory can offer for the construction of fundamental solutions in the case of the three fundamental equations of mathematical physics. This natural line of development of group analysis, passing to the space of distributions, was sketched in [10], giving heuristic considerations and statement of the problems. Yurii Berest [3], my student, recently got remarkable results applying this method to wave equations in Riemannian manifolds with non-trivial conformal group. Some details of infinitesimal group techniques applicable to distributions may be found in [12].

The Laplace Equation. Let us consider the equation

$$\Delta u = \delta(x), \quad x \in \mathbb{R}^n, \tag{16}$$

for a fundamental solution as a boundary-value problem, where at a fixed point, the origin, a Δ-function singularity is given. This boundary problem is invariant under the group of rotations and dilations, generated by the operators

$$X_{ij} = x^j \frac{\partial}{\partial x^i} - x^i \frac{\partial}{\partial x^j}, \quad i, j = 1, \ldots, n,$$

$$Z = x^i \frac{\partial}{\partial x^i} + (2 - n)u \frac{\partial}{\partial u}.$$

A basis of the invariants of this group consists of the single function $J = u|x|^{n-2}$. According to the invariance principle, the fundamental solution is to be sought as an invariant solution, determined by the equation $J = \text{const}$. Thus,

$$u = C|x|^{2-n}. \tag{17}$$

Substituting Eq. (17) into Eq. (16), we find the value of the constant, $C = 1/(2 - n)\Omega_n$, where Ω_n is the measure of the surface of the unit sphere in n-space. Thus, *the fundamental solution was determined from the condition of invariance up to a constant multiple, and the differential equation served only for the normalization.*

The Heat Equation. The equation

$$u_t - \Delta u = \delta(t, x) \tag{18}$$

with n-dimensional Laplace operator in the space of variables x^i is invariant under the group of rotations, Galilei transformations, and dilations, which is generated by

$$X_{ij} = x^j \frac{\partial}{\partial x^i} - x^i \frac{\partial}{\partial x^j}, \quad Y_i = 2t \frac{\partial}{\partial t} - x^i u \frac{\partial}{\partial u},$$

$$Z = 2t \frac{\partial}{\partial t} + x^i \frac{\partial}{\partial x^i} - nu \frac{\partial}{\partial u}.$$

This group has the invariant

$$J = ut^{n/2} e^{|x|^2/4t}.$$

Therefore, an invariant solution has the form

$$u = Ct^{-n/2}e^{-|x|^2/4t}. \tag{19}$$

Equation (18) serves as a normalizing condition: Substitution of Eq. (19) into Eq. (18) yields the value of the constant, $C = (2\sqrt{\pi})^{-n}$.

The Wave Equation. For the equation

$$u_{tt} - \Delta u = \Delta(t, x), \tag{20}$$

the group of symmetries is generated by

$$X_{ij} = x^i \frac{\partial}{\partial x^j} - x^j \frac{\partial}{\partial x^i}, \quad Y_i = t\frac{\partial}{\partial x^i} + x^i \frac{\partial}{\partial t},$$

$$Z = t\frac{\partial}{\partial t} + x^i \frac{\partial}{\partial x^i} + (1-n)u\frac{\partial}{\partial u}, \quad i, j = 1, \dots, n.$$

The operators X_{ij} and Y_i generate the group of rotations and Lorentz transformations and have two invariants: u and $\tau = t^2 - |x|^2$. Therefore, an invariant solution is to be sought of the form $u = f(\tau)$. The condition of invariance under the group of dilations with generator Z takes the form

$$2\tau f'(\tau) + (n-1)f(\tau) = 0 \tag{21}$$

Let us look only at odd n, for in the case of even n we would have to use the method of balayage of Hadamard. Then, setting $n = 2m + 1$, $m = 0, 1, \dots$, we write Eq. (21) as $\tau f'(\tau) + mf(\tau) = 0$, which is known to have the general solution

$$f(\tau) = \begin{cases} C_1\theta(\tau) + C_2, & \text{for } m = 0 \\ C_1\delta^{(m-1)}(\tau) + C_2\tau^{-m}, & \text{for } m \neq 0. \end{cases}$$

Substitution of these formulas in Eq. (20) gives $C_1 = \frac{1}{2}\pi^{-m}$, $C_2 = 0$. In this way, the invariance principle yields the fundamental solution

$$u = \begin{cases} \frac{1}{2}\theta(t^2 - |x|^2), & n = 1 \\ \frac{1}{2}\pi^{(1-n)/2}\delta^{(n-3)/2}(t^2 - |x|^2), & n \geq 3, \end{cases}$$

where θ is the Heaviside function and $\delta^{(n-3)/2}$ is the derivative of order $(n-3)/2$ of the δ-function.

Kepler's Laws

The motion of a material point under a central force with potential $V = \alpha/|x|$ satisfies the obvious law of conservation of angular momentum $M = m\,(x \times v)$, where m is the mass of the particle and x and v are its coordinate and velocity, respectively. This conservation law, known in celestial mechanics as Kepler's Second Law, is a corollary of the invariance of the Lagrange equations of motion under the group of rotations and follows from Noether's theorem. Write the infinitesimal transformation of the rotation group with vector parameter $a = (a^1, a^2, a^3)$ in the form

$$\bar{x} = x + \delta x, \quad \delta x = x \times a. \tag{22}$$

Then from the group point of view, *Kepler's Second Law expresses the invariance of the problem under infinitesimal rotations* (22).

The Kepler problem is also invariant under the inhomogeneous dilation generated by the operator

$$X = 3t\frac{\partial}{\partial t} + 2x^i\frac{\partial}{\partial x^i}.$$

An invariant both of the rotation group and of this dilation is the quantity $J = t^2/r^3$. *The existence of this invariant is called in celestial mechanics Kepler's Third Law.*

Finally, the Kepler problem has a special group of symmetries, which in the notation of Eqs. (22), can be written

$$\bar{x} = x + \delta x, \quad \delta x = (x \times v) \times a + x \times (v \times a). \tag{23}$$

This 3-parameter group differs from ordinary Lie groups of point and contact transformations, being more general; it is called the *group of Lie-Bäcklund transformations* [2]. The computation of the symmetry group (23) is carried out in [9], p. 346. From Noether's theorem one gets a vector integral of the motion

$$A = v \times M + \alpha\frac{x}{|x|},$$

found first by Laplace [14]. Taking the scalar product of Laplace's vector A with the radius vector x, one readily infers that the orbit of Keplerian motion is an ellipse. This is Kepler's First Law. So from the group point of view, *Kepler's First Law expresses the invariance of the problem under the 3-parameter Lie-Bäcklund group with infinitesimal transformation* (23).

Thus, all three of Kepler's Laws of celestial mechanics have a group-theoretic nature.

Concluding Remarks

I could continue in this spirit, for there are many entertaining applications of Lie theory, and nowadays newly developed methods of group analysis are awaiting application. But I hope what I have said is enough to convince you that acquaintance with the classical foundations and modern group-theoretic methods has become an important part of the mathematical culture of anyone constructing and investigating mathematical models of natural problems. For this, one can go to the beautiful books of Lie, Bianchi, and so on, and more recent works (see the References).

In conclusion, I would like to carry over to Lie theory in mathematical physics Einer Hille's remark [8]: "I hail a [semi-]group when I see one and I seem to see them everywhere! Friends have observed, however, that there are mathematical objects which are not [semi-]groups."

A. V. Bäcklund

References

1. W. F. Ames, non-*linear Partial Differential Equations in Engineering*, Vols. I and II, New York: Academic Press (1965, 1972).
2. R. L. Anderson and N. H. Ibragimov, *Lie-Bäcklund Transformations in Applications*, Philadelphia: SIAM (1979).
3. Yu. Berest, Construction of fundamental solutions for Huygens equations as invariant solutions, *Dokl. Akad. Nauk SSR*, 317(4), 786-789 (1991).
4. L. Bianchi, *Lezioni sulla teoria dei groupi continui finiti di trasformazioni*, Pisa: Spoerri (1918).
5. G. Birkhoff, *Hydrodynamics*, Princeton, NJ: Princeton University Press (1950, 1960).
6. G. W. Bluman and S. Kumei, *Symmetries and Differential Equations*, New York: Springer-Verlag (1989).
7. T. Hawkins, Jacobi and the birth of Lie's theory of groups, *Arch. History Exact Sciences* 42(3), 187-278 (1991).
8. E. Hille, *Functional Analysis and Semi-groups*, New York: Amer. Math. Soc. (1948), preface.
9. N. H. Ibragimov, *Transformation Groups Applied to Mathematical Physics*, Dordrecht: D. Reidel (1985).
10. N. H. Ibragimov, *Primer on the Group Analysis*, Moscow: Znanie (1989).
11. N. H. Ibragimov, *Essays in the Group Analysis of Ordinary Differential Equations*, Moscow: Znanie (1991).
12. N. H. Ibragimov, Group analysis of ordinary differential equations and the invariance principle in mathematical physics, *Uspekhi Mat. Nauk*, 47 (1992), 89–156.
13. F. Klein, Theorie der Transformationsgruppen B. III, *Pervoe prisuzhdenie premii N. I. Lobachevskogo, 22 okt. 1897 goda*, Kazan: Tipo-litografiya Imperatorskago Universiteta (1898), pp. 10-28.
14. P. S. Laplace, *Mécanique céleste*, T. I. livre 2, Chap. III (1799).
15. S. Lie, Über die Integration durch bestimmte Integrale von einer Klasse linearer partieller Differentialgleichungen, *Arch. for Math.* VI (1881).
16. S. Lie, Klassifikation und Integration von gewöhnlichen Differentialgleichungen zwischen x, y, die eine Gruppe von Transformationen gestatten, *Arch. Math.* VIII, 187-453 (1883).
17. S. Lie, *Theorie der Transformationsgruppen, Bd. 1 (Bearbeitet unter Mitwirkung von F. Engel)*, Leipzig: B. G. Teubner (1888).
18. S. Lie, Die infinitesimalen Berührungstransformationen der Mechanik, *Leipz. Ber.* (1889).
19. S. Lie, *Vorlesungen über Differentialgleichungen mit bekannten infinitesimalen Transformationen (Bearbeitet und herausgegeben von Dr. G. Scheffers)*, Leipzig: B. G. Teubner (1891).
20. S. Lie, Zur allgemeinen Theorie der partiellen Differentialgleichunen beliebiger Ordnung, *Leipz. Ber.* I, 53-128 (1895).
21. S. Lie, *Gesammelte Abhandlungen*, Bd. 1-6, Leipzig-Oslo.
22. M. Noether, Sophus Lie, *Math. Annalen* 53, 1-41 (1900).
23. P. J. Olver, *Applications of Lie Groups to Differential Equations*, New York: Springer-Verlag (1986).
24. L. V. Ovsiannikov, Group properties of differential equations, Novosibirsk: USSR Academy of Science, Siberian Branch (1962).
25. L. V. Ovsiannikov, *Group Analysis of Differential Equations*, Boston: Academic Press (1982).
26. A. Z. Petrov, *Einstein Spaces*, Oxford: Pergamon Press (1969).
27. E. M. Polischuk, *Sophus Lie*, Leningrad: Nauka (1983).
28. V. V. Pukhnachev, Invariant solutions of Navier-Stokes equations describing free-boundary motions, *Dokl. Akad. Nauk SSSR* 202(2), 302-305 (1972).
29. W. Purkert, Zum Verhältnis von Sophus Lie und Friedrich Engel, *Wiss. Zeitschr. Ernst-Moritz-Arndt-Universität Greifswald, Math.-Naturwiss.* Reihe XXXIII, Heft 1-2, 29-34, (1984).
30. G. F. B. Riemann, Ueber die Fortpflanzung ebener Luftwellen von endlicher Schwingungsweite, *Abh. K. Ges. Wiss. Göttingen* 8 (1860).
31. L. I. Sedov, *Similarity and Dimensional Methods in Mechanics*, 4th ed., New York: Academic Press (1959).
32. H. Stephani, *Differential Equations: Their Solution Using Symmetries*, Cambridge: Cambridge University Press (1989).

18

The Missing Link

Felipe Acker

Introduction

The goal of this article is to change the views of mathematicians throughout the world on three fundamental theorems of elementary analysis: Cauchy-Goursat's Theorem, Stokes' Theorem and the Mean Value Theorem. My claims are the following:

1. Cauchy-Goursat's Theorem is really a mere corollary of Green's.
2. The usual treatment of Stokes' Theorem is misguided and I will do it properly.
3. The Mean Value Theorem does generalize to higher dimensions as an *equality*.

Sophisticated objects like differentiable manifolds and exterior differential forms will not figure in the exposition, lest they discourage potential readers. For a more technical version, see [1] and [2].

To make my points of view clear, I'll begin by summarizing what seems to be received wisdom about these theorems.

Cauchy-Goursat's Theorem states (in a simplified version) that if A is an open subset of the complex plane, f is holomorphic on A, and R is a (closed) rectangle contained in A, then

$$\int_{\partial R} f(z)\, dz = 0$$

(where ∂R represents the boundary of R with positive orientation).

Almost every introductory complex analysis book contains the remark that "with the additional hypothesis that f is C^1," the proof can be carried out using Green's Theorem. So there seems to be a general belief in the reciprocal: Without this additional hypothesis, Green's Theorem wouldn't apply. This is a fallacy, as I'll prove below. The real question is, How could such a fiction persist for one entire century?

For **Stokes' Theorem,** let's restrict ourselves to Green's Theorem: the exposition will be less technical, and the central ideas won't suffer. The simplest version states that if R is a rectangle and

$$P, Q : R \to \mathbb{R}$$

Volume 18, No. 3 (Summer 1996), 4-9

are C^1 functions, then

$$\int_{\partial R} P\,dx + Q\,dy = \int\int_R \left(\frac{\partial Q}{\partial x} - \frac{\partial P}{\partial y}\right) dx\,dy.$$

Although it is obvious that the C^1 hypothesis can be relaxed, the universally accepted proof is based on iterated integration and needs some kind of regularity of *each one* of the partial derivatives $\partial Q/\partial x$ and $\partial P/\partial y$. However, if we look at this theorem as a generalization of the Fundamental Theorem of Calculus, we feel that the natural hypothesis should be the integrability of $(\partial Q/\partial x - \partial P/\partial y)$, even if individually $\partial Q/\partial x$ and $\partial P/\partial y$ are bad. This is, in fact, true: if P and Q are differentiable (approximable by linear functions), it is enough to assume $(\partial Q/\partial x - \partial P/\partial y)$ to be Riemann integrable. The precise hypotheses are more subtle, as I will show, but this version is sufficient in order to get Cauchy-Goursat as a corollary.

Now let's turn to the **Mean Value Theorem** or, should I say, the Mean Value Equality: if

$$f : [a, b] \to \mathbb{R}$$

is continuous on $[a, b]$ and differentiable at each point of $]a, b[$, then there exists a point c in $]a, b[$ such that

$$f'(c) = \frac{f(b) - f(a)}{b - a}.$$

The trouble appears when we try to generalize this result to higher dimensions: the pretty and geometrical equality becomes an inequality. I think the best expression of what everybody seems to believe was given by Jean Dieudonné in his celebrated *Foundations of Modern Analysis* [4]:

> After the formal rules of Calculus have been derived (sections 8.1 to 8.4), the other sections of the chapter are various applications of what is probably the most useful theorem in Analysis, the mean value theorem, proved in section 8.5. The reader will observe that the formulation of that theorem, which is of course given for vector-valued functions, differs in appearance from the classical mean value theorem (for real-valued functions), which one usually writes as an *equality* $f(b) - f(a) = f'(c)(b - a)$. The trouble with that classical formulation is that: 1°. **there is nothing similar to it as soon as f has vector values;** 2°. it completely conceals the fact that *nothing* is known on the number c, except that it lies between a and b, and for most purposes, all one needs to know is that $f'(c)$ is a number which lies between the g.l.b. and l.u.b. of f' in the interval $[a, b]$ (and *not* the fact that it actually is a value of f'). In other words, **the real nature of the mean value theorem is exhibited by writing it as an** *inequality,* **and not as an equality.**

Well, Dieudonné was wrong![1] The Mean Value Theorem does generalize to higher dimensions as an *equality*.[2] This is a key idea: when we prove the Fundamental Theorem of Calculus, we really need the Mean Value Theorem in the equality form. If we try to mimic this proof to get Stokes' Theorem, it becomes clear that a general version of the Mean Value *Equality* would be welcome. In the chain leading from the Fundamental Theorem of Calculus to Stokes' Theorem, this is the missing link.

---

[1] I really appreciate people like Dieudonné (or, on the opposite side, Arnold) who express polemic opinions; polemics is fundamental to intellectual activity. I prefer Arnold, but, as the French say, "il faut de tout pour faire un monde."

[2] And this equality reveals a new aspect of its nature. The theorem referred to by Dieudonné is usually called by French authors *finite increases theorem*. I claim the true *mean value theorem* is the one I will present below.

The Fundamental Theorem of Calculus and Stokes' Theorem

Let me briefly recall the proof of the Fundamental Theorem of Calculus just to emphasize the role played in it by the Mean Value Theorem.

THE FUNDAMENTAL THEOREM OF CALCULUS.   Let $f : [a, b] \to \mathbb{R}$ be continuous on $[a, b]$ and differentiable on $]a, b[$. If f' is (Riemann) integrable, then

$$\int_a^b f = f(b) - f(a).$$

PROOF: Let $P = \{a_0, \ldots, a_n\}$, $a = a_0 < a_1 < \cdots < a_n = b$ be a partition of $[a, b]$. Then writing

$$f(b) - f(a) = \sum_{i=1}^n f(a_i) - f(a_{i-1})$$

and applying the Mean Value Theorem to each subinterval $[a_{i-1}, a_i]$, we get

$$f(b) - f(a) = \sum_{i=1}^n f'(c_i)(a_i - a_{i-1}),$$

where

$$c_i \in \ ]a_{i-1}, a_i[, \quad i = 1, \ldots, n.$$

The right-hand side converges to

$$\int_a^b f'. \qquad \qquad \square$$

Now let us turn to Stokes' Theorem and try to adapt the proof of the one-dimensional case. For simplicity, let us restrict our study to the elementary case of Green's Theorem.

GREEN'S THEOREM (TENTATIVE).   Let $R = [a, b] \times [c, d]$ and $P, Q : R \to \mathbb{R}$ be continuous on R and differentiable in its interior. Let

$$\frac{\partial Q}{\partial x} - \frac{\partial P}{\partial y}$$

be integrable on R. Then

$$\int_{\partial R} P \, dx + Q \, dy = \int \int_R \left(\frac{\partial Q}{\partial x} - \frac{\partial P}{\partial y} \right) dx \, dy.$$

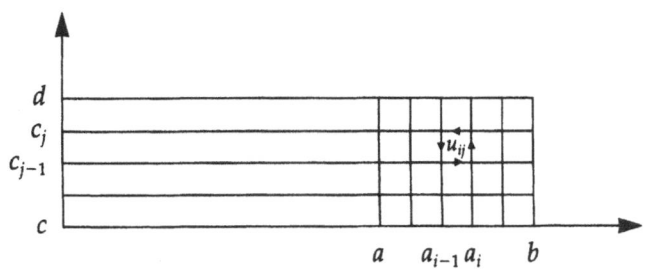

PROOF. The proof should begin by taking two partitions,

$$a = a_0 < a_1 < \cdots < a_n = b, \quad c = c_0 < c_1 < \cdots < c_m = d,$$

and considering the rectangles $R_{ij} = [a_{i-1}, a_i] \times [c_{j-1}, c_j]$. Then just observe that, with positive orientations,

$$\int_{\partial R} P\, dx + Q\, dy = \sum_{i=1,\dots,n} \sum_{j=1,\dots,m} \int_{\partial R_{ij}} P\, dx + Q\, dy.$$

We are now at the point where the Mean Value Theorem is needed. If only we could write, for each i and for each j,

$$\int_{\partial R_{ij}} P\, dx + Q\, dy = \left(\frac{\partial Q}{\partial x} - \frac{\partial P}{\partial y} \right) (u_{ij})(a_i - a_{i-1})(b_j - b_{j-1}),$$

for some u_{ij} in the interior of R_{ij} (just as in the one-dimensional case), the proof would be finished.

Now, what is the trouble if we try to generalize the Mean Value Theorem? The problem seems to be that we must then generalize Rolle's Theorem, whose usual proof is peculiar to real-valued functions.

The Mean Value Theorem

I will give now an alternative proof of the one-dimensional Mean Value Theorem. Refer to Figure 18-1. If we divide $[a, b]$ into three (equal) parts, setting

$$h = \frac{b - a}{3}$$

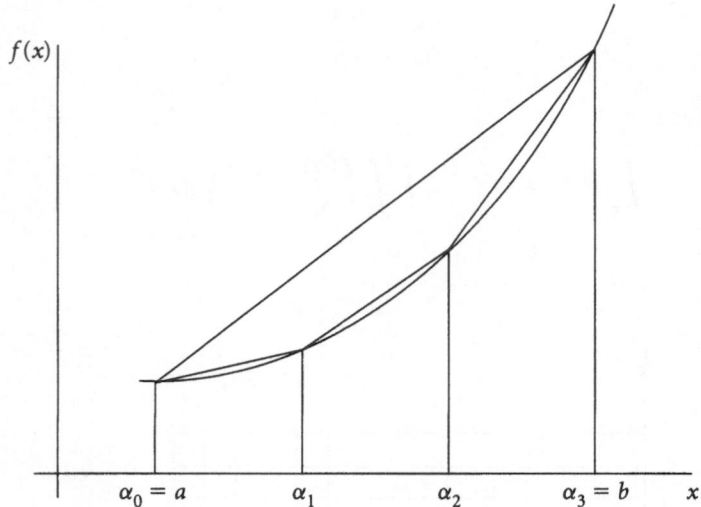

FIGURE 18-1

and letting $\alpha_0, \alpha_1, \alpha_2, \alpha_3$ be the endpoints of the subintervals we get, then we must have, for the slopes

$$\bar{m} = \frac{f(b) - f(a)}{b - a}, \quad m_i = \frac{f(\alpha_i) - f(\alpha_{i-1})}{h}, \quad i = 1, 2, 3,$$

that either $m_i = \bar{m}$ for all the i's or there are two of them (m_1 and m_3, for instance, in Figure 18-1) such that $m_1 < \bar{m} < m_3$. Now look at the (continuous) function

$$m : [a, b - h] \to \mathbb{R}$$

$$x \mapsto \frac{f(x + h) - f(x)}{h}.$$

What we have said implies that, for some point $\bar{x}$ in $]a, b - h[$, we have $m(\bar{x}) = \bar{m}$ (be sure you understand that $\bar{x}$ can be taken in the open interval).

Next, make $a_1 = \bar{x}$ and $b_1 = \bar{x} + h$ and iterate the procedure for the interval $[a_1, b_1]$, and so on. We will get a sequence of nested intervals $[a_n, b_n]$ with

$$b_n - a_n = \frac{b - a}{3^n}, \quad \frac{f(b_n) - f(a_n)}{b_n - a_n} = \frac{f(b) - f(a)}{b - a}.$$

Let c be the intersection point of this sequence of intervals. As c is in $]a, b[$, we are sure that $f'(c)$ exists and

$$f'(c) = \lim_{n \to \infty} \frac{f(b_n) - f(a_n)}{b_n - a_n} = \frac{f(b) - f(a)}{b - a}. \qquad \square$$

Everyone can see that this proof is not simpler than the usual one, and I do not pretend otherwise. The difference is that I am able to generalize it to higher dimensions.

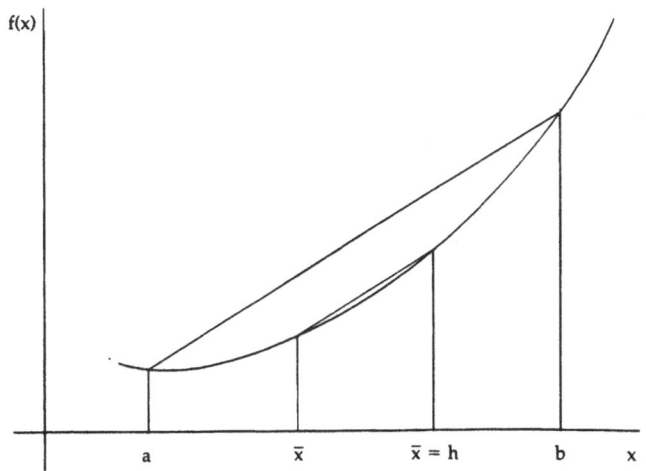

FIGURE18-2

The Generalized Mean Value Theorem and Its Corollaries

Let's begin with what is sometimes called the *physical interpretation of the divergence.* To fix some notation, when I say "rectangle," read "closed rectangle with sides parallel to the coordinate axes"; "$\mu(R)$" should be read "area of R."

PROPOSITION.   Let A be an open subset of the plane, let

$$F : A \to \mathbb{R}^2$$

be continuous on A, and let (R_n) be a sequence of similar rectangles containing the point u_0 such that

$$\lim_{n \to \infty} \mu(R_n) = 0.$$

Let ω be the differential form $P\,dx + Q\,dy$, where $F = (P, Q)$.[3] If F is differentiable at u_0, then

$$\lim_{n \to \infty} \frac{1}{\mu(R_n)} \int_{\partial R_n} \omega = \left(\frac{\partial Q}{\partial x} - \frac{\partial P}{\partial y} \right)(u_0).$$

PROOF.   Let T be the differential of F at u_0, and write, for u in A,

$$F(u) = F(u_0) + T(u - u_0) + \epsilon(u),$$

with

$$\lim_{u \to u_0} \frac{\epsilon(u)}{|u - u_0|} = 0.$$

In other words,

$$F = F_0 + \epsilon,$$

where F_0 is C^1. Let ω_0 be the differential form corresponding to F_0, and ω_ϵ the differential form corresponding to ϵ. Apply the C^1 form of Green's Theorem to ω_0 to get

$$\int_{\partial R_n} \omega_0 = \int \int_{R_n} \left(\frac{\partial Q}{\partial x} - \frac{\partial P}{\partial y} \right)(u_0) = \mu(R_n) \left(\frac{\partial Q}{\partial x} - \frac{\partial P}{\partial y} \right)(u_0)$$

(this could be proved directly with some more effort). Finally, taking into account that the rectangles R_n are similar, we get

$$\lim_{n \to \infty} \frac{1}{\mu(R_n)} \int_{\partial R_n} \omega_\epsilon = 0,$$

and the proof is finished. $\square$

---

[3]To get what is called the *divergence*, we should of course have written $\omega = -Q\,dx + P\,dy$.

The above result motivates the following definition.

DEFINITION.   Let A be an open subset of the plane, and let

$$P, Q : A \to \mathbb{R}$$

be such that, for $\omega = P\,dx + Q\,dy$,

$$\int_{\partial R} \omega$$

is well defined whenever R is a rectangle. The differential form ω will be said to have an exterior derivative $d\omega(u)$ (which will be just a real number) at the point u of A if for any sequence (R_n) of similar rectangles containing u such that $\mu(R_n)$ converges to zero, we have

$$\lim_{n \to \infty} \frac{1}{\mu(R_n)} \int_{\partial R_n} \omega = d\omega(u).$$

The name exterior derivative is taken here, in a somewhat improper sense, as a reminder of what general concept I am trying to put into a "popular form."

Now observe that if, for instance, P and Q are continuous on A and are differentiable at u, then ω has an exterior derivative at u. Moreover, if we change the values of P and Q at u so as to lose continuity, this will not change the values of the integrals and ω will still have the same exterior derivative at u. This is not quite surprising: a differential form assigns numbers to objects (curves, in the present case; manifolds, in higher dimensions), not to points. So, the continuity of ω should be taken in the sense that the integrals of ω over two neighboring curves should differ very little.

DEFINITION.   If $\omega = P\,dx + Q\,dy$ is a differential form defined on some subset A of the plane, ω will be said to be continuous on A if for every rectangle R the function

$$u \mapsto \int_{\partial(u+R)} \omega$$

is continuous (wherever defined), where $u + R = \{u + r; r \in R\}$.

We now have good definitions for proving the generalized Mean Value Theorem. The proof follows the same steps as the one-dimensional case. The basic trick is the following.

LEMMA.   Let $\omega = P\,dx + Q\,dy$ be a continuous differential form on the rectangle R. Then there exists a rectangle R_0 contained in the interior of R such that

$$\frac{1}{\mu(R_0)} \int_{\partial R_0} \omega = \frac{1}{\mu(R)} \int_{\partial R} \omega,$$

the sides of R_0 being one-third of those of R.

PROOF. Let $R = [a, b] \times [c, d]$. Taking

$$h = \frac{b-a}{3}, \quad k = \frac{d-c}{3}, \quad \bar{R} = [0, h] \times [0, k],$$

we get, if we divide the sides of R into three intervals, nine rectangles R_i, $i = 1, \ldots, 9$, congruent to $\bar{R}$.

Now note that

$$\int_{\partial R} \omega = \sum_{i=1}^{9} \int_{\partial R_i} \omega$$

and that, for all i,

$$\mu(R) = 9\mu(R_i).$$

So

$$\frac{1}{\mu(R)} \int_{\partial R} \omega = \frac{1}{9} \sum_{i=1}^{9} \frac{1}{\mu(R_i)} \int_{\partial R_i} \omega.$$

Now, just as in the one-dimensional case, we are sure that either

$$\frac{1}{\mu(R)} \int_{\partial R} \omega = \frac{1}{\mu(R_i)} \int_{\partial R_i} \omega, \quad i = 1, \ldots, 9$$

(and then we can just choose the central rectangle to be R_0) or we have, for two of the rectangles, say R_1 and R_2,

$$\frac{1}{\mu(R_1)} \int_{\partial R_1} \omega < \frac{1}{\mu(R)} \int_{\partial R} \omega < \frac{1}{\mu(R_2)} \int_{\partial R_2} \omega.$$

The differential form ω being continuous, we can, pushing R_1 and R_2 inside R, suppose R_1 and R_2 to be contained in the interior of R and still satisfy the above inequalities. Then the function

$$m :]a, a + 2h[\times]c, c + 2h[\to \mathbb{R},$$

where

$$m(u) = \frac{1}{\mu(\bar{R})} \int_{\partial(u+\bar{R})} \omega - \frac{1}{\mu(R)} \int_{\partial R} \omega,$$

is continuous and takes positive and negative values. It must then vanish for some u_0, and we take $R_0 = u_0 + \bar{R}$.

We are finally ready for

THE MEAN VALUE THEOREM.   Let the differential form $\omega = P\,dx + Q\,dy$ be continuous on the rectangle R and have an exterior derivative at each point of the interior of R. Then there is a point u in the interior of R such that

$$d\omega(u) = \frac{1}{\mu(R)} \int_{\partial R} \omega.$$

PROOF.   Just as in the one-dimensional case, we iterate the lemma to get a sequence of nested rectangles (R_n) such that

1. all the R_n's are contained in the interior of R,
2. the lengths of the sides of R_{n+1} are one-third of those of R_n,
3. $\frac{1}{\mu(R_n)} \int_{\partial R_n} \omega = \frac{1}{\mu(R)} \int_{\partial R} \omega$.

Let u be the intersection point of the sequence (R_n).
 We get now, as an immediate consequence, the following stronger form of Green's Theorem:

THEOREM.   Let the differential form $\omega = P\,dx + Q\,dy$ be continuous on the rectangle R and have an exterior derivative at each point of the interior of R. If $d\omega$ is (Riemann) integrable, then

$$\int_R d\omega = \int_{\partial R} \omega.$$

COROLLARY.   Let R be a rectangle and $P, Q : R \to \mathbb{R}$ be continuous on R and differentiable at each point of the interior of R. If

$$\frac{\partial Q}{\partial x} - \frac{\partial P}{\partial y}$$

is (Riemann) integrable, then

$$\int_R \left(\frac{\partial Q}{\partial x} - \frac{\partial P}{\partial y} \right) dx\,dy = \int_{\partial R} P\,dx + Q\,dy.$$

COROLLARY.   Let A be an open subset of the complex plane and let $f : A \to \mathbb{C}$ be holomorphic. If R is a rectangle contained in A, then

$$\int_{\partial R} f(z)\,dz = 0.$$

Remark. It is usual to present Goursat's Theorem under the hypothesis that f is holomorphic on A, except at a subset D of isolated points such that

$$\lim_{z \to z_0} f(z)(z - z_0) = 0$$

for each z_0 in D. This hypothesis just assures the continuity of the complex differential form $f(z)\,dz$. On the other hand, we have $d\omega(z) = 0$ if z is not in D. To conclude the proof, we just argue that $d\omega(z) = 0$ also if $z \in D$. This is a consequence of the following proposition, which generalizes a well-known result of real analysis (proved using the Mean Value Equality).

PROPOSITION.   Let A be an open subset of the plane and let $\omega = P\,dx + Q\,dy$ be a continuous differential form on A. Suppose ω has an exterior derivative at each point of A, except perhaps at z_0. Then, if

$$\lim_{z \to z_0} d\omega(z) = \eta,$$

we have $d\omega(z_0) = \eta$.

The proof is left as an exercise. We can also prove the Cauchy-Goursat Theorem without this proposition. To make things simpler, let $D = \{z_0\}$. If z_0 isn't in the interior of R, just apply Green's Theorem. If z_0 is in the interior of R, draw a horizontal line and a vertical line through z_0 and apply Green's Theorem to the four rectangles you get (this is also a hint for the exercise!).

Of course the above version of Green's Theorem can be extended in the obvious way to the Divergence Theorem on a block $B = [a_1, b_1] \times \cdots \times [a_N, b_N]$ in $\mathbb{R}^N$. The next step is Stokes' Theorem on chains and on manifolds; see [1] and [2].

Final Remarks

I apologize for having included some proofs, but I had to justify my bombastic claims. I hope everybody is now persuaded. You may say, "Okay, the guy has found a trick," referring to the proof of the Mean Value Theorem. But there is something else to be observed: the definition of the exterior derivative. In the books on differential forms, the exterior derivative is made to look like some kind of algebraic operation on partial derivatives (something like a mixture). One notable exception is Arnold's book [3].

Let me emphasize this: Exterior derivation is a genuine analytic operation, and it is an independent concept—it does not depend on differentiation. The one-dimensional derivative contains both the idea of *linear approximation* and that of *flux* (didn't Newton call it *fluxion?*). The first one generalizes as the differential, the latter as the exterior derivative. People working on what is called *Partial Differential Equations* should perhaps think about this: the physically relevant "partial differential equations" (at least those from classical physics) are, in fact, "exterior derivative equations," and current Sobolev spaces are almost surely not their habitat.

References

1. F. Acker, *The Mean Value Theorem for Differential Forms*, 1997.
2. F. Acker, *Advanced Calculus* (in Portuguese), 1997.
3. V. Arnold, *Mathematical Methods of Classical Mechanics*, New York: Springer-Verlag, 1978.
4. J. Dieudonné, *Foundations of Modern Analysis*, New York: Academic Press, 1960.

APPLIED MATHEMATICS

19

Yeast Oscillations, Belousov-Zhabotinsky Waves, and the Non-retraction Theorem

Steven Strogatz

When you think of scientific applications of topology, you usually think of physics: fiber bundles [1], solitons [2], and instantons in particle physics, for example. But more and more, topology is beginning to play an important role in chemistry and biology. When biochemists discuss the three-dimensional coiling of DNA, differential topology is the language of choice [3–6]. The widespread occurrence of closed *rings* of DNA leads to complex topologies: knots or even networks of inter-locked rings (as in the Trypanosome parasite that causes sleeping sickness) [7, 8]. Amazing enzymes called "topoisomerases" can break DNA rings and rejoin them with altered topology [9, 10]. These feats have only recently been mimicked by man-made technology—after decades of effort, organic chemists [11, 12] have synthesized the first molecular Möbius strip.

In this spirit I present two disparate examples of science unified by a theorem of topology, the so-called non-retraction theorem. The first example should appeal to beer drinkers—it concerns a peculiarity of the yeasts' brewing schedule. Then turning from rhythms in time to patterns in space, the second example deals with chemical waves in a remarkable broth, the Belousov-Zhabotinsky reagent. Unlike the tedious chemistry of high school days, this little dish of reactions seems positively alive. Against a motionless orange backdrop, blue spiral waves rotate and spread like grassfires, each dominating its own patch of reagent, colliding with and annihilating intrusive blue waves from afar. There is much topology in these two-dimensional patterns and even more in their three-dimensional analogues, all organized around the non-retraction theorem.

1. Yeast

Yeast is most famous for its biochemical knack of converting sugar to alcohol. But with some pro-vocation by inquisitive biologists, the cells are capable of topological tricks of timing that may sur-prise you.

Before discussing topology, we need a bit of biology. The full story is told in [13–15]. The rele-vant facts concern the yeasts' timetable. Under contrived laboratory conditions, yeast cells do not metabolize sugar through a series of monotonous steps; instead some of the intermediate reactions

Volume 7, No. 2 (Spring 1985), 9–17

proceed via oscillations (though the end result is still a steady trickle of alcohol). For example, bio-chemists have monitored the levels of a metabolic intermediate, NADH, which conveniently fluo-resces under ultraviolet light. Its blue glow brightens and dims periodically, revealing a *cycle* in the underlying biochemistry. Furthermore, the individual cells are mindful of the others; they pulse in unison, collectively synchronized in a biochemical rhythm which waxes and wanes twice per minute.

It would be extraordinary luck if all the yeast *began* in unison. So how do they manage to get in step and stay there? Presumably the cells communicate by molecular diffusion and then each accel-erates or hesitates to get in step with the stirred mass.

A simple experiment [13] elucidates these adjustments. Two equal volumes of cells, each syn-chronous within itself, yet out of step with the other volume, are mixed together. After mixing, they rapidly resynchronize in the standard cycle, but perhaps at some compromise phase. How does the compromise phase depend on the old phases of the "parent" ingredients?

We could dream up all sorts of rules: "Whichever group peaked first sets the pace, and the other falls in step." But that rule is nonsense—since both groups are periodic, neither can be said to have peaked "first." Another try: "If groups 1 and 2 have progressed through fractions ϕ_1 and ϕ_2 of the cycle, then upon mixing they adopt the average, $1/2(\phi_1 + \phi_2)$." Nor is this rule well-defined, since the fractions depend on an arbitrary selection of the "beginning" of each cycle. In other words, the pur-ported compromise depends unrealistically on our conventions: whether the cycle is said to begin at its peak, at its trough, or wherever.

So fix any reference point of the cycle's beginning but stick to it. Having done so, we can define the useful notion of *phase*: the cell's biochemical rhythm is at phase ϕ when it has progressed through a fraction ϕ of its cycle, as measured in equal time steps from the arbitrary zero point. Notice that phase should be regarded not as a real number, but instead as a real number (mod 1) or equivalently as a point on the unit circle S^1, because ($\phi = 0 =$ beginning) and ($\phi = 1 =$ end) correspond to the same part of the biochemical cycle. Now the question about phase compromise can be restated [16]. The phase ϕ of the mixture has some unknown functional dependence on the phases ϕ_1 and ϕ_2 of the ingredients:

$$\phi = f(\phi_1, \phi_2), \quad f : S^1 \times S^1 \to S^1$$

What can be said, *a priori*, about the function f? (Biologists [13] have actually done the experiments, and thus measured f, but that would be giving away the answer.) It is plausible that:

(A1) interchanging the *names* of groups 1 and 2 will not affect the outcome: $f(\phi_1, \phi_2) = f(\phi_2, \phi_1)$.

(A2) if the separate groups agree in phase initially, they continue at that phase, unaffected by mix-ture with their own kind. That is, if $\phi_1 = \phi_2$, then $\phi = \phi_1 = \phi_2$. (In science, this is usually called a "control experiment.")

(A3) Slight changes in the initial phases alter the outcome only slightly: in other words, f is contin-uous in each argument.

Though reasonable, these axioms probably sound weak—they don't seem to pin down the func-tion's detailed behavior. And yet within them lurks a topological surprise, the one advertised at the outset.

They are inconsistent.

In the proof of inconsistency, the final step invokes the non-retraction theorem, discussed in the next section.

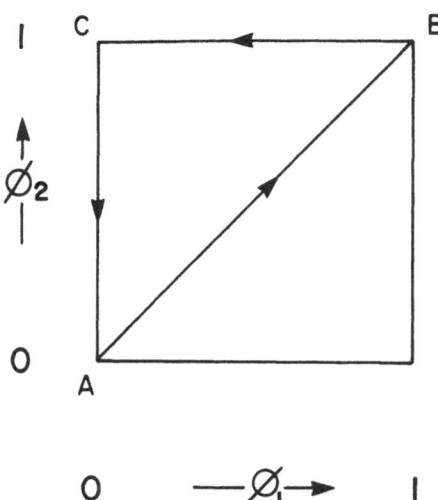

FIGURE 19-1 *A square of initial phases (ϕ_1, ϕ_2) in the phase compromise experiment for yeast glycolysis. Phases 0 and 1 represent the same point in the biochemical cycle. The contours of the compromise phase $\phi = f(\phi_1, \phi_2)$ are not shown here, since this figure is for a preliminary thought experiment: according to axioms A1−A3, $\Delta\phi(AB) = 1$ and $\Delta\phi(BC) = \Delta\phi(CA) = M = $ integer. Hence $W = \Delta\phi(ABCA) = 2M + 1 \neq 0$. By the non-retraction theorem (Section 2), f must be discontinuous within ABCA.*

PROOF:

Step 1:   Instead of thinking of f as a function on the torus $T^2 = S^1 \times S^1$, regard it as a function on the unit square $I^2 = [0, 1] \times [0, 1]$. Keep track of the phase assigned to each point of the closed curve ABCA in the unit square shown in Figure 19-1. In particular, since the curve begins and ends at A, the compromise phase $\phi = f(\phi_1, \phi_2)$ must change by an integer number of full cycles in one counter-clockwise lap around this path. This integer is the *winding number W.*

Step 2:   Compute W. Along arc AB, $\phi_1 = \phi_2$, so axiom (A2) implies $\phi = \phi_1 = \phi_2$. Hence along AB, ϕ increases by one cycle. For convenience we write $\Delta\phi(AB) = 1$. Along arc BC, $\Delta\phi(BC)$ equals some integer number M of cycles (since points B and C represent the same biochemistry and so are assigned the same ϕ.) The number M is unknown, but we *do* know $\Delta\phi(BC) = \Delta\phi(CA) = M$. (This follows from the symmetry (A1).) These contributions imply a winding number of $W = 2M + 1$, around ABCA.

Step 3:   Since M is an integer, W is odd. In particular, $W \neq 0$. That's all we need to derive a contradiction. For if phase is to be assigned continuously (Axiom 3) throughout the triangular region bounded by ABCA, then the non-retraction theorem (next section) implies that $W = 0$ along this boundary.

Which axiom is the culprit? The experiments implicate A3, the continuity axiom. Indeed, the data seem to show the most localized breakdown possible: a "phase singularity" [14], a single point at which all phase contours converge. Elsewhere, all three axioms appear to be obeyed (Figure 19-2).

Near the singularity, the experimental results are extremely sensitive to initial conditions and therefore practically irreproducible. One must sympathize with the pioneering experimenters [13]; at least three times, they meticulously probed the zone here claimed to be singular, each time obtaining a different compromise phase. Their dogged conclusion:

The synchronization does not follow any clear pattern. The reason . . . is not understood at this time. Much more experimentation is probably needed before the effects can be correlated to some variable in the system.

At the time, "more experimentation" may have seemed the only recourse. In retrospect, the yeasts were merely behaving as they had to, single-celled creatures abiding by the non-retraction theorem.

2. The Non-retraction Theorem

The non-retraction theorem forbids certain kinds of mappings from a space to its boundary. In its simplest form it merely states a familiar fact about intervals: A closed interval cannot be mapped

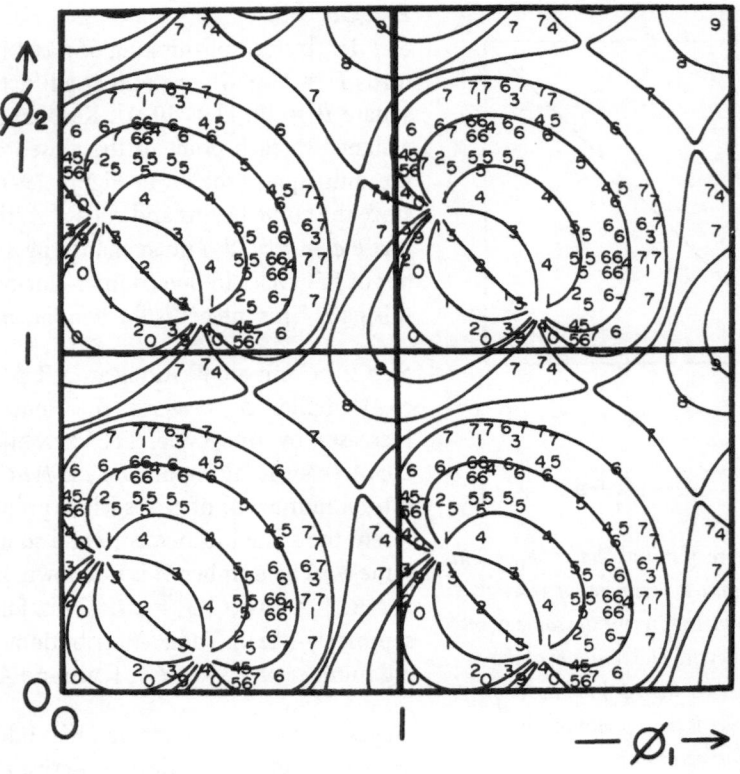

FIGURE 19-2 *Phase compromise experiments, plotting compromise phase as a function of two parent phases. The digits represent observed compromise phase, plotted in one-tenth cycle intervals, using the data of Ghosh et al. [13]. Contour lines link roughly equivalent data. Note that axioms A1–A3 are satisfied, except at the predicted phase singularities, where continuity fails. From Winfree [14], p. 62.*

continuously onto its two endpoints. All the points in the interval must go one way or the other—otherwise, the interval is ripped.

Stated more formally, let $I = [0, 1]$, and let $\partial I = \{0, 1\}$ be its boundary. By a *retraction* of I, we mean a continuous function $r\colon I \to \partial I$ such that $r \mid \partial I =$ identity. In other words, r leaves the boundary pointwise fixed. Of course, there *is* no retraction of I, since I is connected and its image $r(I)$ must be connected as well.

It might appear that the disconnnectedness of the boundary is essential for this result; it is not. Consider, for example, the closed unit disk $D^2 = \{(x, y) \in \mathbf{R}^2 \mid x^2 + y^2 \leq 1\}$ and its boundary $\partial D^2 = S^1$. The non-retraction theorem still holds, even though S^1 is connected. *There is no continuous function $r\colon D^2 \to S^1$ that leaves the boundary pointwise fixed* (Figure 19-3).

We can provide an intuitive proof of this fact using the winding number. Let $C = \partial D^2$. Then r assigns a "phase" $r(P)$ to each point P on C. Now r is the identity on C; when P executes one anticlockwise circuit of C, $r(P) = P$ executes one anticlockwise cycle of phase in S^1. Hence, along C, the winding number is 1.

Now deform C, shrinking it radially (Figure 19-4). The winding number must change continuously throughout the deformation. But since the winding number is an integer, it must be constant and hence is identically 1, even when C becomes arbitrarily small. This is a contradiction: on small circles the winding number is zero, as in Figure 19-4.

The winding number argument suggests a refinement of the theorem. The contradiction arose because the winding number was non-zero on ∂D^2; the stronger assumption that $r \mid \partial D^2 =$ identity

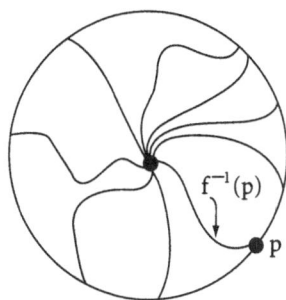

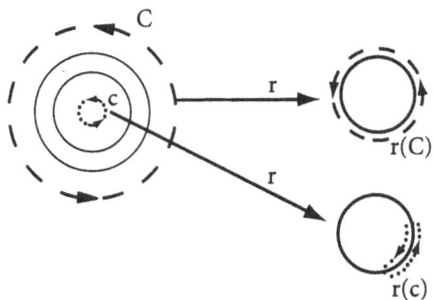

FIGURE 19-3 *Any function $f : D^2 \to S^1$ which leaves the boundary $\partial D^2 = S^1$ pointwise-fixed must be discontinuous. Wiggly curves in the disk represent loci of constant phase, i.e., the contours $\{f^{-1}(p) : p \in S^1\}$. Here the contours converge to a "phase singularity," the most localized discontinuity possible.*

FIGURE 19-4 *The continuous function $r : D^2 \to S^1$ acts as the identity on C, so r(C) winds once around S^1. Hence W = 1 on C. As the large circle C shrinks continuously to a small circle c, W varies continuously and, being integer-valued, remains constant. Hence W = 1 on c. But for c sufficiently small, r(c) is an arc with no net winding around S^1: hence W = 0 on c. This contradiction implies that r is discontinuous somewhere in D^2.*

is needlessly restrictive. Pairing down to essentials, we see that *any continuous function $r : D^2 \to S^1$ must have winding number zero around ∂D^2.* This is most easily proved using fundamental groups [17].

It was this form of the theorem that we invoked when discussing yeast, and we will invoke it twice more when discussing BZ reagent.

We mention in passing only one of the vast generalizations of the results above. If M is *any* compact manifold with boundary, then there is no smooth map $r : M \to \partial M$ that leaves ∂M pointwise fixed. The argument can be found in [18] and uses an elegant idea of Hirsch [19]. Essentially it says the following. If such an r exists, then for any regular value $x \in \partial M$, $r^{-1}(x)$ must be a one-dimensional manifold (with boundary). But the boundary of $r^{-1}(x)$ is precisely $r^{-1}(x) \cap \partial M$ and consists of the point x itself. There are, however, no one-dimensional manifolds with only one boundary component.

3. Belousov-Zhabotinsky reagent

Figure 19-5 shows an example of "Pa Ndau." In this ritual needlework embroidered by one of the Hmong women of Laos, the spirals are said to hold a magic spell; the borders guard the magic of the inner images. All Hmong children begin life strapped to their mothers' backs in carriers elaborately decorated with Pa Ndau, and Pa Ndau honors the dead on their final journeys [20].

Nature, too, stitches the spiral patterns of Pa Ndau. In a dish of motionless BZ reagent (Figure 19-6), a blue spiral wave of "fire" (actually, chemical oxidation) spreads across a medium of orange "grass" (quiescent, unoxidized reagent). The wavefront advances relentlessly, oxidizing its neighbors in front and thus turning them blue. The visual effect is striking: Thanks to the spiral geometry [21], the wave appears to rotate rigidly, slowly turning like the spray from a lawn sprinkler. Meanwhile its passage leaves a "burnt out" refractory region behind. Gradually the exhausted region recovers. Its color returns to deep orange, indicating its renewed susceptibility to another oxidizing pulse from its neighbors. Because of this renewal, the pattern is *self-regenerating*: it propagates round and round, like a dog chasing its own tail, until the energy resources in the medium are spent (after about an hour.)

FIGURE 19-5 *A "snail" motif in Pa Ndau, needlework of the Hmong people of Laos.*

FIGURE 19-6 *The Belousov-Zhabotinsky chemical reagent, seen by blue light transmitted through a 1-mm depth. The bright spirals mark loci of maximum concentration of ferriin (Belousov's blue indicator dye); the intervening darkness is deep orange. Waves of oxidation propagate from the spiral centers at a few mm per minute, the spirals turning in about one minute. The dish is 90 mm in diameter. From Winfree and Strogatz [29], Figure 2.*

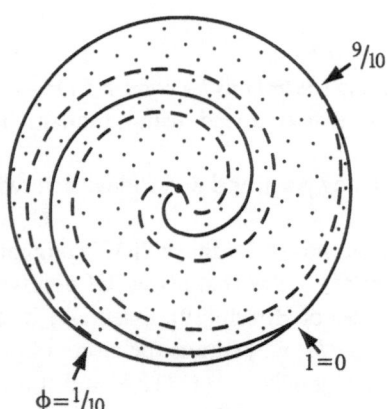

FIGURE 19-7 *Contours of uniform phase on a disk periodically excited by a clockwise rotating wave. During one circuit of the boundary, at fixed time, phase changes through a full cycle. Hence a phase singularity lurks within the disk. From Winfree and Strogatz [29], Figure 5.*

BZ reagent has become a Rorschach inkblot test. Applied mathematicians [22, 23] see it as an analog computer for demonstrating rotating wave solutions to reaction-diffusion equations. Chemists point to it as an oscillating [24] oxidation-reduction reaction. And for physiologists, it mimics pernicious rotating electrical waves on the heart [25, 26], believed to be involved in certain forms of heart attack.

Here it is recruited in the name of topology. The existence of rotating waves in BZ reagent entails some weird topological consequences, in the form of "phase singularities." First we must adapt the earlier notion of phase.

Since every area element in the domain of a spiral wave is periodically excited by its recurrent passage, each element may be assigned a phase. As before, a phase is a point on S^1; now it represents the fraction of a cycle elapsed since the arrival of the previous wavefront (at the point in question).

This phase description breaks down in an informative way: it detects a point to which phase cannot be meaningfully assigned (hence, one which is not periodically excited). The proof relies on the non-retraction theorem, as applied to a closed curve C that encircles the spiral's inner portions (Figure 19-7). The winding number of phase around C is $W = \pm 1$. (Why? Because as time advances, the wave periodically rotates through all locations along C. Hence, at fixed time, one point is just now being excited, those ahead of the wave are in varying stages of anticipation, and those behind are in varying stages of recovery, running through all phases *once* in a circuit of C.)

As before, the key point is that $W \neq 0$ around C. Now if we idealize the assignment of phase throughout the dish of reagent as a mapping $f : D^2 \to S^1$, the non-retraction theorem requires that f be discontinuous. The alternative is that f fails to be defined somewhere in D^2, i.e., some point is not periodically excited. The latter option is physically reasonable; molecular diffusion smears the observed wavefront, and the inner end of the spiral appears to thrash about, or "meander" [27].

Of course, the actual medium is three-dimensional. In thick layers of BZ reagent, the spirals are seen to be slices of a wave surface shaped like a *scroll* [14, 28]. The singular point elongates to a *thread* of phaselessness (Figure 19-8a). For chemical reasons [29, 30], the thread generically closes in a ring, and the scroll closes in a toroidal "scroll ring" wave.

The closure might occur in many topologically distinct ways: the ends could be joined directly (Figure 19-8a), the scroll could first be twisted along its length through 360° (Figure 19-8b), or it could even be knotted. What would these structures look like, and would they be chemically viable?

The simplest is shown in Figure 19-9. It is a surface of revolution swept out by a spiral orbiting a distant axle. The spirals all intersect and terminate at a common kink. (They terminate because of the empirical observation that such colliding waves annihilate, just as colliding grassfires cannot continue through each other's ashes.) These waves have been observed [31] and photographed recently [32]. Their chemical viability is indisputable.

On the other hand, the once-twisted scroll ring (Figure 19-8b) is chemical nonsense. Though *locally* a normal

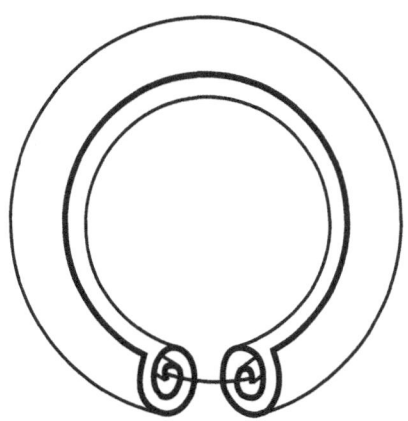

FIGURE 19-8a *A scroll-shaped wave surface formed from spirals. The singular point of a spiral wave is here replaced by a singular filament about which the scroll is coiled. Without violating local chemical continuity, the scroll may be closed in an untwisted ring.*

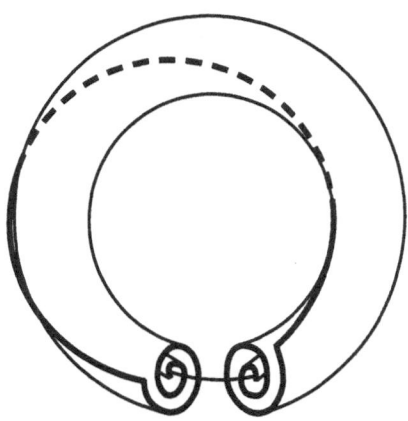

FIGURE 19-8b *As in (a), but a 360° twist is imparted along the length of the scroll before joining its ends. Both figures from Winfree [14], p. 253.*

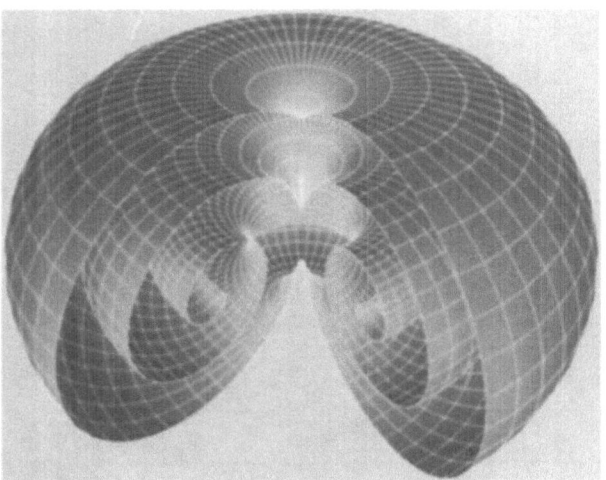

FIGURE 19-9 *Computer image of an "untwisted scroll ring," obtained from Figure 19-8a by adding more turns to the spirals. A wide sector has been removed to make the insides visible. The inner piece of the wave is attached to the singular ring; the outer piece has budded off and is formed by the collision of spirals with their diametrically opposite counter-parts. Real chemical scroll rings [31, 32] look less symmetric but are topologically similar to this figure.*

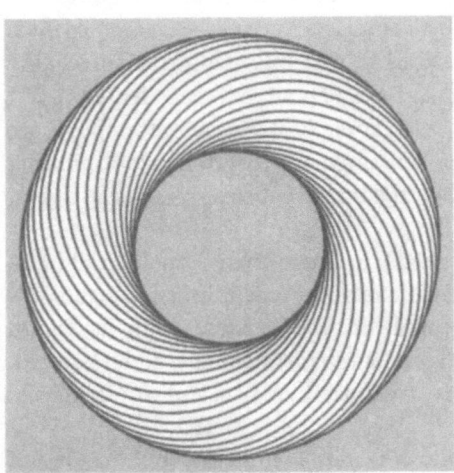

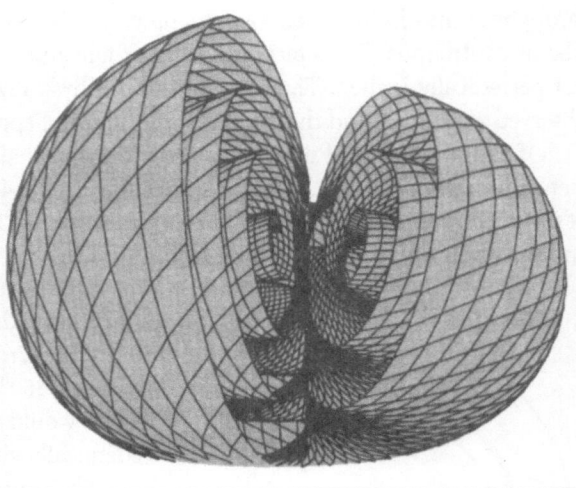

FIGURE 19-10 *Curves of constant phase on an imaginary torus encasing a twisted scroll ring. One of the curves represents the position of the moving wavefront. Because of the scroll's twist, the curves link the toroid's hole. Only half of each curve is visible. From Winfree [14], p. 255.*

FIGURE 19-11a *Similar to Figure 19-9, but obtained instead by extending the twisted scroll ring of Figure 19-8b. Unlike Figure 19-9, this wave is in one piece, attached to the inner singular ring—strictly speaking, the wavefront must be continued to infinity so that it will not have to end along an artifactual edge. No such structure could exist chemically. (In the computer construction, the edge was placed in the invisible sector.) From Winfree [36], Figure 14a.*

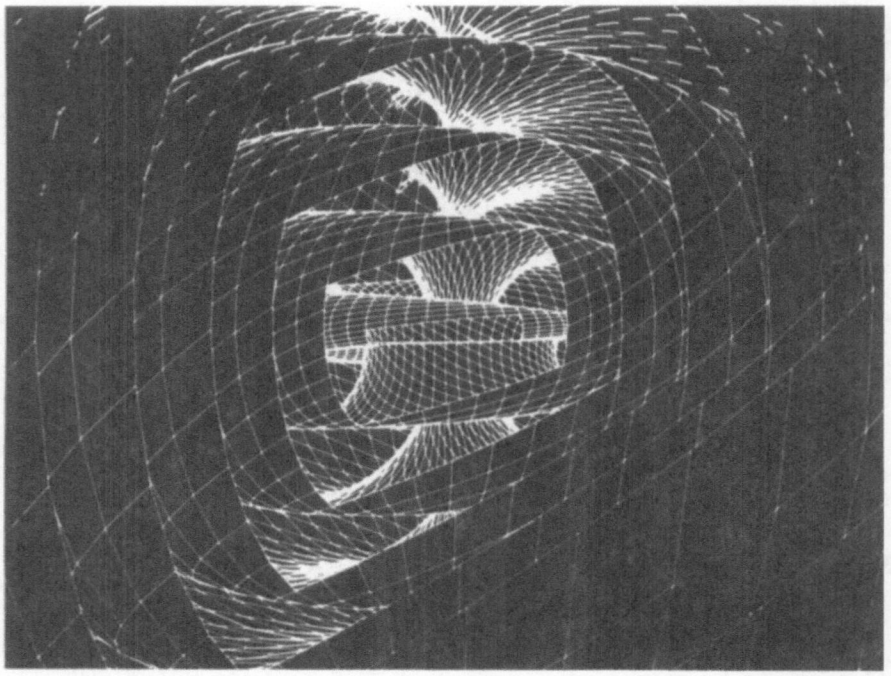

FIGURE 19-11b *The screw-like surface near the vertical axle is viewed through several windows in the fully evolved wave. The original singular filament is horizontal, near the center; note that it is threaded by the axle singularity, as predicted by the non-retraction theorem. From Winfree [36], Figure 14b.*

scroll wave, its *global* behavior is unphysical. (Scroll rings are reminiscent of fiber bundles [1] in this interplay of local and global properties.) To visualize a twisted scroll ring, imagine extending the surface's edge in Figure 19-8b, stopping its development whenever bits of edge collide.

Not easy, is it?

A more indirect approach is instructive. Encase the twisted scroll ring with an imaginary torus. The intersection of the wavefront with the toroid is a ring which threads the hole exactly once (since the generating spiral twists once as it orbits the hole axis.) The past and future positions of this ring constitute the instantaneous contours of constant phase on the torus (Figure 19-10). In a circuit around the toroid's inner equator, the contours are encountered in succession, running through a full cycle after one lap, so that $W = \pm 1$.

Hence, around the inner equatorial circle, $W \neq 0$. This circle bounds a disk that plugs the toroid's hole. As in earlier arguments, there must be a phase singularity within the disk. Here's a new wrinkle: Since the theorem applies to *any* disk-like diaphragm bounded only by the equator, each such diaphragm is pierced by a singular point. Following the singularity from one diaphragm to another, we trace a *new* singularity, an unforeseen filament threading the original singular ring.

Computer graphics replace this existence proof with a picture (Figure 19-11). The wavefront is a screw surface, a gnarled helicoid built from spirals. As in a parking garage, we can move from one level to the next by staying on the surface and circulating around the central axle. Chemically the axle would be a sink, a collision locus for waves emanating from the original singular ring. However, sinks have never been observed in BZ reagent.

The non-retraction theorem spawns a prediction: A twisted scroll ring is not chemically viable. Indeed, further investigation shows that scroll ring topology is *quantized;* the only configurations of linked, twisted, and knotted rings which are potentially realizable are those for which a certain mathematical index vanishes [29, 33–35]. On topological grounds, it seems that waves in BZ reagent are like atoms—they fall into a periodic table, not of chemistry but of shape.

References

1. H. J. Bernstein and A. V. Phillips (1981) *Sci. Amer. 245*(1), 122.
2. G. Rebbi (1979) *Sci. Amer. 240*(2), 92.
3. W. F. Pohl (1980) *Math. Intelligencer 3,* 20.
4. J. H. White (1969) *Amer. J of Math. 91,* 693.
5. F. B. Fuller (1971) *Proc. Natl. Acad. Sci. USA 68,* 815.
6. F. H. C. Crick (1976) *Proc. Natl. Acad. Sci. USA 73,* 2639.
7. P. T. Englund, S. L. Hajduk, and J. C. Marini (1982) *Ann. Rev. Biochem. 51,* 695.
8. M. A. Krasnow, A. Stasiak, S. J. Spengler, F. Dean, T. Koller, and N. R. Cozzarelli (1983) *Nature 304,* 559.
9. J. C. Wang (1982) *Sci. Amer. 247*(1), 94.
10. M. Gellert (1981) *Ann. Rev. Biochem. 50,* 879.
11. D. M. Walba, R. M. Richards, and R. C. Haltiwanger (1982) *J. Amer. Chem. Soc. 104,* 3219.
12. E. Wasserman (1962) *Sci. Amer. 207,* 94.
13. A. K. Ghosh, B. Chance, and E. K. Pye (1971) *Arch. Biochem. Biophys 145,* 319.
14. A. T. Winfree (1980) *The Geometry of Biological Time.* New York: Springer-Verlag.
15. B. Chance, E. Pye, B. Hess, and A. Ghosh (1973) *Biochemical and Biological Oscillators.* New York: Academic Press.
16. A. T. Winfree (1974) *J. Math. Biol. 1,* 73.
17. J. R. Munkres (1975) *Topology: A First Course.* Englewood Cliffs: Prentice Hall.
18. J. W. Milnor (1965) *Topology from the Differentiable Viewpoint.* Charlottesville: University of Virginia Press.

19. M. Hirsch (1963) *Proc. Amer. Math. Soc. 14*, 364.

20. M. Campbell, N. Pongnoi, and C. Voraphitak (1978) *From the Hands of the Hills*. Hong Kong: Media Transasia.

21. A. T. Winfree (1972) *Science 175*, 634.

22. D. S. Cohen, J. C. Neu, and R. R. Rosales (1978) *SIAM J. Appl. Math. 35*, 536.

23. P. S. Hagan (1982) *SIAM J. Appl. Math. 42*, 762.

24. I. R. Epstein, K. Kustin, P. DeKepper, and M. Orban (1983) *Sci. Amer. 248*(3), 112.

25. V. I. Krinsky (1978) *Pharmac. Ther. B 3*, 539.

26. A. T. Winfree (1983) *Sci. Amer. 248*(5), 144.

27. O. E. Rossler and C. Kahlert (1979) *Z. Naturforsch. 34a*, 565.

28. A. T. Winfree (1973) *Science 181*, 937.

29. A. T. Winfree and S. H. Strogatz (1983) *Physcia-8D*, 35.

30. A. T. Winfree (1974) *Sci. Amer. 230(6)*, 82.

31. A. T. Winfree (1974) *Faraday Symp. Chem. Soc. 9*, 38.

32. B. Welsh, J. Gomatam, and A. Burgess (1983) *Nature 304*, 611.

33. A. T. Winfree and S. H. Strogatz (1983) *Physcia-9D*, 65 and 333.

34. A. T. Winfree and S. H. Strogatz (1984) *Physica-13D*, 221.

35. S. H. Strogatz, M. L. Prueitt, and A. T. Winfree (1984) *IEEE Comp. Graphics and Appl. 4(1)*, 66.

36. A. T. Winfree (1984) in *Fronts, Interfaces, and Patterns*, A Bishop, editor. Los Alamos National Laboratory Symposium. Amsterdam: North-Holland.

20

Strings

Yu. I. Manin

This paper is dedicated to the memory of Vadik Knizhnik.

Introduction

Recently I got an invitation to a conference on strings and superstrings to be held in El Escorial, Spain. Enclosed was a poster (see Figure 20-1) representing as a Riemann surface of genus 17 with 4 cusps the famous San Lorenzo Monastery built by Felipe II. The double symbolism of a monstrous castle-surface hanging by a cord had been skillfully rendered by an artist. The rope and prongs were reminiscent of the Inquisition and the sadistic monarch, while at the same time they were a commonplace visualization of the new toys of theoretical physicists—classical and quantum strings.

Well, not quite new. The foundations of string theory were laid back in the 1960s when Veneziano discovered his remarkable dual amplitude in strong interaction physics. It was soon understood that the Veneziano model describes quantum scattering of relativistic one-dimensional objects, i.e., strings, instead of common point-particles. This model was in qualitative agreement with experimental data on the parton-like behaviour of strong interactions. One could imagine a meson as a tube of color flux with quarks attached to its ends. The string length scale then should be roughly 10^{-13} cm. Because a string has internal excitation modes, this could explain the proliferation of strongly interacting particles.

However, the hadronic interpretation of dual string theory was plagued by many quantitative disagreements. To quote just one, it so happened that the quantum theory of relativistic strings seemed to be consistent only in 26-dimensional space-time, while hadrons apparently lived in our 4-dimensional world!

Meanwhile the quantum chromodynamic, i.e., the theory of quantized Yang-Mills fields, gained momentum as the correct theory of string interactions, and strings became outdated. The modern Renaissance of the string theory is based upon its reinterpretation suggested in 1974 by J. Scherk and John Schwarz.

String theory is now considered as a candidate theory of elementary particle physics at the Planck scale ($\sim 10^{-33}$ cm) rather than at the hadronic scale. This romantic leap of twenty orders of magnitude makes the situation in modern theoretical physics extremely bizarre and poses new problems of relating theory to the phenomenology of low-energy (former high-energy) physics. Psychologically this leap was prepared for by a decade of Grand Unification models based upon Yang-Mills fields with a large gauge group and a bold extrapolation of the high-energy behavior of coupling constants of strong and electro-weak interactions.

Volume 11, No. 2 (Spring 1989), 59–65

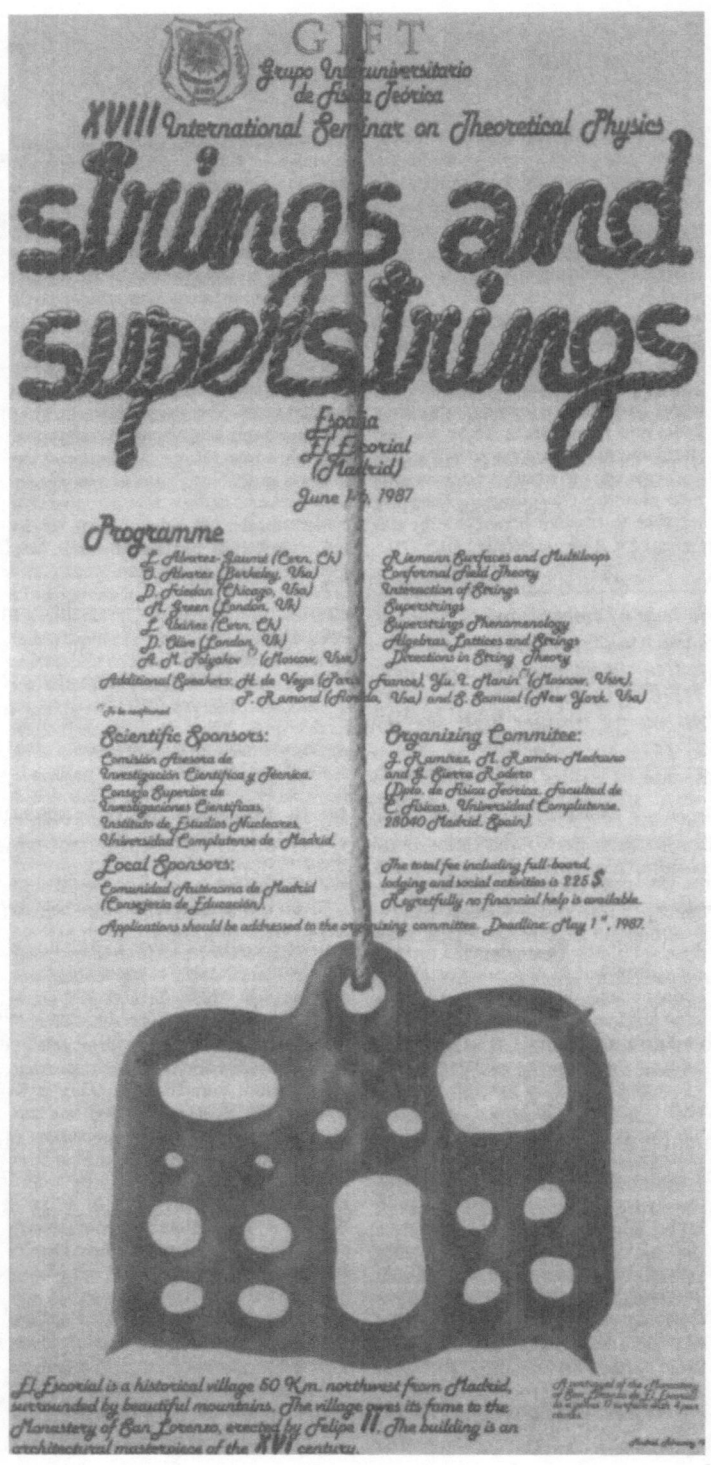

FIGURE 20-1 *Invitation to a conference on strings and superstrings.*

Another essential ingredient of modern string theory, also created in the 1970s, is supersymmetry, i.e., a mathematical scheme that allows us to incorporate bosons and fermions in a multiplet mixed by a symmetry supergroup. At the classical level, this involves an exciting extension of differential and algebraic geometry, Lie group theory, and calculus by introducing anticommuting coordinates that represent the half-integer spins of fermions. A string endowed with such fermionic coordinates is called a superstring. In a sense, supersymmetry implies general covariance and thereby requires unification with gravity.

The quantum theory of superstrings is consistent in 10-dimensional space-time. Because this is still far from our four dimensions, it was suggested, as a revival of an old Kaluza-Klein idea, that the extra six dimensions should be compactified at Planck scale. More specifically, our space-time presumably has a structure of a product $M^4 \times K^6$, where M^4 is the Minkowski space of special relativity while K^6 is a compact Riemannian space of diameter $\sim 10^{-33}$ cm, i.e., a point for all practical purposes. For theoretical purposes, however, it is not a point by any means. In a fantastic paper [16], Ed Witten and collaborators proposed that (in a vacuum state) K^6 is a Calabi-Yau complex manifold with a complicated topology responsible for such exotic properties of our universe as the existence of three (or four) generations of elementary constituents of matter, i.e., leptons and quarks.

In the beginning of the 1980s, Michael Green and John Schwarz discovered that requirements of consistent quantization (the so-called anomaly cancellation) place severe restrictions on the possible gauge groups of the superstring theory. It seems now that a specific superstring model, called the $E_8 \times E_8$ heterotic superstring, could eventually become a "Theory of Everything." Such are Great Physical Expectations.

Mathematically, the (super)string theory is no less interesting. As the Moscow physicist Vadik Knizhnik once remarked, unification of interactions is achieved through unification of ideas. Physics papers devoted to many facets of string theory are now filled with homotopy groups, Kac-Moody algebras, moduli spaces, Hodge numbers, Jacobi-Macdonald identities, and modular forms. A researcher trying to find his or her way in this mixture of seemingly disparate structures and techniques soon discovers that a physicist's intuition often transcends the purely mathematical one.

Some Physics

Let me review the physical content of modern quantum field theory before proceeding to its mathematical scheme.

In the 1920s, fundamental physics consisted of four principal theories: electromagnetism, general relativity (i.e., gravity theory), quantum mechanics, and statistical physics. Broadly speaking, the first three dealt with "elementary" phenomena while the fourth dealt with "collective" phenomena and their general laws.

The scale of elementary phenomena was defined by four fundamental constants: e (electron charge), G (Newton's constant), c (light velocity), $\hbar$ (Planck's constant). The group of dimensions generated by them essentially coincides with the group generated by three classical physical observables: mass, length, and time. In other words, starting from $G, c, \hbar$ one can define the "natural," or Planck, units:

$$M_{Pl} = (\hbar c G^{-1})^{1/2} \sim 10^{-5} g, \quad \ell_{Pl} = \hbar M_{Pl}^{-1} c^{-1} \sim 10^{-33} \text{ cm}, \quad t_{Pl} = \ell_{Pl} c^{-1} \sim 10^{-43} \text{ sec}.$$

The trouble is, we know of no elementary processes of Planck scale. Indeed, modern accelerators allow us to probe space-time at scales down to 10^{-16} cm and 10^{-26} sec only. On the other hand, M_{Pl} is the mass of a macroscopical drop of water about 0.2 mm diameter, and elementary particles of such a large mass do not exist in our world.

This incompatibility of the three fundamental theories was long considered as evidence of a need for a deeper theory combining all of them, for short, a $(G, c, \hbar)$-theory. In fact, two approximations had been discovered: general relativity, which could be considered as a (G, c)-theory, and quantum electrodynamics, i.e., a $(c, \hbar)$-theory. No one has yet succeeded in developing a consistent $(G, \hbar)$-theory, i.e., quantum gravity. The actual history of physics in our century followed an alternative course. Since the discovery of radioactivity and the subsequent construction of the first accelerators, the list of elementary particles and forces tended to grow; immense efforts of several generations of physicists were devoted to the development of quantum field theory explaining the variety of observed phenomena.

In the 1960s we had to be content with the following picture. There are several sorts of matter particles, seemingly point-like ones, i.e., without discernible internal structure; stable matter consists of quarks and electrons. All matter particles are fermions, i.e., they obey Fermi statistics and have spin $1/2$. There are also quanta of four fundamental forces: photons (electromagnetic force), gluons (strong force), vector mesons (weak force), and graviton (?) (gravity). They are bosons, meaning that they obey Bose-Einstein statistics and have spin 1 (or 2 for graviton).

Although elementary particles are point-like, they do have internal degrees of freedom. Mathematically, this means that a wave-function of a quark, say, is not a scalar function on space-time but a section of a vector bundle associated with a principal G-bundle, where G is a Lie group called the gauge group. (The choice of G ideally should have been governed by fundamental laws of nature but in the practice of the 1960s was model dependent). Similarly, a wave function of a quantum of fundamental force is a connection on the corresponding vector bundle, i.e., a matrix-valued differential form describing a parallel transport of internal state vectors along paths in space-time. A (second-quantized) theory of this kind is generally called a Yang-Mills theory.

The highest achievements of this epoch were the so-called standard model describing electroweak and strong interactions by means of the Yang-Mills fields with the gauge group $SU(3) \times SU(2) \times U(1)$ and several projects of Grand Unification based upon some large (preferably simple) group G containing $SU(3) \times SU(2) \times U(1)$. This large group should be a symmetry group of a fundamental theory at a higher energy, which is broken by some mechanism at lower energy leading to the effective Lagrangians of our present-day physics.

In all these developments gravity was neglected, because the gravitational interaction between elementary particles is many orders of magnitude weaker than, say, the electromagnetic one (this is another way of saying that Planck mass is very large). In fact, the overall picture of our universe is determined by different forces at different scales. At the scale about 10^{-13} cm, quarks are bound into protons and neutrons by the strong force. The atomic nucleus consists of protons and neutrons bound by residual forces. The strong interaction, being a short-distanced one, dies out at atomic scale, and the electromagnetic interaction takes its part binding electrons and nuclei into a neutral atom. The electromagnetic interaction is long-distanced and very strong in comparison with gravity, but for some reason positive and negative electric charges cancel each other out in large lumps of matter like stars and planets. Gravitational charge, i.e., mass, never cancels but only adds up, so that at astronomical scale, gravity becomes the main force. The residual electromagnetic forces, in the form of light and radio waves, serve as a source of energy and information for our kind of living matter. (This hierarchy of scales reflected in the hierarchy of physical theories is a characteristic trait of our modern understanding of nature. Any future unified theory will have to explain it.)

Thus, all observed effects of gravity are in fact collective ones. They can become discernible at an elementary interaction level only in sufficiently excited matter, e.g., if an elementary particle is accelerated to the energy $\sim M_{Pl}c^2$, which is far outside the range of any conceivable laboratory. However, such conditions existed in the very early universe, and the physics of these extremal states probably determined the universe's later fate on a cosmological scale.

Given this background, let us now reconsider some features of string models. Their first striking property is a prediction of a definite dimension of the space-time at Planck scale: 26 for bosonic strings and 10 for superstrings. This source of embarrassment in dual models of hadrons now becomes one of the major predictions of the theory. However, this prediction is not directly testable and it poses the problem of explaining the apparent four-dimensionality of the low-energy world. Very loosely speaking, one can imagine that $26 - 10 = 16$ dimensions somehow conspire to accommodate the internal degrees of freedom of fundamental particles (16 being the rank of the gauge groups $E_8 \times E_8$ and $SO(32)$), while the remaining $10 - 4 = 6$ dimensions "spontaneously" compactify at the Planck scale early in the course of cosmological evolution.

Below we shall have more to say about the mathematical origin of these critical dimensions, 26 and 10. It suffices to mention here that their appearance is a pure quantum field theoretical effect.

A second property of string models is the unification of the four known forces, including gravity, in an effective Lagrangian of low-energy approximation to the full theory, which itself is much richer.

String models place very tight constraints on the possible Grand Unification gauge group. It may be $E_8 \times E_8$, one factor being responsible for common matter, with the other one responsible for the so-called dark matter, which interacts with the common one only through gravity.

Finally, string models incorporate supersymmetry into the framework of fundamental physics.

One must add that such a theory actually does not exist yet. We have an ideal picture, a puzzle some of whose fragments have marvelously found their proper places while others still remain a challenge. Besides, all of this may prove someday to be just wishful thinking—as physics. Fortunately, mathematics is less perishable.

Mathematical Structure of Quantum Field Theory

In the following pages I shall try to make explicit some mathematical structures of fundamental physics, stressing specific properties of string models. A model of a physical system starts with a description of a set $\mathcal{P}$ of "virtual classical paths" and an action functional $S: \mathcal{P} \to \mathbb{R}$. Generally speaking, $\mathcal{P}$ is a function space, e.g., a space of maps of manifolds $f: M \to N$, or a product of such spaces. Maps may be subject to some boundary conditions; N may be a bundle over M and $\mathcal{P}$ may consist of sections of this bundle, etc. Usually, M and/or N are (pseudo)Riemannian manifolds (with fixed or variable metrics) and S is obtained by integrating a natural volume form over M or N. Three examples follow.

(i) *General relativity.* Here M is a fixed 4-dimensional C^∞-manifold, $\mathcal{P} =$ space of Lorentz metrics $g = g_{ab}dx^a dx^b$ on M (i.e., sections of $S^2(TM) \to M$ with positivity conditions), and

$$S(M, g) = -(16\pi G)^{-1} \int_M R \, \text{vol}_g \quad \text{(Hilbert-Einstein action)}, \tag{1}$$

where G is the Newton constant, $R =$ Ricci curvature, $\text{vol}_g =$ the volume form of g.

(ii) *Massive point-particle propagating in a space-time (M, g).* Here $\mathcal{P}$ is the set of maps $\gamma : [0, 1] \to M$,

$$S(\gamma) = -m \int_0^1 ds, \tag{2}$$

where $m =$ mass, $ds^2 = \gamma^*(g)$, the induced metric. The image of $[0,1]$ is the particle's virtual world-line.

(iii)   *String propagating in a space-time (M, g).* Here $\mathcal{P}$ is the set of maps $\sigma : N \to M$, where N is a surface whose image is the string's world-sheet,

$$S(\sigma) = -\frac{T}{2} \int_N \text{vol}_{\sigma^*(g)} \quad \text{(Nambu's action)}, \tag{3}$$

where T is the so-called "string-tension" of dimension $(\text{length})^{-2}$ and $\sigma^*(g)$ is the induced metric.

We shall always measure time in length units and action in Planck units (or, as physicists say, put $\hbar = c = 1$).

Classical equations of motion.   These are equations for stationary points of S: $\Delta S = 0$. Finding their solutions or exploring their qualitative properties is the main task of classical mathematical physics.

Quantum expectation value and partition function.   These are given by the following formal expressions (Feynmann integrals):

$$\langle \theta \rangle = Z^{-1} \int_{\mathcal{P}} \theta(p) e^{iS(p)} Dp, \tag{4}$$

$$Z = \int_{\mathcal{P}} e^{iS(p)} Dp, \tag{5}$$

where $\theta : \mathcal{P} \to \mathbb{R}$ is an observable and Dp is a formal measure on $\mathcal{P}$.

Most problems of quantum field theory can be thought of as problems of finding a correct definition and a computation method for some Feynmann path integral. From a mathematician's viewpoint almost every such computation is in fact a half-baked and *ad hoc* definition, but a readiness to work heuristically with such *a priori* undefined expressions as (4) and (5) is necessary in this domain.

There are several standard tricks. First, one tends to work with the so-called Euclidean versions of (4) and (5), where $e^{iS(p)}$ turns into $e^{-S(p)}$. Besides "better convergence" (whatever that means), this makes explicit a basic analogy between quantum field theory in (D space, 1 time) dimensions and statistical physics in $D + 1$ space dimensions, allowing us to use rich technical tools and insight on collective phenomena. Second, one tries to reduce (4) and (5) to finite-dimensional integrals using group invariance and/or approximations. Third, one tries to reduce (4) and (5) to Gaussian integrals, whose theory is the only developed chapter of infinite-dimensional integration, e.g., by means of an appropriate perturbation series.

Here is, for example, a standard heuristic explanation of the correspondence between classical and quantum laws of motion. Assume for simplicity that (with given boundary conditions) the equation $\Delta S = 0$ has only one solution $p = p_o$. Assume also, by analogy with the finite-dimensional case, that the stationary phase approximation is valid, i.e., the quantum expectation value $\langle \theta \rangle = \int_{\mathcal{P}} \theta(p) e^{iS(p)} Dp$ coincides with $\theta(p_o)$ up to a universal factor and a small error. This means that quantum observables practically have their classical values on the classical path.

A necessary condition for the validity of the stationary phase approximation is that $S/\hbar$ be large on $\mathcal{P}$. This is in accordance with the discovery of early quantum theory that the classical regime corresponds to $\hbar \to 0$.

A successful calculation of a path integral often involves one or more limiting processes that differ from the Archimedes-Newton-Lebesgue prescription of adding up an infinity of infinitesimal contributions. In fact, such a calculation usually gives a finite value as a difference or a quotient of two (or more) infinities.

I believe that there is a message in this observation. Each level of reality we become aware of is but a flimsy foam on the surface of an infinitely deep ocean, usually called a vacuum state. It is a state of lowest energy, but its energy is infinite. We are divided by a thin film from an eternal fire, whose first tongues are the flames of the nuclear age. Will a mature string theory start a new auto-da-fé?

Returning to mathematics, it sometimes happens that the infinities of a concrete model apparently do not reduce to a finite number. A notorious example is the Einstein gravity theory (1), which for this reason is called unrenormalizable. Only after an extension to a larger picture, hopefully a stringy one, can gravity become finite.

If a theory is renormalizable (or, as happens with super-symmetric models, even finite), the indeterminacies in choosing "infinite constants" are resolved by experimental data—values of various charges and coupling constants. Ideally, nothing like this should be allowed: a perfect theory must predict everything.

Operator approach.   We described earlier the Lagrangian approach to quantization. There is an alternative approach which in various contexts is called the Hamiltonian, canonical, or operator approach; it takes the following form.

On the space of solutions P of the classical equations of motion $\Delta S = 0$, there is a natural Poisson structure, i.e., a Lie bracket on a space F of functionals on P. For example, in classical mechanics the space of classical paths P can be identified with the phase space, because a classical motion is defined by its initial value of position and momentum. The corresponding Poisson structure is defined by the well-known symplectic form. (This basic example suggests taking for P an appropriate space of boundary values, which is often done in string theory.)

A unitary representation of a subalgebra of F in a Hilbert space H defines an operator quantum description of the system. Of course, such a representation is rarely unique, and the non-uniqueness corresponds to the indeterminacy of path integrals. The so-called geometric quantization is a method of constructing such representations in functions on P, or sections of a bundle on P, or in appropriate cohomology.

An expectation value $\langle\theta\rangle$ of an observable $\theta \in F$ calculated by means of a path integral should coincide with an operator expectation value of the kind $\langle \text{vac}|\zeta(\theta)|\text{vac}\rangle$ (or $\langle\psi|\zeta(\theta)|\phi\rangle$ for appropriate state vectors vac, ζ, $\psi \in H$) defined by a representation ζ. In order that this make sense one should ensure that θ can be considered as a functional on $\mathcal{P}$ uniquely defined by its restriction to $P \subset \mathcal{P}$.

In general, the path integral quantization and operator quantization should be considered as complementing each other rather than being strictly equivalent. This complementarity inherited from classical mechanics reappeared in the guise of the Schroedinger and Heisenberg approaches to quantum mechanics.

In string theory, comparison of the two approaches poses some intriguing problems on the connections between modular forms on Teichmueller spaces and moduli spaces of vector bundles, on the one hand, and representation theory of the Virasoro, Kac-Moody, and similar Lie algebras on the other hand. Before the advent of quantum field theory in this mathematical domain, only genus one modular forms appeared as character series of representations, and even that seemed a mystery.

Symmetries.   The basic structure $(\mathcal{P}, S)$ is often complemented by an action of a group G on $\mathcal{P}$ leaving S invariant (or, infinitesimally, by an action of a Lie algebra $\mathfrak{A}$ which in an infinite-dimensional situation may well be non-integrable). Mathematical aspects of such a picture may have various physical interpretations of which we mention several.

Classical symmetries of a flat space-time acting upon $\mathcal{P}$ give rise to the energy-momentum operators. Noether's theorem on conservation laws is reflected in the structure of the momentum map $\mu\colon P \to \mathfrak{A}^*$, where P is the space of classical motions with an invariant symplectic structure.

Local gauge symmetries in the theory of Yang-Mills fields and diffeomorphisms of the space-time of general relativity give rise to physically indistinguishable states. Thus, in this case one should actually call $\mathcal{P}/G$ the space of virtual paths, choose various G-invariant subspaces as spaces of quantum states, etc. String theory also provides such phenomena.

Path integral (or operator) quantization may break up classical G-invariance of the picture due to the indeterminacies in the regularization scheme.

A precise description of the resulting non-invariance is the subject-matter of the theory of anomalies. In recent years it became clear that the essential aspect of anomalies reflects cohomology properties of G. The disappearance of quantum anomalies is considered to be an important metatheoretical criterion of the consistency of a quantum model. This disappearance led to the discovery of critical dimensions and preferred gauge groups.

Finally, a few words should be said about the conformal group, which is the group of local rescalings of a metric of space-time or string world sheet: $g_{ab} \to e^f g_{ab}$. The conformal invariance of a physical model implies the absence of a natural scale (length, mass, or energy). In the context of statistical physics this happens in the vicinity of phase transitions. One can conjecture that fundamental physics is governed by conformally invariant laws. Anyway, conformal invariance plays an essential role in string theory.

Correspondence principles.   Historically, the correspondence principle is a loosely formulated prescription for deriving classical laws of physics from quantum ones.

Modern fundamental physics is a conglomerate of theories, or models, each of which is applicable at an appropriate scale or is a simplified version of a more adequate but too complex theory. All informal rules of patching together these models at the outskirts of their validity ranges may reasonably be called correspondence principles.

If, as I believe, this openness of physics is its essential characteristic, then correspondence principles themselves may be elevated in status and could eventually become considered as physical laws acting in transition periods as rites of passage. For example, 26, 10, and 4 may have been successive steps of the formation of our space-time from an infinite-dimensional quantum chaos during the first $10^{-?}$ seconds of creation.

Reading Suggestions

Entering string theory is not an easy task for mathematicians. Two recent publications may help to get a general perspective and to choose a particular topic for deeper study: the monograph-textbook [1] and the anthology [2].

Two ICM Berkeley talks [3] and [4] were at least partly devoted to strings. Witten's talk is a beautiful initiation to quantum field theory.

The Belavin and Knizhnik discovery [5], together with previous works by Quillen and Faltings, led to important progress in the theory of determinant bundles [6]–[10], extending Grothendieck's Riemann-Roch theorem. Much remains to be done in this domain, which is simultaneously "the component at infinity" of arithmetical geometry [11], [12].

Path integrals of string theory could eventually be reinterpreted in arithmetical terms via a version of Siegel-Tamagawa-Weil theory. A beautiful recent result by Shabat and Voevodsky, drawing upon previous work by Grothendieck and Bely, shows that natural lattice approximations in string theory are inherently arithmetical ones [13].

For a rich representation-theoretic part of string theory see [14], [15], [10] and many pages of [1] and [2].

Good luck, with a few strings attached!

References

1. M. B. Green, J. H. Schwarz, and E. Witten, *Superstring Theory,* in 2 vols, Cambridge: Cambridge University Press (1987).
2. J. H. Schwarz, ed., *Superstrings: The First Fifteen Years of Superstring Theory,* in 2 vols, Singapore: World Scientific (1986).
3. E. Witten, *Physics and geometry,* Berkeley ICM talk (1986).
4. Yu. I. Manin. *Quantum strings and algebraic curves,* Berkeley ICM talk (1986).
5. A. A. Belavin and V. G. Knizhnik, Algebraic geometry and the geometry of quantum strings, *Phys. Lett.* 168B (1986), 201–206.
6. D. S. Freed, Determinants, torsion, and strings, *Comm. Math. Phys.* 107 (1986), 483–513.
7. J.-M. Bismut, H. Gillet, and C. Soulé, Analytic torsion and holomorphic determinant bundles, preprint, Orsay 87-T8 (1987).
8. P. Deligne, Le déterminant de la cohomologie, Current trends in arithmetical algebraic geometry, 93–177, *Contemp. Math.* 67, Amer. Math. Soc. 1987.
9. A. A. Beilinson and Yu. I. Manin, The Mumford form and the Polyakov measure in string theory, *Comm. Math. Phys.* 107 (1986), 359–376.
10. A. A. Beilinson and V. V. Schechtman, Determinant bundles and Virasoro algebras, *Comm. Math. Phys.* 118 (1988), 651–701.
11. G. Faltings, Calculus on arithmetic surfaces, *Ann. of Math.* 118 (1984), 387–424.
12. Yu. I. Manin, *New Dimensions in Geometry,* Springer Lecture Notes in Math. 1111 (1985), 59–101.
13. V. A. Voevodsky and G. B. Shabat, Equilateral triangulations of Riemann surfaces and curves over algebraic number fields, preprint (1987).
14. G. B. Segal, Unitary representations of some infinite dimensional groups, *Comm. Math. Phys.* 80 (1986), 301–342.
15. B. Feigin and B. Fuchs, Representations of the Virasoro algebra, *Seminar on supermanifolds,* 5 (D. Leites, ed.), Stockholm: Dept. of Math., Stockholm Univ., N 25 (1986).
16. P. Candelas, G. Horowitz, A. Strominger, and E. Witten, Vacuum configurations for superstrings, *Nucl. Phys.* B258 (1985), 46–90.

21

Integrability in Mathematics and Theoretical Physics: Solitons

S. P. Novikov

Mathematical problems cannot always be solved. Sometimes they turn out to be too hard, and their solutions stretch out over decades and even centuries; sometimes they prove to be insoluble "in principle." Alas, we mathematicians are not like those Soviet giants of the thirties who could raise hundreds of calves in one year from two cows or mine a mountain of coal overnight.

Even thousands of years are not long enough to achieve the trisection of the angle and the quadrature of the circle with compass and straightedge; to express the roots of fifth-degree algebraic equations in radicals; to give algorithmic solutions of the problems of group theory and Diophantine equations; to invent an axiom system for arithmetic leaving no statement undecidable within the system. . . .

Generic Laws and Special Cases

Sir Isaac Newton, in his famous "Second letter to Oldenburg" (1676), gave an anagram, i.e., a short encrypted sentence, stating what he regarded as the principal discoveries made in the work. His claim, put in modern terms, was to have found a universal method of solving algebraic and differential equations, by expressing the desired solution as a power series and evaluating its coefficients by substituting it into the equation.

Actually, aside from this simple procedure and its analogs, today there is still no universal theoretical approach. As for a numerical solution on a computer, nobody solves equations this way. No significant quantitative or qualitative information about the behavior of solutions can be obtained in this way, except for an important special case: When the solution you are studying is close to a known solution found in advance by other methods, or when the equation you are studying is close in an appropriate sense to one that is exactly solvable. But to carry this method through, you will need the "exact solution" really *exactly,* at the level of formulas and algebraic or analytic identities, not merely numerical approximations. We will see examples of this in what follows.

In classical physics of the nineteenth and early twentieth centuries, the fundamental laws of nature were expressed by differential equations. In contemporary quantum theory, the picture is

Volume 14, No. 4 (Fall 1992), 13–21

more complicated, but differential equations have not lost their significance. Many special problems are described by differential equations in an approximation depending organically on the particular concrete situation. One sometimes also encounters more complicated types of equation (difference equations, integral equations, equations with delay, etc.).

What differential equations do we need, and which of them do we know how to solve? What can be said in general about the properties of their solutions? How have views of this changed in the last three centuries?

The first important and difficult problem beginning the era of differential equations was solved by Newton: the problem of the motion of two point masses with masses m_1 and m_2 under the action of mutual gravitational attraction. (Let us skip over the question of the role of the apple, and also the question whether Newton wrote down the equations all by himself or with help from another great English physicist, Robert Hooke.) Newton solved the 2-body problem exactly and derived Kepler's laws. To be sure, the 2-body problem is merely an approximate model. In real life, massive bodies are not point masses; the effect of other planets is not so small; and the stability of the full model is dubious.

For more than 150 years many scientists doubted the validity of Newton's gravitational theory of the Solar System. Yet it is valid—even though we are still unable to establish strong stability mathematically for values of the parameters differing as much as our real Solar System does from the exactly solved Newton 2-body model of the Earth and Sun.

This example already gives us an idea of the basic structure of the approach of mathematical physics: to represent the system under study in one sense or another as a small perturbation of an exactly solvable model. Theoretical physicists proceed in this way even when the perturbation is not at all small. What else can they do? There are practically no alternatives. Direct numerical solution is often impossible, and anyway its results may be hard to use theoretically.

The simplest models where all the difficulties can be brought out are what are called autonomous dynamical systems in some "phase space," whose points correspond to states of the system and are given by coordinates $(y) = (y_1, \ldots, y_n)$. The dynamical system is written

$$\frac{dy_j(t)}{dt} = f_j\big(y_1(t), \ldots, y_n(t)\big). \tag{1}$$

It often happens that the phase space has some non-trivial topology, for example in the description of rotations of a body; the coordinates (y) may be only "local" or may have some singularities (as do, for example, polar and spherical coordinates). In Newtonian mechanics, giving the coordinates means specifying positions and velocities of all particles. In many more special problems, giving the coordinates means instead specifying values of a minimal set of parameters needed to characterize the state of the physical system to the degree of approximation relevant to the problem, where this set of parameters may change over time ("dynamical variables").

Systems that can be considered closed in the given approximation have the property that no state loses or gains energy. They are described by so-called conservative (Lagrangian or Hamiltonian) equations.

In dissipative systems, frictional forces are added to the equations of motion—so that energy decreases. In more general systems, the right-hand member of (1) may be more complicated, representing dissipation, outside energy sources, etc.

Such terms as "energy" (and the now forgotten "vis viva") were borrowed at the end of the seventeenth century from the occult sciences. They came to denote in theoretical physics precisely defined mathematical quantities, the so-called integrals of the motion. A general integral of the motion in our case is any function $F(y)$ of the coordinate values (the state of the system) such that its value does not change over time as the system evolves according to (1): $dF(y(t))/dt = 0$, i.e., $F(y(t))$ is constant.

In a typical dynamical system taken at random, there will not be any single-valued continuous integrals of the motion at all. In an autonomous Hamiltonian system, there will be in general only one, the energy. In a Hamiltonian system with a group of symmetries, there are more integrals: as many as there are generators in the group of symmetries. For example, the principle that the laws of physics are invariant under *all* motions of space (translation of time, rotation and translation of space) gives rise to these integrals: energy, momentum (coming from translations), and angular momentum (coming from rotations). One uses these integrals to eliminate some of the coordinates. Thus, from the group of translations, one gets rid of the coordinates and the velocity of the center of mass; the case of rotations is more complicated.

It results from the non-commutativity of the group of rotations that one cannot eliminate all three Euler angles and their rates of change relative to the center of mass; only four variables of relative motion can be eliminated, rather than six as was done for the group of translations.

So there appears in the theory of Hamiltonian systems an algebraic peculiarity, this sensitivity to the algebraic structure of the group of symmetries—in this case, the group of motions of the 3-dimensional Euclidean space $\mathbf{R}^3$ in which we live. (This was the view of the classical mechanics of the seventeenth through the nineteenth centuries. The point of view of the contemporary theory of elementary particles and gravitation is more complicated, and still further complications are in the offing, but here let us not get out of $\mathbf{R}^3$.)

Anyway, we have no hope of finding more general integrals of motion, other than total energy, total momentum, and total angular momentum, for the problem of the motion of n point masses. For $n = 2$, the 2-body problem, these are enough to integrate the equations completely; for $n > 2$, the problem of three or more bodies, they are not. At the end of the nineteenth century, it was even rigorously proved (by Poincaré and others) that in the 3-body problem there are no analytic integrals of the motion except those mentioned already, arising from the group of symmetries of the laws of nature.

At this stage, a prevailing attitude developed toward the problem of integrability, which can be briefly summed up this way: If there is no symmetry, then there are no integrals, at least "as a rule," in sensible problems encountered in natural science. It is necessary to study arbitrary systems of the greatest generality: That is what can describe reality.

Let me give a little commentary on this attitude, in three points, of which the first two are mathematical ("What can we do?") and the third philosophical ("What do we believe?").

(1) *Typical, generic systems certainly should be studied.* Unquestionably, they approximately describe many particular cases. But *how* should they be studied? Already in the 1920s and 1930s, the language of set theory and fractional dimension (fractals) was worked out, and in the 1950s, the concepts of probability in dynamical systems; in the last thirty years, practically no new mathematical concepts and terminology have been added. But how could these things be perceived in mathematical models of actual physical systems? Aside from a few ingenious, specially concocted models in pure mathematics, this did not become possible until the use of modern computers. Ingenuity and analysis are ineffective here, except for systems close to integrable ones. The author is aware of very few examples where analytic methods allowed visualizing stochastic behavior of orbits of an autonomous dynamical system—let alone of general, arbitrarily chosen systems. One very restricted but curious class of such systems I encountered in the early 1970s in the investigation of spatially homogeneous solutions of Einstein's gravitational equations—what are called homogeneous cosmological models. A group of well-known physicists (E. Lifshits, I. Khalatnikov, and their student V. Belinskii) found analytically a very special stochastic régime. But a single exception only confirms the rule. *Without computers you cannot visualize randomness in real systems* (in "general position").

(2) *Integrable systems and systems close to integrable ones can be studied in much greater detail* using and at the same time adding to the arsenal of general algebraic and analytic methods together with computer technology. Computers are not used only for straightforward finding of orbits, but

join with deep mathematics to bring out an incomparably more diverse array of beautiful, intellectually satisfying aspects of the problem.

The general qualitative theory of Hamiltonian systems close to integrable ones was founded by Poincaré. Between 1955 and 1965, Soviet mathematicians, A. N. Kolmogorov and his student V. I. Arnold, and the Western mathematician J. Moser created the remarkable KAM theory, revealing the answer to the question of the behavior of orbits. It must be noted, however, that the theorems of this theory require a degree of closeness to an integrable system which is not found, for example, in the real Solar System, in the problem of motion of a sputnik, etc. (what are called the small parameters, measuring this degree of closeness, have values in reality many orders of magnitude greater than required by the KAM theorems). Nevertheless, the qualitative picture of behavior of orbits is made clear by the KAM theory. Algebraic and functional structures based on it are of decisive significance in theoretical analytic constructions depending on closeness to an integrable system. These properties of systems are ignored by the general qualitative KAM theory, but they take primary roles in the ideas and methods I will be talking about later.

(3) *Physicists and mathematical philosophers of science for the most part do not believe that the laws of nature are to be expressed by the arbitrarily chosen, general equations. Most of them somehow believe "de facto" in a higher reason.* For example, among spherically symmetric potentials the Newtonian potential is distinguished; it is the one which in the 2-body problem admits many periodic motions, those followed (approximately) by the planets revolving about the Sun. This is a specific property just of gravitational forces inversely proportional to the square of the distance. Any perturbation or alteration of the analytic form of the force destroys the periodicity of the motion, and the famous drift of perigee appears. In Newton's law there is hidden an extra symmetry not entailed by any obvious group of symmetries of space-time. But it is exactly this relationship that permits the Solar System to endure! Furthermore, this same hidden symmetry survives in quantum mechanics, and (because electrical and gravitational forces of attraction follow the same law) it determines to first approximation the quantum spectrum of a charged particle and Mendeleev's table.

But is this important now, at the end of the twentieth century? Are we not at the end of the period of development of the mathematical apparatus of natural sciences based on exactly solvable super-symmetric models, going back to Fermat (derivation of the refraction of light from the "divine" variational principle of minimal time) and Newton (solution of the 2-body problem from the laws of mechanics)?

In classical mechanics that period could be considered closed probably by the mid-nineteenth century; in the more modern branches of physics, by the mid-twentieth century. The theory of perturbations, i.e., the study of systems close to exactly solvable systems, remained the basic method in quantum mechanics as well.

How about the yet-undiscovered laws governing the transformation of elementary particles or the evolution of the cosmos in early or later stages at very large size scales? What should we expect—that they will be supersymmetric or arbitrarily chosen? Probably the former; but it would be hard to know in advance just which sort of supersymmetry it will be. *The higher reason is not predictable just from our present knowledge and concepts; we can only make partial predictions. (Lucky for us!)*

Solitons and Scattering

The repertory of admittedly important integrable systems was not added to for a long time. As already noted, rules have exceptions. In the nineteenth century some special systems of mechanics and geometry were found to have further integrals of the motion in the absence of any evident symmetry: Jacobi, Neumann, Weierstrass, Clebsch, Kovalevski, Chaplygin, Steklov, etc. Plainly, some okay people were involved in this! Especially popular in the 1880s was the integrable case of the Kovalevski top in a gravitational field with special values of the inertia tensor. The mathematical investigation of these cases was very beautiful. It relied on deep use of the then-recent theory of Riemann surfaces (its most non-trivial analytic aspects), which until then had not appeared in mathematical physics. With the exception of the simplest case of elliptic functions, it did not appear in physics after that either, for almost one hundred years! Back in the nineteenth century it seemed that these and similar special cases had nothing to distinguish them physically, and consequently everyone outside of a small circle in classical mechanics forgot about them. They did not have at that time any serious influence on the development of physics.

Keep in mind that *the difference between integrable and general Hamiltonian systems appears primarily in quantitative and qualitative characteristics of the orbits of the motion, as long as nothing is known about the system but the shapes of its orbits.*

In the integrable case, the motion can be described (under the very general and easily verifiable hypothesis of "compactness" of the phase surface of constant energy integral) by so-called quasi-periodic functions. The Fourier spectrum reduces to finite number of basic (perhaps incommensurable) frequencies and their integral linear combinations. The spectrum of a generic function is completely different. This is indeed the way to tell the difference between generic and integrable systems experimentally, on the computer—though here one may be misled by computational errors, especially with the computing capabilities which existed 20–40 years ago (and even today it is often very easy to go wrong).

In the early 1950s, the famous quantum physicist (and designer of nuclear reactors) Enrico Fermi, together with the mathematicians Pasta and Ulam, carried out a numerical computer experiment, following the motion of a discrete set of points on the line—a variant of the discretized 1-dimensional string. The aim was to understand how the stochastic equidistribution of energy among degrees of freedom proceeds in the presence of a non-linear term and discretization of the equation. The answer was unexpected: Nothing stochastic was observed; the motion appeared quasi-periodic up to the limits attainable by the best computer available in the United States (run by the best people). Specialists confronted with this did not immediately give a clear-cut response because such computations could not be considered absolutely convincing, and it was the first example of its kind. Other examples of integrability came up in the theory of non-linear waves.

This is a most curious story and is worth recalling. The English (or Scottish) gentleman, applied mathematician, and naval engineer John Scott Russell, as he recounts in a work dated 1844, was observing in 1834 the motion of a barge in a narrow canal while riding on horseback rather fast. The barge for some reason suddenly stopped. The mass of water moving in front naturally did not stop. It continued to move forward by itself and shaped itself into a sort of mound of water, about 30 feet long and 1–1.5 feet high, moving along the canal without noticeable change of shape at a speed of about 8–9 miles an hour. Russell rode after this object (called much later a "soliton") for some time on his horse. It is believed today that this was the canal near the university building in Edinburgh, Scotland. (It is amusing that attempts to reproduce the phenomenon experimentally in the 1980s in that same canal were unsuccessful. The soliton failed to appear when the barge stopped.)

I remark in passing that Russell throughout his scientific teaching and engineering career never had a regular secure long-term job (unlike our scientists), though he once had a chance to build a large ship, and another time he was recommended to the mathematics faculty of Edinburgh University by Hamilton (but did not get the appointment). This information was provided in an article by R. Bullow [1].

Anyway, the well-known scientists Boussinesq (1872) and Rayleigh (1876) found theoretically that a soliton should appear in shallow water and found its analytic form. Then in 1895 the Korteweg–de Vries (KdV) equation was derived, and from that basic equation of shallow-water theory came the already known soliton along with a new object, the cnoidal wave.

Let us continue the story of the fate of these discoveries, leaving out the history of the rigorous derivation of shallow-water solitons from the exact equations of hydrodynamics. The discoveries that concern us came in the 1960s, not from that source but from plasma physics. As a result of the work of the applied mathematicians M. Kruskal and N. Zabusky (United States) and specialists in plasma physics in the USSR and the United States (Sagdeev, Gardner, Morikawa), it became clear that *solitons and the KdV equation can approximately describe the evolution of waves in the most diverse media, under the most simple and crude hypotheses of non-linearity and dispersion, if dissipation is negligibly small.* This equation and its simplest solutions display a certain universality analogous to linear equations like D'Alembert's.

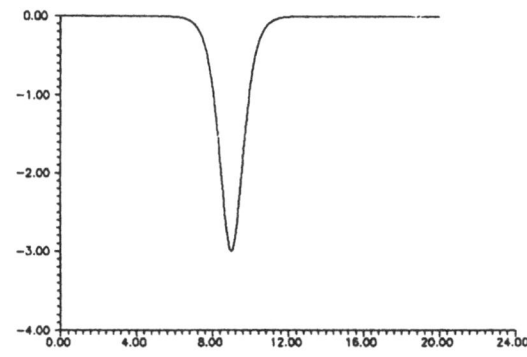

FIGURE 21-1 *Soliton.*

Its traveling-wave solutions $u(x - ct)$, already known in the nineteenth century, occur in the form of solitons (Figure 21-1),

$$u = -2\frac{K^2}{ch^2\big(K(x - 4K^2t)\big)}, \quad c = 4K^2,$$

or combs (Figure 21-2): periodic cnoidal waves,

$$u = 2\wp(x - ct + iw' \mid g_2, g_3) - \frac{c}{6}.$$

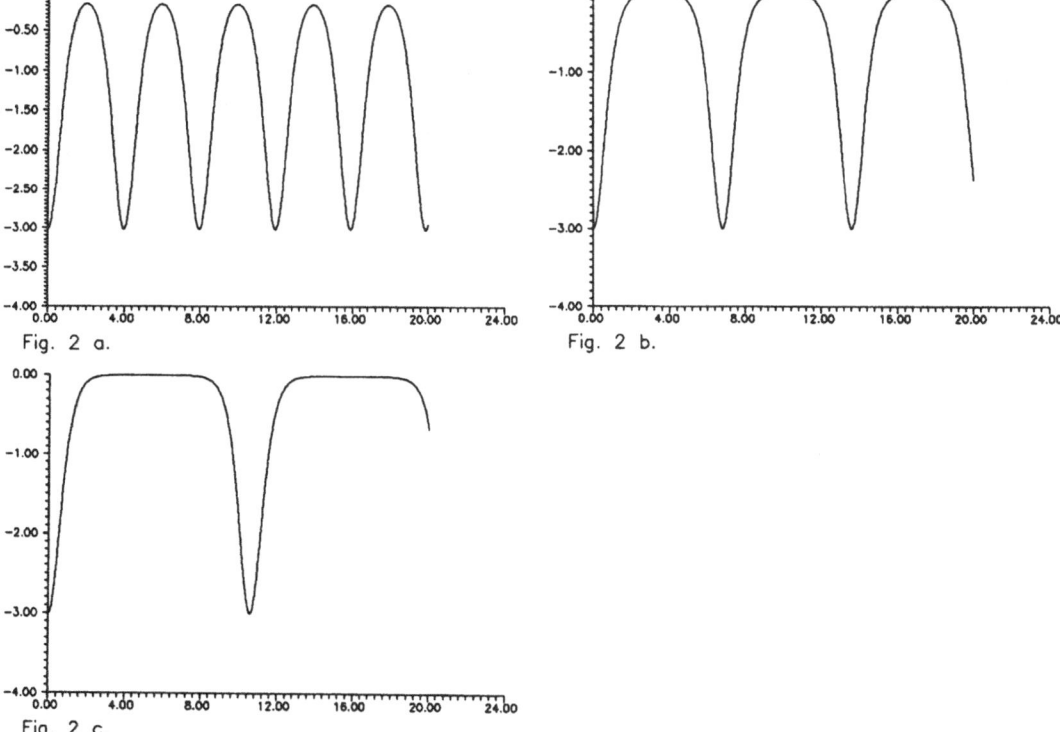

Fig. 2 a.

Fig. 2 b.

Fig. 2 c.

FIGURE 21-2 *Cnoidal waves.*

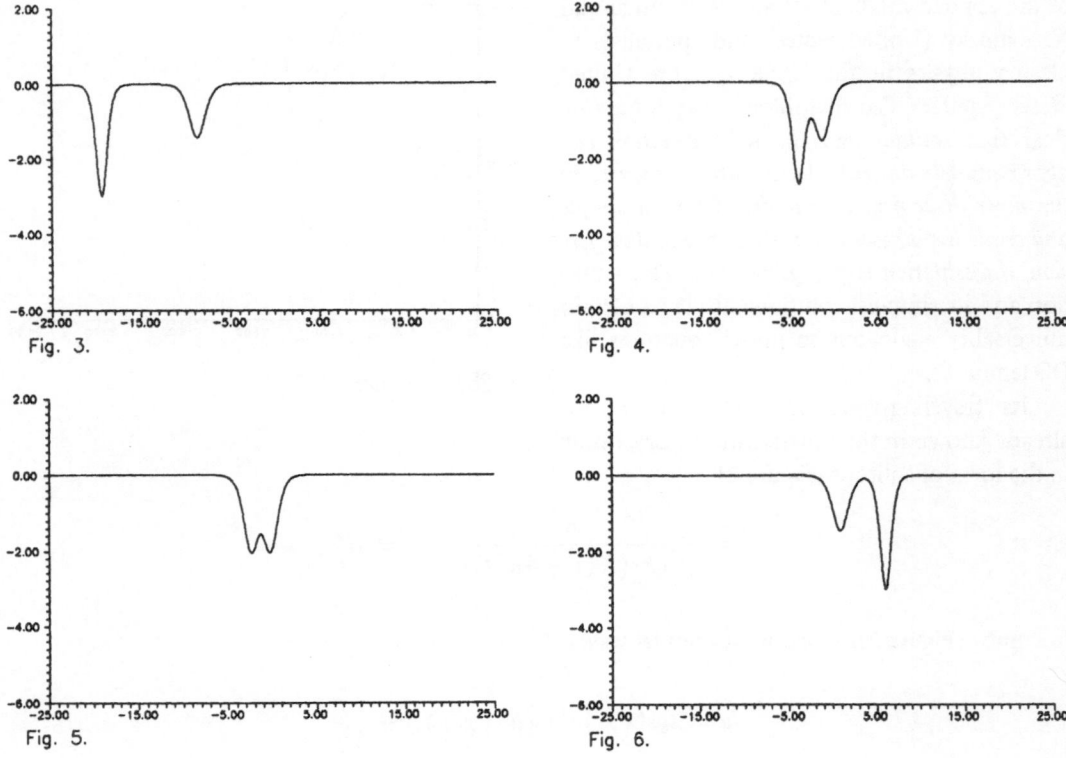

Fig. 3.   Fig. 4.

Fig. 5.   Fig. 6.

FIGURES 21-3–21-6 *Interaction of two solitons.*

(Here $\wp$ denotes the Weierstrass $\wp$-function [2].) Figures 21-2a to 21-2c illustrate the decay of a cnoidal wave to solitons as the parameters g_2, g_3 tend to the degenerate point $g_2^3 = 27g_3^2$.

The maximal height of a soliton is proportional to the square of the phase velocity c. In 1965, Kruskal and Zabusky numerically investigated the evolution of the waveform when the initial situation is the sum of two solitons with speeds $c_1 < c_2$, with the center of the faster soliton starting farther to the left:

$$u = -2\partial_x^2 \ln \det A,$$

$$A = \begin{bmatrix} (e^{2K_1(x-4K_1^2 t)} + 1)/(2K_1) & 1/(K_1 + K_2) \\ 1/(K_1 + K_2) & (e^{2K_2(x-4K_2^2 t)})/(2K_2) \end{bmatrix}$$

How would they evolve? The result turned out most interesting (Figures 21-3–21-6). The left soliton caught up with the right, as expected (it *did* have higher speed), and they flowed together. But quite unexpectedly, after some time had elapsed, the same solitons could again be made out, retaining their individuality completely. The faster soliton, as it were, had jumped over the other (merging with it in the course of the jump), exchanging with it a certain phase-shift. Regarding a soliton as something like a particle, one could say there had been an elastic scattering event, with individuality of particles conserved, though during it they overlapped, producing a complicated waveform. Such behavior is the opposite of stochastic! Kruskal as early as 1965 found a series of other conservation laws (integrals of the motion) for the KdV equation, which somehow makes it clearer how the appearance of the initial shape can be remembered.

Bear in mind that soliton solutions (i.e., isolated traveling waves, perhaps with internal structure) occur for a wide class of systems, some of them stochastic or non-conservative, etc., but that the behavior that has been described for a pair of solitons as in KdV theory is a sign of some internal degeneracy, some hidden symmetry. Two years later the famous 1967 paper of Gardner, Green, Kruskal, and Miura (GGKM) gave, in a certain sense, an exact construction of the general solution of the KdV equation in the soliton case, when $u(x, t)$ falls off with prescribed speed for large $|x|$ (evolution in the class of rapidly decreasing functions).

The method was striking. It depended on the use of results of quantum scattering theory, both the direct and the inverse problems, which had been solved some time earlier by the Soviet mathematicians I. M. Gelfand, B. M. Levitan, and V. A. Marchenko for other purposes. Quantum theory came into the GGKM procedure as a mathematical trick! A year later the American mathematician P. Lax made sense out of the algebra underlying the procedure (see box, p. 247); the connection between the GGKM procedure and the language of Hamiltonian mechanics was established in 1971 by the Soviets L. D. Faddeev and V. E. Zakharov, and also by Gardner; Zakharov together with A. B. Shabat in the USSR, as well as Lamb in the United States, found in 1971 a number of new physically important integrable systems in which the Lax representation and the GGKM procedure work for rapidly decreasing waves "of soliton-like type." Beginning in 1973, systems admitting a Lax representation were found in profusion.

Among such systems there are now some which are 2-dimensional (but not 3-dimensional as yet, it seems—at least no important ones). In none of them would you say there is any evident symmetry in a traditional sense. Among the 2-dimensional integrable systems with hidden symmetry appeared a known equation, which had been introduced by Soviet plasma physicists B. Kadomtsev and his student V. Petviashvili in the study of transverse stability of a 1-dimensional effect described by the KdV equation. This equation, called the KP equation, has a very beautiful mathematical theory.

All in all, integrable systems are numerous among physically important universal systems describing processes in the first non-linear approximation, in one and sometimes in two spatial dimensions. And they all have in common a hidden algebra—some analog of the so-called Lax representation. I will try (below) to give an idea what that is. The point for now is that this representation associates with the non-linear equation a certain auxiliary linear operator L, analogous to the energy operator or Hamiltonian in quantum mechanics, knowledge of whose spectrum can help in solving the non-linear system. For the KdV equation, L is a scalar Schrödinger operator with potential $u(x, t)$, where t is a parameter. In the above-mentioned rapidly-decreasing case, the spectrum is described by scattering theory for L.

I do not want here to get into the details of the mathematics of scattering theory for 1-dimensional Schrödinger or Dirac operators or whatever. What is important here is that the coefficient functions of these fundamental quantum mechanical operators describe states of our non-linear system in soliton theory. In particular, for the Korteweg–deVries equation, we have the Schrödinger operator

$$L = -\frac{d^2}{dx^2} + u(x, t)$$

acting on an arbitrary function $\psi(x)$ by

$$L[\psi(x)] = -\psi'' + u(x, t)\psi.$$

The notion of scattering makes sense when the coefficient function ("potential") $u(x, t)$ is, for every value of t and for $|x| \to \infty$, rapidly decreasing in some good sense. Therefore, sufficiently far away (mathematicians say, for $x \to \infty$ or $x \to -\infty$), the eigenfunctions of the Schrödinger operator, i.e., the solutions of $-\psi'' + u\psi = \lambda\psi$, hardly differ from ordinary exponentials $\exp(ikx)$, or more precisely from linear combinations of $\exp(ikx)$ and $\exp(-ikx)$ (the eigenvalue is positive, being the square of the wave number: $\lambda = k^2$, $k \in \mathbf{R}$). The comparison of the asymptotics of the same eigenfunction for $x \to \infty$ and for $x \to -\infty$ constitutes what mathematicians call "scattering." This connection of scattering theory with exponentials $\exp(\pm ikx)$ makes the method of solving the KdV equation by the GGKM setup in a sense analogous to the well-known Fourier method. Remember that that method, discovered by Bernoulli and D'Alembert in the eighteenth century in connection with oscillations of a string, became later (beginning with Fourier) a universal method of solving linear differential equations with constant coefficients. Exactly this class of integrable systems has remained the basic mathematical framework of twentieth-century theoretical physics.

Numerous works have followed the idea of the GGKM method (or method of inverse scattering problems, whose Russian acronym is MOZR) as a non-linear analog of the Fourier method [3].

New Exact Integrals

However, in the modern theory of solitons this point of view is seen to be incomplete. It too is bound by the "soliton-like" character of the waveform $u(x, t)$, namely, the property of being rapidly decreasing in $|x|$. This point of view fails for the simplest and most important class of periodic waves $u(x, t)$, those having a period T such that $u(x + T, t) = u(x, t)$ for every value of the time t. There are other classes of functions, important in the theory of non-linear oscillations and quantum field theory, for which the KdV equation is still not solved. Nor was any bridge found to join the "nonlinear Fourier method" with the famous integrable cases of classical mechanics, whose complicated and obscure development I have already recalled. There ought to be in anything new some kernel, some essential part, which harks back to something old and "well-forgotten"—that is, forgotten because at the time it could not be understood (even by its authors)!

Then what about the periodic case? Well, the cnoidal wave, a solution of the KdV equation known back in the nineteenth century, is periodic in x. The subject of periodic (and quasi-periodic) solutions of the KdV equation turned out to be complicated mathematically. A successful approach was found in my work in 1974 and developed by me together with my students Dubrovin and Krichever and by others: Leningrad colleagues, especially Its and Matveev, and many Western mathematicians including P. Lax, H. McKean, and P. Van Moerbeke. It starts from the same algebra, but here the mathematical objects which come up are typical of solid-state quantum mechanics: the Schrödinger operator with a periodic potential, the Bloch waves describing quantum states of electrons in crystals. In contrast to the scattering theory associated with rapidly decreasing potentials, the inverse problems of quantum mechanics had never been solved in the periodic case prior to soliton theory. Their solution turned out to be inextricably linked with soliton theory, with the KdV equation; both theories were thereby enriched.

The mathematical technique for the periodic problem unified new results in the non-linear KdV equation with a deeper understanding of spectral theory of 1-dimensional periodic structures (crystals) based on the theory of Riemann surfaces and the related computations with special functions.

This marked the appearance of a bridge joining modern integrable systems of the theory of non-linear waves (soliton theory) with the integrable cases of the top from the nineteenth century which I recalled earlier, citing S. Kovalevski and others. Riemann surfaces finally entered the toolbox of mathematical physics, after a lapse of 100 years.

All this, of course, applies not only to the KdV equation but also to all the other integrable systems of soliton theory discovered in the 1970s and 1980s in the USSR, the United States, England, Italy, France, Japan, etc. The best known of them are the so-called 1-dimensional non-linear Schrödinger equation; the famous "sine–Gordon" equation, which already had come up in Lobachevskii geometry (negative curvature) in the nineteenth century and reappeared in the theory of superconductivity (the Josephson effect) and other physical problems in the 1960s and 1970s; discrete systems (the so-called Toda chains and the discrete analog of KdV); and many, many others

Consider the solution $\Psi_+(x, k)$ such that

$$\Psi_+ \sim e^{ikx}, \quad x \to -\infty,$$
$$\Psi_+ \sim a(k, t)e^{ikx} + b(k, t)e^{-ikx}, \quad x \to +\infty,$$
$$k \in \mathbf{R}, \quad L[\Psi_+] = k^2\psi_+.$$

For the real potential $u(x, t)$, we have

$$|a|^2 - |b|^2 = 1.$$

THEOREM (GGKM).   The KdV equation for $u(x, t)$ $[u_t = 6uu_x - u_{xxx}]$ implies the following simple equations for the "scattering parameters" a, b:

$$\frac{\partial a}{\partial b} \equiv 0, \quad \frac{\partial b}{\partial t} = 8ik^3b.$$

The last equation may be easily deduced from the "Lax representation" for the operators $L = -\partial_x^2 + u$, $A = -\partial_x^3 + \frac{3}{2}u\partial_x + \frac{3}{4}u_x$.

LEMMA.   The Heisenberg equation

$$\frac{\partial L}{\partial t} = [L, A] = LA - AL$$

and the KdV equation

$$u_t = 6uu_x - u_{xxx}$$

are equivalent (please check it by an elementary formal calculation).

(Zakharov, Shabat, Ablowitz, Kaup, Newell, Segur, Hirota, Hénon, Flaschka, Manakov, Moser, Calogero, Degasperis, etc.).

Let me try to explain for the example of the autonomous dynamical system (1) how we are led to the new integrals, spectra of operators, and Riemann surfaces—though it will still be hard to see why these simple ideas have such deep consequences.

Assume that we are given, on the phase space with coordinates $(y) = (y_1, \ldots, y_n)$, two matrix-valued functions depending on an additional parameter: $L(y, \cdot), A(y, \cdot)$.

These matrix functions must be chosen such that the equation of the dynamical system (1)

$$\frac{dy_j}{dt} = f_j(y_1, \ldots, y_n)$$

is completely equivalent to a matrix Heisenberg equation of quantum mechanics

$$\frac{dL(y(t), \lambda)}{dt} = LA - AL = [L, A] \tag{2}$$

(in soliton theory we then say that (1) admits a Lax representation).

The important thing is that this relationship occurs often. It is what determines the new type of hidden symmetry. Let us look at the properties of this equation.

1. The eigenvalues of L are integrals of the system (1); for we see at once that they are independent of t if they satisfy the matrix equation (2). This is a situation familiar from quantum mechanics.
2. If L depends upon the numerical parameter λ, then its eigenvalues μ_j as functions of λ are multivalued functions with branch points, and indeed they define, in the complex domain, a Riemann surface consisting of the roots of the characteristic equation for the eigenvalues of $L(\lambda)$: $\det(L(\lambda) - \mu E) = 0$. Here the determinant of $L - \mu E$ has been set equal to zero. It is a polynomial in two variables λ, μ, and the solutions for μ are being considered as functions of λ.

Example 1.   Let us take the simplest 2-by-2 example. Here $n = 3$, $y_1 = u$, $y_2 = u_x$, $y_3 = u_{xx}$. If we take matrix functions of the form

$$L = \begin{bmatrix} a & b \\ c & d \end{bmatrix}, \quad A = \begin{bmatrix} 0 & 1 \\ u - \lambda & 0 \end{bmatrix},$$
$$a = u_x = -d, \quad b = 2u + 4\lambda, \quad c = -u_{xx} + 2u^2 + 2\lambda u - 4\lambda^2,$$

then the Heisenberg–Lax equation $L' = [L, A]$ reduces to a differential equation for the function $u(x)$ (check it!). A solution of this equation is (up to a constant) the Weierstrass elliptic function. This function $u(x - ct)$ is also a cnoidal wave of the KdV equation (the soliton being a degenerate special case).

The determinant of $L - \mu E$ is a quadratic trinomial in μ with coefficients depending upon λ. Its zeros describe the simplest Riemann surface for a 2-valued function:

$$\mu^2 = \sqrt{R(\lambda)}, \quad R(\lambda) = a^2 + bc = -\det \Lambda(\lambda).$$

The branch points of the Riemann surface are at the zeros of the polynomial $R(\lambda)$ and at $\lambda = \infty$. There are three in all. The coefficients of $R(\lambda)$ are integrals of the differential equation; the solution of the equation is in terms of the well-known Weierstrass elliptic function associated with the Riemann surface. This very Riemann surface turns out to have deep significance in the theory of the quantum Schrödinger operator with periodic potential $u(x)$ which determines the electron spectrum. Even this example gives a methodological indication; things are much the same in more complicated situations, where the Riemann surfaces and their special functions are more complicated.

This, then, is how the higher symmetry of dynamical systems arises, though it was not until the mid-1970s that this was understood.

I will not try to describe the many deep investigations in this field in the last fifteen years. It is perhaps worth repeating, however, that the mutual relationship

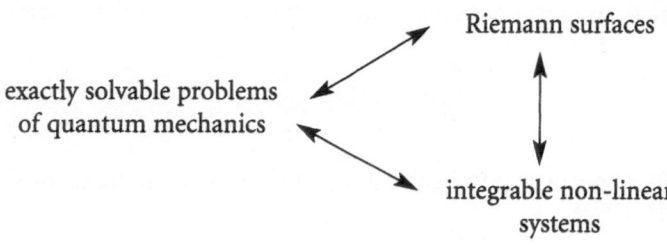

has paid off not only for non-linear systems but also reciprocally for quantum mechanics and theory of Riemann surfaces.

In some interesting models of the solid state, the quantum-mechanical Schrödinger operator (operator of energy) is not at all arbitrary as in other cases, but is specially constructed from the variational principle of "minimal free energy." It can be related by the GGKM–Lax approach to an equation of KdV type, which plays the role of mathematical symmetry of the quantum problem. This approach made it possible to solve the Peierls–Fröhlich model from the 1930s, and it became clear that it described a number of contemporary experiments, in particular the phenomenon of "charge waves" in 1-dimensional (quasi-1-dimensional) substances. This research was done by physicists together with the mathematician Krichever at the Landau Institute of the Soviet Academy, about ten years ago (I. Dzyalashinskii, S. Brosovskii, and others).

In this important example, the situation is a sort of dual or inverse to that of GGKM: It is the quantum problem that has physical content; the Schrödinger equation describes the motion of real electrons, whereas the non-linear solitons of the KdV-like system are only a "symmetry group" of the Peierls–Fröhlich model which allow it to be exactly solved.

In recent years, these ideas of soliton theory have spread to quantum field theory and statistical physics, where curious exactly solvable models are also occasionally found. Hidden algebraic mechanisms of integrability are inherited both from soliton theory and from certain other ideas from within quantum theory, which had its own history of exactly solvable models originating in famous works of the 1930s and 1940s by H. Bethe, L. Onsager, and other leading physicists.

The remarkable and in some sense exactly solvable theory of quantum strings has been a major concern of theoretical physicists for some years now. They expected "everything in the world" from them; no doubt they will at least get *something* useful.

Strings remain a focus of attention. The ideas involved are not infrequently borrowed from soliton theory—but I cannot get into that because it would require a whole separate article. Let me just say that the theory of Riemann surfaces plays a fundamental role both in the theory of strings and in the so-called conformal field theory (CFT), growing out of the work of A. Polyakov and his colleagues A. Zamolodchikov, G. Belavin, and V. Knizhnik, in the early 1980s at the Landau Institute [4].

What moral might I draw for the reader? I do not know if I have succeeded in convincing anybody of anything; but for myself, reflecting on my experiences in science and those of friends and colleagues I worked with or learned from, I am led to this general conclusion: The finding of new exactly solvable models has been centrally involved in the evolution of theoretical physics and mathematics; its role has waxed and waned. It certainly seems that over the last twenty years, after an appreciable break, completely new solvable models came to the place of honor in our fields; and yet it may be that now there will follow an appreciable period of consolidation, centered on the development and application of this body of ideas, without the introduction of essentially new basic models.

References

1. R. Bullow, *Solitons* (M. Lokshmanan, ed.), New York: Springer-Verlag (1987).

2. *Handbook of Mathematical Functions* (M. Abramowitz and I. Stegun, eds), New York: Dover Publications, Inc. (1965).

3. For example, S. Novikov, V. Manakov, L. Pitaevskii, and V. Zakharov, *Theory of Solitons*, New York: Plenum (1984).

4. See *Physics and Mathematics of Strings* (dedicated to the memory of V. Knizhnik) (Brink, Polyakov, and Friedan, eds.), Singapore: World Scientific (1990).

22

On Newton's Problem of Minimal Resistance

Giuseppe Buttazzo and Bernhard Kawohl

1. Introduction

In 1685, Sir Isaac Newton studied the motion of bodies through an inviscid and incompressible medium. In his words (from his *Principia Mathematica*):

> If in a rare medium, consisting of equal particles freely disposed at equal distances from each other, a globe and a cylinder described on equal diameter move with equal velocities in the direction of the axis of the cylinder, (then) the resistance of the globe will be half as great as that of the cylinder. . . . I reckon that this proposition will be not without application in the building of ships.

The problem of finding a body of minimal resistance to motion in a medium seems to be one of the first problems in the calculus of variations (see, for instance, Goldstine [4]). It can be described roughly as follows. Suppose a body moves with a given constant velocity through a fluid, and suppose that the body covers a prescribed maximal cross section (orthogonal to the velocity vector) at its rear end. Find the shape of the body which renders its resistance minimal.

The solution depends on how we define the resistance of a body. The Newtonian pressure law states that the pressure coefficient is proportional to $\sin^2\vartheta$, with ϑ being the inclination of the body contour with respect to the free-stream direction. The sine-squared pressure law can be easily deduced from the assumption that the fluid consists of many independent particles with constant speed and velocity parallel to the stream direction, that the interactions between the body and the particles obey the usual laws governing elastic shocks, and that tangential friction and other effects can be neglected (see Figure 22-1).

So the resistance of a body depends only on its geometry. Suppose that its front end is described by a function $u(x) \geq 0$ defined on the horizontal bottom $\Omega \subset \mathbf{R}^2$. Then it is easy to see that the resistance of the body is proportional to the integral

$$F(u) = \int_\Omega \frac{1}{1 + |Du|^2} dx, \tag{1.1}$$

Volume 15, No. 4 (Fall 1993), 7-12

assuming that the stream direction is vertical downwards. In particular, if the body is a half-sphere of radius R, we have $u(x) = \sqrt{R^2 - r^2}$, and an easy calculation gives the relative resistance

$$C_0 = \frac{F(u)}{\pi R^2} = 0.5$$

as predicted by Newton in 1685. Other bodies with the same value of C_0 are illustrated in Figure 22-2.

Even though Newton's model is only a crude approximation to real physics, it appears to provide good results in the following situations (see, for instance Funk [3]): for a body in a rarefied gas with low speed, for bodies which move in an ideal gas with high Mach number, and for slender bodies. The literature on Newton's problem and on its consequences for low-speed ballistics and hypersonic aerodynamics is very wide; the

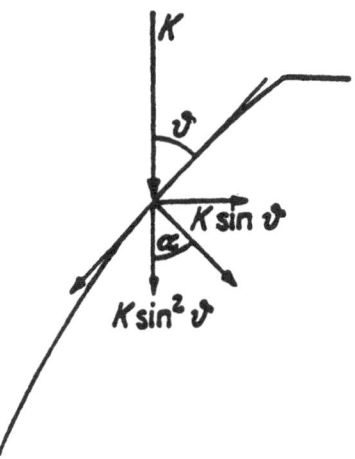

FIGURE 22-1 *Sine-squared pressure law.*

interested reader can find problems related to Newton's model in the books by Miele [9] and Hayes and Probstein [5] and references therein.

However, as far as we know, most of the literature is concerned with the case of rotationally symmetric bodies, for which the analysis reduces to ordinary (i.e., one-dimensional) calculus of variations; in the present article, always within the framework of the Newtonian sine-squared pressure law, we give a first attempt to treat non-symmetric cases.

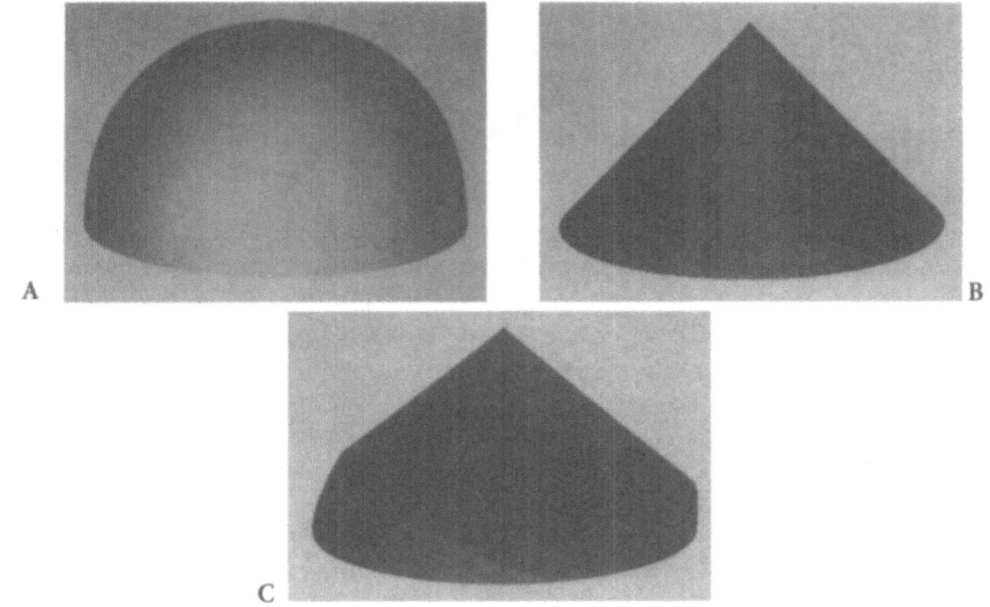

FIGURE 22-2 *(A) Half-sphere; (B) Cone; (C) Pyramid.*

2. The Model

Consider a three-dimensional body E whose horizontal bottom is a subset Ω of $\mathbf{R}^2$, and assume that the upper boundary of E is given by the graph of a function $u \geq 0$ defined on Ω. In other words,

$$E = \{(x, y, z) : (x, y) \in \Omega, 0 \leq z \leq u(x, y)\}.$$

As explained in the Introduction, the resistance of the body E is proportional to the integral (1.1) and so Newton's problem of minimal resistance can be stated as follows: minimize (1.1) over a suitable class of functions u.

Now, the integral functional F above is neither convex nor coercive. Therefore, obtaining an existence theorem for minimizers via the usual direct method in the calculus of variations fails. Moreover, if we do not impose any further constraint on the competing functions u, the infimum of the functional in (1.1) turns out to be zero, as is immediately seen by taking, for instance,

$$u_h(x) = h \operatorname{dist}(x, \partial\Omega)$$

for every $h \in \mathbf{N}$ and letting $h \to +\infty$. Therefore no function u can solve the minimization problem

$$\min \int_\Omega \frac{1}{1 + |Du|^2} dx \tag{2.1}$$

because the integrand is strictly positive on Ω. Even a constraint of the form

$$0 \leq u \leq M \tag{2.2}$$

does not suffice. Indeed, a sequence of functions like

$$u_h(x) = M \sin^2(h|x|)$$

satisfies the constraint (2.2) but is such that

$$\lim_{h \to +\infty} F(u_h) = 0,$$

and again problem (2.1) has no solution.

It is, therefore, necessary to restrict the class of admissible functions for (2.1) beyond the constraint (2.2), and we shall consider only convex bodies E, that is, functions u which are concave on Ω:

$$\min\{F(u) : 0 \leq u \leq M, u \text{ concave on } \Omega\}.$$

Other kinds of constraints can be imposed (see, for instance, Miele [9]): Instead of (2.2), we may consider a bound on the surface area of E

$$\int_\Omega \sqrt{1 + |Du|^2} dx + \int_{\partial\Omega} u \, dH^{n-1} \leq c,$$

or on the volume of E

$$\int_\Omega u \, dx \leq c.$$

From the mathematical point of view, the concavity constraint on u is strong enough to provide an extra compactness which implies the existence of a minimizer; from the physical point of view, a motivation for this constraint is that, thinking of the fluid as of many independent particles, each particle hits the body only once. If E is not convex, it could happen that a particle hits the body more than once. Because $F(u)$ measures only the resistance due to the first shock, it would no longer reflect the total resistance of the body.

Note that convexity of E is sufficient (but not necessary) to guarantee single collision between particles and body; we shall come back to this observation in the last section of the paper.

Other classes of functions u, even if less motivated physically, can be considered from the mathematical point of view; a possible alternative could be the considerably larger class of quasi-concave functions—that is, of functions u whose upper level sets $\{x \in \Omega : u(x) \geq t\}$ are all convex. Note that in the radially symmetric case, a function $u = u(|x|)$ is quasi-concave if and only if it is decreasing as a function of $|x|$. Another intermediate class of admissible functions for which the problem can be studied is the class of superharmonic functions; this will be done in a forthcoming paper [2].

3. The Radial Case

Before studying general convex bodies, it is instructive to investigate radial shapes; this way we shall recover some known results. The first rigorous proof of existence in the radial case is attributed to Kneser [7], even though the explicit form of the solution was already conjectured by Newton himself (see, for instance, Goldstine [4] or Tonelli [12]).

Let Ω be the ball $B(0, R)$, and let us consider only functions of the form $u = u(|x|)$ with u non-increasing as a function of $|x|$. As described in the previous section, the class of such functions coincides with the class of radially symmetric quasi-concave functions. Then, after integration in polar coordinates, the functional F becomes

$$F(u) = 2\pi \int_0^R \frac{r}{1 + |u'(r)|^2} dr,$$

so that the resistance minimization problem can be written in the form

$$\min \left\{ \int_0^R \frac{r}{1 + |u'(r)|^2} dr : u \text{ non-increasing, } 0 \leq u \leq M \right\}. \tag{3.1}$$

It is easy to see that the competing functions u must satisfy the conditions $u(0) = M$ and $u(R) = 0$; in fact, if one of them were violated, the function

$$w_\varepsilon(r) = (1 + \varepsilon)\big(u(r) - u(R)\big)$$

would provide a resistance strictly less than $F(u)$, for some $\varepsilon > 0$. Therefore in (3.1) the conditions $u(0) = M$ and $u(R) = 0$ can be added.

By setting $v(t) = u^{-1}(M - t)$, problem (3.1) can be rewritten in the more traditional form

$$\min \left\{ \int_0^M \frac{vv'^3}{1 + v'^2} dr : v \text{ increasing, } v(0) = 0, v(M) = R \right\}. \tag{3.2}$$

The functional in (3.1) is defined even for a general (discontinuous) non-increasing function as

$$\int_0^R \frac{r}{1 + u_a'^2} dr,$$

where u'_a is the absolutely continuous part of the measure u' with respect to Lebesgue measure; analogously, in (3.2) a general increasing function v provides the value

$$\int_0^M \frac{vv_a'^3}{1+v_a'^2}dt + \int_{[0,M]} vv_s', \tag{3.3}$$

where v_s' is the singular part of the measure v' with respect to Lebesgue measure, and the second integral in (3.3) must be interpreted in a particular BV sense. An equivalent simpler expression for (3.3) is

$$\frac{R^2}{2} - \int_0^M \frac{vv_a'}{1+v_a'^2}dt.$$

We refer to Marcellini [8] and Botteron & Marcellini [1] for a rigorous study of minimization problems of this form in the class of functions which may have a jump at the origin. It is shown there that the minimization problem (3.1) actually admits a solution which is concave and solves the Euler-Lagrange equation

$$ru' = C(1+u'^2)^2 \text{ on } \{u' \neq 0\}$$

for a suitable constant $C < 0$. Therefore, the minimization problems on the classes of concave functions and of quasi-concave functions admit the same solutions.

In the radial case, the solution u can actually be explicitly computed, and—amazingly enough—Newton already gave this solution some 300 years ago; see [4]. Indeed, consider the function

$$f(t) = \frac{t}{(1+t^2)^2}\left(-\frac{7}{4} + \frac{3}{4}t^4 + t^2 - \log t\right) \quad \forall t \geq 1;$$

we can easily verify that f is strictly increasing, so that the following are well defined:

$$T = f^{-1}\left(\frac{M}{R}\right), \quad r_0 = \frac{4RT}{(1+T^2)^2}.$$

With these notations, we get

$$u(r) = M \quad \forall r \in [0, r_0]$$

and, in parametric form,

$$\begin{cases} r(t) = (r_0/4t)(1+t^2)^2 \\ u(t) = M - (r_0/4)\left(-\frac{7}{4} + \frac{3}{4}t^4 + t^2 - \log t\right) \end{cases} \quad \forall t \in [1, T].$$

Note that $|u'(r)| > 1$ for all $r > r_0$ and that $|u'(r_0^+)| = 1$. In fact, as Newton already observed, if u has a small derivative in an interval $[a, b]$, then u can be changed in $[a, b]$ to a function whose slope is first 0 and then -1. This will decrease resistance. In modern terms, we can say the reason is that the convex relaxation f^{**} of the function $f(s) = 1/(1+s^2)$ which appears in (1.1) detaches from f in the interval $(0, 1)$, so that the solution *smells* the concavity of f and stays away from it.

Denoting by C_0 the relative resistance

$$C_0 = \frac{2}{R^2}\int_0^R \frac{r}{1+u'^2}dr,$$

we have $C_0 \in [0, 1]$, and some approximate calculations give

	$M = R$	$M = 2R$	$M = 3R$	$M = 4R$
r_0/R	0.35	0.12	0.048	0.023
C_0	0.37	0.16	0.0082	0.0049

The optimal radial shape for $M = R$ is shown in Figure 22-3. Moreover, we obtain

$$\frac{r_0}{R} \sim \frac{27}{16}\left(\frac{M}{R}\right)^{-3} \quad \text{as} \quad \frac{M}{R} \to +\infty$$

$$C_0 \sim \frac{27}{32}\left(\frac{M}{R}\right)^{-2} \quad \text{as} \quad \frac{M}{R} \to +\infty.$$

For the optimal frustum cone (see Figure 22-4), an elementary calculation gives

$$C_0 = \frac{r_0}{R} = 1 - \frac{M}{2R^2}\left(\sqrt{M^2 + 4T^2} - M\right) \sim \left(\frac{M}{R}\right)^{-2} \quad \text{as} \quad \frac{M}{R} \to +\infty.$$

In particular, for $M = R$ we get $C_0 = r_0/R \sim 0.38$. Note that for a slender cone, it is still $C_0 \sim (M/R)^{-2}$. Summarizing the results which are known in the radial case, we get:

1. problem (3.1) admits a solution u;
2. the modulus of the gradient $|Du|$ stays outside $(0, 1)$;
3. u is concave;
4. $u(R) = 0$;
5. the solution is unique;
6. there is always a *flat region*, that is, there exists $r_0 > 0$ such that $u(r) = M$ for every $r \in [0, r_0]$;
7. u is Lipschitz continuous up to the endpoint.

FIGURE 22-3 *Newton's optimal shape for* $M = R.$

FIGURE 22-4 *The optimal frustrum cone for* $M = R.$

4. The Non-radial Case

Let Ω be a bounded open convex subset of $\mathbf{R}^n$ ($n = 2$ in the physical case) and let $M > 0$ be given. Then, according to Section 2, Newton's problem of least resistance, for convex bodies having given cross section Ω and given height M, reduces to the minimization problem

$$\min\left\{\int_\Omega \frac{1}{1 + |Du|^2}dx : 0 \leq u \leq M, u \text{ concave}\right\}. \tag{4.1}$$

Note that every concave function u is locally Lipschitz continuous in Ω, so that $Du \in L^\infty_{\text{loc}}(\Omega; \mathbf{R}^n)$ and the integral in (4.1) is well defined.

THEOREM 4.1. For every $M > 0$, problem (4.1) admits at least one solution, and every solution u has the property that $|Du| \notin (0, 1)$.

The proof of existence relies on the following compactness result (see Marcellini [8]).

PROPOSITION 4.2. For every $M > 0$ and every $p < +\infty$, the set

$$\mathcal{A}_M = \{u \text{ concave on } \Omega, 0 \leq u \leq M\}$$

is compact with respect to the strong topology of $W^{1,p}_{\text{loc}}(\Omega)$.

By using Proposition 4.2, the existence statement of Theorem 4.1 follows by the usual direct methods of the calculus of variations. More generally, in this way it can be proved that for every $M > 0$, the functional

$$G(u) = \int_\Omega g(x, u, Du)dx$$

admits a minimum point on $\mathcal{A}_M$ provided $g : \Omega \times \mathbf{R} \times \mathbf{R}^n \to [0, +\infty]$ is a Borel function with $g(x, \cdot, \cdot)$ lower semi-continuous on $\mathbf{R} \times \mathbf{R}^n$ for a.e. $x \in \Omega$.

To show that $|Du| \notin (0, 1)$ one can generalize Newton's idea. If $|Du| \in (0, 1)$ on a set B of positive measure, we can replace u by the infimum w of all those planes tangent to the graph of u whose slope is outside $(0, 1)$. The proof that $F(w) < F(u)$ can be carried out by using the co-area formula (we refer to [2] for details). This completes the proof of Theorem 4.1.

In the case $n = 1$, $\Omega = (-R, R)$, the solution is explicitly given by

$$u(x) = \begin{cases} (M/R)(R - |x|) & \text{if } M \geq R \\ (R - |x|) \wedge M & \text{if } M < R. \end{cases}$$

For $n > 1$, however, only these few facts are known about problem (4.1):

1. problem (4.1) admits a solution u;
2. the modulus of the gradient $|Du|$ stays outside $(0, 1)$;
3. u is concave and, therefore, locally Lipschitz continuous.

It would be interesting to prove (or disprove) the following facts which were cited above for the radial case:

4. the solution is zero on $\partial\Omega$;
5. the solution is unique;
6. there is always a flat region, that is, an open set $\Omega_0 \subset \Omega$ such that $u = M$ on Ω_0;
7. the solution is Lipschitz continuous up to the boundary $\partial\Omega$.

Let us briefly address the question of whether minimizers on symmetric domains Ω are necessarily symmetric. Even in the case when Ω is a disc, the usual methods of symmetrization (see Kawohl [6]) fail. Indeed, it is possible to show that the body of Figure 22-2(c) has the property that its resistance increases under Schwarz symmetrization. Nevertheless, we do not know if in the case when Ω is a disc the minimum in (4.1) is attained on the radial solution of Section 3.

Incidentally, among all domains Ω with given area, the disc does not provide the body of least resistance. To see this, let Ω be the disc of radius R, let $M = R$, and consider a long and thin rectangle $\omega = (-\varepsilon/2, \varepsilon/2) \times (0, \pi R^2/\varepsilon)$. Then the function $u(x, y) = (\varepsilon - 2|x|)M/\varepsilon$ defined on ω has resistance of order ε^2. In other words, according to Newtonian mechanics, *cum grano salis* a blade has less resistance than a bullet.

If we enlarge the class of admissible functions by considering bodies having the property that each particle hits the graph of u at most once, even in the radial case we may have minimizers which are not concave. Indeed, the bodies depicted in Figure 22-5 have $M = R$ and $C_0 \sim 0.32$ and so they have less resistance than every convex body of the same height and radius. The slope of the conical parts is $\sqrt{3}/2$.

Another class considerably larger than the class of concave functions, in which the minimum resistance problem is still meaningful, is the class of quasi-concave functions, that is of functions u whose upper level sets $\{u \geq t\}$ are convex. It is not difficult to see, by using the co-area formula, that bounded quasi-concave functions are in $BV(\Omega)$; therefore, in this framework, the minimization problem (4.1) becomes

$$\min\left\{\int_\Omega \frac{1}{1 + |D_a u|^2} dx : 0 \leq u \leq M, u \text{ quasiconcave}\right\}, \qquad (4.2)$$

where $D_a u$ denotes the absolutely continuous part of the measure Du with respect to the Lebesgue measure. We do not know if problem (4.2) admits a solution, and if this solution turns out to be actually concave, even if convex bodies should be expected to be optimal for the resistance analysis arising from the variational point of view.

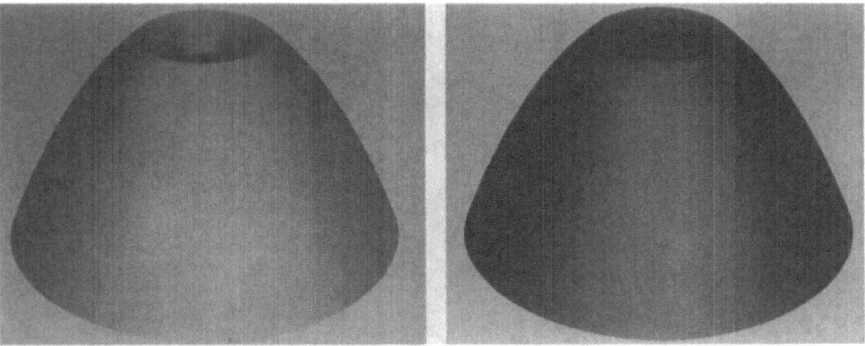

FIGURE 22-5 *Nonconvex bodies with single shock property.*

References

1. B. Botteron and P. Marcellini, A general approach to the existence of minimizers of one-dimensional non-coercive integrals of the calculus of variations, *Ann. Inst. H. Poincaré Analyse Non Linéaire* 8 (1991), 197-223.
2. G. Buttazzo, V. Ferone and B. Kawohl, Minimum problems over sets of concave functions and related questions, *Math. Nachr.* 173(1995), 71–89.
3. P. Funk, *Variationsrechnung und ihre Anwendungen in Physik und Technik*, Grundlehren 94 Heidelberg: Springer-Verlag (1962).
4. H. H. Goldstine, *A History of the Calculus of Variations from the 17th through the 19th Century*, Heidelberg: Springer-Verlag (1980).
5. W. D. Hayes and R. F. Probstein, *Hypersonic Flow Theory*, New York, Academic Press (1966).
6. B. Kawohl, *Rearrangements and Convexity of Level Sets in PDE*, Lecture Notes in Mathematics No. 1150, Heidelberg: Springer-Verlag (1985).
7. A. Kneser, *Ein Beitrag zur Frage nach der zweckmäßigsten Gestalt der Geschoßspitzen. Arch. Math. Phys.* 2 (1902), 267-278.
8. P. Marcellini, *Non-convex Integrals of the Calculus of Variations*, Proceedings of Methods of Nonconvex Analysis; Varenna 1989, A. Cellina, ed. Lecture Notes in Mathematics No. 1446, Heidelberg: Springer-Verlag (1990), pp. 16-57.
9. A. Miele, *Theory of Optimum Aerodynamic Shapes*, New York: Academic Press (1965).
10. I. Newton, *Philosophiae Naturalis Principia Mathematica* (1687).
11. S. Parma, Problemi di minimo su spazi di funzioni convesse, Tesi di Laurea, Università di Ferrara, Ferrara (1991).
12. L. Tonelli: *Fondamenti di Calcolo delle Variazioni*, Bologna: Zanichelli (1923).

23

If Hamilton Had Prevailed: Quaternions in Physics

J. Lambek

This is a nostalgic account of how two certain key results in modern theoretical physics (prior to World War II) can be expressed concisely in the language of quaternions, thus suggesting how they might have been discovered if Hamilton's views had prevailed. In the first instance, *biquaternions* are used to discuss special relativity and Maxwell's equations. To express Dirac's equation of the electron, we are led to replace the complex number i by the *right regular representation* of the quaternion unit i_1. Looked at in this way, it is actually equivalent to the relativistic version of Schrödinger's equation. The complex number i reappears as soon as we consider the electron in an electromagnetic field. When expressed in terms of complex matrices, Dirac's equation turns out to be invariant not only under a projective representation of the Lorentz group and under Weyl's gauge transformation but also under a projective representation of $SU(3)$.

The Invention of Quaternions

Bourbaki introduced the following symbols for various species of numbers:

$$\mathbb{N} \subset \mathbb{Z} \subset \mathbb{Q} \subset \mathbb{R} \subset \mathbb{C} \subset \mathbb{H},$$

referring to the *naturals, integers, rationals, reals, complex numbers,* and *quaternions,* respectively. This sequence expresses a logical development of the number system, but its historical (and pedagogical) development proceeds somewhat differently:

$$\mathbb{N}^+ \subset \mathbb{Q}^+ \subset \mathbb{R}^+ \subset \mathbb{R} \subset \mathbb{C} \subset \mathbb{H},$$

where $\mathbb{N}^+$, $\mathbb{Q}^+$, and $\mathbb{R}^+$ refer to the positive naturals, positive rationals, and positive reals, respectively.

Not many mathematicians can claim to have introduced (invented? discovered?) a new kind of number. Although the positive reals had been effectively used by Thales, as ratios of geometric quantities, it was only after the Pythagoreans discovered, to their great discomfort, that the equation $x^2 = 2$ cannot be solved for rational x, that a rigorous definition of the positive reals was given by Eudoxus, essentially by what we now call Dedekind cuts. As far as we know, the Indian mathematician Brah-

Volume 17, No. 4 (Fall 1995), 7-15

magupta was the first to allow zero and negative numbers to be subject to arithmetical operations, thus permitting the transition from $\mathbb{R}^+$ to $\mathbb{R}$. Cardano, perhaps better known as a physician than as a mathematician, introduced complex numbers, not just to solve equations such as $x^2 + 1 = 0$ but because they were needed to find real solutions of cubic equations with real coefficients.

After Gauss had proved the fundamental theorem of algebra, there was no longer any need to introduce new numbers to solve equations. It was with a different motivation in mind that quaternions were invented by William Rowan Hamilton and, according to Altmann [1986], independently by Olinde Rodrigues. (He also points out that they were already known to Gauss.)

Hamilton had already made important contributions to mathematical physics, the most celebrated one being his reformulation of the Euler-Lagrange equations in a form in which position and momentum appear on the same footing. He was now looking for numbers of the form $x + iy + jz$, with $i^2 = j^2 = -1$, which would do for space what complex numbers had done for the plane. According to Conway [1951], he may have been influenced by the complex number identity

$$(x + iy)(x - iy) = x^2 + y^2$$

when he looked at

$$(x + iy + jz)(x - iy - jz) = x^2 + y^2 + z^2 - (ij + ji)yz.$$

In 1843 he had the sudden insight to abandon the commutative law of multiplication. (It should be noted that matrix multiplication may have come later.) Writing $ij = k$, he found that

$$i^2 = j^2 = k^2 = ijk = -1.$$

Numbers of the form $a + bi + cj + dk$, with $a, b, c, d \in \mathbb{R}$, were called *quaternions*. They were added, subtracted, and multiplied according to the usual laws of arithmetic, except for the commutative law of multiplication.

It will be convenient to replace i, j, and k by i_1, i_2, and i_3, respectively, so that any quaternion can be written as

$$a = a_0 + i_1 a_1 + i_2 a_2 + i_3 a_3 = \sum_{\alpha=0}^{3} i_\alpha a_\alpha,$$

where $i_0 = 1$. The *conjugate* of a is given by

$$a^\dagger = a_0 - i_1 a_1 - i_2 a_2 - i_3 a_3,$$

and one notes that $aa^\dagger$ and $a^\dagger a$ are both equal to the real number

$$N(a) = a_0^2 + a_1^2 + a_2^2 + a_3^2,$$

called the *norm of a*. When $a \neq 0$, one can define $a^{-1} = a^\dagger / N(a)$, so that

$$aa^{-1} = 1 = a^{-1}a.$$

The quaternions form a *skew field* or *division ring*, which is denoted by $\mathbb{H}$ in Hamilton's honour, $\mathbb{Q}$ having been pre-empted for the field of rationals. Note also that $(ab)^\dagger = b^\dagger a^\dagger$.

As far as we know, Hamilton was the first to look at a non-commutative system of numbers. Had matrices been known to him, Hamilton might have defined

$$i_1 = \begin{pmatrix} 0 & 1 \\ -1 & 0 \end{pmatrix}, \quad i_2 = \begin{pmatrix} 0 & i \\ i & 0 \end{pmatrix},$$

where i is the ordinary complex square root of -1, thus forcing

$$i_3 = i_1 i_2 = \begin{pmatrix} i & 0 \\ 0 & -i \end{pmatrix}.$$

If these three matrices are multiplied by $-i$, one obtains the so-called Pauli *spin matrices,* which were to play a role in quantum mechanics later. Thus, the quaternion a could have been identified with the complex 2-by-2 matrix

$$\begin{pmatrix} a_0 + ia_3 & a_1 + ia_2 \\ -a_1 + ia_2 & a_0 - ia_3 \end{pmatrix} = \begin{pmatrix} u & v \\ -v^* & u^* \end{pmatrix},$$

where u and v are complex numbers with complex conjugates u^* and v^*. Note that the quaternion conjugate of such a matrix is

$$\begin{pmatrix} u^* & -v \\ v^* & u \end{pmatrix},$$

which is the same as the complex conjugate of the transposed matrix.

Replacing 0, 1, and i in these complex matrices by

$$\begin{pmatrix} 0 & 0 \\ 0 & 0 \end{pmatrix}, \quad \begin{pmatrix} 1 & 0 \\ 0 & 1 \end{pmatrix}, \quad \begin{pmatrix} 0 & 1 \\ -1 & 0 \end{pmatrix},$$

respectively, one can obtain a representation of quaternions as 4-by-4 real matrices. A more natural way of representing quaternions by real matrices is with the help of linear algebra. Clearly, the mapping $x \mapsto ax$, induced by left multiplication with a, is a linear transformation of the vector space of quaternions x, hence can be represented by a 4-by-4 real matrix $L(a)$. Writing $[x]$ for the column vector associated with the quaternion $x = \sum_{\alpha=0}^{3} i_\alpha x_\alpha$, namely, the transposed of the row vector (x_0, x_1, x_2, x_3), we thus have

$$[ax] = L(a)[x].$$

We may also represent the linear mapping $x \mapsto xb$, induced by right multiplication with b, by a matrix $R(b)$ such that

$$[xb] = R(b)[x].$$

In view of the associative law

$$(ax)b = a(xb)$$

for quaternions, we have

$$R(b)L(a) = L(a)R(b), \quad L(ax) = L(a)L(x), \quad R(xb) = R(b)R(x).$$

Thus, $L : \mathbb{H} \to M_4(\mathbb{R})$ and $R : \mathbb{H}^{\text{op}} \to M_4(\mathbb{R})$ are ring homomorphisms.

Applications to Classical Physics

It seems reasonable to represent a three-dimensional vector (x_1, x_2, x_3) by the headless quaternion

$$\xi = i_1 x_1 + i_2 x_2 + i_3 x_3;$$

but what is the meaning of the fourth coordinate x_0? Influenced by the pre-Socratic philosopher Parmenides, who believed that the flow of time is an illusion, Hamilton might have suspected that x_0 stands for time. But what then is the significance of the norm $N(x)$ when $x_0 \neq 0$? (Of course, when $x_0 = 0$, it stands for the square of the distance from the origin.)

   If we assume that ξ and

$$\eta = i_1 y_1 + i_2 y_2 + i_3 y_3$$

are pure 3-vectors, then

$$\xi\eta = -(\xi \circ \eta) + \xi \times \eta,$$

where

$$\xi \circ \eta = x_1 y_1 + x_2 y_2 + x_3 y_3$$

and

$$\xi \times \eta = i_1(x_2 y_3 - x_3 y_2) + i_2(x_3 y_1 - x_1 y_3) + i_3(x_1 y_2 - x_2 y_1)$$

came to be called the *scalar product* and *vector product*, respectively. It was these two products which were applied to physics in the vector analysis of Gibbs and Heaviside, rather than the quaternion product. In particular, writing

$$\nabla = i_1 \frac{\partial}{\partial x_1} + i_2 \frac{\partial}{\partial x_2} + i_3 \frac{\partial}{\partial x_3},$$

one usually summarizes Maxwell's equations as follows:

$$\nabla \circ \mathbf{B} = 0, \quad \nabla \times \mathbf{E} + \frac{\partial \mathbf{B}}{\partial t} = 0, \quad \nabla \circ \mathbf{E} = \rho, \quad \nabla \times \mathbf{B} - \frac{\partial \mathbf{E}}{\partial t} = \mathbf{j},$$

where

$$\mathbf{B} = B_1 i_1 + B_2 i_2 + B_3 i_3, \quad \mathbf{E} = E_1 i_1 + E_2 i_2 + E_3 i_3$$

represent the magnetic and electric fields, respectively, ρ is the charge density, and $\mathbf{j}$ is the current density. (Units have been chosen to make c, the velocity of light, equal to one light-second per second.) These laws are usually ascribed to Coulomb, Faraday, Gauss, and Ampère, respectively, although it was Maxwell who added the term $\partial \mathbf{E}/\partial t$ to the last equation. This addition is in fact crucial to what follows.

Using the language of quaternions, albeit quaternions with complex components, known as *biquaternions*, Maxwell's four equations may be combined into a single equation:

$$\left(\frac{\partial}{\partial t} - i\nabla\right)(\mathbf{B} + i\mathbf{E}) + (\rho + i\mathbf{j}) = 0.$$

As far as I know, this was first pointed out independently by Conway [1911] and Silberstein [1912], although it might have been realized by Clerk Maxwell himself, when he said in 1869:

The invention of the calculus of quaternions is a step towards the knowledge of quantities related to space which can only be compared, for its importance, with the invention of triple coordinates by Descartes.

It follows from Maxwell's equations that

$$\frac{\partial \rho}{\partial t} + \nabla \circ \mathbf{j} = 0.$$

This is known as the *equation of continuity*, which asserts that the scalar part of

$$\left(\frac{\partial}{\partial t} + i\nabla\right)(\rho + i\mathbf{j})$$

is zero. Another consequence of Maxwell's equations is the existence of a 4-potential, here denoted by the biquaternion $\varphi + i\mathbf{A}$, such that

$$\mathbf{E} = -\nabla\varphi - \frac{\partial \mathbf{A}}{\partial t}, \quad \mathbf{B} = \nabla \times \mathbf{A}.$$

In quaternion notation, this asserts that the vector part of

$$\left(\frac{\partial}{\partial t} + i\nabla\right)(\varphi + i\mathbf{A})$$

is $\mathbf{B} + i\mathbf{E}$. (Sometimes, the scalar part is put equal to zero.)

If Maxwell had faith in quaternions, other physicists despised them. Thus, Oliver Heaviside, as quoted by Conway [1948], asserted that quaternions are "a positive evil of no inconsiderable magnitude," and William Thomson, better known as Lord Kelvin, as quoted by Altmann [1986], said that they "have been an unmixed evil to those who touched them in any way, including Clerk Maxwell." Even Minkowski, who should have known better, rejected quaternions as "too narrow and clumsy." It may be difficult to understand how a mathematical concept can evoke such strong antagonism;

but, even in our day, similar opinions have been expressed about categories, although for different reasons.

Not surprisingly, quaternions never entered the mainstream of physics, yet they had a small but dedicated group of devotees, of whom I shall only mention the few whose articles are cited in the bibliography: Conway, Dirac, Silberstein, Weiss, Gürsey, and Synge. My own interest as a graduate student was raised by the inspiring book by Silberstein [1924] and led to Part I of my thesis, from which, however, all physics was expunged when I realized that my main ideas had been anticipated by Conway [1948].

If we take the quaternion form of Maxwell's equations seriously, we are led to the study of quaternions with complex components, also called *biquaternions*. Extending the representation of quaternions as 2-by-2 complex matrices to biquaternions, we see that the latter must be represented by matrices of the form

$$\begin{pmatrix} u & v \\ -v^* & u^* \end{pmatrix} + i \begin{pmatrix} u' & v' \\ -v'^* & u'^* \end{pmatrix}.$$

But it is easily seen that any 2-by-2 complex matrix is of this form, as we can solve the four equations

$$u + iu' = p, \quad v + iv' = q, \quad -v^* - iv'^* = r, \quad u^* + iu'^* = s$$

for u, v, u', and v'. For example, writing $u = u_0 + iu_1$, $v = v_0 + iv_1$, and so on, and adding the first and last equations, we see that

$$2u_0 + 2iu'_0 = p + s,$$

so that

$$u_0 = \frac{1}{2}(p_0 + s_0).$$

Thus, the algebra of biquaternions is isomorphic to $M_2(\mathbb{C})$.

The complex conjugate

$$(a + ib)^* = a - ib$$

of a biquaternion is then represented by the conjugate matrix, whereas the quaternion conjugate

$$(a + ib)^\dagger = (a_0 + ib_0) - i_1(a_1 + ib_1) - i_2(a_2 + ib_2) - i_3(a_3 + ib_3)$$

is represented by the transposed matrix.

We may also extend the representation L of quaternions as 4-by-4 real matrices to one of biquaternions as 4-by-4 complex matrices. Using the same letter L for the latter, we see that again

$$L(c^*) = L(c)^*, \quad L(c^\dagger) = L(c)^\dagger$$

are the conjugate and transposed matrices, respectively. For the present, we shall favor this representation and identify the biquaternion c with the matrix $L(c)$ in $M_4(\mathbb{C})$, although later we shall look at a representation of biquaternions in $M_4(\mathbb{R})$.

Applications to Special Relativity

The special theory of relativity requires the invariance of the expression $t^2 - x_1^2 - x_2^2 - x_3^2$ under a coordinate transformation passing from a stationary platform to a moving train. This suggests that position in space and time be joined together in a biquaternion of the form

$$x = x_0 + ii_1 x_1 + ii_2 x_2 + ii_3 x_3 = t + ir,$$

where the x_α are real. These biquaternions are characterized by the property $x^* = x^\dagger$ and have been called *hermitian* biquaternions. In fact, the matrices $L(x)$ and $R(x)$ are then hermitian matrices. The differential operator

$$\frac{d}{dx} = \frac{\partial}{\partial t} - i\nabla$$

is also a hermitian biquaternion, but the so-called *six-vector* $\mathbf{F} = \mathbf{B} + i\mathbf{E}$ satisfies $\mathbf{F}^\dagger = -\mathbf{F}$; it is represented by a skew-symmetric matrix.

We may ask which continuous transformations leave the norm of a hermitian biquaternion invariant. It is easily see that this is so for

$$x \mapsto x^*, \quad x \mapsto -x, \quad x \mapsto qxq^{*\dagger} \quad \text{when } N(q) = 1,$$

and for no others. (See, e.g., Lambek [1950].) If we ask, more generally, which continuous transformations leave the norm of the difference of two hermitian biquaternions unchanged, we should add also the *translation*

$$x \mapsto x + a.$$

The group generated by all these transformations is called the *Poincaré group*.

We shall rule out the transformation $x \mapsto x^*$ but admit $x \mapsto -x$, thus following Lewis Carroll, who suggested that time is reversed in a mirror. Transformations of the form $x \mapsto qxq^*$ are called (proper) *Lorentz transformations*; they were originally postulated by Lorentz to account for the Michelson-Morley experiment. This *ad hoc* explanation was later justified by Albert Einstein, who realized that they also described, in addition to rotations, the changes in coordinates when passing from a stationary platform to a uniformly moving train; these are called *boosts* (see Sudbury [1986]) and form the cornerstones of the special theory of relativity.

A Lorentz transformation is given by a biquaternion $q = u + iv$, with $u, v \in \mathbb{H}$, such that $qq^\dagger = 1$, that is,

$$uu^\dagger - vv^\dagger = 1, \quad uv^\dagger + vu^\dagger = 0.$$

q describes a rotation in three-space if $q \in \mathbb{H}$, that is, if $v = 0$; it is a *boost* if it is hermitian, that is, $q^\dagger = u - iv$, which means that u is a scalar and v a 3-vector. It is easily seen that every *Lorentz transformation is a rotation followed by a boost*. Indeed, let

$$\mu^2 = uu^\dagger = 1 + vv^\dagger \geq 1, \quad r = u\mu^{-1}, \quad s = \mu - iuv^\dagger\mu^{-1};$$

then r is a rotation, s is a boost, and $q = sr$.

The space of hermitian biquaternions is called *Minkowski space*; it is generated by 1, ii_1, ii_2, and ii_3. If we put $\lambda\alpha = i$ when $\alpha > 0$ but $\lambda_0 = 1$, we may write these generators as $\lambda\alpha i_\alpha$. Any hermitian biquaternion then has the form

$$x = \sum x'_\alpha i_\alpha = \sum s_\alpha \lambda_\alpha i_\alpha,$$

where the $x_\alpha = x'_\alpha \lambda^*_\alpha$ are real. Applying a Lorentz transformation, we obtain another hermitian biquaternion

$$qxq^{*\dagger} = \sum_{\alpha,\beta} x'_\alpha \lambda^*_\alpha \Lambda_{\alpha\beta} \lambda_\beta i_\beta,$$

where the $\Lambda_{\alpha\beta}$ are real numbers. Putting

$$\lambda^*_\alpha \Lambda_{\alpha\beta} \lambda_\beta = \Delta_{\alpha\beta},$$

we see that

$$L(q)R(q^{*\dagger})[x] = \Delta^\dagger[x].$$

This being so for all $[x]$, we infer that

$$L(q)R(q^{*\dagger}) = \Delta^\dagger,$$

hence, taking the transposed of each side, that

$$R(q^*)L(q^\dagger) = \Delta.$$

Multiplying this by $[x^*]$, for any hermitian biquaternion x, we obtain

$$q^\dagger x^* q^* = \sum_{\alpha,\beta} i_\alpha \Delta_{\alpha\beta} x'^*_\beta = \sum_{\alpha,\beta} i_\alpha \lambda^*_\alpha \Lambda_{\alpha\beta} x^*_\beta.$$

For later reference, we summarize this observation as follows:

LEMMA 1. If a Lorentz transformation transforms $\lambda_\alpha i_\alpha$ into $q\lambda_\alpha i_\alpha q^{*\dagger} = \sum_\beta \Lambda_{\alpha\beta} \lambda_\beta i_\beta$, then $q^\dagger \lambda^*_\beta i_\beta q^* = \sum_\alpha i_\alpha \lambda^*_\alpha \Lambda_{\alpha\beta}$.

Einstein had realized that the mass-momentum $p = m + im\mathbf{v}$ should also transform like x (without using the language of quaternions); hence, he wrote $p = m_0 dx/ds$, where $ds^2 = N(dx)$, and $m_0 = m(dt/ds)^{-1}$ is the *rest mass* of a moving particle, assumed to be invariant. The conservation of momentum is then coupled with the conservation of mass.

Now

$$m = m_0 \frac{dt}{ds} = m_0(1 - v^2)^{-1/2},$$

where $v^2 = \mathbf{v} \circ \mathbf{v}$. If the velocity v of the moving particle is small compared with the velocity $c = 1$ of light,

$$m \doteq m_0\left(1 + \frac{\frac{1}{2}v^2}{c^2}\right),$$

where we have temporarily restored the symbol c to obtain the famous approximation

$$mc^2 \doteq m_0 c^2 + \frac{1}{2} m_0 v^2.$$

Einstein considered this to be the total energy of the particle, the kinetic energy $\frac{1}{2} m_0 v^2$ being augmented by the atomic energy $m_0 c^2$.

The charge-current density $J = \rho + i\mathbf{j}$ may be thought of as $J = \rho_0 dx/ds$, where $\rho_0 = \rho(dt/ds)^{-1}$ is assumed to be an invariant scalar; hence, J also transforms like x, namely $J \mapsto p J p^{*\dagger}$. In our notation, Maxwell's equations are combined into the single equation $d\mathbf{F}/dx + J = 0$. It seems that Henri Poincaré was the first to realize that they are invariant under Lorentz transformations. To see this, we only need $\mathbf{F} \mapsto q^* \mathbf{F} q^{*\dagger}$ and $d/dx \mapsto q(d/dx)q^{*\dagger}$. The former transformation is natural for a 6-vector, as

$$(q^* \mathbf{F} q^{*\dagger})^\dagger = q^* \mathbf{F}^\dagger q^{*\dagger} = -q^* \mathbf{F} q^{*\dagger}.$$

The latter transformation may be justified on general principle: if the column vector of the $x'_\alpha = \lambda_\alpha x_\alpha$ is transformed by the matrix Δ, then the column vector of the $\partial/\partial x'_\alpha = \lambda_\alpha \partial/\partial x_\alpha$ is transformed by the inverse of $\Delta^\dagger$. Now, if Δ leaves $x_0^2 - x_1^2 - x_2^2 - x_3^2$ invariant, the inverse of $\Delta^\dagger$ is Δ; hence,

$$\frac{d}{dx} = \sum_\alpha \lambda_\alpha^* \frac{\partial}{\partial x_\alpha}$$

transforms like x.

In summary, the following hermitian quaternions are transformed by $q(\ )q^{*\dagger}$:

$$x = t + i\mathbf{r} \qquad \text{(position in space-time)},$$

$$\frac{d}{dx} = \frac{\partial}{\partial t} - i\nabla \qquad \text{(partial derivation)},$$

$$p = m + im\mathbf{v} \qquad \text{(energy-momentum)},$$

$$J = \rho + i\mathbf{j} \qquad \text{(charge-current density)},$$

$$\Phi = \varphi + i\mathbf{A} \qquad \text{(4-potential)}.$$

On the other hand, the 6-vector

$$\mathbf{F} = \mathbf{B} + i\mathbf{E} \quad \text{(electromagnetic field)}$$

is transformed by $q(\ )q^\dagger$. The quaternion form of Maxwell's equations is

$$\frac{d}{dx}\mathbf{F} + J = 0.$$

Operating on this equation by $(d/dx)^\dagger$ and noting that $(d/dx)^\dagger(d\mathbf{F}/dx)$ is a 6-vector, we infer that the scalar part of $(d/dx)^\dagger J$ is zero. This is the equation for continuity. We should also add the observation that $(d/dx)^\dagger \Phi + \mathbf{F}$ is a scalar, which is sometimes put equal to zero.

Maxwell's equations describe the electromagnetic field created by a continuous distribution of charge in motion, as expressed by the charge-current biquaternion J. Conversely, a given electromagnetic field also acts on a moving charge, this time viewed as a discrete charge q_0 moving with velocity $\mathbf{v}$, by exerting a force $q_0(\mathbf{E} + \mathbf{v} \times \mathbf{B})$. According to Newton, this should be the rate of change of momentum, but special relativity requires that the rate be measured by d/ds, not by d/dt. Moreover, according to the dictates of special relativity, this should be augmented by a term $q_0(\mathbf{v} \times \mathbf{E} + \mathbf{B})$, expressing the rate of change of energy. Thus, we obtain the relativistic form of the *equation of motion* of an electron, with $q_0 = -e$,

$$\frac{dp}{ds} = e\frac{dx}{ds}\mathbf{F} + ig,$$

where g is some hermitian biquaternion.

There remains a contradiction: we described the charge as continuously distributed in Maxwell's equation, but as discrete in the equation of motion. This contradiction will only be resolved when we pass to the quantum-mechanical treatment of the electron.

Application to Quantum Mechanics

Quantum mechanics prescribes that the momentum $p = m_0 dx/ds$ be replaced by the differential operator $-(h/2\pi i)(d/dx)$, where h is Planck's constant. Choosing units so that $h/2\pi = 1$, we thus expect the equation $pp^\dagger = m_0^2$ to be replaced by the relativistic wave equation

$$\left(\frac{d}{dx}\right)\left(\frac{d}{dx}\right)^\dagger \varphi = -m_0^2\varphi,$$

where $\varphi = \varphi_0 + i\varphi_1$ is a function of position in space-time. This is usually referred to as the *Klein-Gordon equation;* but, as a recent biography of Schrödinger points out (Moore [1989]), Schrödinger himself had considered this form of the wave equation before introducing the time-evolving form which is named after him.

Dirac obtained his celebrated first-order equation by extracting the square root of this second-order differential operator, thus rediscovering the main idea behind Clifford algebras. It does not seem to be widely realized that such *ad hoc* methods are not needed and that Dirac's first-order equations in fact *equivalent* to the Klein-Gordon equation, provided we do not insist that φ remain a scalar. In fact, we shall assume that φ is a biquaternion and write

$$\frac{d\varphi}{dx} = m_0\chi,$$

where $\chi = \chi_0 + i\chi_1$ is another biquaternion. Then

$$m_0\left(\frac{d\chi}{dx}\right)^\dagger = \left(\frac{d}{dx}\right)^\dagger\left(\frac{d}{dx}\right)\varphi = -m_0^2\varphi,$$

so the Klein-Gordon equation is equivalent to the following pair of biquaternion equations:

$$\frac{d\varphi}{dx} = m_0\chi, \quad \left(\frac{d\chi}{dx}\right)^\dagger = -m_0\varphi.$$

How can these be combined into a single first-order equation?

Assume, for the moment, that there is an entity j such that $j^2 = -1$, $ji = -ij$, and $ji_\alpha = i_\alpha j$ for $\alpha = 0, 1, 2,$ or 3. Then we have

$$\left(\frac{d}{dx}\right)(\varphi + j\chi) = \left(\frac{d}{dx}\right)\varphi + \left(\frac{d}{dx}\right)j\chi$$

$$= m_0\chi + j\left(\frac{d}{dx}\right)^\dagger\chi$$

$$= m_0(\chi - j\varphi) - jm_0(\varphi + j\chi).$$

There is certainly no 4-by-4 matrix which anticommutes with the complex number i. But let us pass to real 4-by-4 matrices and identify i_α with the matrix $L(i_\alpha)$ representing it. We shall take $j_\alpha = R(i_\alpha)$, the contravariant representation; then

$$j_1^2 = j_2^2 = j_3^2 = j_3j_2j_1 = -1, \quad j_\alpha i_\beta = i_\beta j_\alpha$$

for $\alpha, \beta = 1, 2,$ or 3. Now replace the complex number i by the real matrix j_1 and identify j with j_2. Putting

$$\Psi = \varphi + j_2\chi = \varphi_0 + j_1\varphi_1 + j_2\chi_0 + j_3\chi_1,$$

we may write the above first-order equation as

$$\frac{d}{dx}\Psi + j_2m_0\Psi = 0,$$

where now

$$\frac{d}{dx} = \frac{\partial}{\partial t} - j_1\nabla,$$

and we have essentially recaptured Dirac's equation for the free electron.

A word of warning. Having replaced i by j_1, we must now write $x = t + j_1\mathbf{r}$, and so on, and even the biquaternion $q = u + iv$ in the Lorentz transformation has now become the matrix $u + j_1v$.

To make sure that our first-order equation is preserved under the Lorentz transformation which sends x to $qxq^{*\dagger}$, hence d/dx to $q(d/dx)q^{*\dagger}$, it suffices to let Ψ be transformed into $q^*\Psi$. (There are other possibilities, namely, sending Ψ to $q^*\Psi q^\dagger$ or $q^*\Psi q^{*\dagger}$, but we shall not consider these.) We note that the Lorentz transformation $q(\ )q^{*\dagger}$ is unchanged when q is replaced by $-q$ but that it corresponds to two distinct ways of transforming Ψ: into $q^*\Psi$ and $-q^*\Psi$, respectively. The transformations of Ψ do not constitute a representation of the Lorentz group but what has been called a *projective representation*. This is the mathematical reason for saying that the electron has spin $1/2$.

It can be shown that the sixteen matrices $1, i_\alpha, j_\beta,$ and $i_\alpha i_\beta$ ($\alpha, \beta = 1, 2,$ or 3) span the space of all 4-by-4 real matrices. The reason is that $\mathbb{H}$ is a central simple algebra of degree 4 over $\mathbb{R}$, hence $\mathbb{H} \otimes \mathbb{H}^{op}$ is isomorphic to $M_4(\mathbb{R})$ (see Jacobson [1980], Theorem 4.6). Thus, Ψ is just an arbitrary 4-by-4 real matrix. However, the transformation rule $\Psi \mapsto q^*\Psi$ permits us to multiply Ψ in the Dirac equation by the column vector $(1, 0, 0, 0)^\dagger$, so we may assume, without loss in generality, that Ψ is the real column vector $(\psi_0, \psi_1, \psi_2, \psi_3)^\dagger$; we may write $\Psi = [\psi]$, where $\psi = \psi_0 + i_1\psi_1 + i_2\psi_2 + i_3\psi_2$ is a real quaternion. There is nothing to prevent us from introducing a complex column vector here; on the other hand, there seems to be no necessity for doing so either, at least as long as we confine attention to the *free* electron, not influenced by an electromagnetic field.

Dirac's equation for the free electron may be written more explicitly in matrix form as follows:

$$\left(\frac{\partial}{\partial t} - R(i_1)\sum_{\alpha=1}^{3} L(i_\alpha)\frac{\partial}{\partial x_\alpha}\right)[\psi] = -m_0 R(i_2)[\psi].$$

Now $L(i_\alpha)[\psi] = [i_\alpha\psi]$ and $R(i_\alpha)[\psi] = [\psi i_\alpha]$, so this is really an equation in real quaternions:

$$\frac{\partial\psi}{\partial t} - \nabla\psi i_1 + m_0\psi i_2 = 0.$$

When written in this way, the equation is less apparently Lorentz-invariant.

When taking $j_1 = R(i_1)$ and $j_2 = R(i_2)$, we made an arbitrary choice. We might just as well have taken $j_1 = R(i_2)$ and $j_2 = R(i_3)$ or, more generally, $j_\alpha = R(ri_\alpha r^\dagger)$ ($\alpha = 1, 2,$ or 3), where r is a real quaternion such that $rr^\dagger = 1$, that is, where $r(\ )r^\dagger$ is a rotation in 3-space. What would happen to Dirac's equation had we chosen a different coordinate frame in 3-space? In its quaternionic form, it would then become

$$\frac{\partial\psi}{\partial t} - \nabla\psi ri_1 r^\dagger + m_0\psi ri_2 r^\dagger = 0.$$

Multiplying by r on the right, we would then obtain

$$\frac{\partial\psi r}{\partial t} - \nabla\psi ri_1 + m_0\psi ri_2 = 0,$$

which is the same as the original equation with ψ replaced by ψr.

I recall telling Dirac in 1949 that I could derive his equation with the help of quaternions. After thinking quietly for several minutes, as was his habit before speaking, he said, "Unless you can do it with *real* quaternions, I am not interested." As I had biquaternions in mind, it was perhaps this remark which finally persuaded me to abandon theoretical physics for pure mathematics. Looking at the problem again almost half a century later, in connection with a project to write an undergraduate textbook on the history and philosophy of mathematics (together with Bill Anglin), I realize that I should have replied: "Yes; but can you do it using a *real* wave vector?" It is only quite recently that I became aware of Dirac's 1945 article, which shows how to express Lorentz transformations with the help of *real* quaternions in a roundabout way. I don't know whether this idea was ever followed up.

The Electron in an Electromagnetic Field

What happens to an electron when an electromagnetic field is present? Classical physics requires that the kinetic energy m should be augmented by the potential energy $-e\,\varphi$, where $-e$ is the charge of the electron and φ is the potential. According to special relativity, the energy-momentum 4-vector p should then be augmented by $-e\Phi$, where $\Phi = \varphi + j_1\mathbf{A}$ is the 4-potential. Finally, quantum mechanics requires that p be replaced by $i(d/dx)$; hence, $p - e\Phi$ by $i(d/dx) - e\Phi$. Multiplying this by $-i$, we see that Dirac's equation becomes

$$\left(\frac{d}{dx} + ie\Phi\right)\Psi + j_2 m_0\Psi = 0,$$

where $\Psi = [\psi]$, ψ being a quaternion.

One is tempted to replace the complex number i by the real matrix j_1 as before, in which case Dirac's equation would become

$$\frac{\partial \psi}{\partial t} - eA\psi - (\nabla \psi - e\varphi \psi)i_1 + m_0\psi i_2 = 0.$$

As long as ψ is non-zero, we could multiply this equation by ψ^{-1} and solve for $\mathbf{A}$. I do not know whether this makes any sense, so I shall allow i to remain a complex number, thus forcing Ψ to be a complex vector, hence ψ a biquaternion.

Now Φ was only determined inasmuch as $(d/dx)^{\dagger}\Phi + \mathbf{F}$ is a scalar. As pointed out by Hermann Weyl [1950], this property is not affected by a so-called *gauge transformation*: replacing Φ by $\Phi + d\alpha/dx$, where α is a real scalar. The same result could have been achieved replacing Ψ in Dirac's equation by

$$\Psi \exp(ie\alpha) = \Psi\big(\cos(e\alpha) + i\sin(e\alpha)\big),$$

for

$$\frac{d}{dx}\big(\Psi \exp(ie\alpha)\big) = \left(\frac{d\Psi}{dx} + ie\frac{d\alpha}{dx}\right)\exp(ie\alpha).$$

[The argument depends on the fact that Ψ commutes with $\exp(ie\,\alpha)$; it would not have worked had we replaced i by j_1.]

Dirac's equation expresses the action of the electromagnetic field, as determined by Φ, on the electron. It replaces the equation of motion discussed earlier. On the other hand, the contribution of the electron to the electromagnetic field was expressed by Maxwell's equations. In terms of Φ these can be written

$$\frac{d}{dx}\left(\frac{d}{dx}\right)^{\dagger}\Phi = J,$$

where J is the charge-current density. Only now we can calculate J with the help of the wavefunction Ψ. The following considerations are adapted from Sudbury [1986], after translation into our language.

Recall the matrix form of Dirac's equation for an electron in an electromagnetic field:

$$\left(\frac{d}{dx} + ie\Phi\right)\Psi = -j_2 m_0 \Psi, \tag{1}$$

where $\Psi = [\psi]$ is a complex column vector, ψ now being a biquaternion. Since the old symbols $*$ and $\dagger$ have changed their meanings in the course of our discussion, we shall now write Ψ^T for the transpose of Ψ and $\Psi^C = [\psi^C]$ for the complex conjugate of Ψ. It will be convenient to invoke the *hermitian conjugate* $\Psi^H = \Psi^{CT} = [\psi^C]^T$. Now

$$\frac{d}{dx} = \frac{\partial}{\partial t} - j_1\nabla, \quad \Phi = \varphi + j_1\mathbf{A}$$

are real symmetric matrices, but j_2 is antisymmetric. Multiplying (1) by Ψ^H, we obtain

$$\Psi^H\left(\frac{d}{dx} + ie\Phi\right)\Psi = -\Psi^H j_2 m_0 \Psi, \tag{2}$$

and the hermitian conjugate of this is

$$\Psi^H\left(\frac{\overleftarrow{d}}{dx} - ie\Phi\right)\Psi = +\Psi^H j_2 m_0 \Psi,\tag{3}$$

where the arrow indicates that differentiation operates leftwards. Adding (2) and (3), we obtain

$$\Psi^H \frac{\overleftrightarrow{d}}{dx}\Psi = 0,$$

that is,

$$\frac{\partial}{\partial t}(\Psi^H\Psi) - \sum_{\alpha=0}^{3}\frac{\partial}{\partial x_\alpha}(\Psi^H j_1 i_\alpha \Psi) = 0.$$

This equation resembles the equation of continuity

$$\sum_{\alpha=0}^{3}\frac{\partial}{\partial x_\alpha}J_\alpha = 0,$$

which suggests that we define

$$J_\alpha = e\Psi^H i_\alpha \lambda_\alpha^* \Psi$$

and consider $J = \sum_{\alpha=0}^{3} J_\alpha \lambda_\alpha i_\alpha$ as a candidate for the charge-current density. Here λ_α is defined as before, except that i has now been replaced by j_1, thus $\lambda_\alpha = j_1 i_\alpha$ when $\alpha > 0$ but $\lambda_0 = 1$. It remains to check that J is transformed by a Lorentz transformation into $qJq^{*\dagger}$. Indeed, J transforms into

$$\sum_\beta e\Psi^H q^\dagger i_\beta \lambda_\beta^* q^* \Psi \lambda_\beta i_\beta,$$

and, by Lemma 1, this is

$$\sum_{\alpha,\beta} e\Psi^H i_\alpha \lambda_\alpha^* \Psi \Lambda_{\alpha\beta} \lambda_\beta i_\beta = \sum_{\alpha,\beta} J_\alpha \Lambda_{\alpha\beta} \lambda_\beta i_\beta = qJq^{*\dagger},$$

as was to be shown.

The wave vector Ψ appearing in the Dirac equation (1) depends on the following three data:

(a) a point in Minkowski space represented by $x = t + j_1\mathbf{r}$. The Lorentz transformation $x \mapsto qxq^{*\dagger}$ corresponds to a transformation $\Psi \mapsto q^*\Psi$.
(b) a choice of the 4-potential Φ, compatible with the electromagnetic field, which itself depends on x. The gauge transformation $\Phi \mapsto \Phi + d\alpha/dx$ is equivalent to the transformation $\Psi \mapsto \Psi \exp(ei\alpha)$.
(c) the choice of j_1 and j_2, which determines $j_3 = j_2 j_1$. We had assumed that $j\alpha = R(i_\alpha)$ but allowed i_α to be replaced by $ri_\alpha r^T$, r being a quaternion with $rr^T = 1$. However, since Ψ is now allowed to be a complex vector, we may as well allow i_α to be replaced by $ri_\alpha r^H$, where r is a biquaternion with $rr^H = 1$. This induces a transformation $\psi \mapsto \psi r$, or $\Psi \mapsto R(r)\Psi$.

The group acting on Ψ according to (a) is the group $SU(2)$ of unitary transformations of determinant 1. The group acting on Ψ according to (b) is the group $U(1)$ of the complex numbers with absolute value 1. The group acting on Ψ according to (c) is a projective representation of $SU(3)$.

The physical theories underlying the foregoing discussion have been known for a long time; they are all discussed in the classic text by Hermann Weyl [1950], a translation of the German edition of 1930. Even some further steps are indicated there, in particular the so-called *second quantization* (accompanied by some hocus pocus to remove unwanted infinities). Following this procedure, one is told to replace the functions Φ and Ψ by operators; but, according to Sudbury [1986], the form of the Dirac equation remains the same. These ideas culminate in quantum-electro-dynamics (QED), which has succeeded in making highly accurate predictions. More recent developments exploit the gauge theories initiated by Weyl. Thus, the strong force acting on a fermion is explained with the help of the group $SU(3)$ and the electro-weak force with the help of the group $SU(2) \times U(1)$. There does not seem to be any significance to the observation that these same groups arise in the above discussion.

What then is the final verdict on the usefulness of quaternions for physics? I am told that they are catching on as a tool for computation, but in the more general framework of Clifford algebras. Indeed, Dirac's original derivation of his equation implicitly used a Clifford algebra argument. One may cling to a feeling that there is something special about quaternions: their 4-dimensionality and the fact that they form a division algebra. These special properties of quaternions might be expected to put restraints on the nature of our universe.

Unfortunately, the 4-dimensionality of Hamiltonian quaternions does not account for the difference between space and time in Minkowski space, and division plays no role in our story. These aspects of quaternions are called upon, however, when one expresses the algebra of 4-by-4 real matrices as a tensor product of $\mathbb{H}$ and $\mathbb{H}^{\mathrm{op}}$, which fact we have exploited here. Be that as it may, I firmly believe that quaternions can supply a shortcut for pure mathematicians who wish to familiarize themselves with certain aspects of theoretical physics.

References

S. L. Altmann, *Rotations, Quaternions and Double Groups*, Oxford: Clarendon Press (1986).

A. W. Conway, On the application of quaternions to some recent developments of electrical theory, *Proc. Roy. Irish Acad.* A29 (1911), 80.

A. W. Conway, The quaternionic form of relativity, *Phil. Mag.* 24 (1912), 208.

A. W. Conway, Quaternions and quantum mechanics, *Ponteacrè Acad. Sci. Acta* 12 (1948), 204-277.

A. W. Conway, Hamilton, his life work and influence, *Proc. Second Canadian Math. Congress*, Toronto: University of Toronto Press (1951).

P. A. M. Dirac, Applications of quaternions to Lorentz transformations, *Proc. Roy. Irish Acad.* A50 (1945), 261-270.

F. Gürsey, Contributions to the quaternionic formalism in special relativity, *Rev. Faculté Sci. Univ. Istanbul* A20 (1955), 149-171.

F. Gürsey, Correspondence between quaternions and four-spinors, *Rev. Faculté Sci. Univ. Istanbul* A21 (1958), 33-54.

N. Jacobson, *Basic Algebra* II, San Francisco: Freeman (1980).

J. Lambek, Biquaternion vectorfields over Minkowski space, Thesis, Part I, McGill University (1950).

J. C. Maxwell, Remarks on the mathematical classification of physical quantities, *Proc. London Math. Soc.* 3 (1869), 224-232.

W. Moore, *Schrödinger, Life and Thought*, Cambridge: Cambridge University Press (1989).

P. J. Nahin, Oliver Heaviside, *Sci. Am.* 1990, 122-129.

S. Silberstein, *Theory of Relativity,* London: Macmillan (1924).

A. Sudbury, *Quantum Mechanics and the Particles of Nature,* Cambridge: Cambridge University Press (1986).

J. L. Synge, Quaternions, Lorentz transformations, and the Conway-Dirac-Eddington matrices, *Commun. Dublin Inst. Adv. Studies* A21 (1972), 1-67.

P. Weiss, On some applications of quaternions to restricted relativity and classical radiation theory, *Proc. Roy. Irish Academy* A46 (1941), 129-168.

H. Weyl, *The Theory of Groups and Quantum Mechanics,* New York: Dover Publications (1950).

Part Five

ARRANGEMENTS AND PATTERNS

24

On the *Problème des Ménages*

Jacques Dutka

1. Introduction

It not infrequently happens that the same mathematical problem appears in different guises in a variety of disciplines and is repeatedly treated by different writers unaware of each other's existence. Such is the case with a classical problem of combinatorial analysis, the *problème des ménages,* which arose more than a century ago in a physical context and was solved anew recently in an actuarial journal.

The problem considered by D. S. Jones and P. G. Moore [1] in 1981 concerned n couples around a circular table, husbands and wives being seated alternately, and the wives are initially assigned places. The number of ways in which the n couples can be seated such that no husband and wife sit next to each other is required.

Let $u(n)$ denote the number of ways. It was shown that

$$u(n) = \sum_{m=0}^{n} \frac{(2n - m - 1)!\, n (-1)^m \pi^{1/2}}{m!\, (n - m - \frac{1}{2})!\, 2^{2n-2m-1}}, \tag{1}$$

some numerical values for $u(n)$ and related quantities for n from 3 to 10 were tabled, and it was also proved that

$$\lim_{n \to \infty} \frac{u(n)}{n!} = e^{-2}. \tag{2}$$

The authors did not discuss the genesis of the problem. Subsequently, in 1983, G. E. Thomas [2] published a short note in which he stated that the problem, under the name *problème des ménages,* is discussed in many texts on combinatorial analysis and quoted a number of results which are known or can readily be derived.

This problem turns out to have connections with a number of diverse areas, and, in the course of its long history, has aroused considerable interest. It is the purpose of this paper to sketch this history in reasonable detail and to supplement earlier accounts.

Volume 8, No. 3 (Summer 1986), 18–25, 33

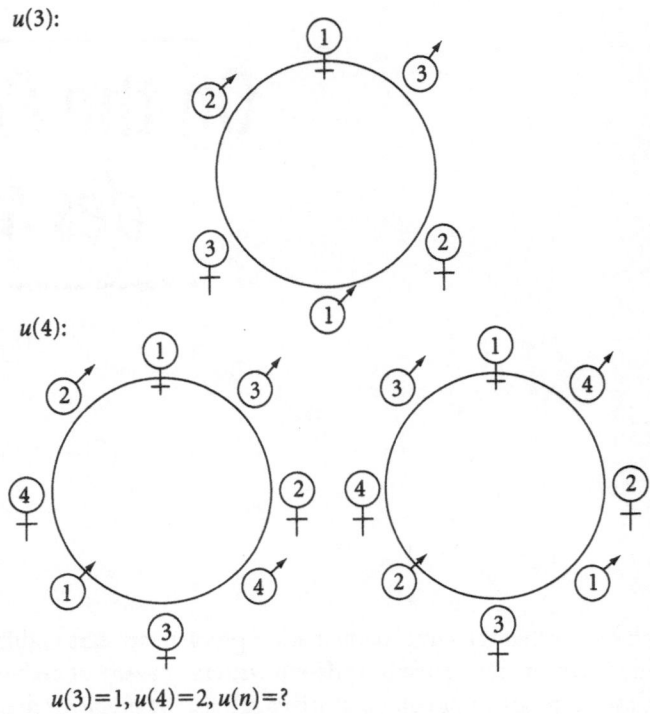

$u(3)=1, u(4)=2, u(n)=?$

FIGURE 24-1 *Problème des ménages.*

2. Explicit Solution of the Problem

The problem essentially as stated in Section 1, appeared in 1891 in a book on number theory by Edouard Lucas [3, 215 and Note III]. He termed it the *problème des ménages,* and showed that it was equivalent to determining the number of permutations discordant with (in his notation)

$$\begin{cases} 1, 2, 3, 4, \dots, (n-1), n \\ 2, 3, 4, 5, \dots, n, 1 \end{cases} \tag{1}$$

In this form, the problem is seen to be an extension of the classic problem of matchings, or "problème des rencontres" of P. R. de Montmort. Here the number of permutations of a row of n objects in which no object remains in its original place is required. It is proved in standard texts on combinatorial analysis that this is

$$d(n) = n! \sum_{k=0}^{n} \frac{(-1)^k}{k!}, \tag{2}$$

the integer nearest to $n!/e$.

Lucas also gave some results obtained by others and a table of associated numerical values. (Further details are given in Section 6.)

The first *explicit* solution of the problem was given in 1934 by J. Touchard [4] without proof.

Lucas (1891):

$u(4)$:
$\begin{cases} 1 & 2 & 3 & 4 \\ 2 & 3 & 4 & 1 \\ 3 & 4 & 1 & 2 \end{cases},\ \begin{cases} 1 & 2 & 3 & 4 \\ 2 & 3 & 4 & 1 \\ 4 & 1 & 2 & 3 \end{cases}$

FIGURE 24-2

Suppose the wives are already seated, which may be done in $2 \cdot n!$ ways since they may occupy "odd" or "even" seats and then be permuted in $n!$ ways. Then the number of ways of seating the husbands so that no husband sits next to his own wife is

$$u(n) = \sum_{k=0}^{n} (-1)^k \frac{2n}{2n-k} \binom{2n-k}{k} (n-k)!. \qquad (3)$$

A proof was given by I. Kaplansky [5] in 1943 which depends on some preliminary lemmas:

A. The number of ways of selecting k objects, no two consecutive, from n objects arrayed in a row is

$$\binom{n-k+1}{k}. \qquad (4)$$

For the selections, $f(n, k)$ in number, consist of two mutually exclusive types:
   (i) those selections, $f(n-2, k-1)$ in number, in which object 1 is included but object 2 is not, and (ii) those selections, $f(n-1, k)$ in number, in which object 1 is not included. Thus

$$f(n, k) = f(n-2, k-1) + f(n-1, k)$$

for $2 \le k \le n-1, n \ge 3$ with the conditions $f(n, 1) = n, f(1, n) = 0$ for $n > 1$. The result follows by mathematical induction.

B. The number of ways of selecting k objects, no two consecutive, from n objects arrayed in a circle is

$$\binom{n-k}{k} \frac{n}{n-k}. \qquad (5)$$

Let $g(n, k)$ denote the number of selections of this type. These consist of $f(n, k)$, the number of selections of the preceding type, *less* $f(n-4, k-2)$, the number of selections of the preceding type which contain objects 1 and n but neither objects 2 nor $n-1$. Thus

$$g(n, k) = f(n, k) - f(n-4, k-2)$$
$$= \binom{n-k+1}{k} - \binom{n-k-1}{k-2} = \binom{n-k}{k} \frac{n}{n-k}.$$

From the statement of the problem, $2n$ conditions are excluded in a seating arrangement. Husband i is in seats i or $i+1, 1 \le i \le n-1$, and husband n is in seats n or 1. Suppose a subset of k conditions is selected from the $2n$ and the number of possible seating arrangements which satisfy these conditions is required. If the k conditions are compatible, this is $(n-k)!$, the number of ways of seating the remaining husbands. If the k conditions are not compatible—no husband may occupy two seats nor may two husbands occupy the same seat, this is zero. Let $v(k)$ denote the number of ways of selecting k compatible conditions from the $2n$, arrayed in a circle. By Lemma B,

$$v(k) = \binom{2n-k}{k} \frac{2n}{2n-k}.$$

On applying the principle of inclusion and exclusion, the ménage number is

$$u(n) = \sum (-1)^k v(k)(n-k)!.$$

(See, e.g., J. Riordan [6, Ch. 3] or L. Comtet [7, Ch. IV].) This proves (3).

To show the equivalence of the result of Jones and Moore [1] and (3), rewrite (1.1) as

$$u(n) = (-1)^n \cdot 2 + n \sum_{m=0}^{n-1} \frac{(-1)^m}{m!} \frac{(2n-m-1)!}{(2n-2m-1)!} \frac{(2n-2m-1)!\sqrt{\pi}}{(n-m-\frac{1}{2})!\,2^{2n-2m-1}}.$$

On applying Legendre's formula for the duplication of the gamma function with $z = 2n - 2m$, one gets

$$\frac{\Gamma(z)\sqrt{\pi}}{\Gamma\left(\frac{z+1}{2}\right)2^{z-1}} = \Gamma\left(\frac{z}{2}\right) = (n-m-1)!$$

and

$$u(n) = (-1)^n \cdot 2 + n \sum_{m=0}^{n-1} \frac{(-1)^m}{m!} \frac{(2n-2m)!}{2n-2m} \frac{(2n-2m)}{(2n-2m)!} \frac{(n-m)!}{(n-m)}.$$

Combine the terms on the right and (3) follows.

3. Sequences in the Genoese Lottery

It does not appear to have been observed previously that Lemmas A and B obtained in Section 2 can be used to solve a classical lottery problem investigated by Leonhard Euler, N. de Beguelin and John (III) Bernoulli in the eighteenth century. A brief history of lotteries, particularly the Genoese lottery, will be sketched.

Lotteries for the disposal of war booty or of valuable objects which could not readily be sold have been known since ancient times. A public lottery for repairing the city of Rome was instituted in the reign of Augustus Caesar. The earliest mention of "lotto" in medieval Italian documents dates from the fifteenth century. The Lotto de Firenze, with money prizes, was established in 1530. In 1620, a

FIGURE 24-3

state lottery was instituted in the Republic of Genoa following a plan formulated by Benedetto Gentile, a member of the Republic's supreme council. Five names were drawn by lot semi-annually from among the ninety members of the council, and these were replaced on the council by others. (At various times the rules were changed—numbers were substituted for names, the total number was increased to one hundred and twenty, etc.) A winner in the lottery received certain multiples of his stake, depending on how many of the numbers drawn he had selected in advance. In the early years a member of the council sometimes insured against the loss of his seat by wagering that his name would be drawn in the lottery. If his name was drawn, he was compensated by a cash prize. Popular superstition held that Gentile, whose name was never drawn, had made a compact with the Devil! (J. Corblet [8, 130–131]).

"The Genoese lottery," as it was called, was soon imitated in the principal cities of Italy and then in other countries. In France, the first drawing of a lottery, with similar rules, was held in 1758 for the benefit of l'École Royale Militaire. In 1774, P. S. Laplace, who was then on the faculty of this institution, discussed the lottery, which he called "loterie de l'École Militaire." This designation afterwards puzzled the historians M. Cantor and L. G. du Pasquier, who expected the usual "loterie genoise."

In Germany, the finances of Prussia were at a low ebb after the Seven Years' War and Frederick the Great sought to increase the state revenues by introducing a "Genoese lottery." (This was in fact instituted in 1763.) Leonhard Euler, who was then a leading member of the Berlin Academy of Sciences, was consulted by Frederick on various occasions regarding mathematical questions concerning the lottery. Euler wrote several memoirs concerning this, including a lengthy paper published in 1767. (See Euler [9], Bernoulli [11] and Todhunter [12, 245–247 and 326–331].)

Euler considers here the probability of obtaining two or more consecutive numbers, termed a sequence, when five numbers are drawn at random from ninety numbered consecutively from 1 to 90 in the Genoese lottery. (Euler regards the numbers drawn as arranged in a row so that 90 and 1 do not form a sequence.) He considers the case when m numbers are drawn at random from n consecutively numbered integers, initially for some numerical values, and then by a remarkable inductive procedure—one is tempted to write, virtually a commonplace with Euler—obtains the result

$$\frac{E(n, m)}{\binom{n}{m}} = \frac{\binom{n-m+1}{m}}{\binom{n}{m}} \tag{1}$$

for the probability that no sequences occur. Evidently, this is essentially equivalent to Lemma A of Section 2.

The problem investigated by Euler was also treated (in the same volume) by N. de Beguelin [10] in two papers, and by Jean (III) Bernoulli [11]. (In contrast to Euler, however, these authors regarded the numbers drawn in the lottery as arranged in a circle so that 90 and 1 form a sequence, etc.) The probability of not obtaining a sequence when m numbers are drawn at random from n consecutively numbered integers was found by Beguelin and Bernoulli to be

$$\frac{B(n, m)}{\binom{n}{m}} = \frac{\dfrac{n}{m}\binom{n-m-1}{m-1}}{\binom{n}{m}}. \tag{2}$$

This is essentially equivalent to Lemma B of Section 2.

A more readily available summary of the memoirs of Euler, Beguelin and Bernoulli was given in a book by Todhunter [12].

4. Tait's Knot Problem

P. G. Tait (1831–1901), a well-known nineteenth-century physicist, was for many years a professor of natural philosophy at Edinburgh. He investigated the theory of vortex atoms in the 1870s and stemming from this formulated a problem whose solution he thought he required in his study of knots: "How many arrangements are there of n letters, when A cannot be in the first or second place, B not in the second or third etc.?" (Cf. Lucas' formulation at the beginning of Section 2.)

Tait suggested the problem to Arthur Cayley (1821–1895), then the foremost mathematician in Britain, and in 1877 (some fourteen years before the publication of [Lucas, 3]), the latter developed a solution based on a systematic application of the principle

> **Muir (1877):**
> $u(4)$ is the number of non-zero terms in the expansion of the determinant,
>
> $$\begin{vmatrix} 0 & 1 & 1 & 0 \\ 0 & 0 & 1 & 1 \\ 1 & 0 & 0 & 1 \\ 1 & 1 & 0 & 0 \end{vmatrix}$$
>
> (Equivalent to : $u(4)$ is the value of the permanent of this array.)

FIGURE 24-4

of inclusion and exclusion (Cayley [13] or E. Netto [14, 75–78]). Subsequently, Thomas Muir [15] formulated Tait's problem as a question concerning the number of terms in the expansion of a particular kind of determinant and obtained a complicated recurrence formula:

$$u(n) = (n - 2)u(n - 1) + (2n - 4)u(n - 2) + (3n - 6)u(n - 3) + (4n - 10)u(n - 4)$$
$$+ (5n - 14)u(n - 5) + (6n - 20)u(n - 6) + (7n - 26)u(n - 7) + (8n - 34)u(n - 8)$$
$$+ \cdots + \frac{1 - (-1)^n}{2}. \tag{1}$$

Cayley wrote an addendum to Muir [15] in which he used (1) formally to obtain a generating function for $u(n)$ of a complicated form. Later Muir [16], starting from (1), obtained the simpler recursion formulas

$$u(n) = nu(n - 1) + 2u(n - 2) - (n - 4)u(n - 3) - u(n - 4) \tag{2}$$

and

$$(n - 2)u(n) = n(n - 2)u(n - 1) + nu(n - 2) + (-1)^{n-1} \cdot 4 \tag{3}$$

with the initial conditions $u(3) = 1$, $u(4) = 2$. (The latter conditions can be verified directly.) This is probably the most useful formula thus far obtained for the calculation of successive values of $u(n)$. (Cf. Section 6.) Muir also verified the generating function obtained by Cayley.

Somewhat anticlimactically, it later turned out that Tait's knot problem was more difficult than that which he had originally formulated. Let P be an arrangement of $\{1, 2, \ldots, n\}$ which satisfies the conditions required above by Tait and C be any one of the n cyclic permutations. Then $P' = C^{-1}PC$ is also a permutation satisfying these conditions. If two arrangements P and Q satisfy these conditions and there is a suitable cyclic permutation C for which $Q = C^{-1}PC$, then P and Q are *equivalent* permutations. The total number of permutations $u(n)$ satisfying Tait's conditions may be grouped into distinct equivalence classes. What was actually required was $T(n)$, the number of such equivalence classes for a given n. (It may be verified that $T(3) = u(3) = 1$, $T(4) = u(4) = 2$, but $T(5) = 5$ while $u(5) = 13$.) The problem of determining $T(n)$ in the general case remained open for many years but was solved in 1956 by E. N. Gilbert [15].

5. The *Problème des Ménages*

When Lucas [3] discussed this problem, he did not know of Tait's problem and of the previous work of Cayley and Muir. One of his collaborators, Col. C. Moreau, obtained a recurrence relation equivalent to (4.3) (Lucas [3, 494–5]). Lucas [3, Note III] also credits another of his collaborators, C. A. Laisant, with the solution presented in his book. Laisant [16] envisaged the problem as determining the number of ways of placing n rooks on an $n \times n$ chessboard with certain restrictions and obtained recurrence relations from which that of Col. Moreau follows.

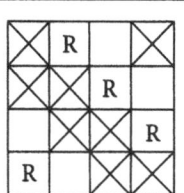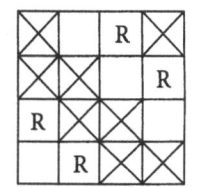

$u(n)$ is the number of ways of placing n non-attacking rooks on an $n \times n$ chessboard if two squares in cyclic order in each row are forbidden

FIGURE 24-5

The problem was discussed in 1902 by H. M. Taylor [17], who did not mention any prior work. (As remarked by Kaplansky and Riordan [18, 115] in their historical discussion: "Alone of all authors in this respect, he chose to seat the men first.") Taylor developed a number of simultaneous recurrence equations (equivalent to those in (Lucas [5, Note III])) and, on elimination,

$$nu(n+2) = (n^2 + n + 1)\big(u(n+1) + u(n)\big) + (n+1)u(n-1), \tag{1}$$

which can be obtained from (4.3).

A number of books on mathematical recreations have since given versions of Taylor's solution.

Another solution is the operational method described by Major P. A. MacMahon [19, Vol. 1, 253–254], which reduces to the determination of the coefficient of a particular term in the product of n linear forms. This result, equivalent to one previously obtained by Laisant, is difficult to apply in practice.

In 1943, at about the same time that the work of Kaplansky [5] appeared, a quite different explicit formula for $u(n)$ was published by Waldemar Schöbe [20]. His starting point was the set of recurrences obtained in Lucas [5, Note III], from which he obtained the result

$$u(n) = 2(-1)^n + n \sum_{v=1}^{n} \frac{(-1)^{v-1}}{(v-1)!} \frac{d^2(n-v)}{(n-v)!} \tag{2}$$

where $d(n)$ is defined by (2.2), and showed that $u(n)/n^2 \to e^{-2}$ as $n \to \infty$.

Laisant (1891). MacMahon (1915):
Let s_r denote the sum of the elements which may occupy the rth position in a row of n different letters, $r = 1, 2, \ldots, n$. Then

$$u(4) = \frac{\partial^4(s_1 s_2 s_3 s_4)}{\partial a \cdot \partial b \cdot \partial c \cdot \partial d} = \frac{\partial^4(c+d)(a+d)(a+b)(b+c)}{\partial a \cdot \partial b \cdot \partial c \cdot \partial d}$$

$$= \text{Coefficient of } abcd \text{ in } (c+d)(a+d)(a+b)(b+c)$$

FIGURE 24-6

In 1953, almost two decades after his result for the *problème des ménages* had been stated, Jacques Touchard [21] published his solution to this as well as to more general problems. In addition to (2.3), he also obtained interesting additional results. E.g.,

$$u(n) = 2 \int_0^\infty T_{2n}(\sqrt{t}/2)e^{-t}dt \tag{3}$$

where the Chebyschev polynomial is defined by

$$2T_n(x) = 2\cos n(\arccos x) = \left(x + \sqrt{x^2 - 1}\right)^n + \left(x - \sqrt{x^2 - 1}\right)^n.$$

Moreover, a generating function for $u(n)$ can be given in terms of a Neumann series

$$\sum_{n=0}^\infty u(n)I_n(2t) = e^{-2t}(1 - t)^{-1} \tag{4}$$

where I_n is the Bessel function of the first kind with imaginary argument.

6. Related Problems

A. W. Joseph [22] in 1947 investigated a variation of the matching problem mentioned at the beginning of Section 2:

"Two persons turn up cards *alternately*, each starting with a similar, but differently shuffled, pack of cards. What is the probability that before all the cards have been dealt two identical cards are showing at the same time?"

He solves this problem in two ways. The latter, "more satisfactory derivation," consists in obtaining a recursion formula for a particular function, which he denotes by u_n. (This is *not* the function $u(n)$ defined in Section 1.) Assume the cards turned up by the first player are, in order, $1, 2, \ldots, n$. Then u_n, the number of ways for no match to occur, may be denoted symbolically, in Joseph's notation by

$$\left\{ \frac{\bar{1}}{2}, \frac{\bar{2}}{3}, \ldots, \frac{\overline{(n-1)}}{\bar{n}}, \bar{n} \right\}$$

which means that the second player does not draw a 1 or a 2 on his first turn, a 2 or a 3 on his second turn, $\ldots$, $(n - 1)$ or n on his $(n - 1)$st turn or n on his nth turn. The required probability is $1 - (u_n/n!)$. A recursion formula for u_n was derived

$$u_n = (n - 1)u_{n-1} + (n - 1)u_{n-2} + u_{n-3} \ (n \geq 4) \tag{1}$$

with $u_1 = u_2 = 0$, $u_3 = 1$, and an asymptotic approximation for $u_n/n!$ was obtained.

Joseph was not aware that the equivalents of (1) had already been given by Lucas [5, Note III, Eq. (8)] and Taylor [17, Eq. (5)] and that the latter had tabulated values of the equivalent of u_n. The relation between u_n and the *ménage* numbers is

$$u_n = \sum_{m=2}^n u(m), \quad n \geq 2. \tag{2}$$

A number of formulas for the *ménage* numbers have been found in addition to those above which are useful under particular conditions. By an analytical investigation, M. Wyman and L. Moser [23] showed that $u(n)$ is the integer nearest to

$$ne^{-2} \sum_{m=0}^{n-1} \frac{(-1)^m (n - m - 1)!}{m!}, \tag{3}$$

and about half of these terms are redundant since their sum is less than one half. Later, W. Schöbe [24] proved by elementary methods that for $n \geq 4$, the upper limit in the summation can be replaced by $[(n - 1)/2]$. Wyman and Moser also supplemented the earlier historical exposition of Kaplansky and Riordan [18] and gave an extensive table of $u(n)$ for $n = 0(1)65$.

S. M. Kerawala [25] derived an asymptotic expression for $u(n)$ by substituting a series of inverse powers of n as a particular solution of (4.3) and obtained the result

$$u(n) = n!e^{-2} \left[1 - \frac{1}{n} - \frac{1}{2n^2} + \frac{1}{3n^3} + \cdots \right] + 4(-1)^n \left[\frac{1}{n^2} - \frac{5}{n^4} + \cdots \right]. \tag{4}$$

The *problème des ménages* is treated from a different and more general viewpoint in the textbook by John Riordan [6, Ch. 8], an outstanding contributor to the modern development of combinatorial analysis. The generating function for R_n, the number of husbands seated next to their wives, is

$$U_n(t) = \sum_{k=0}^{n} \frac{2n}{2n - k} \binom{2n - k}{k} (n - k)!(t - 1)^k \tag{5}$$

and the corresponding function, when the husbands and wives are seated alternately at a straight table, is

$$V_n(t) = \sum_{k=0}^{n} \binom{2n - k}{k} (n - k)!(t - 1)^k \tag{6}$$

with

$$U_n(t) = V_n(t) + (t - 1)V_{n-1}(t). \tag{7}$$

The expression for $U_n(t)$ is essentially due to Touchard [4].

In 1944–46, Riordan pointed out the relation between the number of Latin rectangles of a particular type and the *ménage* numbers. Latin rectangles are $m \times n$ arrays, $m \leq n$, in which each row is a permutation of $\{1, 2, \ldots, n\}$ and no column contains repeated numbers. (Such arrays, particularly square arrays, are important in the design of experiments.)

In what follows, the first row will be assumed to be in normalized order $1, 2, \ldots, n$, so that the total number of Latin rectangles is $L(m, n) = n!K(m, n)$, where $K(m, n)$ denotes the number of normalized Latin rectangles. Evidently, $K(2, n) = d(n)$, where $d(n)$ is defined by (2.2), and the *ménage* number $u(n)$ is the number of $3 \times n$ Latin rectangles whose first two rows are given by (2.1). Riordan derived a result which may be written

$$K(3, n) = \sum_{k=0}^{[n/2]} \binom{n}{k} d(k)d(n - k)u(n - 2k), u(0) = 1 \tag{8}$$

(See Riordan [6, 205ff.].) Only partial results are known for the enumeration of Latin rectangles with $m > 3$.

References

1. Jones, D. S. and Moore, P. G., "Circular seating arrangements," *Journal of the Institute of Actuaries*, Vol. 108 (1981), 405–411.
2. Thomas, D. E., Note, *Journal of the Institute of Actuaries*, Vol. 110 (1983), 396–398.
3. Lucas, E., *Théorie des Nombres*, Paris, 1891. (Reprinted in 1961.)
4. Touchard, J., "Sur un problème des permutations," *Comptes Rendus de L'Acad. des Sciences*, T. 198 (1934), 631–633.
5. Kaplansky, I., "Solution of the 'problème des ménages,' " *Bulletin of the American Mathematical Society*, Vol. 49 (1943), 784–785.
6. Riordan, J., *An Introduction to Combinatorial Analysis*, New York, 1958.
7. Comtet, L., *Advanced Combinatorics*, Dordrecht-Holland, 1974.
8. Corblet, J., "Étude historique sur les loteries," *Revue de l'Art Chrétien*, 5 Année (1861), 12–28, 57–76, 121–137.
9. Euler, L., "Sur la probabilité des séquences dans la Lotterie Genoise," *Histoire de l'académie des sciences de Berlin*, T. 21 (1765), 1767, 191–230, reprinted in *Opera Omnia* Ser. 1, Vol. 7, Leipzig and Berlin, 1923.
10. Beguelin, N. de, "Sur les suites ou séquences dans la Lotterie de Genes," *Histoire de l'académie des sciences de Berlin*, T. 21 (1765), 1767, 231–256, 257–280.
11. Bernoulli, Jean, "Sur les suites ou séquences dans la Lotterie de Genes," *Histoire de l'académie des sciences de Berlin*, T. 25 (1769), 1771, 234–253.
12. Todhunter, I., *A History of the Mathematical Theory of Probability . . .*, London, 1865, reprinted New York, 1949.
13. Cayley, Arthur, "On a problem of arrangements," *Proceedings of the Royal Society of Edinburgh*, Vol. IX (1875–78), 338–342.
14. Netto, Eugen, *Lehrbuch der Kombinatorik*, Zweite Auflage, Berlin, 1927, reprinted New York, 1958.
15. Gilbert, E. N., "Knots and classes of ménage permutations," *Scripta mathematica*, Vol. 22 (1956), 228–233.
16. Laisant, C., "Sur deux problèmes de permutations," *Bulletin de la Société Mathématique de France*, T. 19 (1890–91), 105–108.
17. Taylor, H. M., "A Problem on Arrangements," *Messenger of Mathematics*, Vol. 32 (May 1902–April 1903), 60–63.
18. Kaplansky, I. and Riordan, J., "The problème des ménages," *Scripta Mathematica*, Vol. 12 (1946), 113–124.
19. MacMahon, P. A., *Combinatory Analysis*, 2 vols., Cambridge, 1915–16, reprinted New York, 1960.
20. Schöbe, W., "Das Lucassche Ehepaarproblem," *Mathematische Zeitschrift*, Bd. 48 (1943), 781–784.

21. Touchard, J., "Permutations discordant with two given permutations," *Scripta Mathematica*, Vol. 19 (1953), 109–119.

22. Joseph, A. W., "A problem in derangements," *Journal of the Institute of Actuaries Students' Society*, Vol. VI (1947), 14–22.

23. Wyman, M. and Moser L., "On the *problème des ménages*," *Canadian Journal of Mathematics*, Vol. 10 (1958), 468–480.

24. Schöbe, Waldemar, "Zum Lucasschen Ehepaarproblem," *Mitteilungen der Vereinigung schweizerischer Versicherungsmathematiker*, Bd. 61 (1961), 285–292.

25. Kerawala, S. M., "Asymptotic solution of the 'problème des ménages,' " *Bulletin of the Calcutta Mathematical Society*, Vol. 39 (1947), 82–84.

25

Quasicrystals:
The View from Les Houches

Marjorie Senechal and Jean Taylor

1. Introduction

Soon after the announcement of their discovery in 1984 [1], quasi-crystals hit the headlines. Here was a substance—an alloy of aluminum and manganese—whose electron diffraction patterns exhibited clear and unmistakable icosahedral symmetry (a view along a five-fold axis is shown in Figure 25-1). A clear and unmistakable diffraction pattern of any sort is evidence of "long-range order": The diffraction pattern is a picture of a Fourier transform. Long-range order is usually synonymous with periodicity, and every periodic structure has a translation lattice. But a simple argument shows that five-fold rotational symmetry is incompatible with lattices in $\mathbf{R}^2$ and $\mathbf{R}^3$: every lattice has a minimum distance d between its points, but if two points at this distance are centers of five-fold rotation about parallel axes, the rotations will generate an orbit with smaller distances between them (Figure 25-2). By this chain of reasoning, it appeared that the impossible had occurred.

For the past five years, quasi-crystals have been studied intensively by metallurgists, solid-state physicists, and mathematicians (few crystallographers have shown much interest in them). The problem has gradually been resolved into three more or less separate questions, not necessarily according to the field of the researcher:

1. *Crystallography.* How are the atoms of real quasi-crystals arranged in three-dimensional space?
2. *Physics.* What are the physical properties of substances with long-range order but no translational symmetry?
3. *Mathematics.* What kind of order is necessary and sufficient for a pattern of points to have a diffraction pattern with bright spots?

As Cahn and Taylor pointed out in 1985 [2], to answer Question 3 we must draw on a variety of techniques from many branches of mathematics, including tilting theory, almost periodic functions, generalized functions, Fourier analysis, algebraic number theory, ergodic theory and spectral measures, representations of $\mathrm{GL}(n)$, and symbolic dynamics and dynamical systems.

This article is a report on the current status of the problem. We began our discussions while attending a conference on *Number Theory and Physics* at the Centre de Physique, Les Houches, France,

Volume 12, No. 2 (Spring 1990), 54–64

FIGURE 25-1 *A diffraction pattern of an aluminum-manganese quasi-crystal. Its five-fold rotational symmetry produced shock-waves in the world of solid-state science. Photograph courtesy of John Cahn.*

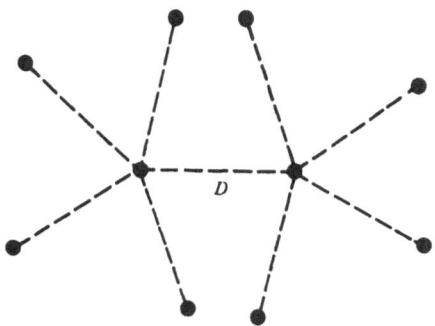

FIGURE 25-2 *Five-fold symmetry is incompatible with periodicity, because it violates discreteness. Two five-fold centers at the minimum distance d generate points whose distance is less than d.*

in March 1989. One of the foci of that conference was quasi-periodicity and quasi-crystals, and during our ten days there we enjoyed extended discussions with a variety of observers and practitioners of this field. But we warn the reader that the view we present here may not be widely shared; in particular, Question 3 is usually not phrased in such generality (see Section 6). And like the view of the Mont Blanc massif from the conference center (Figure 25-3), the general outline and size of the problem is rather clear, but features that are prominent from our perspective may mask others, including the summit.

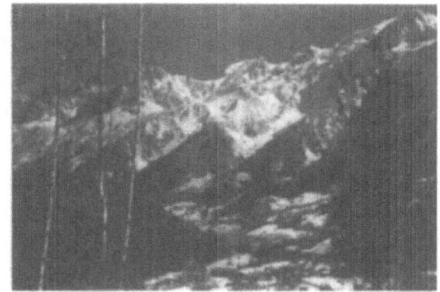

FIGURE 25-3 *The Mont Blanc massif, seen from the Centre de Physique, Les Houches, France. Photo by Pierrette Cassou-Nouges.*

2. What Is a Crystal?

The discovery in 1912 that crystals diffract X-rays lent overwhelming experimental support to the hypothesis that crystalline structure is periodic. What could be a better example of Pierre Curie's banal but widely admired Principle of Symmetry: "When certain causes produce certain effects, the elements of symmetry in the causes ought to reappear in the effects produced"? Since then, the lattice has been taken as the definition of the crystalline state.

For the first year or two after their discovery, the question most hotly debated among solid-state scientists was: Are quasi-crystals crystals? By this was meant, can the structure of these alloys be interpreted within the framework of periodicity (for example, as a mosaic or intergrowth), or is it something truly new? Now, nearly five years later, quasi-crystals of varying compositions (aluminum-lithium-copper, uranium-palladium-silicon, and many others) and high perfection have been grown in laboratories all over the world and have been analyzed in great detail, and the mosaic and intergrowth models have been discarded by essentially everyone but Linus Pauling [3]. It is clear that the question should be interpreted differently. To ask whether quasi-crystals are crystals really means to ask what we mean by "crystal." We cannot see inside the solid state; we know it only through the images provided by diffraction, electron microscopy, and other modern techniques. It might be more appropriate to define a "crystal" to be a structure with sufficient long-range order to exhibit images with properties associated with those we call crystalline, such as a diffraction pattern with sharp spots.

A diffraction pattern for a material is essentially a two-dimensional slice of the square modulus of the Fourier transform of its density distribution; it faithfully records the amplitudes of the transform, but gives no direct record of the phases. Geometrically, we can think of a periodic crystal as an orbit of its symmetry group, which is an infinite discrete group with compact fundamental region. It can be shown that every orbit of such a group is a union of a finite number of congruent lattices. In the first approximation, we can take the density distribution of a periodic crystal to be a sum of weighted delta functions located at the nodes of each of these lattices (with the same weight for each point of a particular lattice). The Poisson summation formula implies that the Fourier transform of a lattice sum of deltas is again such a sum. Thus the diffraction pattern of a lattice is a lattice; it is in fact its dual lattice. The original lattices and their weights can be recovered from the diffraction pattern with the help of some techniques for recovering the "phase factor"; this is the experimental and theoretical framework for crystal structure analysis.

However, the lattice hypothesis is not without its problems. There are crystals with extremely large repeat units, with thousands of atoms in the unit cell. There are crystals that are more or less random stackings of two-dimensional periodic structures. There are crystals in which the lattice is disturbed by a modulation. Finding a periodic framework on which to hang these structures can be likened to the pre-Keplerian problem in astronomy of trying to explain planetary orbits by decorating the circle with the right number and arrangement of epicycles. No number or arrangement will be correct for the quasi-crystals! The quasi-crystal phenomenon shows us that a diffraction pattern can theoretically show sharp spots even if there is a single non-periodic but well-defined geometrical pattern that gives rise to it. And although it is widely assumed that the crystal lattice is a global consequence of the play of local interatomic forces, from the standpoint of physics or mathematics this is an open problem. Indeed, Miekisz and Radin have shown that generically one would expect local forces to generate non-periodic structures [4]. In fact, one can even find crystals almost arbitrarily close to "ideal quasi-crystals," in the same way that irrational numbers can be approximated by rational numbers.

Thus the deeper question is: what local ordering properties are necessary and sufficient to produce orderly images?

Two minimal properties that we might require of a locally ordered point set $L \subset \mathbf{R}^n$ are discreteness and relative density: There is a minimum distance d between any pair of points of L, and a number $\delta > 0$ such that every sphere of radius δ contains at least one point of L; such an L is sometimes called a *Delone system*. (Incidentally, Delone's name is sometimes spelled Delaunay, reflecting the fact that his forebears went to Russia with Napoleon and stayed on.) A finite region of a Delone system with no additional structure roughly describes the centers of the atoms in a monatomic gas in a closed container. Increasing the structure increases the order. Sufficient local order implies periodicity: Delone and his colleagues proved [5] that there is a number $k = k(\delta, n)$, where n is the dimension of the space in which L lies, such that if the sets $\{x \in \mathbf{R}^n : |x - u| \leq k\} \cap L$ are congruent for each $u \in L$, then L is an orbit of a crystallographic group.

The patterns we are interested in have order somewhere between amorphous and periodic. The question is, what intermediate conditions are necessary and sufficient to ensure that L can produce a diffraction pattern with "bright spots"? For example, one condition might be *local isomorphism*: every finite configuration of L occurs in every bounded region of sufficient size. But although local isomorphism is present in all the examples that we know about, there is no proof that it is either necessary or sufficient. Other local ordering conditions can of course be proposed, but not much is known about their effect either. The question remains open.

One obvious difficulty is that "bright spot" is not well defined. We can simplify matters by defining it to mean that there are Dirac deltas in the Fourier spectrum, weighted so that some peaks appear isolated. Then there are two cases to consider: Either the entire Fourier spectrum is a set of deltas (possibly dense), or else the spectrum also contains a continuous component. But from the experimental point of view a bright spot need not represent a "real" delta; it might be due to features of the transform that closely approximate delta functions. Point sets with this property are of theoretical as well as experimental interest (see Section 4).

We can formulate these conditions more precisely. Any Delone system D has a density distribution that is a countable infinite sum of weighted Dirac deltas on the points of D; we can assume as a first approximation that all the weights are equal to 1. Then the distribution $\rho(x)$ can be written $\sum_{v \in D} \delta(x - v)$, where $x \in \mathbf{R}^n$; a distribution of this form is sometimes called a *Dirac comb*. We are looking for Dirac combs whose Fourier transforms $\tilde{\rho}(t) = \sum_{v \in D} \exp(2\pi i t \cdot v)$ are closely related to Dirac combs, where by "closely related" we mean one of the following:

(a) The Fourier transform is precisely a Dirac comb; such a comb is also called a Poisson comb. (An important special case is when the frequencies v at which the delta functions of the Fourier transform occur have a finite basis over the integers; the original density is then said to be *quasiperiodic*.)
(b) The Fourier transform contains a Dirac comb together with a continuous part.
(c) The Fourier transform "looks like" it contains a Dirac comb but does not in fact do so; this can happen when the spectrum has a singular continuous component.

Characterizing the order properties of Dirac combs satisfying (a), (b), or (c), and classifying these combs, is a central problem of quasi-crystallography and involves all of the branches of mathematics listed above. In the absence of a complete answer to our question, we study examples. The two classes of combs that have been studied in most detail are those obtained by projection, and one-dimensional combs. We will also discuss some of the relations between combs and tilings; some interesting recent work is discussed in Section 5.

3. Point Sets Obtained by Projection

Three years before the discovery of quasi-crystals, the English crystallographer Alan Mackay [6], long an advocate for a more general crystallography, devised an ingenious experiment. He arranged for an optical diffraction mask to be constructed whose holes were located at the vertices of a tiling by Penrose rhombs (Figure 25-4). These tilings, which are constructed by juxtaposing the rhombs according to strict matching rules, are non-periodic. Yet they aren't "disordered": one can discern a great deal of local order. Local configurations with five-fold symmetry not only occur, they occur all over the place. Indeed, the pattern of vertices has the local isomorphism property discussed

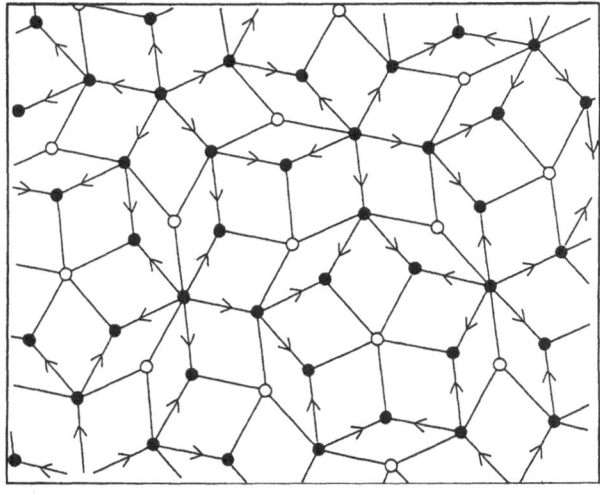

FIGURE 25-4 *Part of a Penrose tiling by rhombs. Vertex colors and edge arrows must be matched.*

above. Moreover, the tilings are self-similar. (These and other properties of the Penrose tilings will be discussed in more detail in Section 5.) As Mackay suspected, the diffraction pattern obtained with this mask was clear and sharp, almost crystalline. And it had the crystallographically forbidden five-fold symmetry.

De Bruijn's construction.   Since the quasi-crystal in question, i.e., the set of the vertices of the Penrose tiles, is not a lattice, how can we explain Mackay's experimental results? The necessary insight was supplied by N. G. de Bruijn in a remarkable set of papers published in 1982 [7]. In these papers, de Bruijn gave a global method for constructing the Penrose tilings that allows us to index the vertices of the rhombs with five integer coordinates $(x_0, \ldots, x_4)$. Thus the vertices can be identified with a subset S of the points of the integer lattice in $\mathbf{R}^5$.

Moreover, de Bruijn showed that

$$1 \le \sum_{k=0}^{4} x_i \le 4 \quad \text{for all } \vec{x} = (x_0, \ldots, x_4) \in S. \tag{1}$$

Since this sum is also the scalar product $(x_0, \ldots, x_4) \cdot (1, \ldots, 1)$, the points of S lie in a region $M \subset \mathbf{R}^5$ which projects to a bounded interval on the line containing $\vec{m} = (1, 1, 1, 1, 1)$; note that this vector is the body diagonal of the unit 5-cube γ_5. The vertices of the Penrose rhombs are the projections of S onto a plane Π, one of the two invariant planes of the five-fold rotation about that diagonal, which cyclically permutes the five coordinate axes. Both of these planes are irrational: their intersections with the lattice are just $\{0\}$. The tile vertices are integral linear combinations of the projections of the five orthonormal unit coordinate vectors of $\mathbf{R}^5$ onto Π. Not all vectors satisfying (1) are in S; S is the set of points M' of M whose projection onto $\Pi^\perp$ lies in the projection of γ_5 onto that subspace. (Any projection of an n-cube is a zonohedron; in this case it is a rhombic icosahedron.)

Katz and Duneau [8], among others, have shown that the projection formalism greatly facilitates the computation of Fourier transforms. The density function of the set S is the product of the density function of the integer lattice in $\mathbf{R}^5$ and the characteristic function of M'. Thus, since the Fourier transform of a product of two functions is the convolution of the Fourier transforms of the individual functions, and since the Fourier transform of the projection of the density function of S is the restriction of the Fourier transform of that function to Π, we can compute the diffraction pattern of the Penrose vertices. (That all of this can be made rigorous has been shown, by rather different arguments, by de Bruijn [9] and by Porter [10].) The Fourier transform is a countable sum of delta functions at a dense set of points in the plane; thus the set of vertices is a Poisson comb. Although the delta functions are dense, we see bright spots in the diffraction pattern, because the amplitude of the transform attains local maxima at isolated endpoints, and at most other points is very small.

The general case.   The Penrose tilings are of course very special. To what extent does the property of being a Poisson comb depend on their remarkable properties? Curiously, this dependence is not very strong. For example, while it is easy to construct Poisson combs by projection, as far as we know most of them are not self-similar. (If we translate M in $\Pi^\perp$, the projected pattern will include local vertex configurations that are forbidden by the matching rules of the Penrose tiles.) Or, we can carry out the analogous construction in $\mathbf{R}^n$, producing plane point sets that are the vertices of tilings with local n-fold symmetry for which no matching rules are known (see Figure 25-5). All of these patterns have the local isomorphism property, however.

More generally, let Λ be a lattice in $\mathbf{R}^n$, and let Π_k be any irrational k-dimensional subspace of $\mathbf{R}^n$ (again, irrational means that $\Pi_k \cap \Lambda = \{0\}$). First, let us see under what conditions we can obtain a Delone system in Π_k by projection. Since Π_k is irrational, the orthogonal projection of all of Λ onto

Π_k will be non-periodic, but it will also be dense. We need to find a subset S of Λ that projects to a discrete set (relative density is guaranteed by the fact that Λ is a lattice). We know that there is a minimum distance d between points of Λ: if $\vec{x}$ and $\vec{y}$ are two vectors of Λ, then $|\vec{x} - \vec{y}| \geq d$. We can decompose $\vec{w} = \vec{x} - \vec{y}$ into its Π_k and $\Pi_k^{\perp}$ components w_k and $w_k^{\perp}$. If we insist that $|w_k^{\perp}| < d - \epsilon$ for some $\epsilon > 0$, then $|w_k| \geq d$. This means that we can obtain a discrete set of points in Π_k by requiring that the image of the set $S \subset \mathbf{R}^n$ under projection to $\Pi_k^{\perp}$ lie in an appropriately chosen compact set $T \subset \Pi_k^{\perp}$; T is sometimes called the *window* for the projection. Thus S lies in the cylinder $M = T \oplus \Pi_k \subset \mathbf{R}^n$. The computation of the Fourier transform then proceeds as in the examples above. The projected set always turns out to be a Poisson comb.

There are many variations on the projection theme. The window need not lie in the orthogonal complement of Π_k; the vector space need not be Euclidean. One technique used quite extensively at the moment is to try to replace Λ by a periodic set of

FIGURE 25-5 *A plane tiling by three kinds of rhombs, projected from $\mathbf{R}^7$. [Courtesy of André Katz.]*

connected surfaces in $\mathbf{R}^n$ and to consider the intersection of Π_k with these surfaces. (Again, de Bruijn has provided a firm mathematical basis for much of this, in a different language [11].) It is true that if one has a Poisson comb and its Fourier transform's delta functions have a finite basis, and if one knows the full complex amplitudes of these deltas, then it is possible to reconstruct a periodic density in $\mathbf{R}^n$ and a plane Π_k such that the restriction of the density in $\mathbf{R}^n$ to Π_k will give the delta functions of the density in $\mathbf{R}^k$. However, it is not at all obvious that there are densities in $\mathbf{R}^n$ that consist of "surfaces" of any particular smoothness. Also, although using arbitrary surfaces, rather than those from projecting a lattice, gives a broader class of Poisson combs, it does not always give a noticeably broader class of diffraction patterns, because these combs may differ from the lattice-projection ones only in their intensities and phases, as de Bruijn has noted [11].

A word about symmetry. The most striking thing about the diffraction pattern of the Penrose vertices is its five-fold rotational symmetry; quasi-crystals might never have been noticed if this symmetry had not been observed. Indeed, successful crystal structure determination depends on finding directions of high symmetry so that bright spots will appear in the diffraction pattern.

The symmetry we observe in the Fourier transform of a projected pattern depends on the symmetry group G of the lattice Λ and on the choice of Π_k. G is a semi-direct product of $\mathbf{Z}^n$ and a finite subgroup $P \subset O(n)$, where P is the stabilizer of $0 \in \Lambda$. If Π_k is invariant under P or under a subgroup of P, the Fourier transform will reflect this. This leads us to the important and interesting problem of determining which lattices in $\mathbf{R}^n$ have invariant subspaces of whatever dimension, and how crystallographic groups built on these lattices act on those subspaces. In short, the projection method has opened an interesting chapter in n-dimensional crystallography. To date, those lattices for which G is or contains the icosahedral group have been studied in the most detail (see, e.g., [12]), because they arise in the theory of the three-dimensional Penrose tiles (see Section 5) and in the interpretation of diffraction patterns of real quasi-crystals.

But from our point of view, it is the bright spots that are fundamental, not symmetry *per se*. Since bright spots are theoretically present in every projected pattern, we know that their occurrence is not dependent on rotational symmetry. Indeed, it seems that rotational symmetry has nothing *a priori* to do with our problem, except that when we find non-crystallographic rotational symmetry in a diffraction pattern, we know that it was produced by a non-periodic Dirac comb. On the other hand, bright spots in a diffraction pattern indicate some sort of long-range order or generalized symmetry. This brings us back to the questions raised in Section 2.

4. Order on the Line

The one-dimensional case is the most tractable; here we find examples of all three types of ordering for non-periodic point sets. We will describe a few of them.

Sequences with average lattices.   The standard example of a one-dimensional quasi-crystal is the "Fibonacci" sequence of points

$$u_n = n + (\tau - 1)[n/\tau], \tag{2}$$

where $\tau = (1 + \sqrt{5})/2$ is the golden number of classical and modern fame and $[x]$ is the greatest integer function. (We will explain the relation of this sequence to the classical Fibonacci sequence below.) This sequence can be obtained by projection from $\mathbf{R}^2$ to a line Π with slope $1/\tau$. Let us consider the more general case in which the line has slope α, where α is an irrational number. Using as our window the projection of a unit square of the integer lattice Λ onto $\Pi^\perp$, the cylinder M is the strip bounded by the lines $y = \alpha x$ and $y = \alpha x + \alpha + 1$ (Figure 25-6) and $M' = M \cap \Lambda$. The points in this strip have coordinates

$$(x, y) = ([n/(\alpha + 1)], n - [n/(\alpha + 1)]),$$

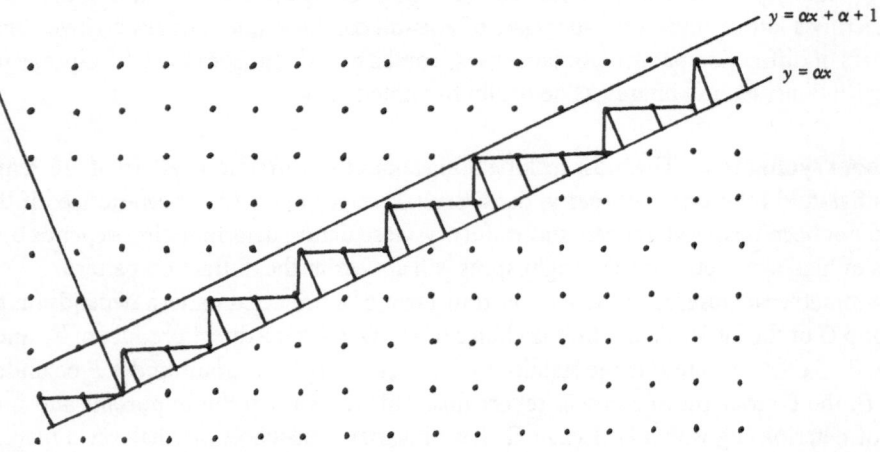

FIGURE 25-6 *A tiling of the line obtained by projection from* $\mathbf{R}^2$. *All tilings of this type have average lattices.*

and project onto the points $(x + \alpha y)\vec{u}$, where $\vec{u}$ is the vector $(1, \alpha)/(1 + \alpha^2)$ along Π. Thus the projected points form a sequence

$$p_n = \alpha n - (\alpha - 1)\left[\frac{n}{\alpha + 1}\right], \tag{3}$$

and $p_{n+1} - p_n = 1$ or α. When $\alpha = 1/\tau$, we obtain the sequence above (if we multiply everything by $1/\tau$). The methods of the preceding section can be used to show that all of these sequences are Poisson combs.

The sequences obtained in this way are often called "one-dimensional quasi-crystals." But they do not really illustrate the quasi-crystal phenomenon, because in fact these sequences are one-dimensional modulated lattices. Modulated crystals were known long before the quasi-crystals were discovered and have been intensively studied for many years.

In what sense are these sequences modulated lattices? Using the equality $[x] = x - \{x\}$, where $\{x\}$ is the fractional part of x, we have

$$p_n = \alpha n - (\alpha - 1)\frac{n}{\alpha + 1} + (\alpha - 1)\left\{\frac{n}{\alpha + 1}\right\} = \frac{\alpha^2 + 1}{\alpha + 1}n + (\alpha - 1)\left\{\frac{n}{\alpha + 1}\right\}, \tag{4}$$

or in the case of the Fibonacci sequence (2), $u_n = (2 - 1/\tau)n + (\tau - 1)\{n/\tau\}$. Thus we see that $\{p_n\}$ is built upon the one-dimensional lattice of points of the form $n(\alpha^2 + 1)/(\alpha + 1)$ for $n \in \mathbf{Z}$, deviating from it by an amount which is at most $|\alpha - 1|$.

This property, of having a limiting average spacing and a bounded modulation away from the lattice with this spacing, is called having an "average lattice."

One can compute the Fourier transforms of sequences of type (4), or indeed of any sequence of the form

$$v_n = \alpha n + \beta\{\gamma n\} \tag{5}$$

in a straightforward manner [13]; the sequences are always Poisson combs.

In fact, any sequence whose elements are of the form $\alpha n + g(n)$, where $g(n)$ is periodic or almost periodic, is also a Poisson comb. For appropriate choices of parameters, these sequences will be non-periodic; it is not known whether they can be obtained by projection.

It is not known which of the sets obtained by projection onto subspaces of dimension greater than 1, considered in Section 2, have average lattices. However, some of them do. Duneau and Oguey have recently shown [14] that the set of vertices of a tiling obtained by projection from $\mathbf{R}^8$ to $\mathbf{R}^2$ has an average lattice; the construction applies to certain other tilings as well.

Sequences obtained by substitution. If we interpret the letters a and b to be line segments of lengths τ and 1, respectively, then the sequence u_n discussed above is the limit

$$\lim_{n \to \infty} T^n(w_0),$$

where w_0 is a word of the two-letter alphabet $\{a, b\}$ and T is the map, or substitution rule,

$$T(a) = ab, \quad T(b) = a.$$

When $w_0 = b$, then the length of the word $T^n(w_0)$ is F_{n+1}, where F_n is the nth Fibonacci number $(F_0 = 0, F_1 = 1, F_{n+1} = F_n + F_{n-1})$; u_n is sometimes called a Fibonacci sequence, although the classical Fibonacci sequence is $T_n(a)$. It is not known which of the more general sequences v_n discussed above can be produced by substitution rules, but obviously, more general substitution rules can be used to produce sequences on the line.

What can be said about the Fourier transforms of substitution sequences? To prove that a Dirac comb is a Poisson comb requires knowing the whole Fourier transform; the only way this has been accomplished so far is to show that the density ρ is the sum of delta functions on a lattice modulated by a periodic or almost periodic function or that ρ is obtained by slicing through a periodic density in a higher-dimensional space. On the other hand, one can show that a density has property (b) as follows. The density $\rho(x) = \sum_{n=0,1,\ldots} \delta(x - v_n)$ has Fourier transform $\tilde{\rho}(q) = \sum_n \exp(2\pi i q \cdot v_n)$. For any frequency q, the sequence of partial sums $\left\{ \sum_{n=0}^N \exp(2\pi i q \cdot v_n) \right\}$ is bounded by $N + 1$. If we can show that for some q the sequence grows like cN for some positive c, then asymptotically the sum is a Dirac delta. It is possible to use this technique for some substitution sequences.

Any composition rule T acting on an alphabet of n letters can be represented by an $n \times n$ matrix M with non-negative integer entries. If there is a $k \in \mathbf{Z}$ such that all the entries of M^k are positive, then the Perron-Frobenius theorem tells us that M has a simple eigenvalue θ that is greater in absolute value than all the others. Bombieri and Taylor [13] showed that if $|\theta| > 1$, while all its conjugates have modulus less than one (i.e., if θ is a Pisot-Vijayaraghavan, or P-V, number) then the Fourier transform of the sequence can be computed as a limit of the Fourier transforms of the words $T^n(w_0)$. The transform contains a Dirac comb, because there are frequencies (forming a dense set) for which the sequence of transforms grows like N. In fact, all of these sequences lie in sets that can be obtained by projection.

Every substitution T on a finite alphabet gives rise to a topological dynamical system. By assigning appropriate lengths to the letters of the alphabet, a fixed point of a sequence $T^n(w_0)$ can be interpreted as the list of the successive differences of an increasing sequence of real numbers, and we can study the order properties of such sequences. The dynamical systems associated with substitutions of constant length, and their spectra, have been studied by Queffélec [15]. (Note, however, that the Fibonacci sequence is *not* of constant length.)

Other one-dimensional Poisson combs.   Aubry, Godrèche, and Luck [18] studied a family of sequences that, for some choices of parameters, appear to be Dirac combs of type (c).

Let Δ be a subinterval of $(0,1)$ and ω be any positive real number. Consider the sequence w_n of 0's and 1's obtained by setting $w_n = 0$ if $\{n\omega\} \in \Delta$ and $w_n = 1$ otherwise. There are two ways to build a sequence of points on the line from the sequence w_n. One can start with a one-dimensional lattice whose nodes are located at the points $n\omega$, $z \in \mathbf{Z}$ and then omit those nodes for which the corresponding w_n is equal to 0. In this way we obtain a lattice with vacancies, which can be shown to be a Poisson comb. Kesten's theorem [17] asserts that this sequence has an average lattice, in the sense defined above, if and only if $\Delta \equiv r\omega \pmod{1}$ for some $r \in \mathbf{Z}$. Thus there exist Poisson combs with no average lattice! The second way to build a sequence is to choose two unequal lengths l_1 and l_2 and let $u_0 = 0$, $u_{n+1} - u_n = l_1 + (l_2 - l_1)w_n$. In this case, it may happen that the Fourier transform has a singular continuous component, i.e., the sequence u_n is a Dirac comb of type (c). As far as we know, property (c) has never been completely established for any example. However, there are cases [18] where the possibility of Dirac peaks can be eliminated using the procedure appropriate to case (b), and the possibility of the transform being absolutely continuous can be essentially eliminated by numerical calculation. The spectrum should therefore contain a singular continuous part; numerical calculations then show it "looks like" a Dirac comb.

5. The Tiling Connection

Crystallographers have used tilings as models for crystal structures for centuries. For example, the lattice can be regarded as a framework for the partition of space into congruent parallelepipedal tiles or "unit cells." These fictitious boxes in turn contain congruent, real, atomic configurations.

The diffraction patterns of the first quasi-crystals looked remarkably like the one obtained earlier by Mackay. Thus it was natural to ask whether the three-dimensional analogue of the Penrose tilings might be a model for quasi-crystals, with the two kinds of tiles playing roles analogous to the unit cells. Further experimental work has shown that this is not the case (see, e.g., [19]). In any case, the tiles in a non-periodic tiling are not analogous to the unit cells of periodic patterns, although it is frequently asserted that they are. There are infinitely many ways to choose the shape and position of the unit cell for a periodic crystal, all equally valid from the abstract point of view. In contrast, in the few cases in which non-lattice point sets can be associated with tilings by copies of one or a few shapes, the choice of cells is usually unique, and it is by no means clear what the relation between the transforms is when masses are placed at vertices or in the tile interiors.

FIGURE 25-7 *A self-similar tiling for which no matching rules are known to exist.* [20]

In fact, it is not clear what aspects of a real structure the tiles in a non-periodic tiling might represent. Like the Big Dipper and other stellar constellations that one learns to identify as a child, the tiles sometimes appear to be highly artificial from a physical point of view, even when they are convex. For example, the minimum distance between vertices in a tiling by Penrose rhombs is the short diagonal of a thin rhomb; in a reasonable structure model one would expect nearest neighbors to be linked in some way.

Still, the possible connection between tilings and quasi-crystal structure continues to be studied, partly because the tilings help us to visualize some kinds of non-periodic order (this is why four of our eight illustrations are tilings). It is easy to produce non-periodic tilings by the projection method. In the special case where T is the projection of the unit n-cube γ_n, the projected points are the vertices of non-overlapping projections of the k-dimensional faces of γ_n. For suitably chosen subspaces, the number of distinct tile shapes (prototiles) will be small ($O(n)$). Thus we can construct many interesting examples.

What do non-periodic tilings have to teach us? The Penrose tilings have three important properties:

1. they have matching rules that force non-periodicity,
2. they can be obtained by a substitution process and they are self-similar,
3. they have strong local order (in fact they are quasi-periodic).

Surprisingly, it appears now that these properties are independent to some extent. There are substitution-produced tilings with matching rules that are not self-similar (several examples are shown in [20]). Figure 25-7 shows a tiling that is self-similar but for which no matching rules seem to exist; recently it has been shown that this tiling is not quasi-periodic (see below). Tilings built with the tiles shown in Figure 25-8, using matching rules, are quasi-periodic but no substitution rule has been found for them.

Matching rules.   Why are matching rules of interest in the study of quasi-crystals? Evidently they are not needed in order for a tiling to be a Poisson comb. Their importance lies instead in our feeling about what features a good model should have. The projection method says nothing about how the quasi-crystal grows—why the atoms order themselves in such a pattern. Some sort of local forcing rules would seem to be an important part of a good model for quasi-crystals, since they are an analogue of the local bonding rules that presumably determine the structure.

The matching rules discovered by Penrose and by Ammann ([20]) were found by trial and error. Is there a more systematic way to do this? De Bruijn showed that his indexing system for the Penrose vertices leads to an unambiguous reconstruction of the Penrose rules, but his arguments do not apply if the set M is translated in $\mathbf{R}^5$. Neither has it proved possible to apply it to any of the other plane tilings projected from $\mathbf{R}^n$. This does not mean that no matching rules exist in these cases. For example, Ammann has found matching rules for certain tilings of the plane by squares and rhombs, projected from $\mathbf{R}^8$ (again, see [20]). But as de Bruijn points out [21], although Ammann's rules are expressed locally, the property of an unmarked tiling to be Ammann-mark-able is not a local property.

More recently, some progress has been made. Using homologous arguments, Katz has developed a method for decorating the tiles of certain projected tilings [22]. He applied it to the "three-dimensional Penrose tiles," thus proving that these tiles can be equipped with matching rules that force non-periodicity (Figure 25-8). However, the construction is not a simple one. When decorations are taken into account the two rhombohedra fall into twenty-two classes.

Recently, Danzer has announced the discovery of a

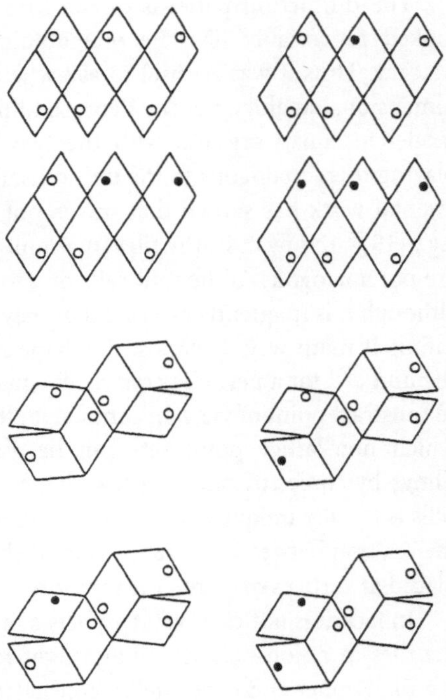

FIGURE 25-8 *The three-dimensional analogue of the Penrose rhombs are two rhombohedra. When decorated with matching rules according to Katz's scheme, the rhombohedra fall into twenty-two classes. Nets for eight of the rhombohedra are shown here; the others can be generated from this set (see Ref. [22]).*

set of four marked tetrahedra [23] that tile $\mathbf{R}^3$ only non-periodically. Although the method by which he found them appears to be less systematic than Katz's, it is of interest because the number of prototiles is small.

Self-similarity. The self-similarity of the Penrose tilings is one of their most remarkable features. But until very recently self-similar tilings have been almost as hard to find as matching rules. In the first place, to be self-similar, a tiling must be a geometric realization of a "fixed point" of a substitution map. Any primitive matrix defines a substitution map, but we do not know of any theory that tells us which substitutions can be realized as tilings. Even when such a realization exists, the tiling need not be self-similar in the sense that the larger configurations into which the tiles are grouped by the action of the substitution map are geometrically similar to the original tiles. Conversely, given a tiling (such as that in Figure 25-5), it may be very difficult if not impossible to determine whether it is invariant under some substitution map T. Recently, Thurston has developed a method for associating self-similar tilings with fractal tile boundaries to a class of algebraic integers [24]. Substitution rules are implicit in the method, but it is not yet clear to us how to extract them.

Nevertheless, tilings invariant under primitive substitution maps are of interest in our context because they necessarily have the local isomorphism property. Moreover, we can sometimes use the substitution map to prove that a tiling is non-periodic, a property that may not be obvious. Notice, for example, that the use of matching rules does not guarantee that a tiling is non-periodic; some other argument must be invoked.

Two different arguments can be used to establish the non-periodicity of a tiling with the substitution property. First, if the grouping of tiles into larger ones is *unique*, the tiling has a hierarchical structure that must be preserved by any translation. But this is impossible, since repeated iteration of this grouping implies that at some hierarchical level the inradius of the tiles will be larger than any specified translation length.

The other argument might be called a "ratio test" for non-periodicity. It involves the eigenvectors of the substitution map. Let T be any primitive, integer $n \times n$ matrix, and let $\{a_1, \ldots, a_n\}$ be any finite alphabet. A word ω of this alphabet contains x_i copies of the letter a_i. We can think of $\vec{x} = (x_1, \ldots, x_n)$ as a vector of the integer lattice in "configuration space." Then $\vec{x}T$ is another vector in this space; its components are the numbers of copies of each of the letters after one application of T. The components of a *left* eigenvector corresponding to its leading eigenvalue θ are the *relative* numbers of the different letters in the infinite word produced by iterating T. We can now state the ratio test. If T has an eigenvalue θ which is a P-V number, then for any initial configuration vector $\vec{x}_0$, the sequence $\theta^{-n}\vec{x}_0 T^n$ will converge to a left eigenvector of θ. If the components of this eigenvector have irrational ratios, then the tiling will be non-periodic, since in a periodic tiling the relative numbers of kinds of prototiles are given by the numbers in a single repeat unit.

If T acts on a tiling, then the prototiles of the tiling play the role of the letters of an alphabet. We assume that they are arranged in such a way that each application of T effects a grouping of the tiles into larger tiles. These tiles need not be similar to the original ones, but if they are, then the relative volumes of the n prototiles after each application of T are the components of a *right* eigenvector of θ. This gives us a way to decide whether a tiling produced by substitution is self-similar; the Penrose tiles pass the test.

Which substitution-invariant tilings produce diffraction patterns with bright spots? There is no definitive answer yet. We have seen that if the tiling can be obtained by projection, then its set of vertices is a Poisson comb. Recently Godrèche and Luck [25] have extended the Bombieri-Taylor method for computing the Fourier transforms of substitution sequences to tilings of the plane by assigning masses to the tiles themselves and expanding the definition of T to take into account the geometry of configurations as well as the numbers of tiles in them. They then showed that Fourier transforms of this density distribution contain Dirac combs even when the matching rules are relaxed. During the Les Houches conference, Godrèche succeeded in showing that the tiling of Figure 25-7 fits into case (b) (but the possibility that the spectrum also contains a continuous component has not been ruled out). It is especially interesting that in this case there is no finite basis for the frequencies of the delta functions of the Fourier transform, so that the tiling is not quasi-periodic [25].

Local order.   The local ordering properties of the Penrose tiles are discussed in [20], so we will not go into detail here. They include local isomorphism, and the fact that the number of different configurations within any finite radius is bounded and grows slowly as the radius increases. These properties hold for all projected and substitution tilings. But it remains an open question to what extent these properties, independently of projection and substitution, can account for the tilings' Fourier transforms.

6. A Word about Definitions

We mentioned at the beginning that we have posed Question 3 more generally than is usually the case.

We did not mention, but many readers will have observed, that we have offered no definition of "quasicrystal." In fact, most other writers define quasi-crystals to be projected (or sliced) structures. There may be some justification for this. As we have shown, the projection/slicing method does produce an extremely large class of Poisson combs. Moreover, the models based on this approach are in very good agreement with experiment. But then, experimentally, it may be impossible to distinguish Poisson combs from the other two cases discussed in Section 2. It seems to us no more appropriate to define quasi-crystals at this stage of our knowledge than to cling to the definition of a crystal as a periodic structure.

Question 3 is non-trivial mathematically, and it is also non-trivial philosophically. The high-dimensional formalism is only a stop-gap to be used until we understand how quasi-crystals grow. String theory notwithstanding, it is reasonable to assume that real quasi-crystals, like real periodic ones, grow in $\mathbf{R}^3$, not in $\mathbf{R}^n$. We need a theory that explains how the patterns that we are interested in can be generated at the local level; it is not clear to what extent the deterministic models we have described are physically meaningful. Modeling growth may require a combination of matching rules, modulations, understanding "the sociological behavior of large groups of atoms" [21], and possibly other ideas. It is too early to know what class of patterns will achieve this. In our view, the definition of quasi-crystal should be left open until the fundamental questions have been answered.

References

1. D. Shechtman, I. Blech, D. Gratias, and J. Cahn, Metallic phase with long-range orientational order and no translational symmetry, *Phys. Rev. Lett.* 53 (1984), 1951–1954.
2. J. Cahn and J. Taylor, An introduction to quasi-crystals, *Contemporary Mathematics* 54 (1987), 265–286.
3. L. Pauling, So-called icosahedral and decagonal quasi-crystals are twins of an 820-atom cubic crystal, *Phys. Rev. Lett.* 58 (1987), 365–368.
4. J. Miekisz and C. Radin, Are solids really crystalline?, *Phys. Rev.* B 39 (1989), 1950–1952.
5. B. Delone, N. Dolbilin, M. Shtogrin, and R. Galiulin, A local criterion for the regularity of a system of points, *Reports of the Academy of Sciences of the USSR* (in Russian) 227 (1976). (English translation: *Soviet Math. Dokl.* 17 (1976), 319–322.)
6. A. Mackay, Crystallography and the Penrose pattern, *Physica* 114A (1982), 609–613.
7. N. G. de Bruijn, Algebraic theory of Penrose's non-periodic tilings of the plane, *Kon. Nederl. Akad. Wetensch. Proc.* Ser. A 84 (*Indagationes Mathematicae* 43) (1981), 38–66.
8. A. Katz and M. Duneau, Quasi-periodic patterns and icosahedral symmetry, *Journal de Physique* 47 (1986), 181–196.
9. N. G. de Bruijn, Quasi-crystals and their Fourier transform, *Kon. Nederl. Akad. Wetensch. Proc.* Ser. A 89 (*Indagationes Mathematicae* 48) (1986), 123–152.
10. R. Porter, The applications of the properties of Fourier transforms to quasi-crystals, M. Sc. Thesis, Rutgers University, 1988.
11. N. G. de Bruijn, Modulated quasi-crystals, *Kon. Nederl. Akad. Wetensch. Proc.* Ser. A 90 (*Indagationes Mathematicae* 49) (1987), 121–132.
12. L. Levitov and J. Rhyner, Crystallography of quasi-crystals; applications to icosahedral symmetry, *J. Phys. France* 49 (1988), 1835–1849.
13. E. Bombieri and J. Taylor, Quasi-crystals, tilings, and algebraic number theory: some preliminary connections, *Contemporary Mathematics* 64 (1987), 241–264.
14. M. Duneau and C. Oguey, Displacive transformations and quasi-crystalline symmetries, *J. Physique* 51 (1990), 5–19.
15. M. Queffélec, *Substitution Dynamical Systems—Spectral Analysis*, Lecture Notes in Mathematics 1294, New York: Springer-Verlag, 1987.
16. S. Aubry, C. Godrèche, and F. Vallet, Incommensurate structure with no average lattice: an example of a one-dimensional quasi-crystal, *J. Physique* 48 (1987), 327–334.

17. H. Kesten, On a conjecture of Erdős and Szüsz related to uniform distribution mod 1, *Acta Arithmetica* 12 (1966), 193–212.

18. S. Aubry, C. Godrèche, and J. M. Luck, Scaling properties of a structure intermediate between quasi-periodic and random, *J. Stat. Phys.* 51 (1988), 1033–1075.

19. M. La Brecque, Opening the door to forbidden symmetries, *Mosaic* (National Science Foundation) 18 (Winter 1987/8), 2–23.

20. B. Grünbaum and G. Shephard, *Tilings and Patterns*, San Francisco: W. Freeman, 1987.

21. N. G. de Bruijn, private communication; see also his preprint "Remarks on Beenker patterns."

22. A. Katz, Theory of matching rules for the 3-dimensional Penrose tilings, *Commun. Math. Phys.* 118 (1988), 263–288.

23. G. Danzer, Three-dimensional analogs of the planar Penrose tilings and quasi-crystals, *Discrete Mathematics*.

24. W. Thurston, Groups, tilings and finite state automata, *Summer 1989 AMS Colloquium Lectures* (preprint).

25. C. Godrèche and J. M. Luck, Quasi-periodicity and randomness in tilings of the plane, *J. Stat. Phys.* 55 (1989), 1–28.

26. C. Godrèche, The Sphinx: a limit-periodic tiling of the plane, *J. Phys. A* 22 (1989), L1163–L1166.

Celtic Knotwork:
Mathematical Art

Peter R. Cromwell

The interlaced ornament produced by Celtic scribes and stone masons has fascinated people for many centuries. The designs, ranging from small individual knots to elaborate panels composed of many motifs, provide the geometrically minded mathematician with a rich source of examples. Many aspects of the interlaced patterns can be studied mathematically, and some of these are explored in this article. We begin with the geometry of the knots.

Constructing Interlaced Patterns

Underlying many of the Celtic knot patterns is a lattice. It is this lattice which imparts the distinctive proportions to Celtic knotwork. It is usually composed of squares, but is occasionally built from 3×4 rectangles. For reference purposes, it is convenient to regard this lattice as the union of two dual lattices whose mesh size is double that of the original lattice (Figure 26-1b). These will be called the *auxiliary grids*. When laying out a design only the vertices of these two grids are drawn in (Figure 26-1c).

The knotwork created on these grids is all related to plaitwork—the basic weaving pattern used in basketwork and other crafts. This is shown in Figure 26-1d. Note how the crossovers in the pattern lie at the intersection points of the two auxiliary grids and that the interlacing has an alternating quality: Each string goes alternately over, then under other strings. This is a characteristic feature in Celtic patterns, too, although one or two anomalies are known (Figure 26-2).

Interlaced patterns formed from portions of plaitwork can be found in Egyptian, Greek, and Roman ornament and in the art of many other cultures. Yet for the Celt, these alternating plaits were merely the raw material on which the artist sets to work. To obtain the more elaborate interlaced designs, the regularity of this primal pattern must be interrupted. This is achieved by breaking two strings at a crossover and rejoining the ends as illustrated in Figure 26-3. (Operations such as this are currently used in combinatorial knot theory.) Note that eliminating crossings in this fashion preserves the alternating property of the interlacing. When sufficiently many crossings are removed, the underlying plait ceases to be the dominant feature in the design and a pattern composed of knot motifs appears. In this way, a bewildering assortment of interlaced designs can be created.

So far the only aids to laying out design are the vertices of the auxiliary grids. Additional construction lines are drawn between these vertices to indicate which crossings are to be removed from

Volume 15, No. 1 (Winter 1993), 36–47

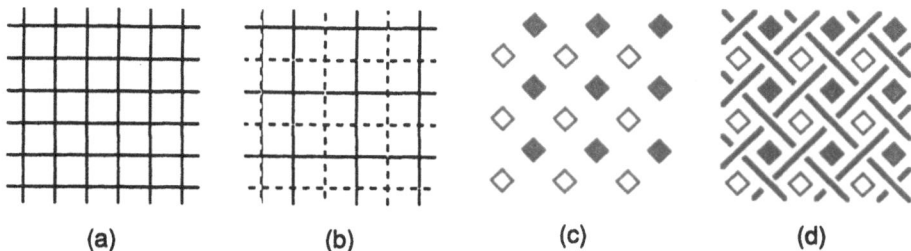

FIGURE 26-1 *How to construct alternating plaitwork.*

FIGURE 26-2 *A non-alternating Celtic pattern.*

FIGURE 26-3 *The rule for eliminating crossovers.*

the plait and how the ends are to be rejoined. These lines are called *break-markers*. Each break-marker is an edge of one of the auxiliary grids. At each crossing point of the plait, there are two such edges, and the edge chosen indicates which one of the two possible reconnections is to be used. A glance at the example in Figure 26-4 will make the convention clear.

To complete the design, the path of the strings is outlined and the background is filled in. This obscures all the construction lines. The pattern is then interlaced to produce the characteristic alternating weave. The construction of some elemental knot motifs is illustrated in Figure 26-5.

Eliminating the crossings according to the simple rule indicated in Figure 26-3 does not always product aesthetically pleasing results. Sometimes a better result is achieved if the path of the strings is allowed to deviate from the path of the plait. At places where two break-markers meet to form a corner, the path of the strings can also be made angular. This has been done in the accompanying figures. In other situations, fairly sharp bends can be replaced with more gentle curves to produce a more graceful, freely flowing design. Figures 26-7 and 26-8d exhibits arcs of several different curvatures. These modifications help to disguise further the underlying plait structure on which the design is based. Another variation produces a very delicate form of interlacing. What are normally taken to

grid and
break-markers

path of string

path outlined and
background filled in

interlacing
added

FIGURE 26-4 *Break-markers aid the construction by indicating how the crossings are to be broken.*

be the two edges of an interlaced ribbon are themselves used as the strings to be interlaced. An example of such "double interlacing" is shown in Figure 26-8j.

Once the technique for constructing Celtic ornament has been understood, the art itself loses some of its mystery. It becomes possible to copy the ancient designs fairly accurately and easily and to create designs of your own. (Do not be surprised, though, if you discover your creations elsewhere.) Any rectilinear area can be filled with knotwork by regarding its boundary as being composed of break-markers on a suitable lattice (see Figure 26-6a). Patterns can also be mapped into curvilinear regions by dividing them into quadrilaterals (Fig. 26-6b).

Simple though this construction procedure may appear, the idea that the Celtic interlaced patterns were related to plaitwork took many years to mature. The methods used by the Celts themselves are no longer known, and the above technique was developed by J. Romilly Allen while he was surveying the patterns in the British Isles at the turn of the century. He records [1] p. xvii:

FIGURE 26-5 *Examples of interlaced motifs together with their construction.*

> The theory of the evolution of Celtic knotwork out of plaitwork . . . is entirely original, and, simple as it appears when explained, took me quite twenty years to think out whilst classifying the patterns.

When we remember the large number of modifications the underlying plait can undergo and the ease with which it is disguised, perhaps this is not so surprising.

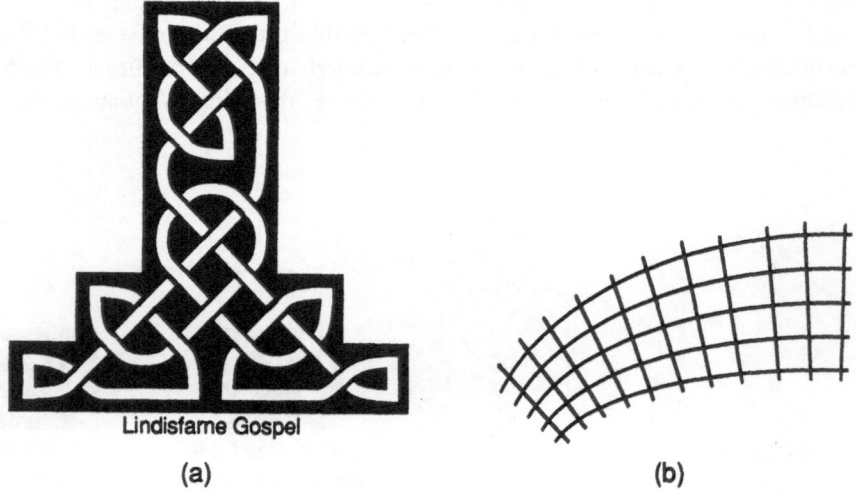

Lindisfarne Gospel

(a)   (b)

FIGURE 26-6 *Irregularly shaped panels can be decorated using the same techniques.*

Interpreting Celtic Designs As Friezes

Much Celtic knotwork is in strip form, either as part of a border or simply as a narrow rectangular panel. In many of these strip patterns, the crossings are eliminated systematically and the pattern of break-markers repeats at regular intervals. This produces a pattern which can be described as "locally periodic": A single motif is repeated side by side. In other forms of ornament, such periodicity gives rise to translational symmetry when the pattern is regarded as a randomly chosen segment of an infinite strip. However, in knotwork, the transition from a small segment to an idealised frieze is not so immediate. Whereas in other forms of decoration, a pattern is simply truncated when it reaches the edge of the available space, knotwork is rarely terminated so abruptly. The pattern is adapted so that the otherwise free ends join up to form continuous strands.

FIGURE 26-7 *This pattern is not based on the standard grid.*

Sutton Hoo Buckle   P1a1

(a)

Rossie Priory Stone   P1a1

(b)

Book of Kells   P2'11

(c)

Book of Durrow   P121

(d)

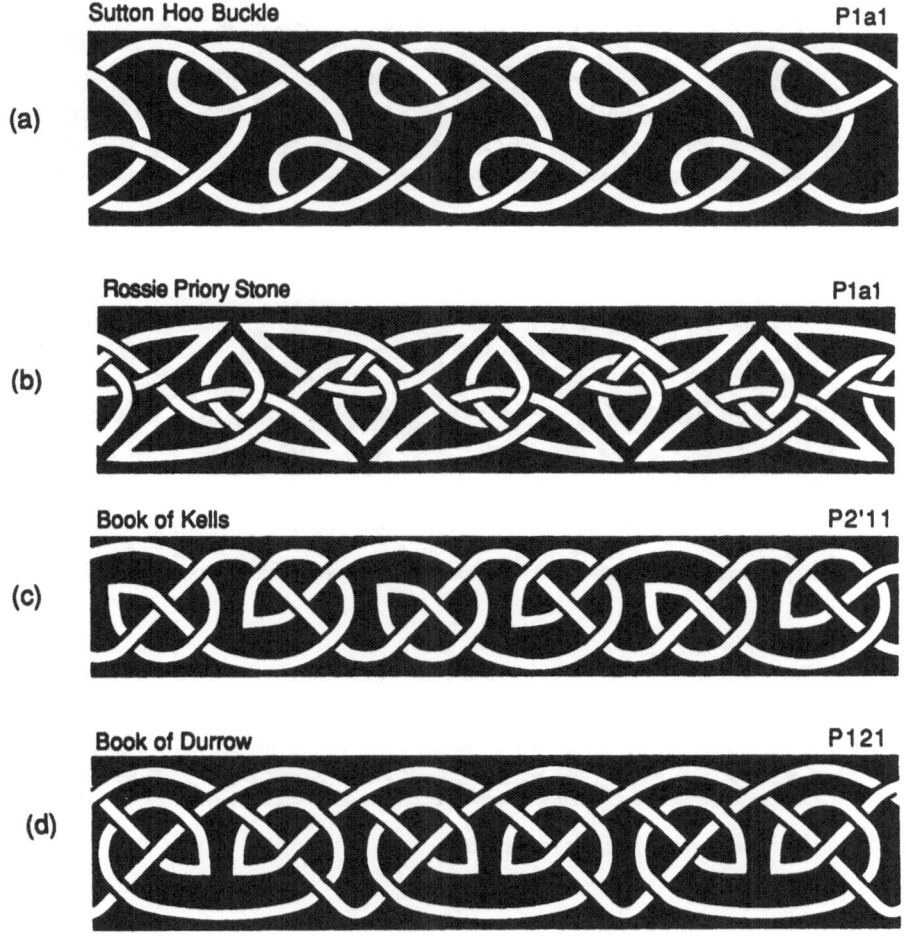

FIGURE 26-8 *Examples of Celtic frieze patterns.*

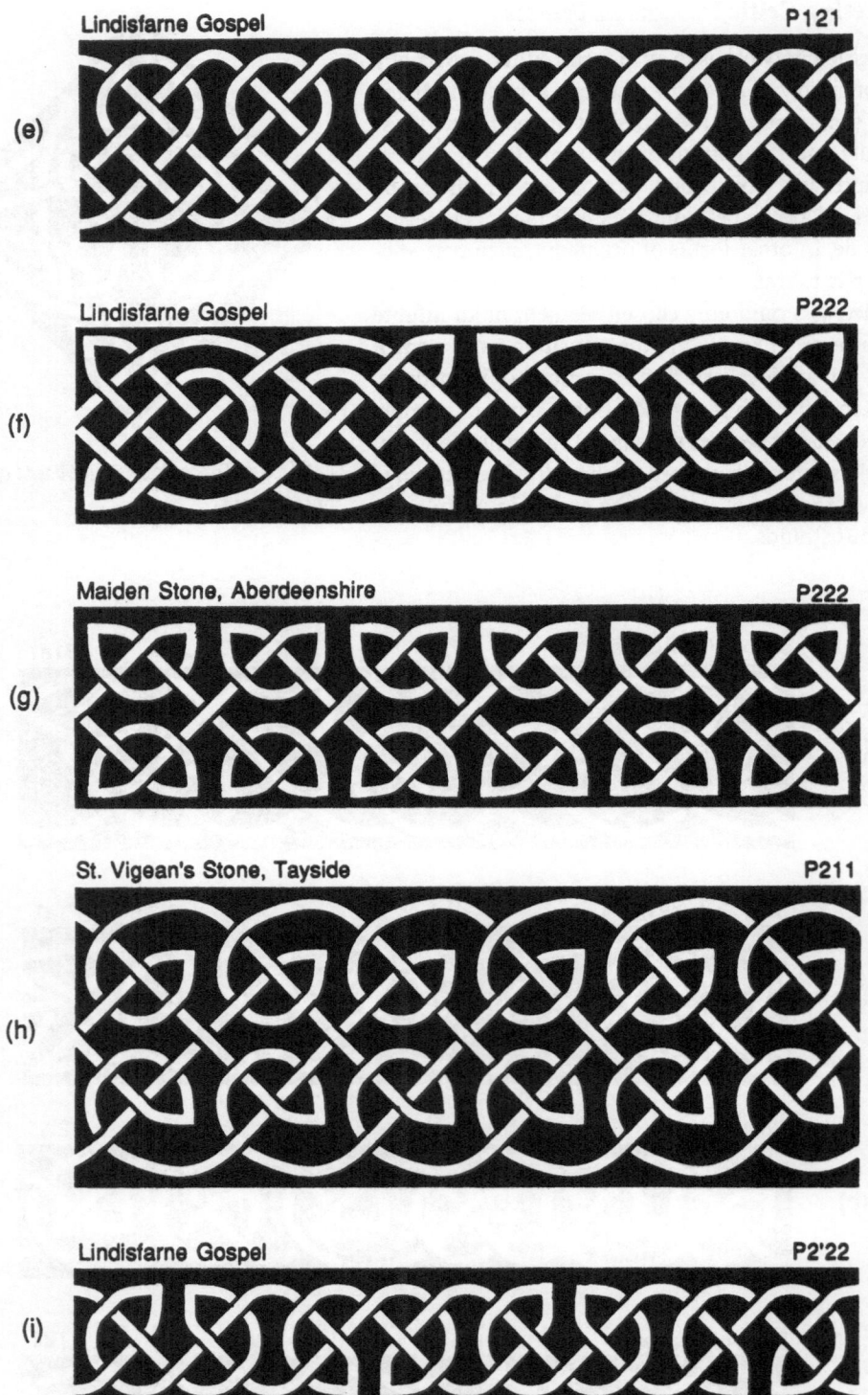

FIGURE 26-8 *Examples of Celtic frieze patterns (continued).*

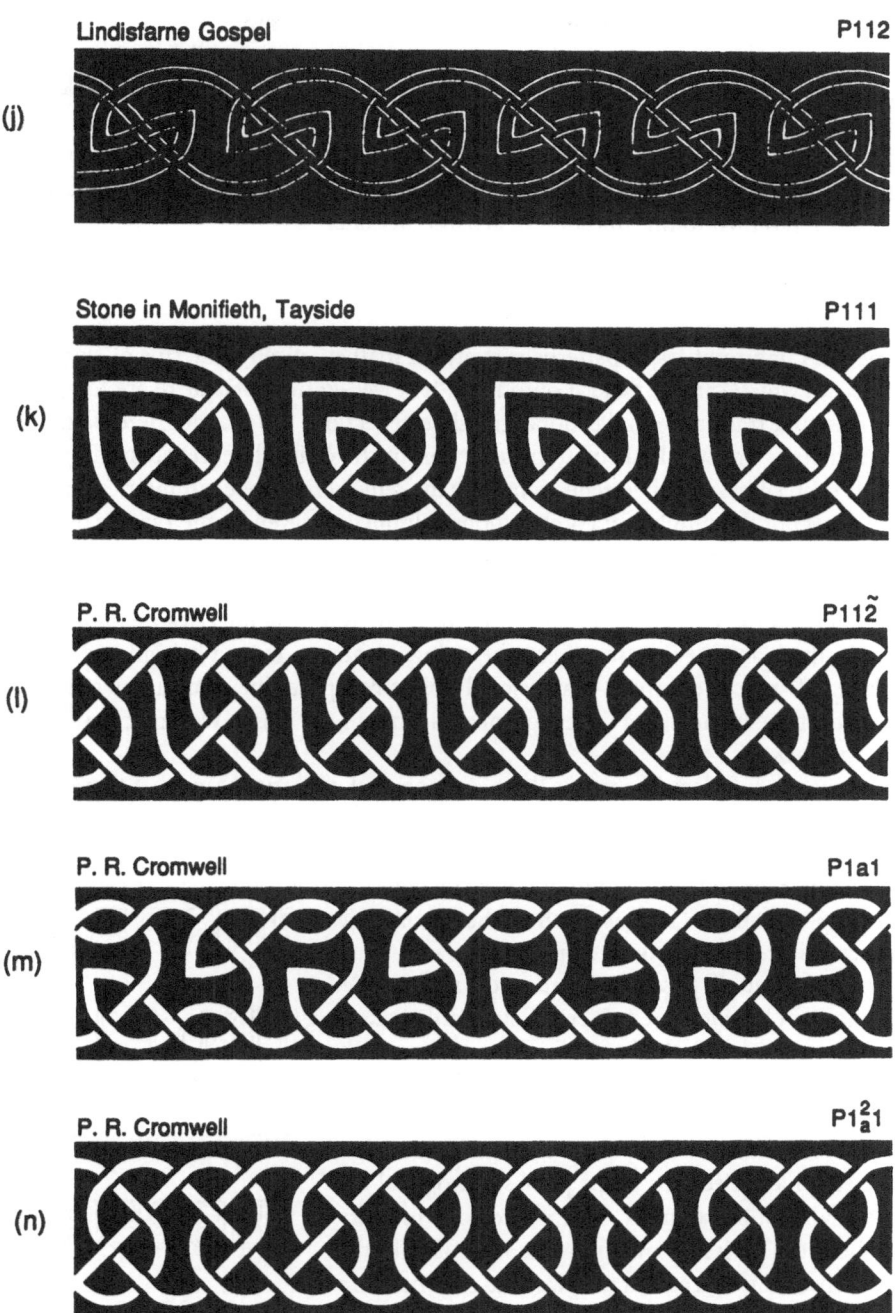

FIGURE 26-8 *Examples of Celtic frieze patterns (continued).*

In this article, the knotwork patterns are regarded as friezes in the conventional mathematical sense: as parts of patterns which extend indefinitely in both directions. In the underlying lattice, the pattern is bounded by two parallel lines of break-markers and the other breaks are arranged so that the pattern as a whole has translational symmetry. Some examples are shown in Figure 26-8. The first two patterns are not constructed on a standard lattice. The pattern in Figure 26-8a is believed to be Scandinavian. Triangular motifs such as that used in Figure 26-8b are normally arranged in set of four to form square patterns. An example based on the same motif is shown in Figure 26-7. I have excluded patterns that can be split up into other patterns. For example, if the knot design in Figure 26-2 is converted into a frieze, the resulting pattern comprises two parallel patterns which are not interlinked. An interlaced pattern is said to be *connected* if the unlaced path (or equivalently, the projection of the link onto the picture-plane) is connected. The frieze constructed from Figure 26-2 is not connected. For the moment, I shall assume that the friezes are connected and that the notion of connectedness captures to some extent our intuitive understanding of when a pattern can be split up. We shall, however, return to this question of separability later.

The Symmetry of Interlaced Friezes

The symmetrical nature of frieze patterns is analyzed mathematically in terms of symmetry operations: isometries which carry a pattern onto itself. It is well-known that there are four isometries of the plane: rotation, translation, reflection, and glide-reflection. For planar frieze patterns, the only possible symmetries are two-fold rotation, reflection in the center-line, reflection in a line perpendicular to the center-line, and glide-reflection along the center-line. These symmetries can be combined in different ways but the number of combinations is limited to seven by the rigidity of the geometry. Examples of patterns exhibiting these seven different symmetry types are shown in Figure 26-9.

When the knotwork friezes are compared with these planar patterns, it becomes apparent that we do not interpret the two kinds of pattern in the same way. The interlaced patterns are not confined to lie in the plane. At the crossovers, the strings appear to extend behind and in front of the picture-plane. We perceive a three-dimensional object composed of continuous strings lying in a neighborhood of the plane, not a collection of arcs lying in the plane. We expect to see the obverse pattern on the back.

Interpreting the interlaced friezes as three-dimensional patterns means that there are additional isometries which can act as symmetries. Two of these are compound motions like the glide-reflection: a *screw* is a rotation followed by a translation along its axis; a *rotatory-reflection* is a rotation followed by reflection in a plane perpendicular to its axis. Both of these isometries can act as symmetries of an interlaced pattern.

The complete set of possible symmetries of two-sided friezes is described by reference to a set of standard axes; the **a**-axis runs along the band, the **b**-axis lies in the plane of the band perpendicular to **a,** and the **c**-axis is normal to the band. The possible symmetries are a two-fold rotation about any of the three axes, reflections in the planes orthogonal to each of the three axes, glide-reflections in the planes orthogonal to **b** and **c,** a screw motion along **a,** and a two-fold rotatory-reflection about **c.** This last symmetry is the same as reflection in a point or inversion (because it is a two-fold symmetry).

There are many different ways in which these symmetries can be combined. As these two-sided friezes are less familiar than the seven one-sided ones, patterns illustrating the thirty-one symmetry types [5] are shown in Figure 26-10. The motif in each of the patterns is a scalene triangle which, in most cases, is colored black on one side and white on the other. When reflection in the plane of the strip is a symmetry, then both sides of the triangle must be the same color: in this case the triangles are colored gray. Next to each pattern is a label of the form $P\square\square\square$ which encodes the symmetries present in

FIGURE 26-9 *The seven one-sided frieze patterns.*

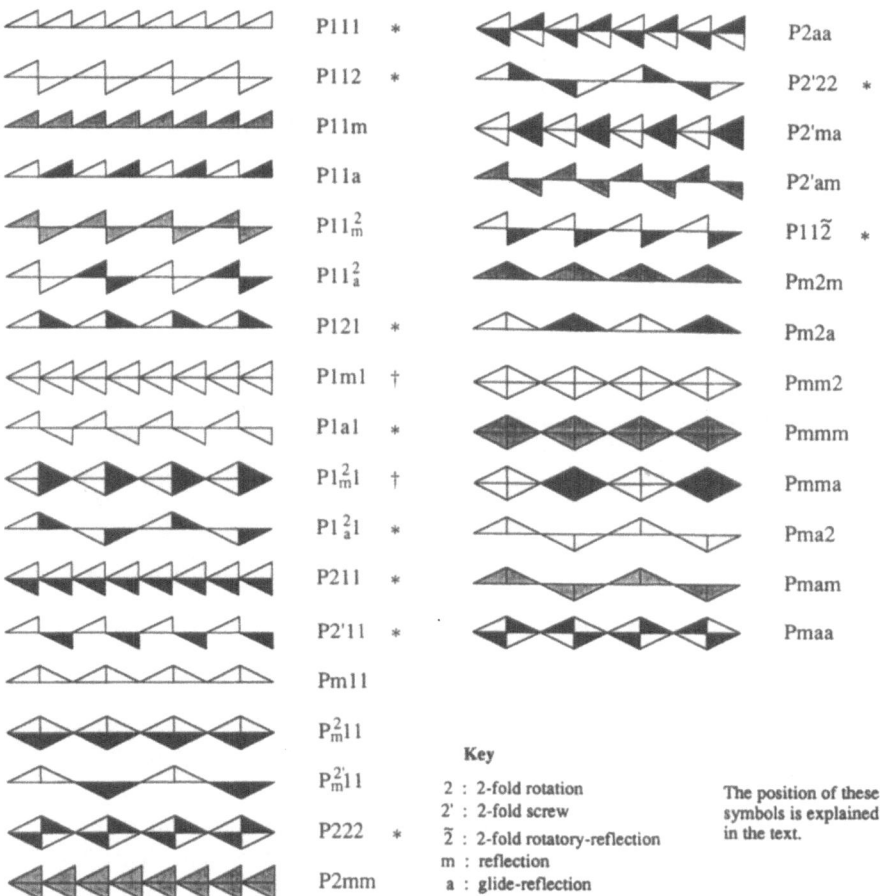

FIGURE 26-10 *The thirty-one two-sided frieze patterns.*

the pattern. After the symbol P (which denotes that the pattern is periodic in one direction) the symbols 2, 2′, $\tilde{2}$, m, and a are used to indicate that an axis of two-fold rotation, two-fold screw, two-fold rotatory-reflection, or the normal vector of a mirror plane or glide plane coincides with one of the reference axes. The first, second, and third symbols after the P correspond to the **a**-axis, **b**-axis, and **c**-axis, respectively. If an axis and a plane of symmetry coincide with the same reference axis, both symbols are given; if no symmetry element corresponds, then the symbol 1 is used as a place marker.

Trying to identify which symmetries are present in a particular knot pattern is not trivial. The reader is encouraged to experiment on the patterns in Figure 26-8. A useful observation is that the crossings in the Celtic friezes all have the same alignment and that they come in two kinds. Furthermore, each kind is stabilized by all the two-fold rotations (see Figure 26-11). Thus, direct symmetries carry a crossing to one of the same kind; indirect symmetries interchange the two kinds.

The Celtic Frieze Groups

After analysing a few examples, one is led to ask how many of the two-sided groups can arise from Celtic knot patterns. Can they all occur, and if not, then which ones? The slightly different question of which ones actually occur in practice can also be considered.

FIGURE 26-11 *Each of these two-fold rotations preserves the crossing.*

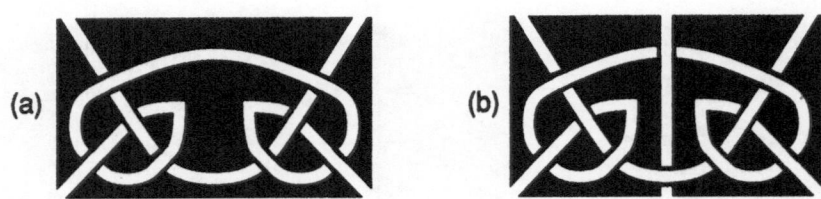

FIGURE 26-12 *A pattern with bilateral symmetry can be made alternating by adding an extra strand.*

The seven "gray groups"—those which have the picture-plane as a symmetry element—cannot occur for knot patterns because the crossing points do not obey this symmetry. Therefore, we can eliminate all the groups whose labels have an *m* as part of the last symbol. The observation that virtually all interlaced ornament are alternating allows further groups to be eliminated.

Suppose that a knot motif has mirror symmetry. An example is shown in Figure 26-12a. To convert this knot into an alternating one, it is necessary to add an extra string which lies in the mirror plane as shown in Figure 26-12b. Those groups which contain reflections in planes orthogonal to the **a**-axis cannot arise from alternating knotwork patterns because the strings that lie in the mirror planes would run directly from the top edge of the band to the bottom—they could never join up with anything else. Thus, those groups whose labels have an *m* in the first position cannot be found in Celtic patterns. The groups whose labels have an *m* as part of the second symbol can arise from alternating patterns: such patterns must have a string that runs straight down the centre-line of the band. There are two of these groups: $P1m1$ and $P1\frac{2}{m}1$. They are marked with a dagger (†) in Figure 26-10. Patterns of this form are not consistent with the standard Celtic grid, however, and so these symmetry groups cannot arise in Celtic patterns either.

There is one more class of group labels that can be eliminated: those which contain an *a* as part of the third symbol (indicating that the picture-plane is also a glide-plane). The reason for this is that to create an alternating design, straight strands must run across the band as in the $Pm\square\square$ case, and these strings can never join up with anything.

There remain ten possibilities. These are marked with an asterisk (*) in Figure 26-10. Examples of Celtic patterns exhibiting each of these symmetry groups are shown in Figure 26-8. Where I have been able to find ancient designs I have used them; a few are my own creations.

The Relative Abundance of Symmetry Types

In my search for examples of Celtic designs, I discovered that the abundance of the different symmetry types varied greatly. Some groups ($P112$, $P222$, $P2'22$) were very common; others were rare. In fact, the only patterns I could find which have group $P1a1$ are not constructed on the standard

FIGURE 26-13 *Alternating braids with five and six strings.*

lattice described above. For the two groups $P11\tilde{2}$ and $P1_a^2 1$, I found no examples at all. Are there any features of these patterns that make them difficult to obtain?

Before resolving this puzzle we shall consider another question. Is there any correlation between the (two-sided) symmetry group of an interlaced design and the (one-sided) symmetry group of its underlying pattern of break-markers? The answer to this is yes. To see why, note that the unlaced path of the strings and the distribution of break-markers have the same symmetry type. The correspondences are listed in Table 26-1. Observe how each of the three rare groups is paired with another group which appears to be the preferred option. In fact, this preference is not a matter of a choice having been made by the designer, consciously or otherwise. It is a consequence of the fact that the ancient patterns terminate and are not true friezes.

Study the two sections of plaited friezes in Figure 26-13. Do you notice any differences? Structurally, they are the same; symmetrically, they are not. This is seen most easily along the edges. In Figure 26-13b, the outermost crossovers are opposite each other; in Figure 26-13a they are staggered with those on one edge lying between those on the other. The symmetry types of these two plaits are, therefore, different.

The symmetry type of a plaited frieze depends on its width. Those plaits that have an even number of lattice cells between the two edges have symmetry type $P222$; those plaits that span an odd number of cells have type $P1_a^2 1$. This observation allows us to resolve the mystery of the missing groups. It also provides an alternative method for enumerating the Celtic groups; because eliminating crossovers can only destroy symmetry, the Celtic groups must be subgroups of $P222$ or $P1_a^2 1$.

Table 26-1 can now be refined to show how the symmetry type depends on the width of the frieze as well as the underlying markers. The result is shown in Table 26-2. We now have at most one group per box of the table: The symmetry type of the interlacing is completely determined by the geometry of the underlying break-markers. Furthermore, all the rare groups are associated with friezes of

Table 26-1.

1-sided	R	b P	A	B	N	V∧	H
2-sided	P111	P1a1	P121	P211	P11$\tilde{2}$	P1$_a^2$1	P222
		P2'11			P112		

Table 26-2.

1-sided	R	b P	A	B	N	V∧	H
2-sided (odd)	P111	P1a1	P121		P11$\tilde{2}$	P1$_a^2$1	
2-sided (even)	P111	P2'11	P121	P211	P112	P2'22	P222

odd width. The only frieze I found that has odd width is shown in Figure 26-8d. Its symmetry type is $P121$—a group which is independent of the width of the frieze.

It is not difficult to discover why odd-width friezes are uncommon. It is not, as has been suggested [1, p. 260], that the patterns look lopsided. Rather it has to do with the fact that the original Celtic patterns are finite designs with no loose ends. At the end of a row of motifs, the strings are paired up and joined—a process which requires an even number of strings. The number of strings is related to the width of the frieze: The parity of the width equals the parity of the number of strings. The Celts could only use odd-width friezes in situations where continuity could be ensured, such as in complete borders. In some places, loose ends occurred naturally. These are often in *zoomorphs*: strange creatures whose elongated tails, limbs, necks, or tongues are intertwined in fanciful patterns. However, these patterns are never large enough to have repeated motifs and so will not provide examples of friezes.

Continuity, Transitivity, and Separability

One problem faced by the designer of interlaced patterns is that of determining the number of components in the completed design. In early forms of Celtic ornament, it is clear that continuity of the path was sought. It was important as a symbol of eternity. Even extremely intricate patterns have just one string arranged in an endless loop. In some examples, the regularity of a pattern has been deliberately abandoned and the pattern modified to ensure that a unique path was obtained. In later times, this rule was followed less strictly, but small rings in a pattern were still avoided. This raises the question of whether there is a simple way to determine the number of strings in a design from the underlying pattern of break-markers.

When the interlacing is a rectangular portion of plaitwork (Figure 26-14), then the answer can be expressed in terms of the bounding rectangle. If the lattice underlying the plait contains $n \times m$ cells, then the pattern will have a single component if h.c.f. $(m, n) \leq 2$. If the pattern is square ($n = m$), then the number of components is $\lceil \frac{1}{2}n \rceil$. In fact, if we count closed loops rather than components, then the number $\frac{1}{2}n$ can be regarded as correct. When break-markers are added and the plaitwork is broken up, these rules are no longer valid. What rules replace them?

How about interlaced friezes: How many strings do they have? The pattern in Figure 26-8f is a chain composed of infinitely many closed loops. All the other friezes in Figure 26-8 have a finite number of infinitely long strings. Most of the patterns from Celtic sources have an even number of strings; only the non-standard patterns a and b, and my designs 1, m, and n of Figure 26-8 have a single strand; Figure 26-8d has three.

For patterns with more than one component, we can investigate whether the symmetry group acts transitively on the strings. Can each string be carried onto any other string by some symmetry

4x4

5x5

3x5

4x6

FIGURE 26-14 *Can the number of components in a rectangular motif be determined from the dimensions of the rectangle?*

of the pattern? In the context of layered patterns and fabrics, a pattern which is string-transitive is said to be *isonemal*. Adopting this terminology, we can say that of the patterns in Figure 26-8, c and e are not isonemal, and all the others except i and j are isonemal by translation. In cases i and j, rotations are required to achieve complete transitivity.

There is a simple test to check whether an interlaced frieze pattern is isonemal. Construct the quotient link by joining the two ends of a fundamental region of the frieze together. If this quotient link has a single component, then the frieze is isonemal by translation.

Another problem arising in the mathematical study of fabrics is that of determining whether a layered pattern falls apart. Can the strings be separated into two or more sets which are not interlinked? We assumed above that friezes are connected. One consequence of this is that Celtic interlaced friezes never fall apart. If an interlaced frieze is alternating, connected, and separable, then so is the quotient link constructed from it. However, alternating diagrams of links represent split links if and only if they are not connected [4].

A Finer Classification

Classifying the multitude of Celtic interlaced patterns into only ten classes is, in some respects, not very satisfactory especially when we realize that the resulting three-dimensional symmetry type is completely determined by the underlying two-dimensional pattern of break-markers. It would be nice to have some kind of classification by pattern type [3] rather than merely by symmetry type.

Intuitively, a pattern is a collection of motifs arranged in a systematic fashion. This regularity is modelled mathematically by requiring that the symmetry group of the pattern acts transitively on the motifs. Classification by pattern type depends on three factors: the symmetry group G of the pattern; the stabilizer $stab_G(M)$ of a motif M; and the set of motif transitive subgroups of G. Two patterns are said to have the same *pattern type* if all these features coincide.

When we try to apply this kind of analysis to Celtic friezes, however, we immediately run into difficulties: The patterns are not discrete, so there is no obvious or natural way to split them up into motifs. The continuity of the strings makes it impossible to choose a motif in an unambiguous way, and the choice made will affect the resulting pattern type. Ironically, the arrangement of break-markers associated with an interlaced frieze pattern is discrete but not necessarily transitive.

Coda

In this study of Celtic knotwork I have concentrated on one particular theme: symmetry. That an analysis in terms of symmetries is possible is due partly to the lattice structure underlying the construction of the designs. This inbuilt regularity and the imposition of periodicity by the artist means that many of the patterns have non-trivial symmetry. The reader may feel that this same rigidity would lead to a dull and sterile art form. A non-mathematician confronted with the mathematical classification of patterns according to their symmetry type wrote [2, p. 70],

> There is no danger that the resources of the pattern maker will be exhausted by the constraints of geometry.

The remark seems appropriate in this context, too. A glance at any of the illuminated manuscripts produced by Celtic scribes will easily convince you that a geometric framework in no way hinders the artist. There is still room for imagination and creativity to express themselves.

References

1. J. Romilly Allen, *Celtic Art in Pagan and Christian Times*, London: Methuen (1904).
2. E. H. Gombrich, *The Sense of Order: a study in the psychology of decorative art*, Ithaca, NY: Cornell University Press (1979).
3. B. Grünbaum and G. C. Shephard, *Tilings and Patterns*, New York: Freeman (1987).
4. W. W. Menasco, Closed incompressible surfaces in alternating knot and link complements. *Topology* 23 (1984), 37–44.
5. A. V. Shubnikov and V. A. Koptsik, *Symmetry in Science and Art* (translated from the Russian by G. D. Archard), New York: Plenum (1974).

Further Reading

Celtic Knotwork

G. Bain, *Celtic Art: the methods of construction*, London: Constable (1977).
I. Bain, *Celtic Knotwork*, London: Constable (1986).
A. Meehan, *Celtic Design: knotwork*, London: Thames and Hudson (1991).

Related Topics

H. Arneberg, *Norwegian Peasant Art: men's handicrafts*, Oslo: Fabritius & Son (1951).
K. M. Chapman, *The Pottery of San Ildefonso Pueblo*, School of American Research, monograph 28, Albuquerque: University of New Mexico Press (1970).
D. W. Crowe and D. K. Washburn, Groups and geometry in the ceramic art of San Ildefonso, *Algebras, Groups and Geometries* (2) 3 (1985), 263–277.
B. Grünbaum, Periodic ornamentation of the fabric plane: lessons from Peruvian fabrics. *Symmetry* 1 (1990), 48–68.
B. Grünbaum and G. C. Shephard, The geometry of fabrics, *Geometrical Combinatorics* (F. C. Holroyd and R. J. Wilson, eds), Pitman (1984), 77–97.
B. Grünbaum and G. C. Shephard, Interlace patterns in Islamic and Moorish art, *Leonardo* (1993).
B. Grünbaum, Z. Grünbaum, and G. C. Shephard, Symmetry in Moorish and other ornaments, *Comp. & Maths. with Appls.*, vol 12B, Nos. 3/4 (1986), 641–653.
A. Hamilton, *The art workmanship of the Maori race in New Zealand*, Wellington: New Zealand Institute (1896).
I. Hargittai and G. Lengyel, The seven one-dimensional space-group symmetries in Hungarian folk needlework, *J. Chem. Educ.* 61 (1984), 1033.
G. H. Knight, The geometry of Maori art. Part I: rafter patterns, *New Zealand Math. Mag.* (3) 21 (1984), 36–40.
G. H. Knight, The geometry of Maori art. Part 2: weaving patterns, *New Zealand Math. Mag.* (3) 21 (1984), 80–86.
D. K. Washburn and D. W. Crowe, *Symmetries of Culture: Theory and Practice of Plane Pattern Analysis*, Seattle: University of Washington Press (1988).

27

The Sacred Cut Revisited: The Pavement of the Baptistery of San Giovanni, Florence

Kim Williams

The pavement of the Baptistery of San Giovanni in Florence has been metaphorically compared to a carpet that has been unrolled in a space. It is true in the case of the Baptistery that the motifs in the floor were inspired by the designs in Arabic carpets. But a decorated pavement can do more than merely embellish a monument. The design of a pavement may reveal an underlying ordering principle or philosophy in the architecture.

The Baptistery of San Giovanni is the oldest church in Florence. The walls of the octagonal building date from about the seventh century after Christ. Its foundations are even earlier. During the 1912–1915 renovation, which returned the Baptistery to its "original" state, an excavation was undertaken inside the Baptistery which revealed the older Roman substructure under the apse. Fragments of the substructure and this mosaic decoration may still be seen through grills in the floor. By the time she became the queen of the Renaissance, almost all of Florence's Roman and Paleo-Christian structures had been destroyed. The Baptistery serves as a reminder of Florence's ancient past. Was the Baptistery intended from its conception to be a baptistery? Probably not. Despite its central plan which is characteristic of baptisteries, it probably served as a chapel. San Giovanni became a baptistery officially, however, when the font was transferred from the church of Santa Reparata in 1128. The pavement is conventionally dated 1207. It is likely that the decision to pave San Giovanni in the late twelfth or early thirteenth century coincided with a desire to heighten the symbolic character of the building as a backdrop to the rite of baptism. The geometry which I have found to be key in the layout of the Baptistery floor provided a particularly rich and apt symbolic base for the program of architectural ornamentation.

The pavement is a patchwork quilt of patterns. Three strips of pavement define the north–south and east axes. The pavement of the west axis was destroyed in 1576 when Grand Duke Francesco I of Tuscany refurbished the Baptistery. A center medallion with a centripetal motif was laid in 1752 but was removed during the 1912 restoration. Today, the west axis and central octagon are paved in simple brown terrazzo. The north–south and east axes radiate out from the center octagon to the three doors. The strips of pavement which define the axes separate the four quadrants of pavement. Each quadrant in its turn is partitioned by decorated borders into compartments, each filled with a different motif. Of the four quadrants, the southeast quadrant is by far the richest, presenting more complex geometrical

Volume 16, No. 2 (Spring 1994), 18–24

patterns as well as figural patterns.[1] Of the axes, the east axis is the most elaborate, the east portal being on axis with the Duomo and so being the ceremonial entrance. (See Figure 27-1.)

Initially drawn to study the wealth of geometric patterns in the Baptistery pavement (some sixty of them), once I began work I became more and more intrigued by the layout of the pavement as a whole. An Italian scholar named Carocci wrote in 1897, "The pavement, as far as I am concerned, wasn't conceived as a systematically arranged whole because it is so irregular in its partitioning and

FIGURE 27-1 *The pavement of the Baptistery of San Giovanni in Florence, showing the central octagonal medallion and the altar on the west side. The southeast quadrant is in the lower left-hand corner. Photograph taken before the 1912 restoration of the Baptistery.*

[1]Enrico Madoni has drawn each motif in the Baptistery pavement in his monograph, *Il Pavimento del Battistere di Firenzo, 1913.*

doesn't present us with a symmetric whole as does, for example, the cathedral of Siena. . . ."[2] I found the partitioning of the Baptistery floor to be very regular, especially when each individual quadrant is compared to the others. The existence of an underlying order was pointed to by the pavement having been so regularly laid out in each of the quadrants, by the layout having been so deliberately emphasized by decorated borders which separated the segments from each other, and by each segment having been filled with a different geometric motif. With the help of a geometrical construction called the "Sacred Cut," I found the ordering mechanism for which I was looking.

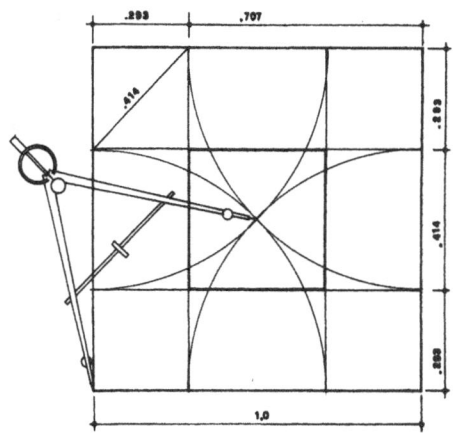

FIGURE 27-2 *The construction and proportions of the Sacred Cut geometry (Figures 27-2–27-10 were drawn by Kim Williams).*

The Sacred Cut is the fundamental construction which forms the basis for Danish engineer Tons Brunés' system of geometry, which he claims has governed the construction of monuments in every period from the Egyptian to the Medieval [1]. Let us examine the construction.

To begin, one takes a reference square of a given dimension. The reference square is divided by placing the compass at a corner of the square and striking an arc which intersects the two adjacent sides and passes through the center of the square. Thus, each of the two sides is divided into segments of proportions $\sqrt{2}/2 \approx .707$ and $1 - (\sqrt{2}/2) \approx .293$ of the length of the side. When similar arcs are struck from the other three corners, the operation is complete, and the reference square is subdivided into a nine-module grid with modules of different proportions. The four corner modules are smallest, .293 square; the middle modules are .293 × .414 rectangles; the inner module is the largest, $\sqrt{2} - 1 \approx .414$ square, and Brunés calls it the "sacred square." It may be seen that the diagonal of each corner square is also equal to $\sqrt{2} - 1 \approx .414$, thus completing a regular octagon. The Sacred Cut also provides a close approximation for squaring the circle: The length of the arc used to construct the Sacred Cut is equal to within 0.6% to the length of the diagonal of one-half the reference square (Figure 27-2).

There is no direct evidence that the Sacred Cut was actually used in antiquity. For example, the Sacred Cut is not mentioned specifically by Vitruvius, who wrote the only surviving document on architectural practice which has come down to us. However, Vitruvius does recommend the use of geometric constructions as devices to ensure proportional harmony in an architecture. Modern-day architectural historians have regarded it as a clue to how ancient monuments were constructed.

Tons Brunés found the Sacred Cut to have played an important part in the proportioning of the Pantheon [2]. Professor Brunés placed a circle within the section of the Pantheon so that it followed the curve of the dome, then circumscribed a square about the circle, which he used as the first reference square. Sacred Cuts performed on this square resulted in horizontals which coincided with the springing line for the dome and the top of the niches between the columns which encircle the main hall. Verticals produced by the Cuts coincided with the upper limit of the coffers in the dome. The sacred square of the first reference square then became the second reference square, and its Sacred Cuts provided verticals which coincided with the diameter of the oculus in the dome and horizontals which coincided with the top of the upper row of windows. The Sacred Cut was useful in analyzing the ground plan of the building as well. It again provided horizontal and vertical lines which coincided with major elements of the building, most notably the width of the porch.

[2]G. Carocci, Il Pavimento nel Battistero Fiorentino, in *Arte Italiana Decorativa e Industriale,* 1897, p. 34, ". . . il pavimento, secondo me, non e stato ideato come cosa organica, perché è assai irregolare nella sua divisione e non presenta un assieme simmetrico come per esempio quallo del Duome di Siena . . ."

More recently, the Sacred Cut has been found by Donald and Carol Watts to have governed the layout of a second-century A.D. housing complex called the "Garden Houses" in ancient Ostia [3]. The Wattses found that three successive Sacred Cuts determined the geometrical order to the Garden House complex. They placed a circle which touched the corner of the courtyard within the complex and constructed a square which circumscribed that circle. Sacred Cuts performed on the east and west sides of this reference square coincided with the position of the outer walls of the courtyard buildings. Similar Sacred Cuts on a second reference square determined by the width of the courtyard and the position of the fountains coincide with the party walls along spines of the courtyard buildings. The sacred square of the second reference square becomes the third reference square, and its Cuts coincide with innermost walls of the courtyard building. The unfolding of all the Sacred Cuts from a common center reveals an emphasis on the major east–west axis of the complex.

The success with which the Sacred Cut has been used in studying Roman architecture recommends its use in analyzing medieval architecture. The step from a Roman housing project to a medieval Tuscan sacred structure is not the large leap it may appear to be. Italian contemporaries of Romanesque architecture believed themselves to be carrying forward Roman building techniques and hence Roman greatness. I was drawn to study the Sacred Cut geometry in relation to the Baptistery not only because of the link between Roman and Romanesque architectures, but particularly because I recognized the kinship between Sacred Cut geometry and the Baptistery's octagonal form. Although the octagon played no part in the analyses of either the Pantheon or the Garden Houses, the Sacred Cut is the method by which one constructs an octagon with equilateral sides from a given square.

Here is how it is applied to the Baptistery floor: We construct a first square. Using the Sacred Cut procedure to construct an octagon, as discussed, we obtain the perimeter walls of the space, referred to as the "outer octagon." Next, by joining alternate midpoints we construct within the outer octagon two superimposed squares, one turned 45° with respect to the other. This forms an eight-point star. Sacred Cuts are next performed on each of the two superimposed squares, and the resulting smaller eight-point star which is formed determines the octagon in the center of the pavement, referred to as the "inner octagon." (See Figures 27-3–27-6.) Originally, the baptismal font occupied the entire inner octagon. After the demolition of the font, a strongly centripetal design in black, white, and red marble took its place.

The Wattses found in the Garden Houses a series of integer dimensions which occurred as ratios of segments constructed by Sacred Cuts. Beginning with a reference square with a side of 41 and per-

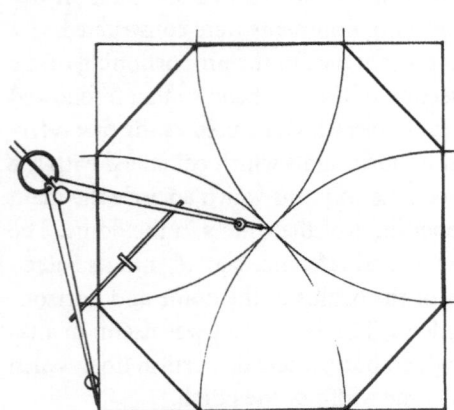

FIGURE 27-3 *Using the Sacred Cut to form the outer octagon.*

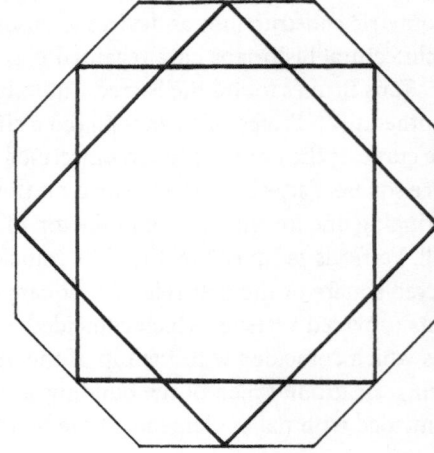

FIGURE 27-4 *The eight-point star found within the outer octagon.*

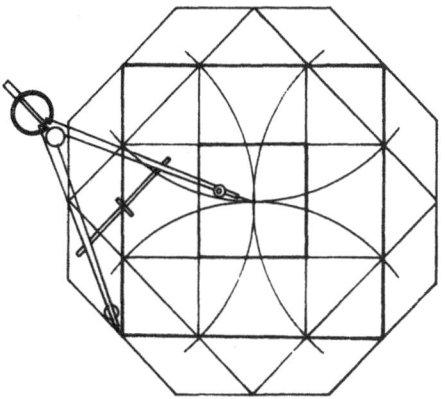

FIGURE 27-5 *Generating the inner octagon using the Sacred Cut.*

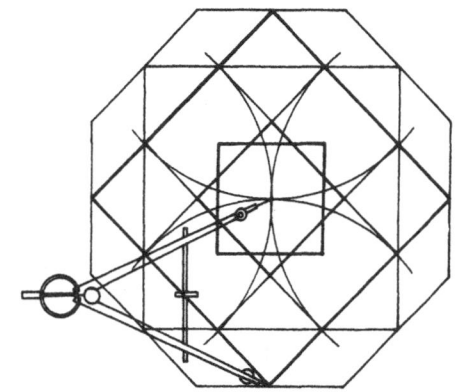

FIGURE 27-6 *Two superimposed Sacred Cuts form the inner octagon.*

forming successive Sacred Cut, using each resulting sacred square as the next reference square, one obtains approximately the following integer series:

$$41 \quad 29 \quad 17 \quad 12 \quad 7 \quad 5 \quad 3 \quad 2 \quad 1.$$

These integers and their multiples were found to figure significantly in the apartments of the Garden Houses complex. The windows in the largest rooms are based on a 7-foot module, the window itself measuring 5 Roman feet and the interval between windows measuring 2 Roman feet; the width of the public corridor and stairway is 17 Roman feet; the inside width of the courtyard buildings is 58 Roman feet.

The Wattses observed that these integers can be obtained from a familiar Pythagorean construction. The construction begins with a square with side 1, whose diagonal, of exact length $\sqrt{2}$ accord-

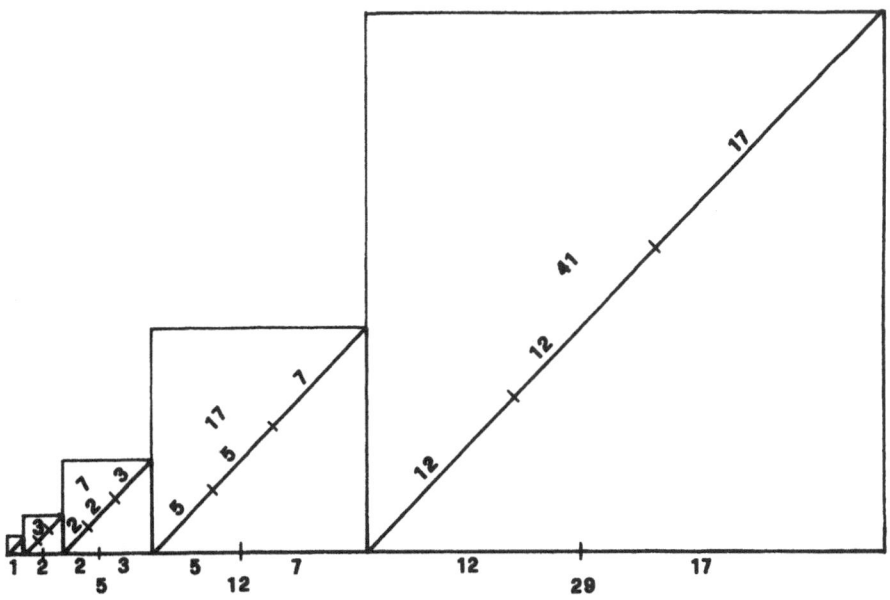

FIGURE 27-7 *The Pythagorean construction for approximating the value of $\sqrt{2}$.*

ing to the Pythagorean theorem, is approximated by integer 1. Subsequent approximations are obtained by recursion: if a_n and b_n are the integers used at the nth stage, then the next stage uses

$$a_{n+1} = a_n + b_n, \quad b_{n+1} = 2a_n + b_n.$$

(See Figure 27-7.) The sequence so produced is

$$1 \quad 1 \quad 2 \quad 3 \quad 5 \quad 7 \quad 12 \quad 17 \quad 29 \quad 41 \quad 70 \ldots$$

as mentioned above. It is easy to prove that the ratios b_n/a_n have the desired property of approaching $\sqrt{2}$.

We have obtained the sequence as two sequences interleaved. The first,

$$(a_n): 1 \quad 2 \quad 5 \quad 12 \quad 29 \quad 70 \ldots,$$

represents the successive values assigned to the sides of the squares; the second,

$$(b_n): 1 \quad 3 \quad 7 \quad 17 \quad 41 \quad 99 \ldots,$$

represents those assigned to the diagonals. Each of these is a Pell's series, with the same recursion $c_{n+1} = 2c_n + c_{n-1}$. The ratios c_{n+1}/c_n approach $\theta = 1 + \sqrt{2}$.[3] This ratio comes naturally into the Sacred Cut construction. It is the exact ratio of the side of the reference square to that of the inner module or "sacred square." If the side of the reference square is taken as $\theta\sqrt{2}$, the modules are simply expressed (Figure 27-8).[4]

P. H. Scholfield points out another valuable quality of this division to the architect. A rectangle divided by two horizontal and two vertical lines generically gives thirty-six different shapes of sub-rectangle. The Sacred Cut construction limits the number of shapes to five [4]. The Baptistery floor contains many rectangles conforming to the shapes occurring in Figure 27-8. (See Figure 27-9.) Such repeated proportionality has been studied under the name of the "principle of repetition of ratios."[5]

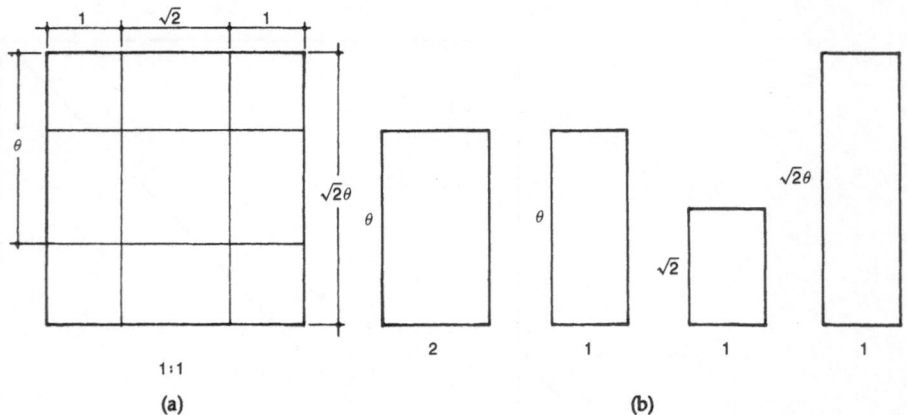

FIGURE 27-8 (a) The Sacred Cut division of the $\theta\sqrt{2}$ square. (b) The four rectangles generated by the Sacred Cut division of the $\theta\sqrt{2}$ square.

[3]The θ value was given its name and extensively analyzed by P. H. Scholfield in The Theory of Proportion in Architecture, Cambridge: Cambridge University Press (1958), 138–144.
[4]This particular division of the square is illustrated in Scholfield, The Theory of Proportion, p. 133, Fig. 3f.
[5]For more on the "principle of repetition of ratios," see Jay Kappraff, Connections: The Geometric Bridge between Art and Science, New York: McGraw-Hill (1991), 19.

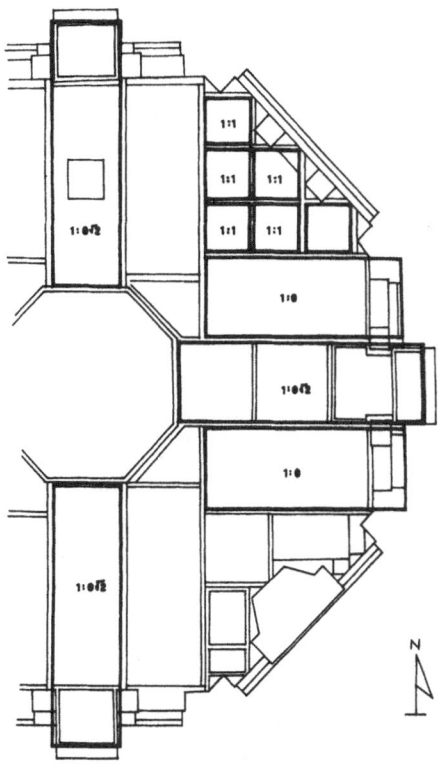

FIGURE 27-9 *The proportions of rectangles in the Baptistery pavement.*

Unlike the Wattses, I was unable to find within the Baptistery any linear dimensions which conformed exactly to the Pythagorean series or the Pell's series.[6] Rather, my having found the proportions of the pavement compartments to be related to θ and $\sqrt{2}$ points to the use of a geometric system rather than the use of a system of integer ratios to lay out the pavement. The Sacred Cut is an ideal example of such a system.

What remains enigmatic in the Baptistery pavement is an apparent irregularity in the pentagonal pieces along the diagonal axes of the inner octagon. To maintain radial symmetry, they should be circumscribable by a square. Instead, they are circumscribed by a rectangle. Similar irregularities in another pavement have been ascribed to craftsmen who, in articulating a system whose ordering principles were unknown or lost, made alterations to the pattern as they worked it out and unintentionally interrupted the system. Knowing that trade secrets in the guilds of the middle ages were jealously guarded, exposing them being punishable by death, this is a plausible explanation. (See Figure 27-10.)

To seek the symbolism in the pavement of San Giovanni, we ask the significance of the geometric figure produced by the Sacred Cut geometry. The superimposed reference squares in the Baptistery create an eight-point star. The eight-point star as an ecclesiastical

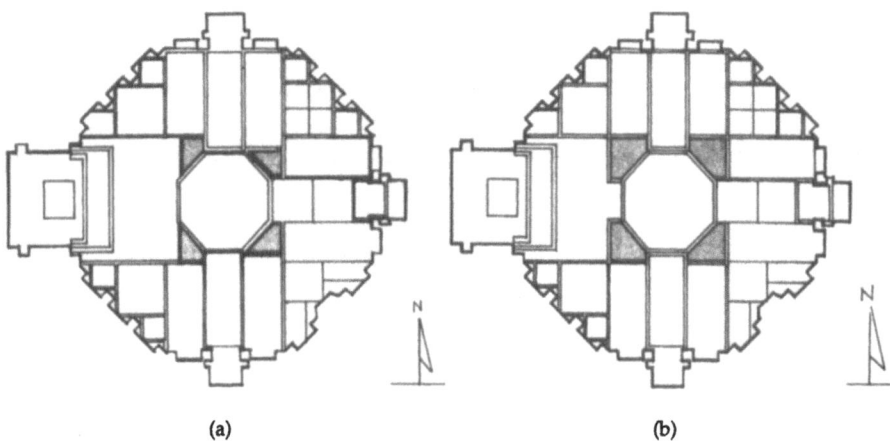

(a)   (b)

FIGURE 27-10 *(a) The existing irregular trapezoids (shaded) which surround the octagonal center of the Baptistery pavement. (b) Ideal arrangement of irregular pentagons surrounding the inner octagon.*

[6]After having taken all the measurements and produced a measured drawing of the pavement, I began to study the measure of the building. The Roman foot was most probably the standard unit of measure used in the Baptistery. Though it has been found to vary, the Roman foot is conventionally considered to be equal to 29.5 cm or 11.625 in. The value I have used to obtain the integers used in my analysis is 28.49 cm. This results in dimensions which correspond roughly to the values in the Pythagorean construction discussed earlier. It must be noted that the Baptistery is not a perfect octagon, being some 41 cm shorter on the east–west axis than on the north–south axis. There is accordingly some disagreement between dimensions in the various quadrants.

emblem signifies resurrection. In medieval number symbolism, eight signified cosmic equilibrium and immortality. It was considered to be particularly closely related to resurrection: Christ was resurrected eight days after he triumphantly entered Jerusalem on Palm Sunday, for example. As a symbol of baptism, the number eight, and by extension the eight-point star and the octagon, represent rebirth or resurrection of the catechumen in Christ. Further, the octagon mediates between a square, which represents things terrestrial, and a circle, which represents things celestial [5]. In fact, in wall intarsia in San Giovanni, the two eight-point stars on the ends have an octagonal center, whereas the two inside stars each have a circular center. I believe the unknown designer of the Baptistery pavement was making a similar statement when he based the layout of the floor on the same geometric figure.

The use of irrational values, or incommensurables, is linked philosophically to the symbolism of the circle and the square. A circle was indefinite, its circumference and area based on the irrational π, whereas the circumference and area of a square were rational values. Philosophically the use of irrational numbers such as θ shows an attempt to rationalize that which is irrational, or in other words, to make sensible that which is divine or only achievable through the intellect. In this context, let's turn again to the Sacred Cut. According to Professor Brunés, its special quality, its "sacredness," lies in its very nearly solving the riddle of how to square the circle. The Wattses believe that the Roman designer of the Garden Houses used the Sacred Cut to represent the squaring of the circle in order to integrate sacred and profane, earthly and celestial.

I believe likewise that the designer of the pavement in San Giovanni intended to imbue his pavement with symbolism to represent this integration. The octagon mediates between square and circle in the Baptistery, thus expressing the purpose of the baptismal rite. The catechumen undergoes death to sin and rebirth in Christ, coming in this way as near to the divine as a human can.

References

1. Tons Brunés, *The Secrets of Ancient Geometry and Its Use*, Copenhagen: Rhodos (1967), Vol. 1, 72–80.
2. Tons Brunés, *The Secrets of Ancient Geometry and Its Use*, Copenhagen: Rhodos (1967), Vol. 2, 38–56.
3. Donald J. and Carol Martin Watts, A Roman apartment complex, *Scientific American*, 255(6) (1986), 132–140.
4. P. H. Scholfield, *The Theory of Proportion*, Cambridge: Cambridge University Press, p. 132.
5. Jean Chevalier and Alain Gheerbrant, *Dizionario dei Simboli*, Milano: Biblioteca Universale Rizzoli (1986), Vol. 2 175.

28

Symmetrical Combinations of Three or Four Hollow Triangles

H. S. M. Coxeter

A *hollow triangle* is the planar region bounded by two homothetic and concentric equilateral triangles, that is, a flat triangular "ring." The edges of the inner triangle are half as long as those of the outer triangle. We will look at two striking constructions in three-space made from such figures.

The Australian sculptor John Robinson assembled *three* such triangular rings to form a structure (entitled *Intuition*) in which certain points on two outer edges of each ring fit into two inner corners of the next, in cyclic order (see Figure 28-1). The topology of the assembly is that of the "Borromean rings," and its symmetry group is C_3, cyclically permuting the three hollow triangles.

Quite independently, the American artist George Odom assembled *four* such triangular rings to form a rigid structure in which the midpoints of the three outer edges of each ring fit into inner corners of the three remaining rings. The four rings are mutually interlocked, and the symmetry group is the octahedral group O or S_4 : 24 rotations permuting the four hollow triangles in all the 4! possible ways. The positions of the twenty-four vertices of the four pairs of homothetic equilateral triangles are most naturally indicated by means of four-dimensional Cartesian coordinates involving the integers

$$0, \quad \pm 1, \quad \pm 2.$$

Felix Klein's Polyhedral Groups

In the opening chapter of his beautiful *Lectures on the Icosahedron* [5], pp. 14–20, Klein identified the rotation groups of the regular tetrahedron $\{3, 3\}$, octahedron $\{3, 4\}$, and icosahedron $\{3, 5\}$, with the alternating and symmetric groups

$$A_4, \quad S_4, \quad A_5.$$

For the tetrahedral and octahedral groups, the four objects which he permuted were the four diameters ("diagonals") that join pairs of opposite vertices of the cube $\{4, 3\}$ or of Kepler's *stella octangula* consisting of two reciprocal tetrahedra (Klein's "tetrahedron and counter-tetrahedron") whose 4 +

Volume 16, No. 3 (Summer 1994), 25–30

4 vertices coincide with the eight vertices of the cube. In the case of the icosahedral group A_5, the five objects which he evenly permuted were the five triads of mutually perpendicular lines (like Cartesian coordinate axes) which join the midpoints of pairs of opposite edges of the icosahedron (or of its reciprocal, the pentagonal dodecahedron {5, 3}); (compare [3], pp. 273–275).

By holding up a model of the icosahedron, with two opposite edges in one's two hands, one soon recognizes these edges as belonging to a *golden rectangle* whose two longer sides are diagonals of pentagons. In such a rectangle [6], p. 12, the ratio of lengths of the sides (longer to shorter) is the "golden ratio"

$$\tau = 2\cos(\pi/5) = 1 + 1/\tau = 1 + 1/1 + 1/1 + \cdots \approx 1.6180339887$$

[3], pp. 140–143. The convention used here is that each solidus dominates whatever follows: $1/1 + 1$ means $1/2$. This convention facilitates the printing of continued fractions.

The equation $\tau = 1 + 1/\tau$ shows that a golden rectangle with sides τ and 1 can be dissected into a square with side 1 and a smaller golden rectangle with sides 1 and τ^{-1}. In Figure 28-2, the smaller rectangle with diagonal BC has been rotated by a quarter-turn and placed inside the square so that the two congruent diagonals AB and BC are perpendicular. The angle $2\phi = BOC$, between the diagonals of the rotated rectangle, is given by

$$\tan 2\phi = BC \Big/ \frac{1}{2} AB = 2.$$

Thus, *the acute angle between the diagonals of a golden rectangle* is arctan 2. This perspicuous proof of a familiar result was kindly supplied by Jan van de Craats.

In terms of the icosahedron, 2ϕ *is* the angle between two adjacent diameters ([3], p. 156). These diameters of the icosahedron are perpendicular to two planes containing adjacent faces of the reciprocal dodecahedron, providing an easy proof for a still more familiar fact: *The dihedral angle of the dodecahedron is*

$$\pi - \text{arctan } 2.$$

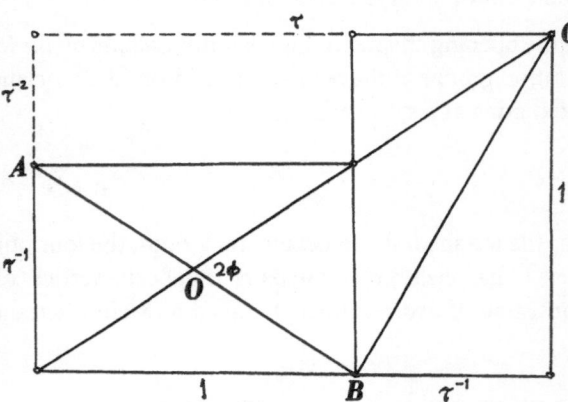

FIGURE 28-1 *The 4 triangles are permuted by 4! rotations.*

FIGURE 28-2 *The diagonals of golden rectangles.*

Max Brückner's Five Tetrahedra

For the icosahedral group

$$(5, 3, 2) \cong A_5, \tag{1}$$

Klein's five Cartesian frames may conveniently be replaced by five tetrahedra whose 5 × 4 vertices coincide with the 20 vertices of the dodecahedron {5, 3} ([2], Plate IX, Figure 11). A simple way to see how this compound arises is to observe the square that lies behind any edge of the dodecahedron. This square and the antipodal square are two opposite faces of a cube whose eight vertices belong to the dodecahedron. Either of the two tetrahedra inscribed in the cube can serve as one of the five tetrahedra inscribed in the dodecahedron; the other four can be obtained by rotation about any pentagonal axis.

Because the 5 × 4 faces of the five tetrahedra lie in the twenty face-planes of an icosahedron, this compound is one of the fifty-eight stellations of the icosahedron, namely, Ef_1 or Ef_1 [4], pp. 5, 25–26 and Plate XVI. A solid model, with five colors for the five tetrahedra, provides a perspicuous demonstration of the isomorphism (1). Each rotation of the model yields an even permutation of the five colors.

The problem of stellating the icosahedron inspired George Odom's discovery of a new construction for the golden ratio ([1], p. 23; [7]):

Let A and B be the midpoints of the sides EF and ED of an equilateral triangle DEF. Extend AB to meet the circumcircle (of DEF) at C. Then B divides AC according to the golden section.

Symmetrical Lattices

When circular discs are closely packed in the Euclidean plane so that each is surrounded by six others, their lattice of centers is most conveniently coordinatized by working in space and taking the plane to have the equation

$$x_1 + x_2 + x_3 = 0.$$

For then, if the discs have diameter $\sqrt{2}$, the centers are simply all the points (x_1, x_2, x_3) whose coordinates are integers satisfying that equation. In particular, the origin $(0, 0, 0)$ is surrounded by a regular hexagon given by the six permutations of $(1, 0, -1)$. In Figure 28-3, for convenience, $(1, 0, -1)$ is contracted to $10\bar{1}$. Note that when the vertices of the hexagon are interpreted as vectors, each is the sum of its two neighbors. This fact is not surprising when we recall that the set of vectors belonging to any lattice is closed with respect to the operation of subtraction.

Analogously in four-dimensional space, the points (x_1, x_2, x_3, x_4) whose coordinates are integers satisfying the equation

$$x_1 + x_2 + x_3 + x_4 = 0 \tag{2}$$

form the lattice of centers of congruent balls (of diameter $\sqrt{2}$) in their most symmetrical close-packing. Now the centers nearest to the origin $(0, 0, 0, 0)$ are the twelve vertices of a cuboctahedron, given by the twelve permutations of $(1, 0, 0, -1)$ or $100\bar{1}$. As we see in Figure 28-4, its edge-length is $\sqrt{2}$. The four planes

$$x_1 = 0, \quad x_2 = 0, \quad x_3 = 0, \quad x_4 = 0$$

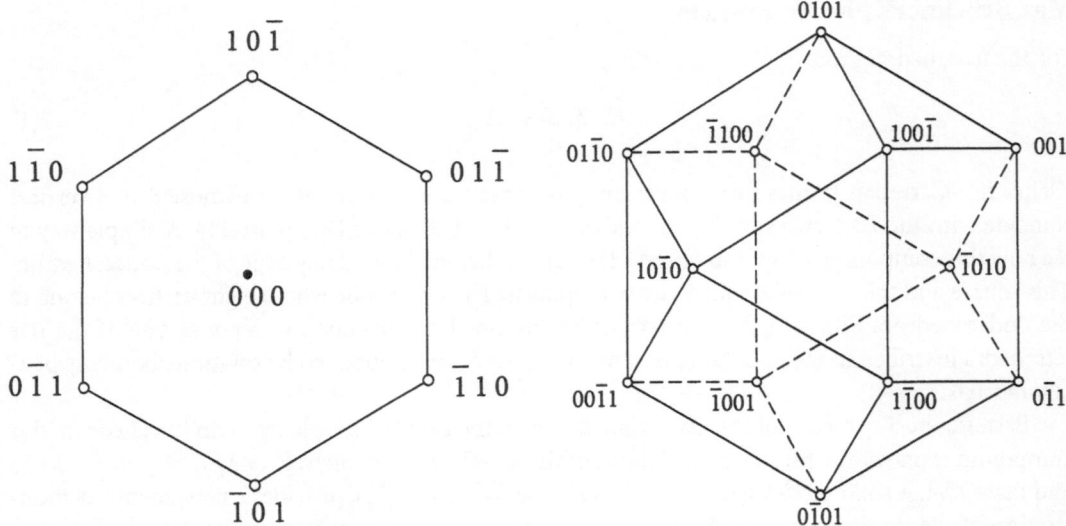

FIGURE 28-3 *A regular hexagon.*   FIGURE 28-4 *A cuboctahedron.*

(by which we mean "sections of four hyperplanes by the special hyperplane (2)") each contain a two-dimensional lattice of the kind we considered earlier.

George Odom's Four Hollow Triangles

Doubling the coordinates, we obtain the twelve vertices of a larger cuboctahedron: the permutations of $(2, 0, 0, -2)$ or $200\bar{2}$. The twenty-four vertices of the two homothetic cuboctahedra can be differently joined by new edges (twelve of length $\sqrt{6}$ and twelve of length $2\sqrt{6}$) so as to form four pairs of homothetic triangles, as in Figure 28-5. The twelve vertices of the four inner triangles are easily seen to coincide with the midpoints of the twelve sides of the four outer triangles. It follows that when the space between each pair of homothetic triangles is filled in with a cardboard (or steel or wooden) lamina, the four interlocked "hollow triangles" form a rigid structure. Such a model, with four colors for the four hollow triangles, was made by George Odom as a piece of abstract sculpture.

As the outer vertices are given by the permutations of $(2, 0, 0, -2)$, the whole sculpture can be enclosed by the cube of edge 4 whose faces lie in the three pairs of parallel hyperplanes

$$x_\nu + x_4 = \pm 2 \quad (\nu = 1, 2, 3).$$

For instance, the two parallel planes

$$x_1 + x_2 = \mp 2, \quad x_3 + x_4 = \pm 2$$

contain the squares

$$(-2, 0, 2, 0) \quad (-2, 0, 0, 2) \quad (0, -2, 0, 2) \quad (0, -2, 2, 0)$$

and

$$(0, 2, 0, -2) \quad (0, 2, -2, 0) \quad (2, 0, -2, 0) \quad (2, 0, 0, -2),$$

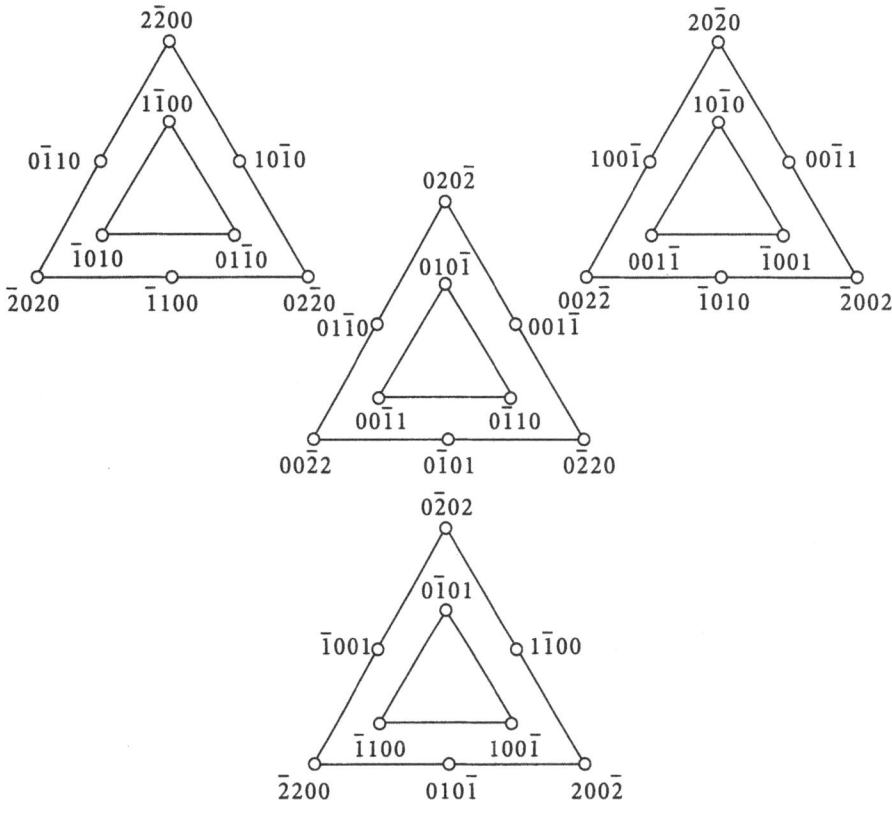

FIGURE 28-5 *Coordinates for the four hollow triangles.*

whose distance apart is 4. In other words, a model will fit neatly into a cubical box of edge 4. Because of its symmetry, it can be placed in the box in twenty-four ways, one for each element of the octahedral group (4, 3, 2), which is the rotation group of the cube {4, 3} (and of the octahedron {3, 4} and of the cuboctahedron $\{{}^4_3\}$). Each way of placing the model in the box provides a different permutation of the four hollow triangles, that is, of the four colors. The sculpture thus yields a strikingly perspicuous demonstration of the isomorphism

$$(4, 3, 2) \cong S_4.$$

John Robinson's Three Hollow Triangles

It is interesting to compare Figures 28-6 and 28-7, bearing in mind that Odom (in the United States) and Robinson (in England) had no communication, though both used hollow triangles of almost exactly the same shape. The differences are as striking as the resemblances. Every two of Odom's four hollow triangles are interlocked, and their assembly is inherently rigid. No two of Robinson's three hollow triangles are interlocked, but paradoxically the whole assembly is inseparable in the manner of "Borromean rings" ([8], pp. 66, 67, 119).

When a model is made, using thin laminae of cardboard, and the lowest vertices of the three outer triangles are allowed to slide freely on a tabletop, the assembly collapses under its own weight so as to form the planar pattern of Figure 28-8. The blue hollow triangle lies over the yellow, the yel-

FIGURE 28-6 *Odom's sculpture.*

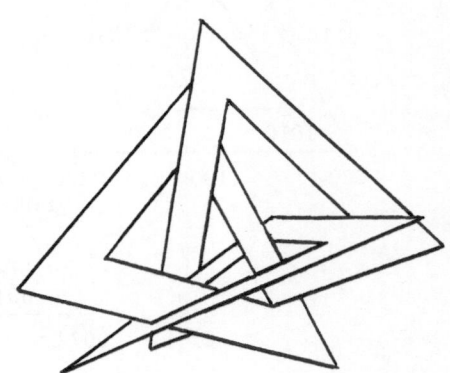

FIGURE 28-7 *Robinson's sculpture.*

low over the red, and the red over the blue. Conversely, when the outermost edges are lifted while trigonal symmetry is preserved, a complicated twisting motion takes place. This continues smoothly for a while and then stops abruptly. It remains true that certain points on two outer edges of each ring fit "into two inner corners of the next, in cyclic order," but these "certain points" slide along those "outer edges" and do not remain midpoints. Robinson fixed his rings at the stage when the lifting motion abruptly stops.

At Marjorie Senechal's "Regional Geometry Institute" in Smith College (funded by NSF), a workshop organized by Doris Schattschneider discovered that the "twisting motion" is an illusion, caused by the modelling material's flexibility. In fact, Robinson's hollow triangles are slightly narrower than Odom's. If the inner equilateral triangle has side 1, the side of the outer triangle is not 2 but $(2\sqrt{6} + 1)/3 \approx 1.9663265$. The details are as follows.

When the Borromean assembly in Figure 28-8 is turned over, it looks like Figure 28-9. An upside-down version of Robinson's sculpture is obtained by lifting the corner A of one hollow triangle

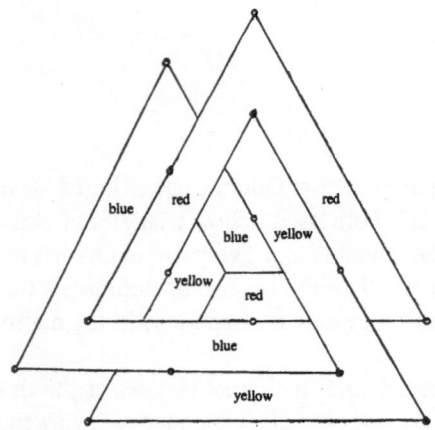

FIGURE 28-8 *The collapsed "Intuition."*

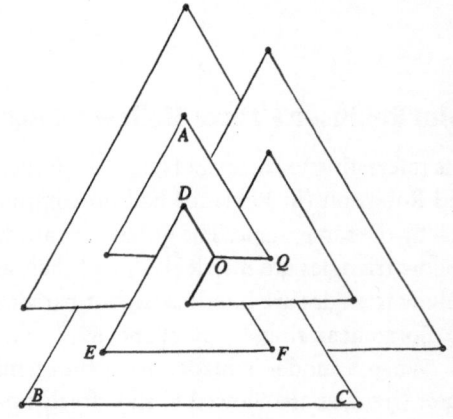

FIGURE 28-9 *The other side of Figure 28-8.*

ABCDEF and the corresponding corners (one initially at *E*) of the other two. The point *O* on the inner side *DF* of the first hollow triangle is seen to coincide with analogous points on the other two hollow triangles. The trigonal symmetry ensures that the *OQ* and *OD* of Figure 28-10, being two of three congruent segments, have the same length, say *x*.

PQ is an inner edge of the third hollow triangle. If *PQ* = *DE* = 1, the length of the outer edge *AB* = *t* can vary from 2 (in the collapsed position) to a slightly smaller value. Extensions of *ED* and *FD* form, with *A*, a rhomb of side $(t-1)/3$, whereas *OD* and *OQ* span another rhomb, of side *x*. Let *β* denote its angle at *Q* and *D*.

In terms of *β* and *t*, we can calculate *OD* = *x*, *AQ* = *y*, and *AP* = *z* by observing that

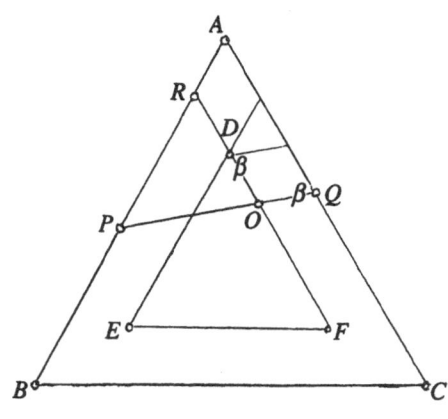

FIGURE 28-10 *The intersection PQ of two hollow triangles.*

$$y = \frac{AQ}{PQ} = \frac{RO}{PO} = \frac{x + (t-1)/3}{1 - x},$$

whereas, from the same triangle *APQ*,

$$\frac{y}{\sin(120° - \beta)} = \frac{z}{\sin \beta} = \frac{1}{\sin 60°}$$

Thus,

$$y = \cos \beta + \frac{\sin \beta}{\sqrt{3}}, \quad z = \frac{2 \sin \beta}{\sqrt{3}} \tag{3}$$

and, since (by similar triangles)

$$\frac{t-1}{2\sqrt{3} \sin \beta} = \frac{(t-1)/3}{z} = \frac{x}{1} = \frac{y - (t-1)/3}{y+1}$$

$$= \frac{\cos \beta + (1/\sqrt{3}) \sin \beta - (t-1)/3}{\cos \beta + (1/\sqrt{3}) \sin \beta + 1},$$

it follows that

$$t = 1 + \frac{2 \sin \beta (\sqrt{3} \cos \beta + \sin \beta)}{\cos \beta + \sqrt{3} \sin \beta + 1} \tag{4}$$

and

$$x = \frac{\sqrt{3} \cos \beta + \sin \beta}{\sqrt{3}(\cos \beta + \sqrt{3} \sin \beta + 1)}. \tag{5}$$

In the collapsed arrangement (Figs. 28-8 and 28-9) we have *β* = 60°, yielding

$$t = 2, \quad x = \tfrac{1}{3}, \quad y = 1, \quad z = 1.$$

In John Robinson's sculpture itself, Elizabeth Whitcomb observed that DQ appears to be one edge of a regular tetrahedron with centre O, so that the angle QOD is $\mathrm{arcsec}(-3)$ and

$$\beta = \mathrm{arcsec}\ 3 \approx 70°31'44''.$$

Setting $\cos \beta = \frac{1}{3}$ and $\sin \beta = \frac{2}{3}\sqrt{2}$ in Eqs. (3)–(5) we obtain

$$t = \frac{1}{3}(2\sqrt{6}+1) \approx 1.9663265,$$

$$x = \frac{1}{2} - \frac{\sqrt{6}}{12} \approx 0.2958809,$$

$$y = \frac{1}{3} + \frac{2\sqrt{6}}{9} \approx 0.8776780,$$

$$z = \frac{4\sqrt{6}}{9} \approx 1.0886893,$$

in good agreement with Robinson's artistic "Intuition."

References

1. Albert Beutelspacher, *Der Goldene Schnitt*, Wissenschaftsverlag, Mannheim, 1988.
2. Max Brückner, *Vielecke und Vielflache*, Leipzig, 1900.
3. H. S. M. Coxeter, *Introduction to Geometry* (2nd ed.), Wiley, New York, 1969.
4. H. S. M. Coxeter, Patrick Du Val, H. T. Flather, and J. F. Petrie, *The Fifty-nine Icosahedra* (2nd ed.), Springer-Verlag, New York, 1982.
5. Felix Klein, *Lectures on the Icosahedron and the Solution of Equations of the Fifth Degree* (2nd ed.), Kegan Paul, Trench Trübner, London, 1913.
6. George Markowsky, Misconceptions about the Golden Ratio, *College Math. J.* 23 (1992), 2–19.
7. George Odom and Jan van de Craats, Problem E 3007, *Amer. Math. Monthly* 90 (1983), 482.
8. John Robinson, *Symbolic Sculpture*, Edition Limitée, Carouge-Geneva, 1992.

Part Six

GEOMETRY AND TOPOLOGY

29

Instantons and the Topology of 4-Manifolds

Ronald J. Stern

Geometric topology is the study of metric spaces which are locally homeomorphic to Euclidean n-space $\mathbf{R}^n$; that is, it studies topological (TOP) n-manifolds. The customary goal is to discover invariants, *usually* algebraic invariants, which classify all manifolds of a given dimension. This is separated into an existence question—finding an n-manifold with the given invariants—and a uniqueness question—determining how many n-manifolds have the given invariant. As is (and was) quickly discovered, TOP manifolds are too amorphous to study initially, so one adds structure which is compatible with the available topology and which broadens the available tools. Presumably, the richer the structure imposed on a manifold, the fewer objects one is forced to study.

As our first example consider $M = \mathbf{R}^n$ (or open subsets thereof). One could study continuous functions on M, but there are many more tools available for studying *smooth* functions. To extend the concept of smoothness to more general manifolds, one specifies a smoothly compatible set of coordinate charts. A *differentiable atlas* on a metric space M is a cover of M by open sets $\{ U_\alpha \}$ (called *coordinate charts*) and homeomorphisms $\Phi_\alpha : U_\alpha \to \mathbf{R}^n$ (so far this is just the definition of a TOP n-manifold) so that the *transition functions* $\Phi_\beta \circ \Phi_\alpha^{-1} : \Phi_\alpha(U_\alpha \cap U_\beta) \to \Phi_\beta(U_\alpha \cap U_\beta)$ are smooth. Note that both the domain and range of the transition functions are open subsets of $\mathbf{R}^n$, so smoothness makes sense. Now the ordinary calculus of $\mathbf{R}^n$ can be patched together via the transition functions to allow one to do calculus on all of M. For instance, $f : M \to \mathbf{R}$ is smooth if $f \circ \Phi_\alpha^{-1} : \Phi_\alpha(U_\alpha) \to \mathbf{R}$ is smooth for each U_α.

There are some obvious redundancies in this notion of differentiation; we say that two atlases $\{ U_\alpha, \Phi_\alpha \}$ and $\{ V_\beta, \Psi_\beta \}$ are equivalent if their union is an atlas on M; that is, if $U_\alpha \cap V_\beta \neq \Phi$, then $\Phi_\alpha \circ \Psi_\beta^{-1}$ is smooth. A smooth (DIFF) structure on M is an equivalence class of atlases on M.

Poincaré promoted another type of structure on a manifold, where the above transition functions $\Phi_\alpha \circ \Phi_\beta^{-1}$, instead of being smooth are required to preserve the natural combinatorial structure of $\mathbf{R}^n$ (piecewise straight-line segments are mapped to piecewise straight-line segments). Such a maximal atlas on M is called a piecewise linear (PL) structure on M; we say M is combinatorially triangulated. A PL n-manifold can be shown to be PL homeomorphic to a simplicial complex that is a so-called combinatorial n-manifold (the link of every vertex is PL homeomorphic to the $(n - 1)$-sphere S^{n-1} with its standard triangulation). These various concepts of a manifold were introduced toward the end of the nineteenth century in order to better understand solutions to differential equations. It is ironic that the field of geometric topology developed, ignoring its roots, and how, as we

Volume 5, No. 3 (Summer 1983), 39–44

shall shortly point out, some of the most pressing questions concerning manifolds today are being answered using modern analysis and geometry.

A basic problem for topologists became, then, to determine when a TOP manifold admits a PL structure and, if it does, whether there is also a compatible DIFF structure. In the 1930s amazingly delicate proofs of the triangulability of DIFF manifolds were given by Whitehead [24] (so that indeed DIFF $\subset$ PL). By the mid 1950s, it was known that every TOP manifold of dimension less than or equal to 3 admits a *unique* DIFF structure [10, 15, 18]. It was unthinkable that for a given TOP (or PL) manifold M there were two distinct calculuses available; that is, that M admitted more than one DIFF structure. In 1956 Milnor showed that there are 28 distinct DIFF structures on S^7. (The examples were provided by some "well-understood" S^3 bundles over S^4 [12].) Other spheres were then discovered to possess exotic DIFF structures. Later work by a host of major mathematicians (e.g., Thom, Kervaire, Milnor, Munkres, Hirsh, Mazur, Poenaru, Lashof, Rothenberg, Haefliger, Smale, Novikov, Browder, Wall, Sullivan, Kirby, and Siebenmann) during the period from 1956 to 1970 began to sort things out in dimensions 5 and greater—a golden era for manifold topology. In the end, the obstructions to putting a PL or DIFF structure on a TOP manifold M of dimension at least 5 became a "lifting" problem (a reduction of the structure group of the tangent bundle) and thus a discrete problem.

In 1968 Kirby and Siebenmann [11] determined that for a TOP manifold of dimension *at least* 5, there is a *single* obstruction $k(M) \in H^4(M; \mathbf{Z}_2)$ to the existence of a PL structure on M; if $k(M) = 0$, there is a PL structure, otherwise there isn't. Moreover, if there is one PL structure, then there are $|H^3(M; \mathbf{Z}_2)|$ distinct PL structures. Earlier, when it was first realized that structure problems really should be lifting problems, it was discovered that (without dimension restrictions) there are further discrete obstructions to "lifting" from PL to DIFF; these involve the homotopy groups of spheres [9]. In particular every PL n-manifold, $n \leq 7$, admits a compatible DIFF structure which, if $n \leq 6$, is unique up to DIFF isomorphism! As one might expect (or hope), our ordinary Euclidean spaces $\mathbf{R}^n$, $n \neq 4$, are indeed ordinary, in that they possess but one DIFF or PL structure.

Surprisingly, virtually none of the techniques developed during these decades has had any impact on the dimensions in which we live and operate, dimensions 3 and 4. In dimension 3, Thurston has made impressive progress in the last six years—much more remains to be done. In dimension 4, there has been equally dramatic progress in the last two years. Perhaps the most striking fact to surface is that, unlike all other Euclidean spaces, $\mathbf{R}^4$ is not so ordinary; there is an exotic DIFF structure on $\mathbf{R}^4$! The existence of this exotic $\mathbf{R}^4$, denoted $\mathcal{R}^4$, is proved using a combination of topology (the work of M. Freedman [7]), geometry, and analysis (the work of S. Donaldson [4]). This $\mathcal{R}^4$ is truly bizarre in that there exists a compact set K in $\mathcal{R}^4$ such that no smoothly embedded 3-sphere in $\mathcal{R}^4$ contains K. Since $\mathcal{R}^4$ is homeomorphic to $\mathbf{R}^4$, there are certainly continuously embedded 3-spheres in it containing K. Thus the horizon of $\mathcal{R}^4$ is extremely jagged. (After looking in the mirror some mornings I am convinced I live in $\mathcal{R}^4$).

At this time it is not known how many such exotic $\mathbf{R}^4$'s exist, although three have been found [8]. Because of the nature of the constructions, topologists speculate that there are uncountably many distinct DIFF structures on $\mathbf{R}^4$—an appealing possibility. The classification of DIFF structures, which in higher dimensions is a discrete problem, could (will) wander into the realm of geometry—a whole moduli space of DIFF structures on 4-manifolds! Whether or not this is true, the work of Donaldson points out the impossibility of characterizing DIFF structures in terms of characteristic classes; i.e., it is not a discrete problem. Taking it to the limit could have so many meanings in dimension 4! Enough mystical wanderings. Why is there such an $\mathcal{R}^4$?

The Intersection Form

We said at the beginning of this article that the goal of geometric topology is to discover algebraic invariants which classify (at least partially) all manifolds in a given dimension. Historically one of the most important of these invariants has been the *intersection form*.

Perhaps it's best to start with two-dimensional manifolds where the intersection form and intersection numbers are more familiar. We can represent one-dimensional homology classes on a smooth surface S by smooth oriented curves.

FIGURE 29-1 *Intersection number zero.*

Suppose $\alpha, \beta \in H_1(S; \mathbf{Z})$ are represented by curves A and B. By slightly perturbing the curves we can assume they intersect transversally in isolated points. That means that at each point of intersection a tangent vector to A, together with a tangent vector to B (in that order!), forms a basis for the tangent space of S. To each point of intersection we assign $+1$ if the orientation of this basis agrees with the orientation of S; otherwise we assign -1. The (oriented) intersection number $A \cdot B$ is defined to be the algebraic sum of these numbers over all points of intersection, and the intersection form is the induced bilinear pairing defined by $I_S(\alpha, \beta) = A \cdot B$. It's easy to see that I_S is skew-symmetric $[I_S(\alpha, \beta) = -I_S(\alpha, \beta)]$ and unimodular. In fact, for any such form we can choose a basis so that the matrix of the form is

$$\begin{pmatrix} 0 & -I \\ I & 0 \end{pmatrix}.$$

Intersection numbers and the intersection form for a smooth 4-manifold M are defined similarly. This time we suppose *two*-dimensional homology classes $\alpha, \beta \in H_2(M; \mathbf{Z})$ are represented by smooth, oriented surfaces A and B and that the surfaces intersect transversally in isolated points. Again we assign $+1$ to a point of intersection if an (oriented) basis for the tangent space of A together with an (oriented) basis for the tangent space of B agrees with the orientation for M; otherwise we assign -1. The intersection number $A \cdot B$ is the algebraic sum of these over all points of intersection, and the intersection form is the bilinear pairing $I_M(\alpha, \beta) = A \cdot B$. This time, however, the pairing is symmetric $[I_M(\alpha, \beta) = I_M(\alpha, \beta)]$. It is still unimodular—the matrix for the form has determinant ± 1.

For smooth manifolds there is another way to define the intersection form. By Poincaré duality we can define the pairing in cohomology rather than homology. If we use DeRham cohomology $H_{DR}^\star(M)$, then $\alpha, \beta \in H_{DR}^2(M)$ can be represented by 2-forms a and b. We simply let

$$I_M(\alpha, \beta) = \int_M a \wedge b.$$

Defining the intersection form on cohomology allows us to extend the definition to *all* 4-manifolds, smooth or not. If $\alpha, \beta \in H^2(M; \mathbf{Z})$ and $[M] \in H_4(M; \mathbf{Z})$ is the fundamental class of M (given by an orientation on M), then $I_M(\alpha, \beta) = (\alpha \cup \beta)[M]$, where " $\cup$ " is the cup product in cohomology.

Here are some examples.

1. The 4-sphere S^4. Since $H_2(S^4; \mathbf{Z}) = 0$, the intersection form is trivial: $I_S^4 = \varnothing$.
2. The complex projective plane $\mathbf{C}P^2$. Here $H_2(\mathbf{C}P^2; \mathbf{Z}) = \mathbf{Z}$, and so the matrix for $I_{\mathbf{C}P^2}$ is (1).
3. The product of spheres $S^2 \times S^2$. In this case $H_2(S^2 \times S^2; \mathbf{Z}) = \mathbf{Z} \oplus \mathbf{Z}$, and we can represent the generators by the embedded surfaces $A = S^2 \times \{pt\}$ and $B = \{pt\} \times S^2$. Since A and B intersect in a single point, and each of them can be "pushed off" themselves, the matrix for $I_{S^2 \times S^2}$ is

$$\begin{pmatrix} 0 & 1 \\ 1 & 0 \end{pmatrix}.$$

4.　The Kummer surface

$$K = \{[Z_0, Z_1, Z_2, Z_3] \in \mathbf{C}P^3 \mid Z_0^4 + Z_1^4 + Z_2^4 + Z_3^4 = 0\}$$

This time things are much more complicated. The rank of $H_2(K; \mathbf{Z})$ is 22, and one can show that the matrix for I_K is given by $E_8 \oplus E_8 \oplus 3 \begin{pmatrix} 0 & 1 \\ 1 & 0 \end{pmatrix}$, where

$$E_8 = \begin{pmatrix} 2 & -1 & 0 & 0 & 0 & 0 & 0 & 0 \\ -1 & 2 & -1 & 0 & 0 & 0 & 0 & 0 \\ 0 & -1 & 2 & -1 & 0 & 0 & 0 & 0 \\ 0 & 0 & -1 & 2 & -1 & 0 & 0 & 0 \\ 0 & 0 & 0 & -1 & 2 & -1 & 0 & -1 \\ 0 & 0 & 0 & 0 & -1 & 2 & -1 & 0 \\ 0 & 0 & 0 & 0 & 0 & -1 & 2 & 0 \\ 0 & 0 & 0 & 0 & -1 & 0 & 0 & 2 \end{pmatrix}$$

(In fact, E_8 is the Cartan matrix for the exceptional Lie algebra e_8.)

The intersection form is indeed a basic invariant for compact 4-manifolds. In 1958 Milnor [13] showed that the homotopy type of a compact, simply-connected 4-manifold is completely determined by the isomorphism class of the intersection form.

The classification (up to isomorphism) of integral unimodular symmetric bilinear forms starts with three things: the rank (the dimension of the space on which the form is defined), the signature (the number of positive eigenvalues minus the number of negative values when considered as a real, rather than integral, form), and the type (the form is even if all the diagonal entries in its matrix are even, otherwise it's odd). A form is positive (negative) definite if all eigenvalues are positive (negative); otherwise it is indefinite.

For indefinite forms the rank, signature, and type form a complete set of invariants [14]. The classification of definite forms, however, is much more difficult. There is only one non-trivial restriction on an even definite form—its signature must be divisible by 8. In fact, it is known that E_8 (mentioned above) is the unique positive definite even form of rank 8; there are two even positive definite forms of rank 16 ($E_8 \oplus E_8$ and E_{16}); twenty-four such forms of rank 24; and many thousands of rank 40.

Given this complicated classification, it was natural to ask which forms could actually occur as the intersection form of a 4-manifold. For example, does there exist a manifold with intersection form E_8? or even $E_8 \oplus E_8$? The Kummer surface comes close for the second, but until two years ago no one knew the answer.

Topological 4-Manifolds

History had taught us that one first understands DIFF and PL manifolds and then proceeds to the more delicate questions concerning TOP manifolds. (Recall that DIFF = PL in dimensions ≤ 6). Imagine the shock when in the summer of 1981 Michael Freedman announced that a compact simply-connected (i.e., every map of the circle extends to a map of the disk) TOP 4-manifold is completely and faithfully classified by two elementary pieces of information!

The first piece of information is the *intersection form* on a 4-manifold M which we have just discussed.

The second piece of information required for Freedman's classification theorem is the Kirby-Siebenmann obstruction $\alpha(M) \in \mathbf{Z}_2$. It is completely characterized by the statement that $\alpha(M) = 0$ if and only if $M \times S^1$ admits a DIFF structure.

FREEDMAN'S THEOREM. Compact, simply-connected TOP 4-manifolds are in 1-to-1 correspondence with pairs $\langle I, \alpha \rangle$, where I is an integral unimodular symmetric bilinear form, $\alpha \in \mathbf{Z}_2$, and if I is even $\sigma(I)/8 \equiv \alpha \pmod 2$.

In particular, (existence): For an integral unimodular symmetric bilinear form I, there is a TOP 4-manifold M_I realizing I as its intersection form; (uniqueness): If I is even, the homeomorphism type of M_I is uniquely determined by I. For odd I, there are exactly two non-homeomorphic M_I realizing I as their intersection form, characterized by the fact that one $M_I \times S^1$ admits a DIFF structure, the other does not. So for instance, there are two manifolds realizing the form (1), namely, $\mathbf{CP}^2$ and another manifold Ch with the property that Ch is homotopy equivalent to $\mathbf{CP}^2$ but Ch admits no DIFF structure.

How beautifully simple—yet so sweeping! The uniqueness statement for $\langle \emptyset, 0 \rangle$ is the four-dimensional topological Poincaré conjecture—that a 4-manifold of the homotopy type of S^4 is homeomorphic to S^4. There also is the existence of a unique TOP 4-manifold with the intersection pairing E_8—a manifold long sought after by topologists.

Comments on Freedman's Proof

For the purposes of our $\mathfrak{R}^4$ we view Freedman's result as the ability to do TOP surgery. In particular, recall the Kummer surface K whose intersection form I_K had a matrix representation as $E_8 \oplus E_8 \oplus 3\left(\begin{smallmatrix}0&1\\1&0\end{smallmatrix}\right)$, where $\left(\begin{smallmatrix}0&1\\1&0\end{smallmatrix}\right)$ is the intersection form for $S^2 \times S^2$. In 1973 A. Casson, in an effort to "surger" K, showed how to represent each of the three $\left(\begin{smallmatrix}0&1\\1&0\end{smallmatrix}\right)$ part of the homology of K by so-called embedded Casson (flexible, kinky) handles. These take the form of disjointly embedded DIFF manifolds $CH_i, i = 1, 2, 3$, each of the proper homotopy type of $S^2 \times S^2 - B^4$ and such that each CH_i contained one of the $\left(\begin{smallmatrix}0&1\\1&0\end{smallmatrix}\right)$ part of the homology of K. If each CH_i were in fact diffeomorphic to $S^2 \times S^2 - B^4$, we could remove the CH_i, leaving a manifold with S^3 boundaries, and then cap off each S^3 by a B^4 to obtain a DIFF manifold whose intersection pairing is the $E_8 \oplus E_8$ form. By a "dual" construction of Casson, each CH_i embeds in the standard $S^2 \times S^2$, in fact, $CH_i = S^2 \times S^2 - X_i$, where X_i is a compactum which is a suspension of an iterated Whitehead continuum. R. D. Edwards then observed that in fact CH_i was DIFF isomorphic to $S^2 \times S^2 - B^4$ if and only if X_i is smoothly cellular in $S^2 \times S^2$; i.e., X_i is the nested intersection of smooth 4-balls in $S^2 \times S^2$. The crux of Freedman's proof is that indeed each CH_i is (TOP) *homeomorphic* to $S^2 \times S^2 - B^4$. Thus one can topologically do the surgery and hence get a TOP manifold whose intersection form is $E_8 \oplus E_8$. Freedman proves that CH_i is standard by first reimbedding the CH_i with delicate geometric control so that the "frontier" of CH_i is geometrically small. Armed with the classical decomposition theoretic fact that, although $\mathbf{R}^3$ mod a Whitehead continuum is not a manifold, but crossed with $\mathbf{R}^1$ is indeed a manifold, Freedman artfully constructs a collar of the frontier of the CH_i, using (and developing) powerful techniques from decomposition space theory (and with a little help from R. D. Edwards, our premier shrinker).

(Freedman's original theorem had a further hypothesis that M with a point deleted admits a DIFF structure. However, F. Quinn [17], during the summer of 1982, showed that any non-compact TOP 4-manifold admits a DIFF structure, along with many other important facts, such as the four-dimensional annulus conjecture!)

The proof of Freedman's theorem itself blends two historically independent schools of topology—the surgery and decomposition space schools. The proof utilizes the most powerful techniques and results from each school. (See the *Freedman's Proof* box.)

DIFF 4-Manifolds

After such a complete understanding of compact simply connected TOP 4-manifolds, one is amazed at how little was still known about DIFF 4-manifolds in the *immediate* post-Freedman era. There were no new DIFF 4-manifolds (although there were some new candidates) or no old candidates that were eliminated. The earliest indication that DIFF 4-manifolds are peculiar is a result of Rochlin in 1952 [20].

ROCHLIN'S THEOREM.   If a simply-connected DIFF 4-manifold has an even intersection form I, then $\sigma(I)$ is divisible by 16.

Recall that the algebraic restriction on such an I is that $\sigma(I)$ be divisible by 8, so the topology of a DIFF manifold restricts the possible intersection pairings. Now Freedman's theorem guarantees the existence of a compact simply-connected TOP 4-manifold M_{E_8} with $I_M = E_8$ and E_8 is even and $\sigma(E_8) = 8$, so that M_{E_8} does not admit a DIFF structure!

Until very recently, Rochlin's theorem and related signature invariants were the only tools available to study DIFF 4-manifolds. (It's amazing we got so far with so little!) After a blow to the head from Freedman's work, rumors were afloat at the end of 1981 that the mathematical physicists were able to detect a new obstruction to the smoothability of 4-manifolds. Then during the summer of 1982 at the first year of the AMS Summer Research Conference Series in New Hampshire (an event planned well before Freedman's work), 4-manifolds topologists were dealt a blow from the forgotten roots of our subject—we were treated to a strong dose of geometry and analysis to explain the remarkable theorem of Simon Donaldson [4], a graduate student at Oxford.

DONALDSON'S THEOREM.   Suppose M is a compact simply-connected DIFF 4-manifold with positive definite intersection pairing I. Then I is equivalent over the integers to the standard diagonal form $\mathrm{diag}(1, 1, \ldots, 1)$.

In particular, $E_8 \oplus E_8$ is *not* diagonalizable over the integers, so that Freedman's TOP 4-manifold $M_{E_8 \oplus E_8}$ does not admit a DIFF structure—a fact that cannot be detected by characteristic classes!

The proof of Donaldson's theorem is a tight combination of topology, differential geometry, and analysis—the non-discrete ingredient that topologists have been missing. (See the *Riemannian 4-manifold* box.)

The existence of $\mathfrak{R}^4$ is now proved indirectly. Freedman provides a topological construction of $M_{E_8 \oplus E_8}$ from the Kummer surface K. It was noticed that, if $\mathbf{R}^4$ has a unique differentiable structure, then this construction can be carried out differentiably to produce a differentiable manifold. Since Donaldson showed such a manifold *cannot* exist, we must conclude that $\mathbf{R}^4$ does *not* have a unique differentiable structure. (See the *construction of $\mathfrak{R}^4$* box for more details.)

The subject of TOP and DIFF 4-manifolds is alive. There is the problem of characterizing non-simply-connected TOP 4-manifolds (Freedman has made some recent progress). The entire subject of DIFF 4-manifolds is wide open—I hope wide open enough to assimilate the new techniques provided by analysis and geometry. We still have left unsettled the smooth 4-dimensional Poincaré conjecture (does S^4 admit exotic DIFF structures?). What is a reasonable characterization of simply-connected DIFF 4-manifolds? Perhaps all that exist are the ones we already know, namely (up to orientation) connected sums of S^4, CP^2, and algebraic surfaces. As a starting point, can we realize the intersection form

$$E_8 \oplus E_8 \oplus 2 \begin{pmatrix} 0 & 1 \\ 1 & 0 \end{pmatrix}$$

by a DIFF 4-manifold?

We have indicated only a flavor of the recent developments in 4-manifold topology. For further, deeper, and more complete information concerning Freedman's theorem, I recommend the reading sequence [3, 21, 7, 22]. For background material relevant to Donaldson's theorem, I recommend the reading sequence [19, 16, 2, 5, 6].

Riemannian 4-Manifolds: The Proof of Donaldson's Theorem

Although the statement of Donaldson's theorem is topological in nature (i.e., a statement topologists can understand), its proof is anything but topological in nature (i.e., a proof a topologist cannot easily digest). Differential geometers have long been aware that four-dimensional space does have some remarkable properties which distinguishes it from spaces of other dimensions. Perhaps at the cornerstone is that the rotation group $SO(n)$ is a simple Lie group for all $n \neq 4$ and that $SO(4)$ is locally isomorphic (in fact double-covers) $SO(3) \times SO(3)$; i.e., $SPIN(4) = SU(2) \times SU(2)$. In terms of Lie algebras this decomposition can be given as follows. The six-dimensional space $\Lambda^2(\mathbf{R}^4)$ of 2-forms on the inner product space $\mathbf{R}^4$ can be viewed as the Lie algebra of $SO(4)$. As $SO(4) = SO(3) \times SO(3)$, $\Lambda^2(\mathbf{R}^4)$ decomposes as the sum of three-dimensional spaces $\Lambda^2(\mathbf{R}^4) = \Lambda^2_+ \oplus \Lambda^2_-$. An alternate description of this decomposition is given in terms of the Hodge star operator $* : \Lambda^2(\mathbf{R}^4) \to \Lambda^2(\mathbf{R}^4)$. If $(e_1, \ldots, e_4)$ is an oriented orthonormal basis for $\mathbf{R}^4$, then $*(e_i \wedge e_j) = e_k \wedge e_l$, where (i, j, k, l) is an even permutation of $(1, 2, 3, 4)$. As $(*)^2 = 1$, $\Lambda^2(\mathbf{R}^4)$ decomposes as the ± 1 eigenspaces $\Lambda^2_\pm$ of $*$. Thus, if M admits a Riemannian metric, $\Lambda^2(T^*M) = \Lambda^2_+(M) \oplus \Lambda^2_-(M)$, and this decomposition is an invariant of the conformal class of the metric on M. An element of $\Lambda^2_+(M)(\Lambda^2_-(M))$ is called a *self dual* (*antiself dual*) 2-form. This decomposition is significant in differential geometry because curvature is a 2-form (with values in a Lie algebra) so that the curvature decomposes into its self dual and antiself dual components. So, given a principal $SU(2)$ bundle ξ over a Riemannian 4-manifold M with a connection ∇, the curvature R^∇ of ∇ is an element of $\Lambda^2(T^*M) \otimes \mathfrak{g}$ [(where $\mathfrak{g}$ is a $SU(2)$ bundle associated to ξ, the *adjoint bundle* of ξ) and $R^\nabla = R^\nabla_+ + R^\nabla_-$. If $R^\nabla_-(R^\nabla_+)$ vanishes, then ∇ is called a *self dual* (*antiself dual*) connection; i.e., $*R^\nabla = R^\nabla(-R^\nabla)$. The existence of such self dual connections has garnered much interest in recent years, as the differential equation

$$*R^\nabla = R^\nabla$$

is the self dual Yang–Mills (YM) equation—a non-abelian [as the group is $SU(2)$, not $SO(2)$] version of Maxwell's equation. (A connection whose curvature satisfies YM is sometimes called an *instanton*.)

Let's consider a specific example. Suppose ξ is the principal SU(2) bundle over S^4 with (real) first Pontrjagin class $p_1(\xi) = 4$. Does ξ have *any* self dual connections, and if so how many; i.e., what is the moduli space $\mathcal{M}$ of solutions to YM? This problem has been completely solved, and in [1] it is shown that $\mathcal{M}$ is an open 5-manifold with a collared copy of S^4 itself near infinity. In fact, $\mathcal{M}$ is the five-dimensional hyperbolic space SO(5, 1)/0(5), the open 5-ball!

What about other manifolds? We now come to the outline of the proof of Donaldson's theorem. Suppose M is a manifold with ξ the principal SU(2) bundle on M with $p_1(\xi) = 4$. C. H. Taubes [23] has shown that, if M has a positive definite intersection form, then there is an (irreducible) solution to YM; i.e., ξ admits a self dual connection that does not reduce to a SO(2) connection. The Atiyah–Singer index theorem and some deformation theory of self dual forms guarantees that $\mathcal{M}$ is a 5-manifold, except there may be a closed set $C \subset \mathcal{M}$ such that in a neighborhood of any point of C, $\mathcal{M}$ is modeled as the zero set of some real analytic map $f: \mathbf{R}^{p+5} \to \mathbf{R}^p$. Furthermore, there are inherent point singularities in $\mathcal{M}$ whose neighborhoods are cones in $\mathbf{CP}^2$ one for each "reducible" self dual connection in ξ, i.e., one for every reduction of the group of ξ to SO(2), i.e., one for every two solutions of $I_M(\alpha, \alpha) = 1$.

Utilizing the work of K. Uhlenbeck and the nature of the construction of the self dual connections by Taubes, Taubes and Uhlenbeck, and (independently) Donaldson have shown that the ends of $\mathcal{M}$ contain a collared copy of M itself. Furthermore, if $\pi_1(M) = 0$ [or no non-trivial representations of $\pi_1(M)$ in SU(2)], there is only one end to $\mathcal{M}$. Suppose $\mathcal{M}$ were indeed a manifold off the inherent singular points so that there would be an orientable DIFF cobordism from M to a disjoint union of $\mathbf{CP}^2$'s. *Again* utilizing the assumption that the intersection pairing I_M on M is positive definite, the Gram–Schmidt process allows the conclusion that I_M is equivalent to the intersection pairing of the other end of the cobordism, namely, the diagonal form diag(1, 1, . . ., 1). The bulk of Donaldson's proof is to show that, although $\mathcal{M}$ may not be a manifold off the inherent singular points, it can be perturbed (rel its end) in the space of connections to be a manifold. (Uhlenbeck [6] has recently given an argument where she perturbs the metric so that $\mathcal{M}$ itself is a manifold off the inherent singular points. In short, some new information is obtained about the topology of M by a study of the space of solutions of a particular PDE, namely, YM. Poincaré would be pleased.

The Construction of $\mathcal{R}^4$

As we pointed out above, $M_{E_8 \oplus E_8}$ exists as a TOP but not DIFF 4-manifold. In the previous section we showed that $M_{E_8 \oplus E_8}$ was obtained by "surgering" the Kummer surface K, that is, by showing that each CH'_i, $i = 1, 2, 3$, was homeomorphic to $S^2 \times S^2 - B^4$. Since $M_{E_8 \oplus E_8}$ does not exist as a DIFF manifold, one of these CH_i must not be DIFF isomorphic to $S^2 \times S^2 - B^4$, say $CH = CH_1$. As R. D. Edwards pointed out, $CH = S^2 \times S^2 - X$ is DIFF (TOP) isomorphic to $S^2 \times S^2 - B^4$ if and only if X is smoothly (topologically) cellular. Thus X is the nested intersection of topological, but not smooth, 4-balls. Thus the interior of one of these topological 4-balls with the DIFF structure inherited from $S^2 \times S^2$ as an open subset is certainly homeomorphic but not DIFF isomorphic to $\mathbf{R}^4$. This is our $\mathcal{R}^4$! Note that this $\mathcal{R}^4$ cannot smoothly embed in S^4, for if it did one could perform the DIFF surgery that is not allowed. In fact, by a more judicious construction $\mathcal{R}^4$ does not smoothly embed in S^4, $\mathbf{CP}^2$ or any positive definite manifold. (A second $\mathcal{R}^4$, denoted $\mathcal{R}_1^4$, was constructed by Gompf [8] by a similar technique, however, $\mathcal{R}_1^4$ is distinct in that it embeds in $\mathbf{CP}^2$ (but not $-\mathbf{CP}^2$), so $-\mathcal{R}_1^4$ is yet a third exotic $\mathbf{R}^4$.)

References

1. M. Atiyah, N. Hitchin and I. Singer (1978) Self duality in four-dimensional Riemannian geometry. *Proc. R. Soc. London Ser. A 362*, 425–461.

2. J. P. Bourguignon and H. B. Lawson, Jr. (1982) Yang-Mills theory: Its physical origins and differential geometric aspects. In *Seminar on Differential Geometry*, S. T. Yau, ed., Annals of Mathematics Studies 102. Princeton University Press, Princeton, N.J., pp. 395–422.

3. A. Casson (1980) Three lectures on new infinite constructions in 4-dimensional manifolds. Notes prepared by L. Guillov, Prepublications Orsay 81T06.

4. S. K. Donaldson (1983). An application of gauge theory to four dimensional topology. *J. Diff. Geom. 18*, 279–315.

5. T. Eguchi, P. Gilkey and A. Hanson (1980) Gravitation, gauge theories and differential geometry. *Phys. Rep. 66*, 213–393.

6. D. Freed, M. Freedman and K. K. Uhlenbeck (1982) Gauge theories and 4-manifolds, MSRI Berkeley Preprint.

7. M. Freedman (1983) The topology of four dimensional manifolds. *J. Diff. Geom. 17*, 357–454.

8. R. Gompf (1983) Three exotic $\mathbf{R}^4$'s and other anomalies. *J. Diff. Geom. 18*, 317–328.

9. M. Hirsch and B. Mazur (1974) *Smoothings of Piecewise Linear Manifolds*, Annals of Mathematics Studies 80. Princeton University Press, Princeton, N.J.

10. B. Kerékjártó (1923) *Vorlesungen über Topologie*, Vol. I, *Flachen Topologie*, Springer, New York.

11. R. Kirby and L. Siebenmann (1977) *Foundational Essays on Topological Manifolds, Smoothings, and Triangulations*, Annals of Mathematics Studies 88. Princeton University Press, Princeton, N.J.

12. J. Milnor (1956) On manifolds homeomorphic to the 7-sphere. *Ann. Math. 64*, 399–405.

13. J. Milnor (1958) On simply connected 4-manifolds. Symposium International Topologia Algebraica, Mexico, pp. 122–128.

14. J. Milnor and D. Husemoller (1973) *Symmetric Bilinear Forms*. Springer-Verlag, New York.

15. E. Moise (1952) Affine structures on 3-manifolds. *Ann. Math. 56*, 96–114.

16. T. Parker (1982) Gauge theories on four dimensional Riemannian manifolds. *Commun. Math. Phys. 85*, 1–40.

17. F. Quinn (1982) Ends III. *J. Diff. Geom. 17*, 503–521.

18. T. Rado (1925) Über den Begriff der Riemannschen Fläche. *Acta Litt. Sci. Univ. Szeged 2*, 101–121.

19. J. H. Rawnsley (1981) *Differential Geometry of Instantons*. Communication of Dublin Institute for Advanced Studies, Series A (Theoretical Physics), No. 25.

20. V. A. Rochlin (1974) New results in the theory of 4-dimensional manifolds (Russian). *Dokl. Akad. Nauk. SSSR 84*, 221–224.

21. L. Siebenmann (1978–1979) Amorces de la chirurgie en dimension quatre: Un $S^3 \times \mathbf{R}$ exotique (d'après A. Casson and M. Freedman). Sem. Bourbaki, No. 536.

22. L. Siebenmann (1981–1982) La conjecture de Poincaré topologique en dimension 4 (d'après M. Freedman). Sem. Bourbaki, No. 588.

23. C. H. Taubes (1982) Self-dual Yang–Mills connections on non-self-dual 4-manifolds. *J. Diff. Geom. 17*, 139–170.

24. J. H. C. Whitehead (1940) On C^1 complexes. *Ann. Math. 41*, 809–832.

30

The Computer-Aided Discovery of New Embedded Minimal Surfaces

David Hoffman

Dedicated to the memory of Karel De Leeuw and Gene Frankel

In 1984, Bill Meeks and I established the existence of an infinite family of complete embedded minimal surfaces in $\mathbf{R}^3$. For each $k > 0$, there exists an example which is homeomorphic to a surface of genus k from which three points have been removed. Figure 30-1 is a picture of the genus-one example. The equations for this remarkable surface were established by Celsoe Costa in his thesis, but they were so complex that the underlying geometry was obscured. We used the computer to numerically approximate the surface and then construct an image of it. This gave us the clues to its essential properties which we then established mathematically. The programming expertise of James T. Hoffman, who is mainly responsible for the quality of the illustrations here, was a central ingredient in our research use of computer graphics. Without the use of a new programming environment, of which he is the principal creator, we would not have made the discoveries I will attempt to describe.

A Brief History of Minimal Surfaces

In order to explain the significance of these new minimal surfaces, I will sketch the history of minimal surface theory by giving six equivalent definitions of a minimal surface, in their approximate historical order. We certainly don't need more than one or two of them to discuss the new example, but they give us a convenient structure in which to describe the history of the subject.

DEFINITION I.   Each point on the surface has a neighborhood which is the surface of least area with respect to its boundary.

A minimal surface in $\mathbf{R}^3$ is the two-dimensional generalization of a straight line in the plane. Any piece of a straight line is the unique curve of minimum length spanning (i.e., connecting) its boundary points. However, while all straight lines in the plane are essentially the same, the collection of all minimal surfaces is very rich and not yet completely understood.

The flat plane is the simplest and oldest known minimal surface. What are some others?

In the 1740s Euler solved the following problem. Consider a surface of revolution defined by a profile curve $y = f(x)$, $a < x < b$, and let $A(f)$ be the area of the surface. Which functions f are

Volume 9, No. 3 (Summer 1987), 8–21

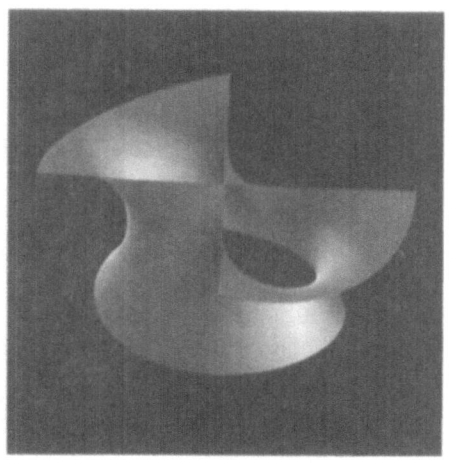

FIGURE 30-1a *The genus-one example: a complete embedded minimal surface, conformally the square torus from which three points have been removed. The surface pictured here has been cut off by a round sphere.*

FIGURE 30-1b *The half of the surface in Figure 30-1a which lies below the (x_1, x_2)-plane.*

critical in the sense that the first derivative of A with respect to f is zero? The answer turns out to be that the graph of $f(x)$ is essentially a catenary:

$$f(x) = \cosh(x) \tag{1}$$

The surface of revolution is called a catenoid. (See Figure 30-2.)

DEFINITION II.   If the surface is written locally as the graph of a real-valued function $z = f(x, y)$, then f satisfies the following (nonlinear elliptic) partial differential equation:

$$f_{xx}(1 + f_y^2) - 2f_{xy}f_xf_y + f_{yy}(1 + f_x^2) = 0. \tag{2}$$

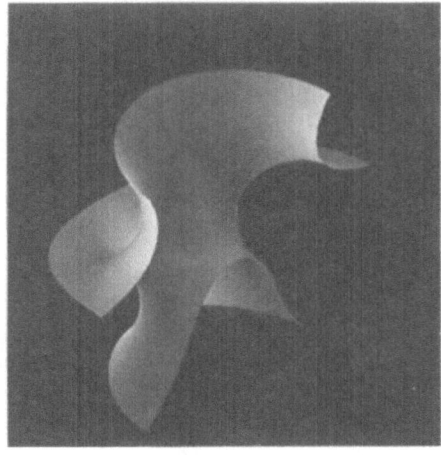

FIGURE 30-1c *The half of the surface in Figure 30-1a which lies in the half-space $x_2 \geq 0$.*

FIGURE 30-2 *The catenoid, the only non-planar minimal surface of revolution.*

This is in fact the Euler-Lagrange equation for the area functional of a graph. The problem of minimizing area is one of the oldest problems in the Calculus of Variations.

In the 1770s, Meusnier gave the following geometric interpretation of equation (2).

DEFINITION III.   The mean curvature H of the surface vanishes identically:

$$H = \frac{1}{2}(k_1 + k_2) = 0. \tag{3}$$

(See Figure 30-3 for an explanation.)

Meusnier also showed that the helicoid which is the image of the mapping

$$X(t, \tau) = \big(t \cos(\tau), t \sin(\tau), \tau\big), (t, \tau) \in \mathbf{R}^2$$

is a minimal surface. (See Figure 30-4.) Except for the flat plane, it is the only ruled minimal surface.

The minimal surface equation (3) was difficult to solve in the eighteenth century. Notice that the helicoid and catenoid both have a one-parameter group of symmetries. In fact these surfaces were discovered by assuming symmetry, thereby reducing the partial differential equation (2) to an ordinary differential equation. The minimal surface equation was so difficult to solve using the techniques available at the end of the eighteenth century that no other minimal surface was discovered, to my knowledge, until 1835 when Scherk found many new examples, including a one-parameter family of isometric non-congruent minimal surfaces containing the helicoid and catenoid. This work

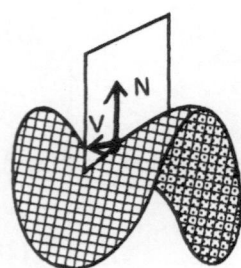

Slice the surface by a vertical plane containing the normal $\vec{N}$ at point p and a tangent vector V to the surface.

The resulting plane curve of intersection has a well-defined curvature κ at p. This is the *normal curvature at p in the direction V,* denoted by κ_V^N

$$\kappa_V^N = \kappa$$

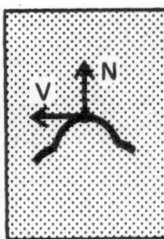

κ_V^N is well defined for any nonzero V and only depends on $V/|V|$. Its extreme values $k_1 \geq k_2$ are the principal curvatures which are assumed in orthogonal directions V_1, V_2.

FIGURE 30-3 *Normal curvature.*

FIGURE 30-4 *The helicoid, discovered by Meusnier in 1776, is the only ruled minimal surface.*

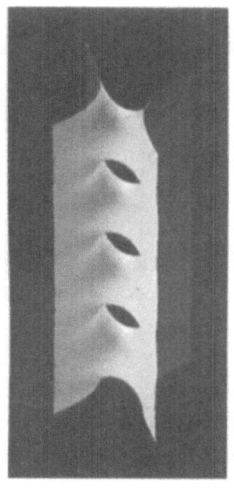

FIGURE 30-5 *A repres-entative piece of Scherk's Second Surface.*

won a prize from the Paris Academy. Scherk's most famous surface can be described as the points in $\mathbf{R}^3$ where

$$\cos(x)e^z - \cos(y) = 0.$$

Another surface he found, locally isometric to the one above, is referred to nowadays as Scherk's Second Surface. It consists of those points in $\mathbf{R}^3$ satisfying

$$\sin(z) - \sinh(x)\sinh(y) = 0.$$

(See Figure 30-5.) Scherk's Second Surface looks like two orthogonal planes whose line of intersection has been replaced by a sequence of tunnels burrowing through in alternating directions. It is a regular surface, free of self-intersections. It should be better known than it is and will figure in our description of new results later on.

While few new solutions to the minimal surface equation were found in the first half of the nineteenth century, some important connections were made that set the stage for future development. The connection between minimal surfaces and harmonic functions is the most important of these.

DEFINITION IV.   If the surface is given by a conformal immersion $X: M \rightarrow \mathbf{R}^3$, then the components of X are harmonic functions.[1]

Some progress was made with this insight in integrating the minimal surface equation by exploiting the connection with analytic functions. This set the stage for what has been described as the first Golden Age of the theory of minimal surfaces, 1850–1880. Much of the work during this period focused on the boundary value problem: find the surface of least area which spans a given contour. This is usually referred to as the Problem of Plateau (after J. F. Plateau, the Belgian experimentalist who did fundamental research on problems involving surface tension and thin films). Soap films, the kind that can be produced by dipping a wire frame into a solution of dishwashing liquid, are stable when their surface tension cannot be reduced by small perturbations of the film. It is a fact (not hard to show) that this means that their surface area is a relative minimum among nearby surfaces with the same boundary. Thus soap films spanning wire contours are actually minimal surfaces in the sense of all our previous definitions and also are really (at least relative) minima in a global sense. We could modify Definition I to read as follows:

DEFINITION V.   (Experimentalist's definition) A sufficiently small piece of the surface may be modelled by the soap film spanning a contour which coincides with the boundary of the piece.

As late as the middle of the last century it was not possible to describe analytically the solution to any boundary value problem for the minimal surface equation unless the boundary curve lay on one of the few explicitly known examples. Then in 1865, H. A. Schwarz announced the solution to the problem of finding the surface of least area whose boundary was a polygon consisting of four edges of a tetrahedron. (For more details see Hildebrandt and Tromba [4].) This solution was also found by Riemann several years earlier but published only after his death. Schwarz, Riemann, Weier-

---

[1]The invariant way of stating this is that the Laplace-Beltrami operator defined by the metric on X annihilates any coordinate projection of X.

strass, and others were able to solve the boundary value problem for a variety of polygonal boundaries. The critical tool in this work was complex function theory.

There is a deep connection between minimal surfaces and analytic functions. Certainly, this follows from Definition IV. But there is another connection that involves the Gauss map. (For a regular oriented surface in $\mathbf{R}^3$, the map associating to a point on the surface its oriented unit-normal, a point in the unit two-sphere, is called the Gauss map.)

DEFINITION VI.   The Gauss map G is anticonformal: Wherever it is non-singular, it preserves the magnitude of angles while reversing their orientation.

It is a fact that any sufficiently smooth surface in $\mathbf{R}^3$ can be parametrized locally by a map from the plane which is regular and angle preserving, i.e., conformal. Thus a regular surface in space has a conformal structure which comes from the induced metric. If these local parameters are denoted by (u_1, u_2), we may introduce the local complex parameter $z = u_1 - iu_2$, making the surface into a Riemann surface. Composing the Gauss map G with stereographic projection σ produces a map, $g = \sigma G$, from the Riemann surface to $\mathbf{C} \cup \{x\}$ which is meromorphic.

Conversely, given a meromorphic function, g, on a simply connected domain in $\mathbf{C}$, it is always the Gauss map of a minimal immersion of the domain into $\mathbf{R}^3$. This minimal surface is not unique. In fact one can specify an arbitrary analytic function, $f \not\equiv 0$, on the domain and the pair will determine a minimal surface up to translation. The details are given in Figure 30-6a. This is the famous Enneper-Weierstrass representation formula, which allows one to write down great numbers of minimal surfaces. It was an equivalent formulation of this representation that was used by Schwarz to solve the boundary value problem. One can write the Schwarz surface as the minimal surface with $f(z) = (1 - 14z^4 - z^8)^{-1/2}$ and $g(z) = z$ on an appropriate domain in the complex plane. Another example is Enneper's surface (Figure 30-7) which can be constructed from the representation by letting $g = z$ and $f = 1$ on the entire complex plane. The representation theorem plays a central role in our construction of the new example.

Let D be a simply connected domain in $\mathbf{C}$, f an analytic function, and g a meromorphic function on D. Then

(6.1a) $X(z) = \operatorname{Re} \int_{p0}^{p} \phi\, dz,$

where

(6.2a) $\phi = ((1 - g^2)f,\ i(1 + g^2)f,\ 2fg)$

is a conformal minimal immersion of D into $\mathbf{R}^3$. The immersion is regular provided the poles of g coincide with the zeros of f, and have at each point exactly half the order of the zeros. The stereographic projection of the Gauss map of X is equal to g.

Let M be a Riemann surface, g a meromorphic function, and η a holomorphic one-form on M. Then

(6.1b) $X(p) = \operatorname{Re} \int_{p0}^{p} \phi$

where

(6.2b) $\phi = (\phi_1, \phi_2, \phi_3) = (1 - g^2,\ i(1 + g^2),\ 2g)\eta,$

is a conformal minimal immersion of M into $\mathbf{R}^3$ provided

(6.3) $\operatorname{Re} \int_{\alpha} \phi_i = 0 \qquad i = 1,2,3$

for any closed curve α on M. It is complete provided

(6.4) $\int_{\beta} |\phi| = \infty$

for any divergent curve β on M.

FIGURE 30-6a *The local Enneper-Weierstrass representation.*

FIGURE 30-6b *The global Enneper-Weierstrass*

Let $M = \overline{M}_k - \{p_1, \ldots, p_r\}$, where $\overline{M}_k$ is a compact surface of genus k and $r > 0$. Suppose g is a meromorphic function on M and η a holomorphic one-form on $\overline{M}_k$ for which (6.3) and (6.4) hold. Then the minimal surface given by (6.1b) has all of its ends separately embedded if and only if

(6.5)   degree of $g = k + r - 1$

and those ends are parallel if and only if, after a rotation,

(6.6) $g(p_i) = \sigma \cdot G(p_i) = 0$ for $i = 1, \ldots, r$.

FIGURE 30-6c *Additional conditions necessary for embeddedness in the finite total curvature case.*

Examples

	M	g	η
Plane	C	0	dz
Enneper's Surface	C	z	dz
Catenoid	$C\backslash\{0\}$	z^{-1}	dz
Helicoid (1 full wind)	$C\backslash[0,\infty)$	iz^{-1}	dz
Scherk's Second Surface (basic piece)	Unit Disk $\backslash\{z\|z^4 = 1\}$	iz	$4(1 - z^4)^{-1}dz$
Trinoid	$C\backslash\{z\|z^3 = 1\}$	z^2	$(z^3 = 1)^{-2}dz$
Genus-one Example	$C/Z \times Z$	$\dfrac{2\sqrt{2\pi}P(1/2)}{P'}$	Pdz
Genus-k Example $k \geqslant 1$	$"w^{k+1} = z^k(z^2 - 1)"$ $\backslash\{(\pm 1,0),(\infty,\infty)\}$	c_k/w	$\left(\dfrac{z}{w}\right)^k dz$

FIGURE 30-6d *Gauss maps g and one-forms η which will produce the surfaces mentioned in this article. They are not unique.*

The Schwarz solution is important to this account for another reason. The solution to the boundary value problem for the quadrilateral may be used as a building block to create a complete minimal surface that is periodic and has a symmetry group containing a three-dimensional lattice of translations. Its fundamental group is infinitely generated. The construction uses a basic observation which has come to be known as the Schwarz Reflection Principle for minimal surfaces: A minimal surface which contains a straight line on its boundary may be analytically extended by reflection across the line. The resulting surface contains the line in its interior. Schwarz also established that a minimal surface with a boundary curve that lies on a plane and meets that plane orthogonally can be extended analytically across the plane by reflection. These

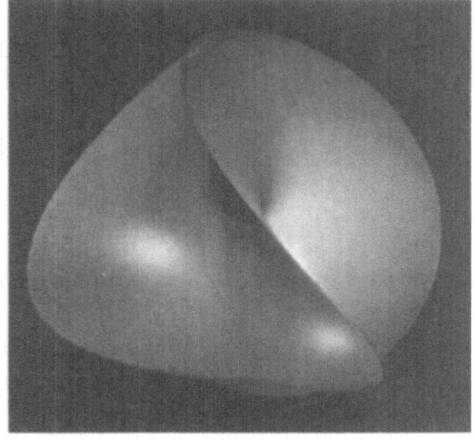

FIGURE 30-7 *Enneper's surface.*

properties are consequences of the Reflection Principle for analytic functions and further illustrate the fundamental connection between minimal surface theory and complex function theory.

Versions of the representation theorem were used by Riemann, Schwarz, and others to build triply periodic complete minimal surfaces out of basic pieces whose boundaries consist of straight lines or plane curves. By arranging the boundaries of the basic piece correctly it could be insured that the resulting minimal surface was **embedded:** that is, free of self-intersections. In recent years, new embedded complete minimal surfaces have been discovered by A. Schoen and by Bill Meeks, all constructed out of basic building blocks. However, this procedure can never produce a complete embedded minimal surface that is homeomorphic to a compact surface with a finite number of points removed. In fact, the only examples of complete embedded minimal surfaces with finitely generated topology that were known before our discovery were the plane, the helicoid, and the catenoid. Topologically, the plane and the helicoid are both once-punctured spheres and the catenoid is a twice-punctured sphere.

To complete this historical survey it is necessary to weave in one more thread: Osserman's work on the global properties of complete minimal surfaces. It began with his work generalizing the Bernstein Theorem. The original theorem of Bernstein stated that equation (2) has no solutions valid in the whole plane except for the trivial ones, where $f(x, y)$ is a linear function $ax + by + c$. An entire graph is a complete surface with the property that its Gauss map takes values in a hemisphere. Osserman proved: If the Gauss map of a complete minimal surface omits an open set, the surface must be a plane.[2] He then went on to develop a general theory of complete minimal surfaces in which an essential tool was the Enneper-Weierstrass representation formula. He recognized that the Enneper-Weierstrass representation was valid for Gauss maps, g, and holomorphic one-forms η defined on Riemann surfaces. It was necessary to make sure that the integrals involved (see Figures 30-6a and 30-6b) were well defined on the given Riemann surface; that is, all the periods were purely imaginary. For completeness, it was required to check that the integrands had certain singular behavior at the boundary points of the Riemann surface, and for regularity that η vanished precisely at the poles of g with twice the order. This is not the proper place to go into careful mathematical details. (For that, see [2], [11], or [14]). For our purposes it is important to note that this is, in principle, an analytic method for construction of complete minimal surfaces with a prescribed finite topology. Another product of Osserman's general investigations was a much clearer understanding of complete minimal surfaces of finite total curvature. The Gauss curvature K is the product of the principal curvatures, so by Definition II, $K \leq 0$ on a minimal surface. The total curvature is $C(S) = \int K dA$. Since K is also the determinant of the differential of the Gauss map, the area of the Gaussian image is equal to the total curvature.

If the total curvature is finite, the differential of the Gauss map, which is meromorphic, must go to zero in some controlled way as one goes off to the boundary of the underlying Riemann surface. I hope that these sketchy details are enough to motivate the following characterization due to Osserman. Let S be a complete minimal surface in $\mathbf{R}^3$. Then S has finite total curvature if and only if

1. S is conformally equivalent to a compact Riemann surface M of genus k with $r > 0$ points removed, the immersion of M is proper, and
2. the Gauss map of S, a meromorphic map, extends meromorphically to M.

In particular, if a complete minimal surface has finite total curvature, it has finite topology. It also follows from (2) that the total curvature is an integer multiple of 4π.

Osserman went on to show that $C(S)$, the total curvature of S, satisfies

$$C(S) \leq -4\pi(k + r - 1), \tag{4}$$

which is a sharpening of the Cohn-Vossen estimate for the total curvature of a complete surface.

Embedded Minimal Surfaces

In view of these results it is natural to replace the requirement of finite topology with the (stronger) assumption that the total curvature is finite. The short list of classical embedded examples of complete minimal surfaces (plane, catenoid, helicoid) have total curvatures equal to 0, -4π, and $-\infty$, respectively. In the category of finite total curvature we therefore have only the plane and the

---

[2]For a non-technical survey of subsequent results, see [3]. For more, see [11], or the Appendix to [14].

catenoid, both of genus $k = 0$ with $r = 1$ and $r = 2$, respectively. This is a rather thin class upon which to make conjectures. On the other hand, no new surfaces were showing up. It gradually became a question of increasing interest to classify the complete embedded minimal surfaces of finite total curvature. There are several reasons for this.

First of all, embedded surfaces are more natural; they correspond to our primitive notion of a surface, the boundary of a solid object. They are also useful. For example, the recent work of Meeks and Yau has used embedded solutions to the Plateau problem to solve some hard problems in topology (their work on the Dehn Lemma and the Smith Conjecture, for example). Moreover, some recent results indicated that there were some real obstructions to the existence of these sorts of surfaces. Rick Schoen proved in

genus \ #ends	0	1	2	3	4	5	6	7
0	X	P	Ca	X	X	X	?	?...
1	X	X	X	?	?	?	?	?...
2	X	X	X	?	.	.		
3	X	X	X	?	.	.		
4	X	X	X	?	.	.		
5	X	X	X	?	.	.		

X   Denotes a topology which cannot support an example.
P = The flat plane.
Ca = The catenoid.
? = No known example.

The examples described in this article remove the question marks in the 3-ends column.

FIGURE 30-8 *A scorecard of known complete embedded minimal surfaces of finite total curvature as of April 1984.*

1982 that the catenoid was the unique complete embedded minimal surface with finite total curvature which was homeomorphic to a compact surface with two points removed [15]. There is no restriction on the genus. It was already known that the plane was the unique embedded example with one end. A little earlier, Meeks and Luquesio Jorge (in Fortaleza, Brazil) had proved that it was impossible to find a complete embedded minimal surface which had finite total curvature and was also homeomorphic to the sphere with 3, 4, or 5 points removed. There had also been some incorrect proofs that the plane and the catenoid were the only examples. These existence and non-existence results, representing the state of things in the spring of 1984, are represented schematically in Figure 30-8.

The Weierstrass-Enneper representation theorem and Osserman's extension of it to Riemann surfaces provide a tool for the construction of complete minimal surfaces of finite total curvature. However, the analytic nature of the construction (look again at Figures 30-6a and 30-6b) leaves one very much in the dark as to the geometry of the surface. If one is interested in constructing an embedded example, the prospect of working from this representation is not very appealing. How can one check for embeddedness?

Several people, including Jorge, F. Gackstatter, and C. C. Chen, had constructed complete minimal surfaces using elliptic functions on Riemann surfaces. No one had a really clear idea what they looked like. Moreover, some recent observations of Jorge and Meeks showed that none of these examples could be embedded. These observations are geometric and also easy to describe.

The Geometry of "Embedded Ends"

Look at the picture of the catenoid (Figure 30-2) and remember that the picture shows only the part of the catenoid inside a relatively small sphere. Imagine you are located in the plane defined by the waist circle and start homeothetically shrinking $\mathbf{R}^3$ while you stay put. What you see after a while looks like two parallel planes that are nearly coincident (Figure 30-9). If you do the same thing for the "trinoid" (see Figure 30-10), you will begin to see three planes, with one line in common, each meeting the other two at an angle of 60 degrees. For Enneper's surface (Figure 30-7) the thought-experiment is a little more difficult: what you see eventually is one triply covered plane. Now recall from the above discussion that a complete minimal surface of finite total curvature is a properly

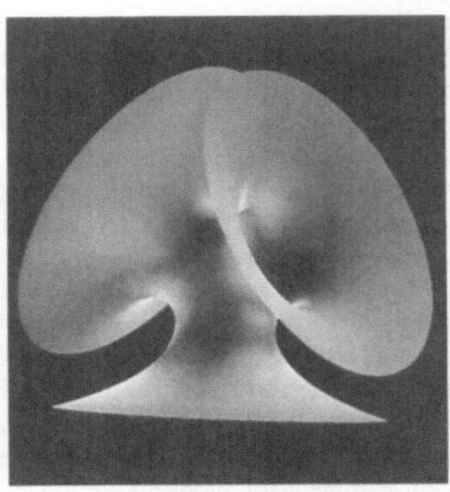

FIGURE 30-9 *The catenoid from far away.*   FIGURE 30-10 *The trinoid.*

immersed surface which is conformally a compact surface from which a finite number of points have been removed. As you travel on the surface toward one of those points, you travel toward infinity (you eventually leave any compact set) in $\mathbf{R}^3$ and the tangent plane to the surface has a well-defined limit. If we call a punctured neighborhood of an omitted point an "end," then the observation of Meeks and Jorge can be stated mathematically as follows. Asymptotically, each end of a complete minimal surface of finite total curvature looks like d coincident copies of a plane (specifically the plane orthogonal to the limiting value of the Gauss map at the end), where d is a positive integer. In the case of the catenoid, there are two ends and the corresponding planes are parallel and each is covered once ($d = 1$ for each end). For the trinoid, there are three ends; the corresponding planes are each covered once ($d = 1$ for each end) but they are not parallel. For Enneper's surface, there is one end and the limiting plane is covered three times ($d = 3$).

Suppose now you are looking at a complete embedded minimal surface of finite total curvature. Since the surface has no intersections, surely each end is itself intersection-free. This forces d to equal 1 at each end. This is the case for the catenoid and the trinoid but not for Enneper's surface. The implication of this observation is (with the help of the Gauss-Bonnet formula) that the contribution to total curvature at each end is as small as possible and the total curvature of the surface is as small (in absolute value) as possible. From (4), we have that the total curvature is equal to $-4\pi(k + r - 1)$. Since the total curvature is precisely -4π times the degree of the Gauss map, we have determined the degree of the meromorphic function g (which is the stereographic projection of the Gauss map) in the Weierstrass representation. Moreover, the ends of an embedded surface do not intersect one another, which means that the limiting tangent planes are parallel. This is the case for the catenoid, but not for the trinoid. In terms of the Gauss map, G, it must be that after (if necessary) a rotation of the surface, $G = (0, 0, \pm 1)$ at each of the ends. Therefore we would expect to see $g = \sigma G = 0$ or ∞ at each end. These conditions are all conditions on the meromorphic function g in the Enneper-Weierstrass representation and are written as such in Figure 30-6c.

All of the known constructions of complete minimal surfaces violated one or the other of these conditions on the Gauss map. All, that is, except one.

The Discovery of the First New Embedded Example

In the fall of 1983, I started to think seriously about using computer graphics to do research in differential geometry. I was interested in using it to work on several problems involving surfaces of constant mean curvature. Also, I had become intrigued by the various sorts of research being done in what is known as computer vision. Through the good graces of the VISIONS research group at the University of Massachusetts, I was given access to modern computing equipment. Setting up the sort of graphics software that I needed proved to be much harder than I had anticipated. Standard packages, usually designed for engineers, could not handle even some of the simplest tasks that I anticipated. After several false starts which consumed many months of time, I had the extreme good fortune to meet Jim Hoffman, a superb graphics programmer who thinks of the computer as an instrument for the creation of beautiful things. He also had a nascent interest in mathematics as well as some ideas, more or less already worked out, about an interactive programming system that could serve to coordinate the computations and the graphics that I wanted to explore. It is called VPL for visual programming language.

James T. Hoffman

At about this time, I found out about an example that met all the known necessary criteria for embeddedness. Bob Osserman had heard from Luquesio Jorge about the unpublished thesis of Celsoe Costa at IMPA in Rio de Janeiro. During a phone conversation, Osserman told me of the existence of this example in the context of a discussion of possible graphics projects. (It is still the case that the first graphics project that I intended to do remains undone to this day.) His thought was that, since one already knew that this example had embedded parallel ends, it might only be necessary to check a compact piece of the surface for self-intersections. I obtained a copy of the thesis and was immediately absorbed by the possibility of seeing the example. Of course, there had to be some estimate of how large a compact set needed to be looked at. Moreover, one thing had to be verified first mathematically. Even though each end was separately embedded, and the ends were asymptotically parallel, it could well happen that they nonetheless intersected one another. It was necessary to show that this did not happen. Fortunately, this was easy to do for this example. Still, the best one could hope for at that time is reflected in the following sentences that I wrote in November 1983.

> There are very few known examples of complete embedded minimal surfaces of finite total curvature and none of such a surface with genus greater than zero. Recently, an example has been found of a complete surface of finite total curvature which has (the) Riemann surface structure of a torus with three points removed. . . . However, since the surface is known only through its Weierstrass representation, it is difficult to tell whether or not it is embedded. It should be possible to look at a large enough piece of the surface using computer graphics and "see" whether or not this is the case: since it is well behaved at the ends this would give strong evidence about self-intersection. If it appeared to be embedded, then a serious effort would be made to prove that it was in fact so. If self-intersections appeared, it would be possible to locate them with good precision on the torus domain and then prove easily that there was a self-intersection. In either case, how this surface looks in space is not known and would be of interest.

Costa's Construction

Let me describe what Costa had done in his thesis. Consider the torus produced by identifying the opposite sides of the unit square. Think of this torus as $\mathbf{C}$ modulo the integer lattice; it is a Riemann surface of genus one with a highly symmetric conformal structure. Remove the three points corresponding to 0, 1/2, and $i/2$. These will be the ends at infinity. For the Gauss map, take g to be c/P'

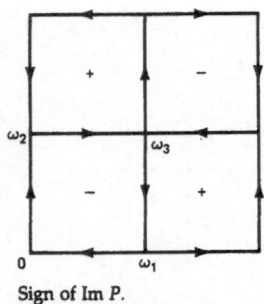
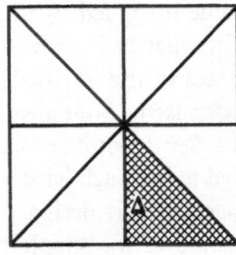
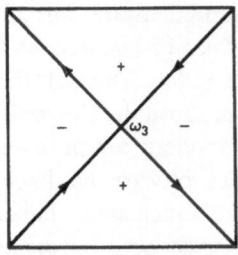

Sign of Im P.
P real on lines indicated.
Arrows in direction of
increasing Re P.

Eight fundamental triangles
corresponding to congruent
pieces of the surface.

Sign of Re P.
P imaginary on lines indicated.
Arrows in direction of
"increasing" Im P.

FIGURE 30-11 *The Weierstrass P-function on the unit square.*

where c is a real constant and P' is the derivative of the Weierstrass P-function. (See Figure 30-11.) For the auxiliary one-form, let $f dz = P dz$. Costa showed that the conformal minimal immersion X produced by the Enneper-Weierstrass representation is well defined on T for precisely one value of c, and the resulting minimal surface is complete. Refer now to Figures 30-6b and 30-6c. Since P' has degree three, the total curvature of $X(t)$ is -12π, exactly what is needed for the surface to have each of its ends separately embedded. Also, P' has either a zero or a pole at each of the three omitted points. Therefore the Gauss map, G, equal to c/P' composed with inverse stereographic projection, is vertical at these points; the ends are parallel. A simple calculation shows that the end corresponding to the point ω_0 is asymptotic to a plane $x_3 = $ constant, while the ends corresponding to ω_1 and ω_2 grow logarithmically in the x_3 direction with $x_3 \to \mp\infty$ respectively. In particular, they do not intersect one another.

The idea of having one "flat" end between two "catenoid" ends occurred to Costa in an unusual way. He was having trouble achieving the minimum total curvature in an example, and was assuming that all of the ends should grow like catenoids. This forces some analytic conditions that he could not meet. Then, at the movies at Rio, watching a documentary about the preparations of some "Samba Schools" for the dance competitions at Carnival time, he saw a dancer with an outlandish headdress that was made to look like two crows—one head up, the other head down—with their wings meeting in an expanding circle in the middle. This gave him the idea to try to create a plane-like end between the catenoid ends. This ultimately worked out to the example we are now discussing. As it turns out, the surface itself does vaguely resemble, according to Costa, the outlandish crow chapeau of an exuberant Carioca at Carnival. See Seymour Papert's *Mindstorms* [18] for another quite different reference to Samba Schools and mathematics.[3]

It Is Embedded!

"I want to reach that state of condensations which constitutes a picture."

Henri Matisse, quoted in *The Visual Display of Quantitative Information*, by E. R. Tufte

[3]Papert proposed the "Samba School" as a model for a new educational environment, where computers are not alienating machines, but are the technological means to create an integrated social context for learning. Thinking about problems and programs is likened to learning new dance—techniques are passed, like dance steps, among participants. The computer is experienced as an extension of the body, and the activity takes on a tactile, kinetic quality.

Computing the coordinates of the surface took a lot longer than expected. At the same time as I was writing the program to do this, I was digging into Abramowitz and Stegun to try to understand a bit more about the ways of computing P' and P. (Hauling the book back and forth between my office and home may be partly responsible for the serious back trouble I experienced the following summer.) Some simple reductions allowed me to eliminate P' from the formulae. Then I came face-to-face with something that at first shocked me, but is a well-known fact of life among people who do serious numerical simulations: Library routines cannot be trusted. Things proceeded smoothly once I got a reliable way to compute P.

William H. Meeks III.

In February of 1984, I heard that Bill Meeks was at the Institute for Advanced Study. I invited him to come to the University of Massachusetts at Amherst and we spent many long enjoyable hours talking about this problem and various other ways to construct an embedded example. Bill was excited by the idea of using computer graphics to attack this problem.

In early March, I was able to compute accurately the coordinates of the surface. Jim's software was not quite in place, so I printed out some values on the surface in tabular form. I was interested to know the spatial values of the surface on the diagonal lines where P is imaginary. (Figure 30-11 again.) It was easy to calculate that $x_3 = 0$ on the image of these curves (that is, under the convention of beginning the integration at $\omega_3 = (1 + i)/2$). What a surprise, in fact a shock, it was to find that these curves were mapped into the straight lines $x_1 = \pm x_2$ in the plane $x_3 = 0$ and that the lines met at right angles. I repeated this computation several times before I allowed myself to believe it. Then I went back to the equations, and much to my surprise it was extremely easy to verify this fact mathematically. It was really the case.

The symmetry of the square is reflected in the symmetry of the P-function. Now it seemed that it might be the case that the surface had the same symmetries. It follows from the Schwarz Reflection Principle for minimal surfaces, mentioned above, that is if a minimal surface contains a straight line, it is invariant under rotation by π about that line. At the very least, the example of Costa had rotational symmetries! I called Meeks who was in Chicago at the time to tell him the news. He found it hard to believe at first but after repeating the calculation became quite excited. We started to believe that the surface had to be embedded.

Within a week or so of this discovery we were able to create pictures of the surface. They were imperfect because, to avoid singularities, we at first had to draw the surface on a subsquare of T, creating spurious edges. Also, we had not yet implemented the ability to cut off a square by a sphere and only draw what was inside. However, Jim Hoffman and I could see after one long night of staring at orthogonal projections of the surface

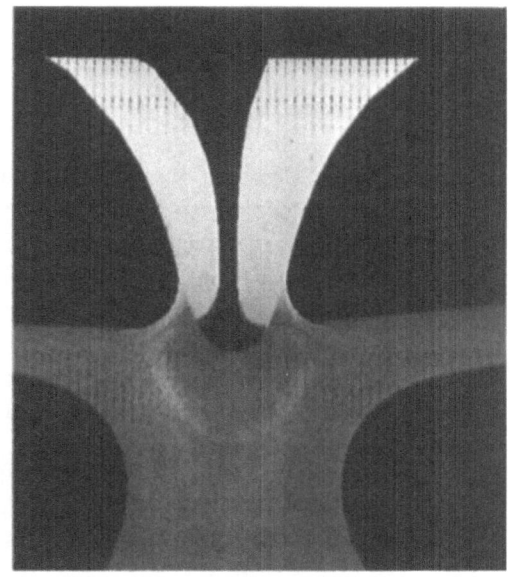

FIGURE 30-12 *One of the original computer images of the genus-one example.*

from a variety of viewpoints that it was free of self-intersections. (See Figure 30-12.) Also it was highly symmetric. This turned out to be the key to getting a proof of embeddedness. Within a week, the way to prove embeddedness via symmetry was worked out. During that time we used computer graphics as a guide to "verify" certain conjectures about the geometry of the surface. We were able to go back and forth between the equations and the images. The pictures were extremely useful as a guide to the analysis.

The Geometry of the Surface

What we were able to prove, with the aid of graphic images, was that the surface in question had the symmetry of the square: Its symmetry group is the dihedral group $D(4)$, a group with eight elements. In terms of the parameter domain T, the group is generated by a rotation ρ by $\pi/2$ about the center point $\omega_3 = (1 + i)/2$ and reflection κ across the horizontal line through ω_3. It contains in particular $\kappa\rho^{-1}$ which is reflection about the positive diagonal through ω_3. The group $D(4)$ moves the triangle Δ in Figure 30-11 into each of the other seven triangles that make up T. The rigid motions of $\mathbf{R}^3$ which correspond to ρ and κ are R = "rotation by $\pi/2$ about the x_3-axis followed by reflection in the (x_1, x_2)-plane" and K = "reflection in the (x_1, x_3)-plane," respectively. The Euclidean motions R and K generate a group with eight elements that moves the positive octant into each of the other octants. It is the symmetry group of the seam on a baseball, "D-2-D" in the parlance of the crystallographers. (See Figure 30-1a and 30-1b, which show the pieces of the genus-one example corresponding to the symmetries R and K.)

Using this information, the surface maybe decomposed into eight congruent pieces, corresponding to the eight triangles which make up T. It can be shown that the boundary of a given triangle Δ is mapped into the boundary of a particular octant Ω. By Definition III of a minimal surface, the components are harmonic, and by an application of the maximum principle, it can be shown that the triangle Δ is mapped into the octant Ω. To prove the surface is embedded is now reduced to the question of whether each piece is embedded. In fact, we were able to show that each piece is a graph over an appropriate plane in $\mathbf{R}^3$. The computer graphics were quite helpful in checking the computation concerning which planes were the correct planes for projection. Proving that each piece is a graph uses, in an essential manner, the fact that the Gaussian image of each triangle is restricted. The proof of embeddedness was completed in this way. (For the complete story, see [8].)

Reaction

At the first public presentations of these results, the reaction was enthusiastic. However, few who looked at the original pictures could see the correspondence between the images and the properties we were claiming for the surface. Polite incredulity is probably an accurate description of the reaction we got. (Compare Figure 30-12 with Figure 30-1!) This was initially surprising to me. Having spent dozens of hours staring at images of this surface as they were being generated on a cathode ray tube and knowing all the little glitches in the pictures that corresponded not to properties of the surface but to our as-yet-imperfect way of making it visible, I had little trouble "seeing" what was there. As the list of powerful mathematicians who admitted to difficulty in understanding the image grew, it became clear that it was necessary to produce more realistic images. In this Jim Hoffman has succeeded wonderfully. This capability has been of great use to our subsequent research.

One of the first public lectures I gave on this discovery was in Berlin, in June of 1984. Herman Karcher was in the audience. From information in the images and descriptions in the lecture, he was able to create a model that evening using typewriter cleaning gum. He showed it to me the next day—presented in a cigar tin! It was accurate enough to help others see the correspondence with the slides that I had shown. Two weeks earlier, I had tried to make a model out of clay borrowed from

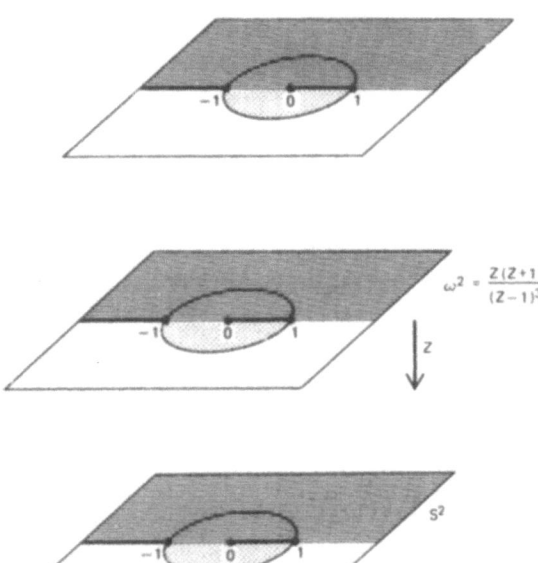

$$\omega^2 = \frac{Z(Z+1)}{(Z-1)^3}$$

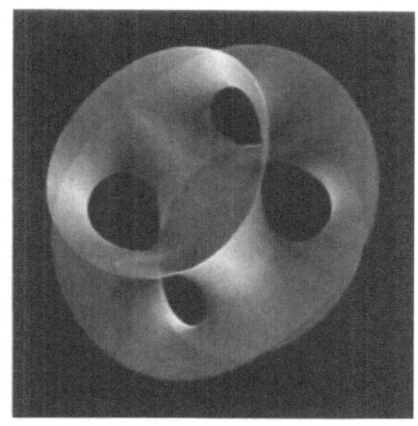

FIGURE 30-14a *The genus-two example.*

FIGURE 30-13 *Schematic representation of the genus-one example as a Riemann surface.*

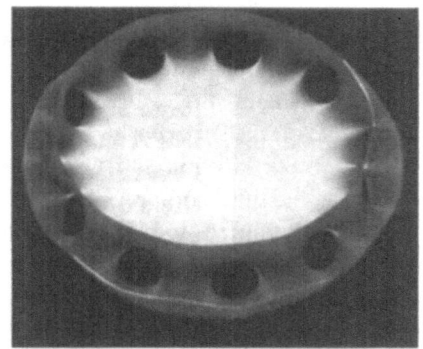

FIGURE 30-14b *The genus-nine example.*

my son's art drawer. The clay in my inept hands proved to be too stiff for the middle part of the surface to stay together when it was moved. It sat, lumpy and sagging, on my desk. I preferred the computer images.

In that talk in Berlin, I discussed the conjecture that Bill and I had made about the existence of higher-genus minimal surfaces with three ends and similar dihedral symmetry. During that summer we proved the existence of a complete embedded minimal surface with three ends of every genus greater than one. Pictures of two of these surfaces, made in the early summer of 1985, appear in Figure 30-14. In the announcement of this discovery, submitted to the *Bulletin of the A.M.S.*, we listed some of the properties of the surfaces which generalize the properties of the genus-one example:

- The surface M_k of genus k has symmetry group which is isomorphic to the dihedral group $D(2k + 2)$.
- The intersection of M_k with the plane $x_3 = 0$ consists of $r + 1$ straight lines meeting at equal angles at the origin.
- The intersection of M_k with the plane $x_3 = c \neq 0$ consists of a single closed curve.

This last statement was more than the reviewer could believe. There were no proofs in the short announcement. It is a bit difficult to imagine a surface of positive genus embedded in space with a height function all but one of whose level sets is a Jordan curve. Of course, three points are missing, one of which is actually at that exceptional level set. Again in this case, pictures helped to clarify the situation and the announcement appeared on schedule in the *Bulletin* [7].

The Higher-Genus Examples

The process of understanding how to construct higher-genus examples actually led to a simplification of the construction of the genus-one example.

The *P*-function and its derivative on the square torus satisfy the differential equation

$$P'^2 = 4P(P^2 - e_1^2), \quad e_1 = P(1/2). \tag{5}$$

This equation actually defines the surface. The Riemann surface corresponding to $\{(z, w) \mid w^2 = 4z(z^2 - 1)\}$ is conformally the square torus T. Moreover, the meromorphic functions z and w produced by projection onto the coordinate planes are equal—up to a real multiplicative constant—to P and P', respectively. Therefore we may construct the surface using the Weierstrass representation with $g = c/w$ as the Gauss map and $\eta = (z/w)dz$ as the auxiliary one-form. The omitted points are then the branch points where $z = \infty$ or $z = \pm 1$. To get a simple picture of this surface, we first perform the linear fractional transformation $z = (\zeta - 1)/(\zeta - 1)$. The defining equation of the Riemann surface becomes $w^2 = -4\zeta(\zeta - 1)/(\zeta - 1)^3$, and the omitted points are the branch points on the surface where $\zeta = \infty$, 0, or -1. The Gauss map may be taken to be $g = cw$ and $\eta = (1/2\zeta w)d\zeta$. Substitution in the Enneper-Weierstrass formula will show that the third integrand is $cd\zeta/\zeta$. Hence, the third component of the immersion X is $x_3 = c \ln |\zeta|$.

Consider the surface to be a double-sheeted covering of the extended complex ζ-plane with the branch cut as indicated in Figure 30-13. The curves lifting the unit circle have $x_3 = c \ln |1| = 0$, so they are mapped into the plane $x_3 = 0$. They turn out to be straight lines emanating from the origin in $\mathbf{R}^3$. The origin is the image of the point on the surface over -1. Similarly, each circle $|z| = \epsilon$ in the z-plane lists to a curve lying in the plane $x_3 = c \ln(\epsilon)$. This means that the region of the Riemann surface above the unit disk is mapped onto the region of the minimal surface below the $x_3 = 0$ plane; the covering of the exterior of the unit disk is embedded onto the region above the $x_3 = 0$ plane. The point P_0 above 0 corresponds to one catenoid end, the point P_∞ above ∞ corresponds to the other. The inversion $z \to 1/\bar{z}$ lifts to a conformal diffeomorphism of the surface that is actually an isometry induced by (the symmetry of the surface which is) rotation about the line $x_1 - x_2 = x_3 = 0$. Complex conjugation $z \to \bar{z}$ lifts to an isometry of the surface that is actually induced by reflection in the plane $x_2 = 0$.

To produce the higher-genus example, start with the Riemann surface $w^{k+1} = z^k(z^2 - 1)$, and let the Gauss map be $g = cw$ and the auxiliary one-form be $\eta = 1/2(z/w)^k dz$. (The real constant c is determined by the need to kill off the periods.) The same linear fractional transformation of the z coordinate that we used above will allow us to view the surface as a $(k - 1)$-sheeted covering of the ζ-plane with the same branch points and cuts as in the genus-one example. Moreover, the third component of the immersion is again equal to $c \ln |\zeta|$. The unit circle in the ζ-plane lifts to $k + 1$ straight lines meeting at $0 = X(P_{-1})$ and diverging to the flat "planar" end at P_1. The mappings $\zeta \to 1/\bar{\zeta}$ and $\zeta \to \bar{\zeta}$ lift to isometries of the surface, which are the restriction of symmetries of the surface: $\zeta \to 1/\bar{\zeta}$ corresponds to rotation by $\pi/(k + 1)$ about the x_3-axis followed by reflection in the plane $x_3 = 0$; $\zeta \to \bar{\zeta}$ corresponds to reflection in the plane $x_2 = 0$.

Not only does this way of constructing the example simplify the proof of embeddedness, it also provides a simple means of drawing these surfaces. First of all, we need only draw the half of the surface that lies over the unit disk; the other piece can be produced by rotation and reflection. Moreover, we can draw the curve on the surface over $\zeta = \epsilon$ by first lifting $k + 1$ circuits of it to the surface and then integrating the one-forms in the Enneper-Weierstrass formula along this lift. It doesn't matter where we start since the resulting curve must be symmetric with respect to the origin; we can simply make its average value equal to zero. The one problem with this procedure is that as ϵ approaches 1, the lift approaches a collection of curves on the surface that get mapped into divergent lines. Thus the length of the lift blows up. Some adjustments must be made to avoid this computational obstacle. These surfaces are illustrated in Figure 30-14.

A natural question about these surfaces is their limiting behavior as the genus goes to infinity. It turns out that if one leaves the parametrization as described in the above paragraph, then as the genus goes to infinity, the surfaces converge to the union of the plane and the catenoid: the tunnels get more numerous but smaller as the genus increases, and, in the limit, they collapse to the circle of intersection of the plane with the catenoid. On the other hand, if each example is homothetically expanded so that the maximum absolute value of Gauss curvature is a constant, say 1, then the ring of tunnels does not collapse. Instead, it expands in the limit to an

FIGURE 30-15 *Plane* $\cup$ *catenoid. One limiting surface of the family. The other is Scherk's Second Surface in Figure 30-5.*

infinite stack of tunnels. The limiting surface is in fact Scherk's Second Surface. (See Figure 30-15 and Figure 30-5.)

Reflections

The discovery of these examples and the working out of their geometric properties have led to a number of theoretical insights and advances in understanding the class of properly embedded minimal surfaces. It is tempting to philosophize about the role of examples in the development of mathematical theory. I will resist. Rather, let me conclude with some personal comments about the role of examples and, more to the point, models of examples in the appreciation and creation of mathematics.

On a recent visit to Columbia I saw a large collection of nineteenth-century plaster models of mathematical objects, manufactured in Leipzig and sold by Martin Schilling. These are exhibited in a glass case that runs along an entire wall in the lounge. The collection includes a model of the real cubic surface which contains twenty-seven lines, solid "graphs" of some elliptic functions, Delaunay surfaces of constant mean curvature, cyclides of Dupin, and many other interesting objects (including some, once well known, that have, for a while at least, receded beyond the horizon of current mathematical consciousness). Also in the case are some artifacts—gas mask, taped hand clubs—of the 1968 student protest at Columbia, discovered in an old desk drawer. Their presence in this minimathematical museum is oddly appropriate, making me think that the mathematicians who first saw these models may well have been involved in similar activities at some point in their lives. (For example, Scherk was on the wrong side of one of the many uprisings in 1848 and subsequently lost his position.)

I have two of these Schilling models in my office, a consequence of being in the mathematics library in Ann Arbor at precisely the time they were being thrown out to make room for more books and journals. I find them enchanting. When I view them I often wonder how doing mathematics then is different from doing it now and how computer modelling of examples is related to these marvelous plaster constructions.[4]

---

[4]A two-volume edition of photographs of these models, along with mathematical explanations, has recently appeared. It was produced by Gerd Fischer and published by Vieweg.

1. The model of an example, if it is a successful one, makes understanding that example easier. If the example is a "good" one, the model has helped in the comprehension of an abstract idea. But in the process of helping to explain an idea, the model itself has become attached to the idea itself in a concrete way. I will never be able to think of the minimal surfaces described above without concurrently reimagining the pictures. After seeing the model of the cubic containing twenty-seven lines, I will forever associate that dusty white model with the equations involved.

2. The process of producing the computer images was a construction which, although done on modern high-speed computing equipment (currently a Ridge 32/110 and a Raster Technologies Model One 380), had a very physical feel to it and a tactile, bodily quality that is a bit hard to convey.[5] It was, in that way, like the making of the molds for these German models. Of course the critical difference was that we could only see what the example was like at the very end of the process. Indeed, for the first example, we didn't know until the end whether this was a model of a significant example.

3. The images produced along the way were the objects that we used to make discoveries. They are an integral part of the process of doing mathematics, not just a way to convey a discovery made without their use. This has a very unusual consequence. When a mathematician understands or discovers something new, the way in which he or she understood things before is rapidly obliterated. Mathematics devours its own history in the process of creating itself. Because of this, historical accounts of mathematics are often anachronistic; for example, a paper of Euler is described in terms of the modern theory of partial differential equations. This may account in part for the depressing fact that new results may rapidly seem trivial and obvious to their discoverers; what it was like not to know these things is lost in the process of finding them out. Notes or old papers usually do not succeed in recreating the previous state of understanding.

However, these visual images can evoke for me the state of understanding that I had at the time they were created and I can remember now (two busy years later) how we were thinking about this group of problems at the time various pictures were produced. Perhaps this has to do with the way in which the human brain stores visual information. These pictures, especially the early imperfect ones, thought of as partial models of particular mathematical examples, are coupled in my mind quite closely with the understanding which they helped bring about.

4. Computer-generated images have obvious advantages over plaster. Images can be created of examples that simply cannot be molded. The range of easily constructible objects is vastly larger: there are no dusty white models of a limit set associated with a Kleinian group, for example. Moreover, methods of computing new or classical objects enter into the process. (This can be simultaneously enriching and maddening!) The computer-created model is not restricted to the role of illustrating the end product of mathematical understanding, as the plaster models are. They can be part of the process of doing mathematics.

Bibliography

1. C. Costa, "Imersões minimas completas em $\mathbf{R}^3$ de gênero um e curvatura total finita," Doctoral thesis, IMPA, Rio de Janeiro, Brasil, 1982. (Example of a complete minimal immersion in $\mathbf{R}^3$ of genus one and three embedded ends. *Bull. Soc. Bras. Mat.,* 15 [1984] 47–54.)
2. L. Barbosa and G. Colares, *Minimal Surfaces in* $\mathbf{R}^3$, Springer Lecture Notes series 1195, Springer-Verlag, Berlin-Heidelberg-New York, 1987.

[5]See Papert [18] for more about the tactile dimension of computing.
All computer-generated images © 1985/1986 David Hoffman and James T. Hoffman. These images were created using a Ridge 32/110 computer, a Raster Technologies Model One/380 graphics controller, and a high-resolution Sony monitor.

3. J-P. Bourguignon, H. B. Lawson and C. Margenn, "Les surfaces minimales," *Pour la Science,* January 1986.

4. S. Hildebrandt and A. J. Tromba, *The Mathematics of Optimal Form,* Scientific American Library, 1985, W. H. Freeman, New York.

5. D. A. Hoffman, *The discovery of new embedded minimal surfaces: elliptic functions; symmetry; computer graphics,* Proceedings of the Berlin Conference on Global Differential Geometry, Berlin, June 1984.

6. D. A. Hoffman, *The construction of families of embedded minimal surfaces,* Proceedings of the Stanford Conference on Variational Methods for Free Surface Interfaces, September 1985, Springer, New York-Berlin, 1987.

7. D. A. Hoffman and W. Meeks III, "Complete embedded minimal surfaces of finite total curvature," *Bull. A.M.S.,* 12, 1985, 134–136.

8. D. A. Hoffman and W. Meeks III, "A complete embedded minimal surface in $\mathbf{R}^3$ with genus one and three ends," *Journal of Differential Geometry,* 21, 1985, 109–127.

9. D. A. Hoffman and W. Meeks III, *The global theory of embedded minimal surfaces.*

10. L. Jorge and W. Meeks III, "The topology of complete minimal surfaces of finite total Gaussian curvature," *Topology,* 22 No. 2, 1983, 203–221.

11. H. B. Lawson Jr., *Lectures on Minimal Submanifolds,* Publish or Perish Press, Berkeley, 1971.

12. J. C. C. Nitsche, *Minimal surfaces and partial differential equations,* in *Studies in Partial Differential Equations,* Walter Littman, ed., MAA Studies in Mathematics, Vol. 23, Mathematical Association of America, 1982.

13. R. Osserman, "Global properties of minimal surfaces in $\mathbf{E}^3$ and $\mathbf{E}^n$," *Ann. of Math.,* 80 (1984), 340–364.

14. R. Osserman, *A Survey of Minimal Surfaces,* 2d Edition, Dover Publications, New York, 1986.

15. R. Schoen, "Uniqueness, symmetry, and embeddedness of minimal surfaces," *J. Diff. Geom.* 18, 1983, 791–809.

16. I. Peterson, "Three Bites in a Doughnut," *Science News* 127, No. 11, 16 March 1985.

17. Dr. Crypton, "Shapes that eluded discovery," *Science Digest,* April 1986.

18. S. Papert, *Mindstorms: Children, Computers, and Powerful Ideas.* Basic Books, New York, 1980.

31

What Is the Difference between a Parabola and a Hyperbola?

Shreeram S. Abhyankar

1. Parabola and Hyperbola

The *parabola* is given by the equation

$$Y^2 = X;$$

we can parametrize it by

$$X = t^2 \text{ and } Y = t.$$

The *hyperbola* is given by the equation

$$XY = 1;$$

we can parametrize it by

$$X = t \text{ and } Y = \frac{1}{t}.$$

Thus the parabola is a *polynomial curve* in the sense that we can parametrize it by polynomial functions of the parameter t. On the other hand, for the hyperbola we need rational functions of t that are not polynomials; it can be

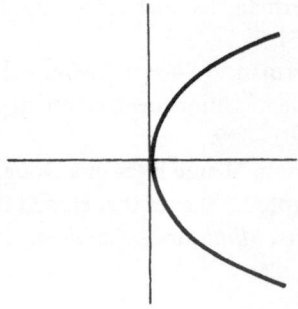

FIGURE 31-1 *Parabola.*

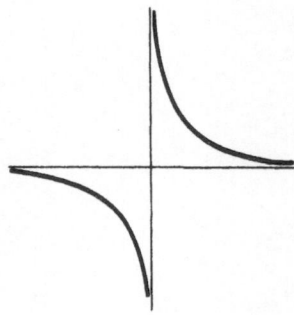

FIGURE 31-2 *Hyperbola.*

Volume 10, No. 4 (Fall 1988), 36–43

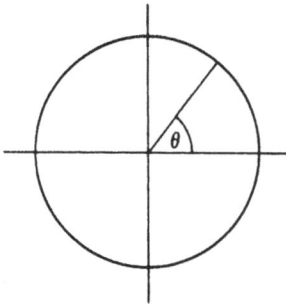

FIGURE 31-3 *Circle.*

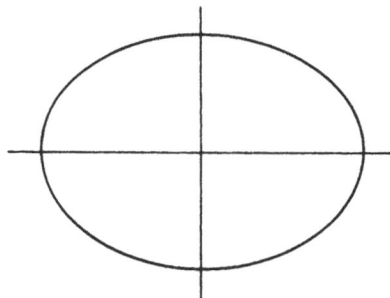

FIGURE 31-4 *Ellipse.*

shown that no polynomial parametrization is possible. Thus the hyperbola is not a polynomial curve, but it is a *rational curve.*

To find the reason behind this difference, let us note that the highest degree term in the equation of the parabola is Y^2, which has the only factor Y (repeated twice), whereas the highest degree term in the equation of the hyperbola is XY which has the two factors X and Y.

2. Circle and Ellipse

We can also note that the *circle* is given by the equation

$$X^2 + Y^2 = 1;$$

we can parametrize it by

$$X = \cos\theta \text{ and } Y = \sin\theta.$$

By substituting $\tan\theta/2 = t$ we get the *rational parametrization*

$$X = \frac{1 - t^2}{1 + t^2} \quad \text{and} \quad Y = \frac{2t}{1 + t^2},$$

which is not a *polynomial parametrization.* Similarly, the *ellipse* is given by the equation

$$\frac{X^2}{a^2} + \frac{Y^2}{b^2} = 1$$

and for it we can also obtain a rational parametrization that is not a polynomial parametrization. I did not start with the circle (or ellipse) because then the highest degree terms $X^2 + Y^2$ (respectively, $(X^2/a^2) + (Y^2/b^2)$) do have two factors, but we need complex numbers to find them.

3. Conics

In the above paragraph we have given the equations of parabola, hyperbola, circle, and ellipse in their *standard form*. Given the general equation of a conic

$$aX^2 + 2hXY + bY^2 + 2fX + 2gY + c = 0,$$

by a linear change of coordinates, we can bring it to one of the above four standard forms, and then we can tell whether the conic is a parabola, hyperbola, ellipse, or circle. Now, the nature of the factors of the highest degree terms remains unchanged when we make such a change of coordinates. Therefore we can tell what kind of a conic we have, simply by factoring the highest degree terms. Namely, if the highest degree terms $aX^2 + 2hXY + bY^2$ have only one real factor, then the conic is a parabola; if they have two real factors, then it is a hyperbola; if they have two complex factors, then it is an ellipse; and, finally, if these two complex factors are the special factors $X \pm iY$, then it is a circle. Here we are assuming that the conic in question does not degenerate into one or two lines.

4. Projective Plane

The geometric significance of the highest degree terms is that they dominate when X and Y are large. In other words, they give the behavior at infinity. To make this more vivid, we shall introduce certain fictitious points, which are called the "points at infinity" on the given curve and which correspond to factors of the highest degree terms in the equation of the curve. These fictitious points may be considered as "points" in the "projective plane." The concept of the projective plane may be described in the following two ways.

A point in the *affine* (X, Y)-*plane*, i.e., in the ordinary (X, Y)-plane, is given by a pair (α, β) where α is the X-coordinate and β is the Y-coordinate. The idea of points at infinity can be made clear by introducing *homogeneous coordinates*. In this set-up, the old point (α, β) is represented by all triples $(k\alpha, k\beta, k)$ with $k \neq 0$, and we call any such triple $(k\alpha, k\beta, k)$ homogeneous (X, Y, Z)-coordinates of the point (α, β). This creates room for "points" whose homogeneous Z-coordinate is zero; we call these the *points at infinity*, and we call their totality the *line at infinity*. This amounts to enlarging the affine (X, Y)-plane to the *projective* (X, Y, Z)-*plane* by adjoining the line at infinity.

More directly, the projective (X, Y, Z)-plane is obtained by considering all triples (α, β, γ) and identifying proportional triples; in other words, (α, β, γ) and $(\alpha', \beta', \gamma')$ represent the same point if and only if $(\alpha', \beta', \gamma') = (k\alpha, k\beta, k\gamma)$ for some $k \neq 0$; here we exclude the zero triple $(0, 0, 0)$ from consideration. The line at infinity is now given by $Z = 0$. To a point (α, β, γ) with $\gamma \neq 0$, i.e., to a point not on the line at infinity, there corresponds the point $(\alpha/\gamma, \beta/\gamma)$ in the affine plane. In this correspondence, as γ tends to zero, α/γ or β/γ tends to infinity; this explains why points whose homogeneous Z-coordinate is zero are called points at infinity.

To find the points at infinity on the given conic, we replace (X, Y) by $(X/Z, Y/Z)$ and multiply throughout by Z^2 to get the homogeneous equation

$$aX^2 + 2hXY + bY^2 + 2fXZ + 2gYZ + cZ^2 = 0$$

of the projective conic. On the one hand, the points of the original affine conic correspond to those points of the projective conic for which $Z \neq 0$. On the other hand, we put $Z = 0$ in the homogeneous equation and for the remaining expression we write

$$aX^2 + 2hXY + bY^2 = (pX - qY)(p^*X - q^*Y)$$

to get $(q, p, 0)$ and $(q^*, p^*, 0)$ as the points at infinity of the conic that correspond to the factors $(pX - qY)$ and $(p^*X - q^*Y)$ of the highest degree terms $aX^2 + 2hXY + bY^2$.

In the language of points at infinity, we may rephrase the above observation by saying that if the given conic has only one real point at infinity, then it is a parabola; if it has two real points at infinity, then it is a hyperbola; if it has two complex points at infinity, then it is an ellipse; and, finally, if these two complex points are the special points $(1, i, 0)$ and $(1, -i, 0)$, then it is a circle. At any rate, all the conics are rational curves, and among them the parabola is the only polynomial curve.

5. Polynomial Curves

The above information about parametrization suggests the following result.

THEOREM.   A rational curve is a polynomial curve if and only if it has only one place at infinity.

Here *place* is a refinement of the idea of a point. At a point there can be more than one place. To have *only one place at infinity* means to have only one point at infinity and to have only one place at that point. So what are the places at a point? To explain this, and having reviewed conics, let us briefly review cubics.

6. Cubics

The *nodal cubic* is given by the equation

$$Y^2 - X^2 - X^3 = 0.$$

It has a *double point* at the origin because the degree of the lowest degree terms in its equation is two. Moreover, this double point at the origin is a *node*, because at the origin the curve has the two *tangent lines*

$$Y = X \text{ and } Y = -X$$

(we recall that the tangent lines at the origin are given by the factors of the lowest degree terms). Likewise, the *cuspidal cubic* is given by the equation

$$Y^2 - X^3 = 0.$$

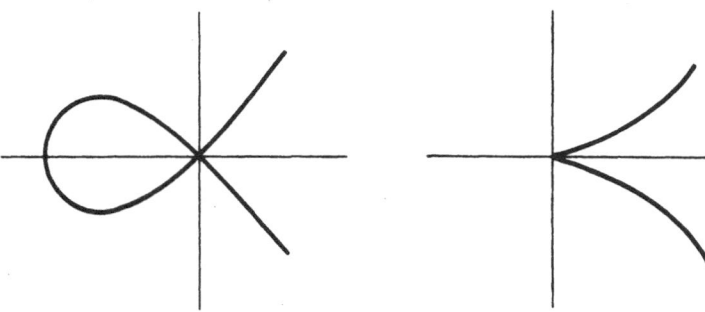

FIGURE 31-5 *Nodal cubic.*   FIGURE 31-6 *Cuspidal cubic.*

It has a double point at the origin. Moreover, this double point at the origin is a *cusp*, because at the origin the curve has the only tangent line

$$Y = 0.$$

A first approximation to places is provided by the tangent lines. So the nodal cubic has two places at the origin, whereas the cuspidal cubic has only one. More precisely, the nodal cubic has two places at the origin because, although its equation cannot be factored as a polynomial, it does have two factors as a power series in X and Y; namely, by solving the equation we get

$$Y^2 - X^2 - X^3 = \left(Y - X(1+X)^{1/2}\right)\left(Y + X(1+X)^{1/2}\right),$$

and by the binomial theorem we have

$$(1 + X)^{1/2} = 1 + (1/2)X + \cdots + \frac{(1/2)\big[(1/2) - 1\big] \ldots \big[(1/2) - j + 1\big]}{j!} X^j + \cdots .$$

7. Places at the Origin

Thus the number of places at the origin is defined to be equal to the number of distinct factors as power series, and in general this number is greater than or equal to the number of tangent lines. For example, the *tacnodal quintic* is given by the equation

$$Y^2 - X^4 - X^5 = 0,$$

which we find by multiplying the two opposite parabolas $Y \pm X^2 = 0$ and adding the extra term to make it irreducible as a polynomial. The double point at the origin is a *tacnode* because there is only one tangent line $Y = 0$ but two power series factors

$$\left(Y - X^2(1 + X)^{1/2}\right)\left(Y + X^2(1 + X)^{1/2}\right).$$

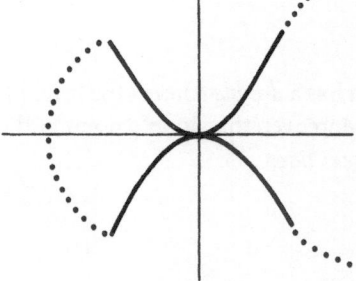

FIGURE 31-7 *Tacnodal quintic.*

So, more accurately, a cusp is a double point at which there is only one place; at a cusp it is also required that the tangent line meet the curve with *intersection multiplicity* three; i.e., when we substitute the equation of the tangent line into the equation of the curve, the resulting equation should have zero as a triple root. For example, by substituting the equation of the tangent line $Y = 0$ into the equation of the cuspidal cubic $Y^2 - X^3 = 0$, we get the equation $X^3 = 0$, which has zero as a triple root.

8. Places at Other Points

To find the number of places at any finite point, translate the coordinates to bring that point to the origin.

To find the number of places at a point at infinity, *homogenize* and *dehomogenize*. For example, by homogenizing the nodal cubic, i.e., by multiplying the various terms by suitable powers of a new variable Z so that all the terms acquire the same degree, we get

$$Y^2Z - X^2Z - X^3 = 0.$$

By putting $Z = 0$ we get $X = 0$; i.e., *the line at infinity* $Z = 0$ meets the nodal cubic only in the point P for which $X = 0$. By a suitable dehomogenization, i.e., by putting $Y = 0$, we get

$$Z - X^2Z - X^3 = 0.$$

Now, P is at the origin in the (X, Z)-plane; the left-hand side of the above equation is *analytically irreducible*; i.e., it does not factor as a power series. Thus the nodal cubic has only one place at P.

Consequently, in view of the above theorem, the nodal cubic may be expected to be a polynomial curve. To get an actual polynomial parametrization, substitute $Y = tX$ in the equation $Y^2 - X^2 - X^3 = 0$ to get

$$t^2X^2 - X^2 - X^3 = 0;$$

cancel the factor X^2 to obtain $X = t^2 - 1$ and then substitute this into $Y = tX$ to get $Y = t^3 - t$. Thus

$$X = t^2 - 1 \quad \text{and} \quad Y = t^3 - t$$

is the desired polynomial parametrization.

As a second example, recall that the nodal cubic $Y^2 - X^2 - X^3 = 0$ has two places at the origin, and the tangent line T given by $Y = X$ meets this cubic only at the origin. Therefore "by sending T to infinity" we could get a new cubic having only one point but two places at infinity; so it must be a rational curve that is not a polynomial curve. To find the equation of the new cubic, make the rotation $X' = X - Y$ and $Y' = X + Y$ to get $-X'Y' - (1/8)(X' + Y')^3 = 0$ as the equation of the nodal cubic and $X' = 0$ as the equation of T. By homogenizing and multiplying by -8 we get $8X'Y'Z' + (X' + Y')^3 = 0$ as the homogeneous equation of the nodal cubic and $X' = 0$ as the equation of T. Labeling (Y', Z', X') as (X, Y, Z), we get $8ZXY + (Z + X)^3 = 0$ as the homogeneous equation of the new cubic and T becomes the line at infinity $Z = 0$. Finally, by putting $Z = 1$, we see that the new cubic is given by the equation

$$8XY + (1 + X)^3 = 0.$$

By plotting the curve we see that one place at the point at infinity $X = Z = 0$ corresponds to the parabola-like structure indicated by the two single arrows, whereas the second place at that point corresponds to the hyperbola-like structure indicated by the two double arrows. Moreover, $Z = 0$ is the tangent to the parabola-like place, whereas $X = 0$ is the tangent to the hyperbola-like place. So this new cubic may be called the *para-hypal cubic*. To get a rational parametrization for it, we may simply take the vertical projection. In other words, by substituting $X = t$ in the above equation, we get $Y = -(1 + t)^3/8t$. Thus

$$X = t \quad \text{and} \quad Y = \frac{-(1 + t)^3}{8t}$$

is the desired rational parametrization; it cannot be a polynomial parametrization.

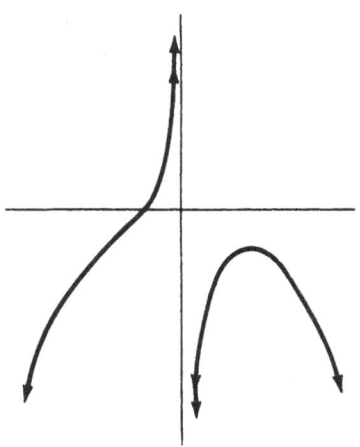

FIGURE 31-8 *Para-hypal cubic.*

9. Desire for a Criterion

In view of the above theorem, it would be nice to have an algorithmic criterion for a given curve to have only one place at infinity or at a given point. Recently in [7] I have worked out such a criterion. See [2] to [6] for general information and [7] for details of proof; here I shall explain the matter descriptively. As a first step let us recall some basic facts about resultants.

10. Vanishing Subjects

In the above discussion I have often said "reviewing this" and "recalling that." Unfortunately, reviewing and recalling may not apply to the younger generation. Until about thirty years ago, people learned in high school and college the two subjects called "theory of equations" and "analytic geometry." Then these two subjects gradually vanished from the syllabus. "Analytic geometry" first became a chapter, then a paragraph, and finally only a footnote in books on calculus.

"Theory of equations" and "analytic geometry" were synthesized into a subject called "algebraic geometry." Better still, they were collectively called "algebraic geometry." Then "algebraic geometry" became more and more abstract until it was difficult to comprehend. Thus classical algebraic geometry was forgotten by the student of mathematics.

Engineers are now resurrecting classical algebraic geometry, which has applications in computer-aided design, geometric modeling, and robotics. Engineers have healthy attitudes; they want to solve equations concretely and algorithmically, an attitude not far from that of classical, or high-school, algebra. So let us join hands with engineers.

11. Victim

Vis-a-vis the "theory of equations," one principal victim of this vanishing act was the resultant. At any rate, the Y-*resultant* $\text{Res}_Y(F, G)$ of two polynomials

$$F = a_0 Y^N + a_1 Y^{N-1} + \cdots + a_N \quad \text{and} \quad G = b_0 Y^M + b_1 Y^{M-1} + \cdots + b_M$$

is the determinant of the $N + M$ by $N + M$ matrix

$$\begin{bmatrix} a_0 & a_1 & \cdots & a_N & 0 & \cdots & \cdots & 0 \\ 0 & a_0 & \cdots & \cdots & a_N & 0 & \cdots & 0 \\ \cdots & \cdots & \cdots & \cdots & \cdots & \cdots & \cdots & \cdots \\ \cdots & \cdots & \cdots & \cdots & \cdots & \cdots & \cdots & \cdots \\ b_0 & b_1 & \cdots & b_M & 0 & \cdots & \cdots & 0 \\ 0 & b_0 & \cdots & \cdots & b_M & 0 & \cdots & 0 \\ \cdots & \cdots & \cdots & \cdots & \cdots & \cdots & \cdots & \cdots \\ \cdots & \cdots & \cdots & \cdots & \cdots & \cdots & \cdots & \cdots \end{bmatrix}$$

with M rows of the *a*s followed by N rows of the *b*s. This concept was introduced by Sylvester in his 1840 paper [10]. It can be shown that if $a_0 \neq 0 \neq b_0$ and

$$F = a_0 \prod_{j=1}^{N}(Y - \alpha_j) \quad \text{and} \quad G = b_0 \prod_{k=1}^{M}(Y - \beta_k),$$

then

$$\text{Res}_Y(F, G) = a_0^M \prod_j G(\alpha_j) = (-1)^{NM} b_0^N \prod_k F(\beta_k) = a_0^M b_0^N \prod_{j,k} (\alpha_j - \beta_k).$$

In particular, F and G have a common root if and only if $\text{Res}_Y(F, G) = 0$.

12. Approximate Roots

Henceforth let us consider an algebraic plane curve C defined by the equation

$$F(X, Y) = 0,$$

where $F(X, Y)$ is a monic polynomial in Y with coefficients that are polynomials in X, i.e.,

$$F = F(X, Y) = Y^N + a_1(X)Y^{N-1} + \cdots + a_N(X),$$

where $a_1(X), \ldots, a_N(X)$ are polynomials in X. We want to describe a criterion for C to have only one place at infinity. As a step toward this, given any positive integer D such that N is divisible by D, we would like to find the Dth root of F. We may not always be able to do this, because we wish to stay within polynomials. So we do the best we can. Namely, we try to find

$$G = G(X, Y) = Y^{N/D} + b_1(X)Y^{(N/D)-1} + \cdots + b_{N/D}(X),$$

where $b_1(X), \ldots, b_{N/D}(X)$ are polynomials in X, such that G^D is as close to F as possible. More precisely, we try to minimize the Y-degree of $F - G^D$. It turns out that if we require

$$\deg_Y(F - G^D) < N - (N/D),$$

then G exists in a unique manner; we call this G the *approximate Dth root of F* and we denote it by $\text{app}(D, F)$. In a moment, by generalizing the usual decimal expansion, we shall give an algorithm for finding $\text{app}(D, F)$. So let us revert from high-school algebra to grade-school arithmetic and discuss decimal expansion.

13. Decimal Expansion

We use decimal expansion to represent integers without thinking. For example, in decimal expansion,

$$423 = (4 \text{ times } 100) + (2 \text{ times } 10) + 3.$$

We can also use binary expansion, or expansion to the base 12, and so on. Quite generally, given any integer $P > 1$, every non-negative integer A has a unique *P-adic expansion*, i.e., A can uniquely be expressed as

$$A = \sum A_j P^j \quad \text{with non-negative integers } A_j < P,$$

where the summation is over a finite set of non-negative integers j. We can also change bases continuously. Namely, given any finite sequence $n = (n_1, n_2, \ldots, n_{h+1})$ of positive integers such that

$n_1 = 1$ and n_{j+1} is divisible by n_j for $1 \le j \le h$, every non-negative integer A has a unique *n-adic expansion;* i.e., A can uniquely be expressed as

$$A = \sum_{j=1}^{h+1} e_j n_j,$$

where $e = (e_1, \ldots, e_{h+1})$ is a sequence of non-negative integers such that $e_j < n_{j+1}/n_j$ for $1 \le j \le h$.

In analogy with P-adic expansions of integers, given any

$$G = G(X, Y) = Y^M + b_1(X)Y^{M-1} + \cdots + b_M(X),$$

where $b_1(X), \ldots, b_M(X)$ are polynomials in X, every polynomial $H = H(X, Y)$ in X and Y has a unique *G-adic expansion*

$$H = \sum H_j G^j,$$

where the summation is over a finite set of non-negative integers j and where H_j is a polynomial in X and Y whose Y-degree is less than M. In particular, if N/M equals a positive integer D, then as G-adic expansion of F we have

$$F = G^D + B_1 G^{D-1} + \cdots + B_D,$$

where $B_1, \ldots, B_D$ are polynomials in X and Y whose Y-degree is less than N/D. Now clearly,

$$\deg_Y(F - G^D) < N - (N/D) \quad \text{if and only if } B_1 = 0.$$

In general, in analogy with Shreedharacharya's method of solving quadratic equations by completing the square, for which reference may be made to [8] (and assuming that in our situation $1/D$ makes sense), we may "complete the Dth power" by putting $G' = G + (B_1/D)$ and by considering the G'-adic expansion

$$F = G'^D + B_1' G'^{D-1} + \cdots + B_D',$$

where $B_1', \ldots, B_D'$ are polynomials in X and Y whose Y-degree is less than N/D. We can easily see that if $B_1 \ne 0$, then $\deg_Y B_1' < \deg_Y B_1$. It follows that by starting with any G and repeating this procedure D times, we get the approximate Dth root of F.

Again, in analogy with n-adic expansion, given any sequence $g = (g_1, \ldots, g_{h-1})$, where g_j is a monic polynomial of degree n_j in Y with coefficients that are polynomials in X, every polynomial H in X and Y has a unique *g-adic expansion*

$$H = \sum H_e \prod_{j=1}^{h+1} g_j^{e_j}, \quad \text{where } H_e \text{ is a polynomial in } X$$

and where the summation is over all sequences of non-negative integers $e = (e_1, \ldots, e_{h+1})$ such that $e_j < n_{j+1}/n_j$ for $1 \le j \le h$.

14. Places at Infinity

As the next step toward the criterion, we associate several sequences with F as follows. The case when Y divides F being trivial, we assume the contrary. Now let

$$d_1 = r_0 = N, g_1 = Y, r_1 = \deg_X \operatorname{Res}_Y(F, g_1),$$

and

$$d_2 = \text{GCD}(r_0, r_1), g_2 = \text{app}(d_2, F), r_2 = \deg_X \text{Res}_Y(F, g_2),$$

and

$$d_3 = \text{GCD}(r_0, r_1, r_2), g_3 = \text{app}(d_3, F), r_3 = \deg_X \text{Res}_Y(F, g_3),$$

and so on, where we agree to put

$$\deg_X \text{Res}_Y(F, g_i) = -\infty \text{ if } \text{Res}_Y(F, g_i) = 0$$

and

$$\text{GCD}(r_0, r_1, \ldots, r_i) = \text{GCD}(r_0, r_1, \ldots, r_j)$$

if $r_0, r_1, \ldots, r_j$ are integers and $j < i$ and $r_{j+1} = r_{j+2} = \cdots = r_i = -\infty$.

Since $d_2 \geq d_3 \geq d_4 \geq \cdots$ are positive integers, there exists a unique positive integer h such that $d_2 > d_3 > \cdots > d_{h+1} = d_{h+2}$. Thus we have defined the two sequences of integers $r = (r_0, r_1, \ldots, r_h)$ and $d = (d_1, d_2, \ldots, d_{h+1})$ and a third sequence $g = (g_1, g_2, \ldots, g_{h+1})$, where g_j is a monic polynomial of degree $n_j = d_1/d_j$ in Y with coefficients that are polynomials in X. Now, for the curve C defined by $F(X, Y) = 0$, we are ready to state the criterion.

CRITERION for having only one place at infinity. C has only one place at infinity if and only if $d_{h+1} = 1$ and $r_1 d_1 > r_2 d_2 > \cdots > r_h d_h$ and g_{j+1} is degree-wise straight relative to (r, g, g_j) for $1 \leq j \leq h$ (in the sense we shall define in a moment).

To spell out the definition of degree-wise straightness, for every polynomial H in X and Y we consider the g-adic expansion

$$H = \sum H_e \prod_{j=1}^{h+1} g_j^{e_j},$$

where H_e is a polynomial in X and where the summation is over all sequences of non-negative integers $e = (e_1, \ldots, e_{h+1})$ such that $e_j < n_{j+1}/n_j$ for $1 \leq j \leq h$. We define

$$\text{fing}(r, g, H) = \max\left(\sum_{j=0}^{h} e_j r_j\right) \text{ with } e_0 = \deg_X H_e,$$

where the max is taken over all e for which $H_e \neq 0 = e_{h+1}$; here fing is supposed to be an abbreviation of the phrase "degree-wise formal intersection multiplicity," which in turn is meant to suggest some sort of analogy with intersection multiplicity of plane curves.

For $1 \leq j \leq h$, let $u(j) = n_{j+1}/n_j$ and consider the g_j-adic expansion

$$g_{j+1} = g_j^{u(j)} + \sum_{k=1}^{u(j)} g_{jk} g_j^{u(j)-k},$$

where g_{jk} is a polynomial in X and Y whose Y-degree is less than n_j. We say that g_{j+1} is *degree-wise straight relative to* (r, g, g_j) if

$$\big(u(j)/k\big)\text{fing}(r, g, g_{jk}) \le \text{fing}(r, g, g_{ju(j)}) = u(j)\big[\text{fing}(r, g, g_j)\big]$$

for $1 \le k \le u(j)$; the adjective *straight* is meant to suggest that we are considering some kind of generalization of Newton Polygon (for Newton Polygon, see [9], Part II, pp. 382–397, where it is called Newton Parallelogram).

15. Places at a Given Point

To discuss places of the curve C defined by $F(X, Y) = 0$ at a given finite point, we may suppose that the point has been brought to the origin by a translation and rotation of coordinates and that neither X nor Y divides F. By the Weierstrass Preparation Theorem (see [1], p. 74), we can write

$$F(X, Y) = \delta(X, Y)F^*(X, Y),$$

where $\delta(X, Y)$ is a power series in X and Y with $\delta(0, 0) \ne 0$ and F^* is a distinguished polynomial; i.e.,

$$F^* = F^*(X, Y) = Y^{N^*} + a_1^*(X)Y^{N^*-1} + \cdots + a_{N^*}^*(X)$$

and $a_1^*(X), \ldots, a_{N^*}^*(X)$ are power series in X that are zero at zero. By ord_X of a power series in X we mean the degree of the lowest degree term present in that power series. We also note that in the present situation, the approximate roots of F^* are monic polynomials in Y whose coefficients are power series in X. Now let

$$d_1 = r_0 = N^*, g_1 = Y, r_1 = \text{ord}_X \text{Res}_Y(F^*, g_1),$$

and

$$d_2 = \text{GCD}(r_0, r_1), g_2 = \text{app}(d_2, F^*), r_2 = \text{ord}_X \text{Res}_Y(F^*, g_2),$$

and

$$d_3 = \text{GCD}(r_0, r_1, r_2), g_3 = \text{app}(d_3, F^*), r_3 = \text{ord}_X \text{Res}_Y(F^*, g_3),$$

and so on, where we agree to put

$$\text{ord}_X \text{Res}_Y(F^*, g_i) = \infty \text{ if } \text{Res}_Y(F^*, g_i) = 0$$

and

$$\text{GCD}(r_0, r_1, \ldots, r_i) = \text{GCD}(r_0, r_1, \ldots, r_j)$$

if $r_0, r_1, \ldots, r_j$ are integers and $j < i$ and $r_{j+1} = r_{j+2} = \cdots = r_i = \infty$.

Since $d_2 \geq d_3 \geq d_4 \geq \cdots$ are positive integers, there exists a unique positive integer h such that $d_2 > d_3 > \cdots > d_{h+1} = d_{h+2}$. Thus we have defined the two sequences of integers $r = (r_0, r_1, \ldots, r_h)$ and $d = (d_1, d_2, \ldots, d_{h+1})$ and a third sequence $g = (g_1, g_2, \ldots, g_{h+1})$, where g_j is a monic polynomial of degree $n_j = d_1/d_j$ in Y with coefficients that are power series in X. For the curve C defined by $F(X, Y) = 0$, we are ready to state the main result of this section.

CRITERION for having only one place at the origin. C has only one place at the origin if and only if $d_{h+1} = 1$ and $r_1 d_1 < r_2 d_2 < \cdots < r_h d_h$ and g_{j+1} is straight relative to (r, g, g_j) for $1 \leq j \leq h$ (in the sense which we shall define in a moment).

To spell out the definition of straightness, first note that in the present situation, the coefficients of a g-adic expansion are power series in X. Now for every polynomial H in Y with coefficients that are power series in X, we consider the g-adic expansion

$$H = \sum H_e \prod_{j=1}^{h+1} g_j^{e_j},$$

where H_e is a power series in X and where the summation is over all sequences of non-negative integers $e = (e_1, \ldots, e_{h+1})$ such that $e_j < n_{j+1}/n_j$ for $1 \leq j \leq h$. We define

$$\text{fint}(r, g, H) = \min\left(\sum_{j=0}^{h} e_j r_j\right) \text{ with } e_0 = \text{ord}_X H_e,$$

where the min is taken over all e for which $H_e \neq 0 = e_{h+1}$; here fint is supposed to be an abbreviation of the phrase "formal intersection multiplicity," which in turn is meant to suggest some sort of analogy with intersection multiplicity of plane curves.

For $1 \leq j \leq h$, let $u(j) = n_{j+1}/n_j$ and consider the g_j-adic expansion

$$g_{j+1} = g_j^{u(j)} + \sum_{k=1}^{u(j)} g_{jk} g_j^{u(j)-k},$$

where we note that in the present situation, the coefficients g_{jk} are polynomials of degree less than n_j in Y whose coefficients are power series in X. We say that g_{j+1} is *straight relative to* (r, g, g_j) if

$$(u(j)/k)\text{fint}(r, g, g_{jk}) \geq \text{fint}(r, g, g_{ju(j)}) = u(j)\big[\text{fint}(r, g, g_j)\big]$$

for $1 \leq k \leq u(j)$; again, the adjective *straight* is meant to suggest that we are considering some kind of generalization of Newton Polygon.

16. Problem

Generalize the above criterion by finding a finitistic algorithm to count the number of places at infinity or at a given point.

17. Example

To illustrate the criterion for having only one place at the origin, let us take

$$F = F(X, Y) = (Y^2 - X^3)^2 + X^p Y - X^7,$$

where p is a positive integer to be chosen. Now

$$F^* = F \quad \text{and} \quad d_1 = r_0 = N^* = N = 4 \quad \text{and} \quad g_1 = Y$$

and hence

$$\text{Res}_Y(F, g_1) = F(X, 0) = X^6 - X^7 \quad \text{and} \quad r_1 = \text{ord}_X \text{Res}_Y(F, g_1) = 6.$$

Therefore,

$$d_2 = \text{GCD}(r_0, r_1) = \text{GCD}(4, 6) = 2$$

and hence

$$g_2 = \text{app}(d_2, F) = Y^2 - X^3 = (Y - X^{3/2})(Y + X^{3/2}).$$

Consequently,

$$\begin{aligned}
\text{Res}_Y(F, g_2) &= F(X, X^{3/2}) F(X, -X^{3/2}) \\
&= (X^{p+(3/2)} - X^7)(-X^{p+(3/2)} - X^7) \\
&= -X^{2p+3} + X^{14},
\end{aligned}$$

and hence

$$r_2 = \text{ord}_X \text{Res}_Y(F, g_2) = \begin{bmatrix} 14 & \text{if } p > 5 \\ 2p+3 & \text{if } p \leq 5. \end{bmatrix}$$

Therefore,

$$d_3 = \begin{bmatrix} 2 \text{ if } p > 5 \\ 1 \text{ if } p \leq 5 \end{bmatrix} \quad \text{and} \quad h = \begin{bmatrix} 1 \text{ if } p > 5 \\ 2 \text{ if } p \leq 5 \end{bmatrix}$$

and

$$r_1 d_1 = \begin{bmatrix} 24 < 26 = (2p+3)d_2 = r_2 d_2 \text{ if } p = 5 \\ 24 \geq 22 \geq (2p+3)d_2 = r_2 d_2 \text{ if } p < 5. \end{bmatrix}$$

Now, if $p = 5$, then

$$g_{11} = 0, \quad \text{and} \quad g_{21} = 0$$

and

$$g_{12} = X^3 \quad \text{and} \quad \text{fint}(r, g, X^3) = 3r_0 = 12 = 2r_1$$

and

$$g_{22} = X^5Y - X^7 \quad \text{and} \quad \text{fint}(r, g, X^5Y - X^7) = 5r_0 + r_1 = 26 = 2r_2,$$

and hence g_{j+1} is straight relative to (r, g, g_j) for $1 \leq j \leq 2$.

Thus we see that if $p > 5$, then $h = 1$ and $d_{h+1} = 2$, whereas if $p < 5$, then $h = 2$ and $d_{h+1} = 1$ and $r_1d_1 > r_2d_2$; finally, if $p = 5$, then $h = 2$ and $d_{h+1} = 1$ and $r_1d_1 < r_2d_2$ and g_{j+1} is straight relative to (r, g, g_j) for $1 \leq j \leq 2$. Therefore, by the criterion we conclude that C has only one place at the origin if and only if $p = 5$.

References

1. S. S. Abhyankar, *Local Analytic Geometry*, New York: Academic Press (1965).
2. ———. A Glimpse of Algebraic Geometry, *Lokamanya Tilak Memorial Lectures*, University of Poona, Pune, India (1969).
3. ———. Singularities of algebraic curves, *Analytic Methods in Mathematical Physics, Conference Proceedings* (Gordon and Breach, eds.) (1970), 3–14.
4. ———. Historical ramblings in algebraic geometry and related algebra, *American Mathematical Monthly* 83 (1976), 409–440.
5. ———. On the semi-group of a meromorphic curve (Part I), *Proceedings of the (Kyoto) International Symposium on Algebraic Geometry*, Tokyo: Kinokuniya Book-Store (1977), 249–414.
6. ———. Generalizations of ancient Indian mathematics and applications (in Hindi), *Second Anniversary Souvenir of Bhaskaracharya Pratishthana*, Pune, India (1978), 3–13.
7. ———. Irreducibility criterion for germs of analytic functions of two complex variables, *Advances in Mathematics* 74 (1989), 190–257.
8. Bhaskaracharya: Beejganita (algebra) (in Sanskrit), India, 1150.
9. G. Chrystal, *Algebra*, Parts I and II, Edinburgh (1886).
10. J. J. Sylvester, On a general method of determining by mere inspection the derivations from two equations of any degree, *Philosophical Magazine* 16 (1840), 132–135.

32

How to Build Minimal Polyhedral Models of the Boy Surface

Ulrich Brehm

Introduction and History

In the middle of the last century A. Möbius gave a combinatorial description of a closed, one-sided, polyhedral surface, which was soon recognized as a topological model of the real projective plane. It also turned out to be the surface that J. Steiner had defined geometrically. Soon, algebraic definitions followed which were used to construct plaster models of the cross-cap and the Roman surfaces. Until the Klein bottle, none of these non-orientable closed surfaces, whether smooth or polyhedral, was known to be "immersible" in $\mathbf{R}^3$. A topological *immersion* $i : M \to \mathbf{R}^3$ is a locally injective continuous mapping. An immersion $i : M \to \mathbf{R}^3$ of a compact 2-manifold (without boundary) is called *polyhedral* if the image of i is contained in the union of finitely many planes.

In 1903 D. Hilbert's student W. Boy proved in [3] that the real projective plane $\mathbf{R}P^2$ allows an immersion in $\mathbf{R}^3$ (with an axis of symmetry of order 3). Several efforts have been made to give an explicit description of such an immersion. A survey of explicit combinatorial, analytic, and algebraic descriptions of such immersions is included in F. Apéry's recent book on the subject [1]. In [5] polyhedral immersions of $\mathbf{R}P^2$ with eighteen vertices were described. The polyhedral immersions given in [1] have even more vertices. In this paper the existence of symmetric polyhedral versions of the Boy surface with only nine vertices and ten facets is shown and an easy recipe for building cardboard models of these objects is given. Formal definitions of the terms "vertex," "facet," and "edge" will be given near the end of the introduction.

T. Banchoff showed in [2] that an immersion of $\mathbf{R}P^2$ in $\mathbf{R}^3$ in general position must have a triple point. A polyhedral immersion can always be perturbated so that the vertices are in "very general" position (for example, with the coordinates of the vertices being algebraically independent). Thus we get the generic case with at least one triple point in the relative interior of three triangular facets. The intersection of any two of these triangles is a line segment which cannot contain a common vertex because a polyhedral immersion is injective in some neighborhood of any vertex. Thus the three triangles containing the triple point have together nine different vertices, so nine is a lower bound for the number of vertices of a polyhedral immersion of $\mathbf{R}P^2$. We will show that this lower bound can indeed be attained.

The nicest way to prove this is to construct a three-dimensional model of the object wanted. This paper contains an easy recipe for building your own cardboard models of minimal polyhedral ver-

---

Volume 12, No. 4 (Fall 1990), 51–56

sions of the Boy surface. We also give the coordinates of the vertices together with the combinatorial structure. Using these data you can also construct a computer model.

Definitions. Let M be a compact 2-manifold (without boundary) and $i : M \to \mathbf{R}^3$ a polyhedral immersion; (a) a *facet* is a connected component of a non-empty set of the form $\text{int}(i^{-1}[H])$ where $H \subset \mathbf{R}^3$ is a plane and int denotes the interior of a set; (b) a *vertex* is a point of M that is in the intersection of the closures of (at least) three facets; (c) the connected components of the set of points of M that are neither vertices nor contained in some facet are called *edges*.

Thus if two vertices happen to be mapped onto the same point in $\mathbf{R}^3$, they are still counted as different vertices. On the other hand, the intersection of (the relative interior of) the images of a facet and an edge is not regarded as an additional vertex. If each facet is a topological open disc, then Euler's formula $f_2 - f_1 + f_0 = \chi(M)$ holds, where f_2, f_1, f_0 denote the numbers of facets, edges, vertices, respectively, and $\chi(M)$ denotes the Euler characteristic of M.

If no misunderstandings can occur, we call the image of a vertex, edge, or facet also a vertex, edge, or facet, respectively. In particular, by coordinates of a vertex we mean always the coordinates of the image point in $\mathbf{R}^3$.

If M is triangulated such that i is piecewise linear, then the local injectivity of i has to be checked only in a neighborhood of each vertex of (the simplicial complex) M.

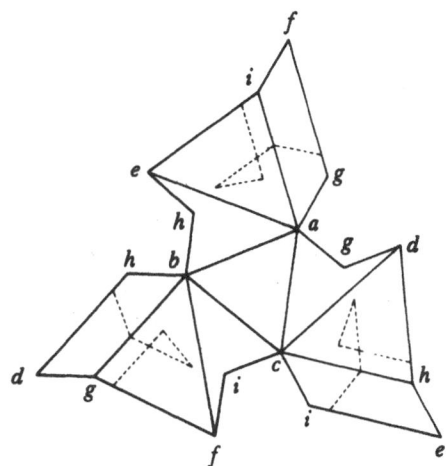

FIGURE 32-1 *A net of a polyhedral immersion P1 of* $\mathbf{R}P^2$. *In Figure 32-2 and Figure 32-3 we show orthogonal projections of P1 in the direction of the axis of symmetry from "above" and from "below." We have indicated the self-intersection lines by dotted lines. Visible lines, dotted or solid, are drawn much thicker than invisible lines. For any two edges with intersecting projections we indicate which of the two edges is above the other one.*

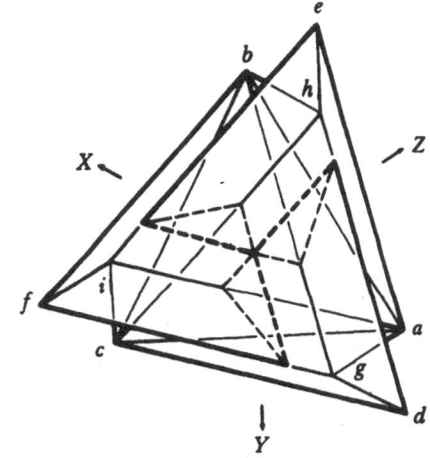

FIGURE 32-2 *The orthogonal projection in the direction of the axis of symmetry from above.*

Polyhedral Immersion of $\mathbf{R}P^2$ with Nine Vertices and Ten Facets

We describe three combinatorially different symmetric polyhedral immersions $P1, P2, P3$ of $\mathbf{R}P^2$ with nine vertices and ten facets. $P1$ has six quadrangular and four triangular facets, whereas each of $P2$ and $P3$ has three pentagonal and seven triangular facets. (The coordinates of the vertices of $P1$ are

a	$(-2, 0, 0)$	b	$(0, -2, 0)$	c	$(0, 0, -2)$
d	$(-1, 2, 1)$	e	$(1, -1, 2)$	f	$(2, 1, -1)$
g	$(-1, 1, 0)$	h	$(0, -1, 1)$	i	$(1, 0, -1)$

In Figure 32-1 we give a net of our immersed poly-hedron $P1$. The dotted lines indicate the self-intersection lines. The list of coordinates of the vertices of $P1$ shows that the mapping $(x, y, z) \rightarrow (z, x, y)$ is a rotation by $2\pi/3$ with axis $\mathbf{R}(1,1,1)$ inducing the permutation (a, b, c) (d, e, f) $(g, h,, i)$ of the vertices. Because this per-mutation induces an automorphism of the net (see Figure 32-1), $P1$ has an axis of symmetry of order 3. The cell-complex defined by Figure 32-1 (ver-tices and edges being identified in the obvious way) clearly is an $\mathbf{R}P^2$ with $f_0 = 9, f_1 = 18, f_2 = 10$.

In Figure 32-10 you can see some pictures of a cardboard model of $P1$. It is easy to check that Fig-ure 32-2 is correct and that a, c, d, g are affinely dependent and $g - a = f - i$. With the symmetry this implies that $P1$ is indeed an immersed polyhe-dron with the combinatorial structure given in Fig-ure 32-1. So we get the following result.

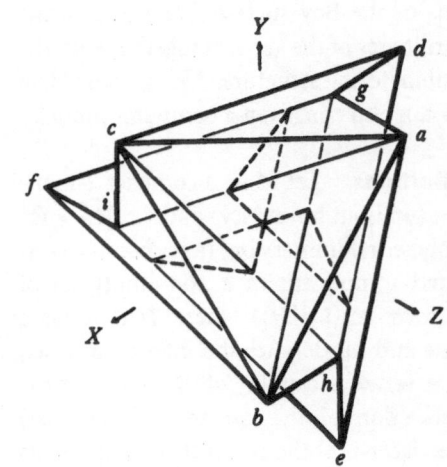

FIGURE 32-3 *The orthogonal projection in the direction of the axis of symmetry from below.*

THEOREM 1: $P1$ is a symmetric polyhedral immersion of $\mathbf{R}P2$ into $\mathbf{R}^3$ with nine vertices, eight-een edges, and ten facets, six of which are quadrangles.

Next we describe two minimal polyhedral versions of the Boy surface containing three pentago-nal facets. The coordinates of the vertices of $P2$ and of $P3$ are

a	$(0, 1, -1)$	b	$(-1, 0, 1)$	c	$(1, -1, 0)$
d	$(2, 2, 0)$	e	$(0, 2, 2)$	f	$(2, 0, 2)$
g	$(1, 1, 0)$	h	$(0, 1, 1)$	i	$(1, 0, 1)$

In Figure 32-4 and 32-5 we give nets of our immersed polyhedra $P2$ and $P3$. The dotted lines indicate the self-intersection lines. The list of coordinates of the vertices shows that the map-ping $(x, y, z) \rightarrow (z, x, y)$ is a rotation by $2\pi/3$ with axis $\mathbf{R}(1, 1, 1)$ inducing the permutation (a, b, c) (d, e, f) (g, h, i) of the vertices. Because this per-mutation induces an automorphism of the nets (see Figure 32-4 and Figure 32-5), $P2$ and $P3$ have an axis of symmetry of order 3. The cell-com-plexes defined by Figure 32-4 and Figure 32-5 (vertices and edges being identified in the obvious way) clearly are $\mathbf{R}P^2$s with $f_0 = 9, f_1 = 18, f_2 = 10$.

$P2$ and $P3$ are combinatorially different because $P3$ contains 6-valent vertices, whereas $P2$ does not contain such vertices.

Note that the Möbius strips arising from $P2$ and $P3$ when omitting the four triangles that do

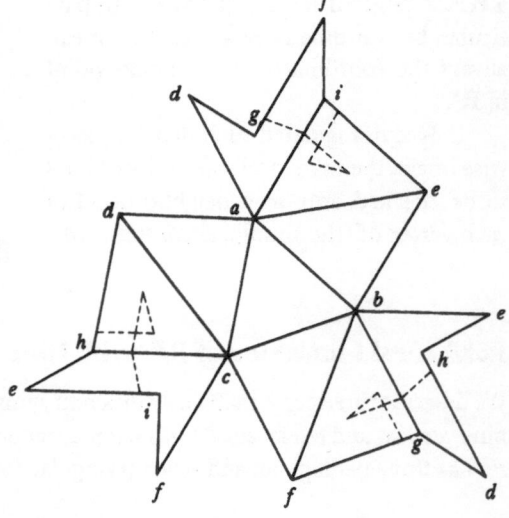

FIGURE 32-4 *A net of a polyhedral immersion P2 of* $\mathbf{R}P^2$.

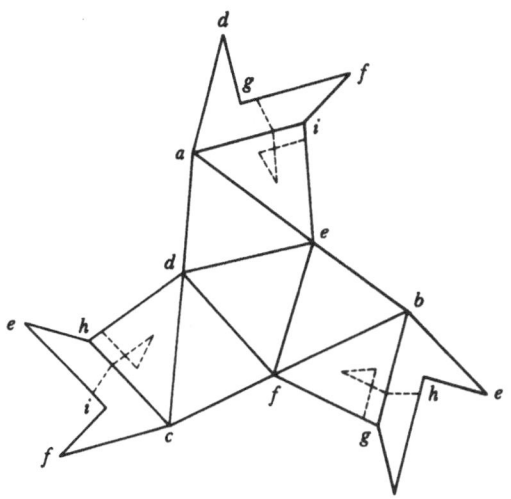

FIGURE 32-5 *A net of a polyhedral immersion P3 of* **R**P^2.

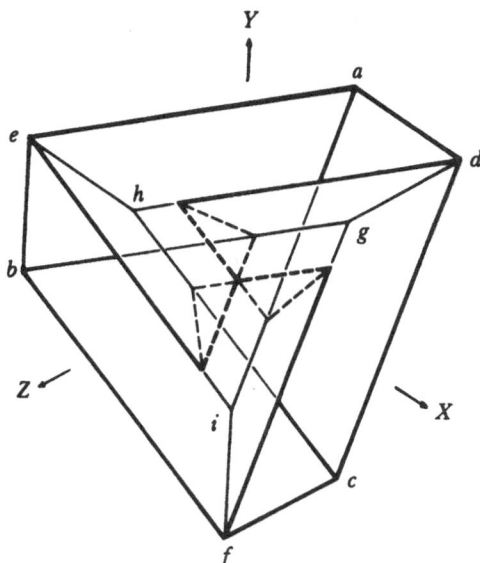

FIGURE 32-6 *Orthogonal projection of the Möbius strip from above.*

not contain the triple point are identical. In Figure 32-6 and Figure 32-7 we show orthogonal projections of this Möbius strip in the direction of the axis of symmetry from "above" and from "below." We have indicated the self-intersection lines by dotted lines. Visible lines, dotted or solid, are drawn much thicker than invisible lines. Note that the three pentagons form a symmetric polyhedral Möbius strip (without self-intersections).

In Figure 32-11 and Figure 32-12 you can see some pictures of *P3* and *P2*, respectively. It is easy to check that Figure 32-6 is correct, that a, d, g, f, i are affinely dependent, and that g lies in the interior of the convex hull of the vertices. With the symmetry this implies that *P2* and *P3* are indeed immersed polyhedra with the combinatorial structure given in Figure 32-4 and Figure 32-5, respectively. So we get the following result.

THEOREM 2: *P2* and *P3* are combinatorially different polyhedral immersions of **R**P^2 into **R**^{3} with nine vertices, eighteen edges, and ten facets, three of which are pentagons.

Now let us modify *P1* by adding a new vertex $j = (-2, -2, -2)$ and replacing the triangle *abc* by the triangles *abj, acj, bcj*. Because *abj* and *abh* are coplanar, we can omit the edge *ab* and get a non-convex pentagon *aehbj*. Symmetrically we omit the edges *ac* and *bc*. Thus we get a polyhedral immersion *P1'* with $f_0 = 10, f_1 = 18, f_2 = 9$.

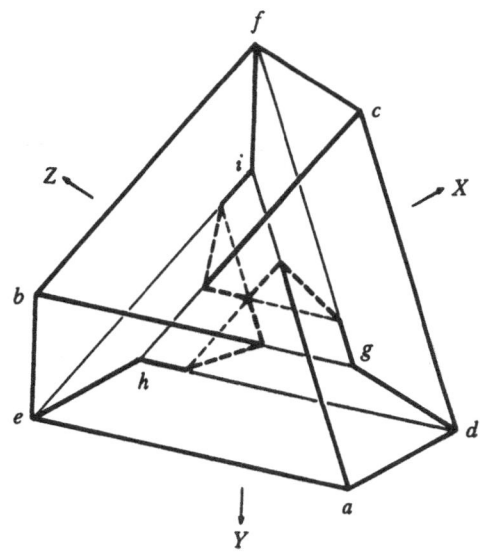

FIGURE 32-7 *Orthogonal projection of the Möbius strip from below.*

Similarly, we can modify $P2$ and $P3$, getting combinatorially different polyhedral immersions $P2'$ and $P3'$ with $f_0 = 10, f_1 = 18, f_2 = 9$. Thus we have shown:

THEOREM 3:   There exist symmetric polyhedral immersions of $\mathbf{R}P^2$ into $\mathbf{R}^3$ with ten vertices, eighteen edges, and nine facets.

Remarks.

1.   One gets a symmetric modification of $P1$ such that the quadrangle $a\,c\,d\,g$ is convex if a, b, c, d, e, f are chosen as the vertices of a regular octahedron with diagonals af, bg, ch and g, h, e are chosen as the midpoints of the edges ad, be, cf, respectively, and the quadrangle $a\,g\,f\,i$ is split into the triangles $a\,g\,i$ and $f\,g\,i$ (splitting $c\,i\,e\,h, d\,g\,b\,h$ similarly).

How To Build Your Own Models of the Boy Surface

We call the symmetric parts of Figure 32-8 and Figure 32-9 consisting of four triangles the large parts and the other parts of Figure 32-8 and Figure 32-9 the small parts of the figure.

1.   Make a sufficiently large copy of Figure 32-8 (for $P1$) or/and of Figure 32-9 (for $P3$), for example by running the figure (several times) through an enlarging copier; the lengths of the edges ab, de, respectively, should be at least 12 cm.
2.   Make a copy of the large part and three copies of the small part of the enlarged Figure 32-8 (resp., Figure 32-9) on cardboard by piercing the vertices with a pin.
3.   Draw the self-intersection lines on both sides of each of the three copies of the small part.
4.   Scratch the edge ai on each of the three copies on one side and the three edges of the regular triangle on the reverse side for $P1$ (resp., the same side for $P3$) of the cardboard.
5.   Cut out the four parts of the net along the contours. Also cut out the windows. In order to link the three windows, make a short cut from the edge ae to the edge of the window near u as indicated in Figure 32-8 (resp., Figure 32-9).
6.   Fold the three copies of the small part along the edge ai and fold the large part in the opposite direction for $P1$ (resp., the same direction for $P3$).
7.   Put the three congruent pieces together, creating the triple point; Figures 32-2, 32-3, 32-6, 32-7, 32-10, 32-11, 32-12 may be helpful.
8.   Glue the three pieces together along the corresponding edges using a self-adhesive (transparent) strip and similarly close the cuts in the triangles. In the case of $P3$ you get the Möbius strip which $P2$ and $P3$ have in common. Note that some of the dihedral angles are quite sharp (between 23° and 30°), namely, the angles at the edges dh, ei, fg (for $P1, P2$ and $P3$) and ag, bh, cf (for $P1$ and $P2$), whereas all other dihedral angles of $P1, P2$, and $P3$ are between 58° and 110°.
9.   Add the large part to finish $P1$ (resp., $P3$) and glue the two pieces together (along all pairs of corresponding edges).

To build $P2$ you have to construct the central part of the net of $P2$ (cf. Figure 32-4) with a circular window in the equilateral triangle and glue this part to the Möbius strip you got in step 8 from the construction of $P3$.

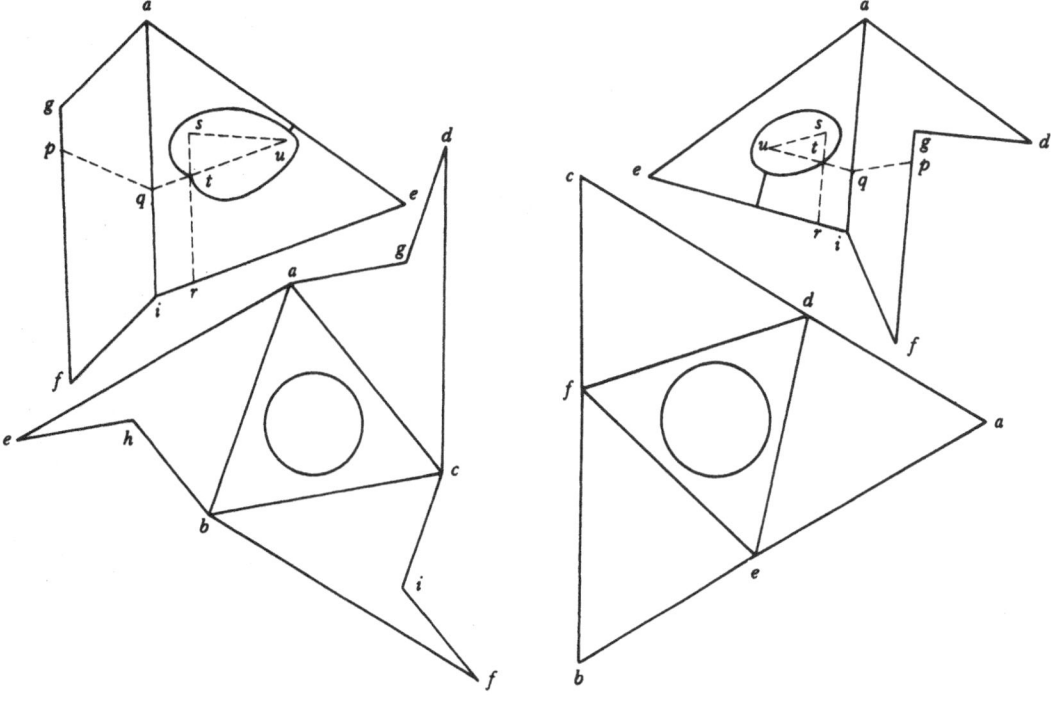

FIGURE 32-8 *Part of P1.*   FIGURE 32-9 *Part of P3.*

2.   A triangulation of the Möbius strip with nine vertices, whose boundary forms a triangle and whose automorphism group has order 6, but which cannot be immersed in $\mathbf{R}^3$ was first described by the author in [4].

3.   In order to build nice models, I suggest cutting a circular "window" into the regular triangle and cutting curved "windows" into the triangles containing the triple point, such that no material self-intersections occur but such that the full self-intersection figure, and in particular the triple point, are still visible (and marked by lines on the model on both sides).

In Figure 32-8 we show the central part of the net of $P1$ and a third of the net of the self-intersecting part of $P1$ (cf. Figure 32-1) indicating the self-intersection lines and the suggested windows. The lengths of the line segments are $ab = ac = bc = 2\sqrt{2}$, $ag = dg = bh = eh = ci = fi = \sqrt{2}$, $ad = be = cf = gi = \sqrt{6}$, $fg = ai = ei = \sqrt{10}$, $ae = bf = cd = \sqrt{14}$, $gp = ir = qt = st = \frac{2}{13}\sqrt{10}$, $tr = tu = iq = \frac{5}{13}\sqrt{10}$.

In Figure 32-9 we show the central part of the net of $P3$ and a third of the net of the Möbius strip which $P2$ and $P3$ have in common (cf. Figure 32-5) indicating the self-intersection lines and the suggested windows. The lengths of the line segments are $be = cf = ad = ai = di = ei = fg = \sqrt{6}$, $ag = dg = gi = fi = \sqrt{2}$, $ae = bf = cd = \sqrt{10}$, $de = df = ef = 2\sqrt{2}$, $gp = ir = qt = st = \frac{1}{7}\sqrt{6}$, $tr = tu = iq = \frac{2}{7}\sqrt{6}$, ($ab = ac = bc = \sqrt{6}$ for P2).

Acknowledgment: I wish to thank D. Ferus for taking the photos of the models (Figures 32-10, 32-11, 32-12).

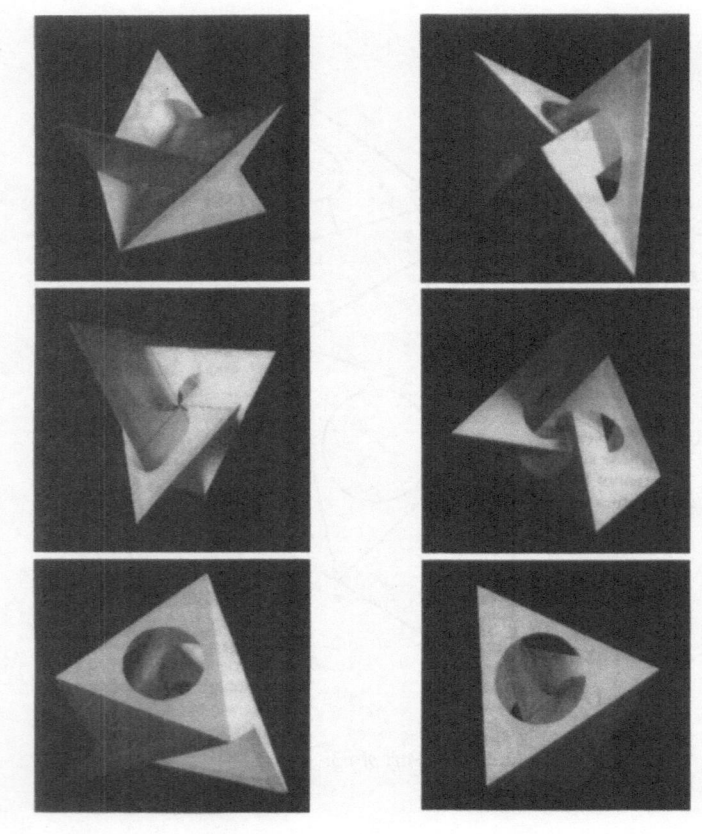

FIGURE 32-10 *A model of P1.*   FIGURE 32-11 *A model of P3.*   FIGURE 32-12 *A model of P2.*

References

1. F. Apéry, *Models of the real projective plane,* Braunschweig: Vieweg (1987).
2. T. Banchoff, Triple points and surgery of immersed surfaces, *Proc. Amer. Math. Soc.* 46 (1974), 407–413.
3. W. Boy, Über die Curvatura integra und die Topologie geschlossener Flächen, *Math. Ann.* 57 (1903), 151–184.
4. U. Brehm, A non-polyhedral Möbius strip, *Proc. Amer. Math. Soc.* 89 (1983), 519–522.
5. K. Merz and P. Humbert, Einseitige Polyeder nach Boy, *Comm. Math. Helv.* 14 (1941–42), 134–140.

33
Recent Developments in Braid and Link Theory

Joan S. Birman

This article is about the theory of braids and the geometry of links in the 3-sphere and new interconnections between them. We are particularly interested in the family of polynomial invariants of link type discovered by Vaughan Jones in 1984 and its recent generalizations.

The discovery of new powerful and easily computed invariants of oriented links in oriented 3-space in 1984 was a huge surprise. The hard work had been done already in what seemed to be totally unrelated studies of Von Neumann algebras. The proofs of the topological invariance of the new polynomials, first by Jones [14], [15] and later by others [12], [21], [17] gave essentially no insight into the geometric meaning of the new tools. The surprises deepened when it was realized that Jones' link polynomials were related in a bewildering variety of ways to areas of physics where there had been no previous hints that knotting or linking might be involved [3], [23], [27]. Physicists who had studied the (quantum) Yang-Baxter equations seemed to have a machine ready to grind out link polynomials in such profusion that it seemed as if one needed invariants to distinguish the invariants, yet they had not known that the Yang-Baxter equation was related in any way to link theory. None of the new polynomials was previously known to topologists. More recently it has even become likely that the new invariants of links in S^3 generalize to new invariants of closed 3-manifolds and of link complements in 3-manifolds [27], [10].

Braid groups had been introduced into mathematical literature by Emil Artin in 1925 [2]. Artin's motivation in studying them was the interesting relationship between braiding and knotting. While braid groups were of clear interest in their own right, they had contributed little that was new to knot theory until 1984, when it soon became clear that the existence of some kind of braiding was a connecting thread between knot theory, operator algebras, and the various areas of physics involved in the discoveries. That is the theme we wish to explore in this article. The full story, whose implications are certain to be deep and far-reaching, is only beginning to be understood.

Links and Closed Braids

A *link* **K** is a finite collection of pairwise disjoint oriented circles embedded in oriented 3-space $\mathbb{R}^3$ or the 3-sphere S^3. If **K** consists of only one embedded circle, it is called a *knot*. We restrict ourselves to piecewise linear or (equivalently) smooth embeddings to avoid links that have pathological local behavior. Links **K** and **K**′ are said to determine the same link *type* if the oriented link **K** can be

Volume 13, No. 1 (Winter 1991), 52–60

deformed onto the oriented link **K'** by an isotopy of S^3. The equivalence class of all links with the same type as **K** is denoted by K.

The nine examples depicted in Figure 33-1 determine fewer than nine distinct link types. How can we tell? Our problem is known as the *knot problem*. That it is a non-triviality, even in the special case of recognizing the "unknot," will be obvious to anyone who has attempted to restore order in a box used to save leftover bits of string. There seems to be no systematic way to do it.

Near the turn of this century there was a flurry of interest in our problem on the part of physicists, notably Lord Kelvin and Peter Guthrie Tait, who thought that the arrangement of distinct elements in the periodic table might be related to knotting in the ether. Their ideas led to the gathering of a mass of experimental data in the form of tables of presumably distinct knot types, organized according to the number of double points in a planar projection. An enormous amount of patience and the liberal use of the eraser evidently dominated that early work. The expressed hope [25] was that computable invariants of knot type would be revealed in the natural course of assembling such data, but to the disappointment of those early workers that did not happen. The tables, instead, served two other equally important purposes. First, they gave convincing evidence of the non-triviality of the link problem. Second, they provided a rich set of examples for the more sophisticated studies that followed. The tables are used to this day and have had a strong influence on all of the work in the area. They can be seen in all their beauty and with surprisingly few corrections in current graduate-level textbooks on the subject, for example [24]. A sample page from the tables of 10-crossing knots is given in Figure 33-2.

Faced with the problem that there are just too many different ways to depict a single link type, one might try to reduce the number of representatives by adding extra structure. James Alexander did just that in 1923 [1]. His contribution is of fundamental importance to our story, so we pause to describe it. Let K be a knot or link type, and let **K** be a representative parametrized by cylindrical coordinates (r, θ, z) $r \neq 0$, in $\mathbb{R}^3$. Let t denote arclength on **K**. Our representative **K** is called a *closed n-braid* if $d\theta/dt > 0$ at all points of **K**. The integer n is the number of times **K** meets a half-plane through the z-axis. (This number is necessarily independent of the choice of the half-plane). The

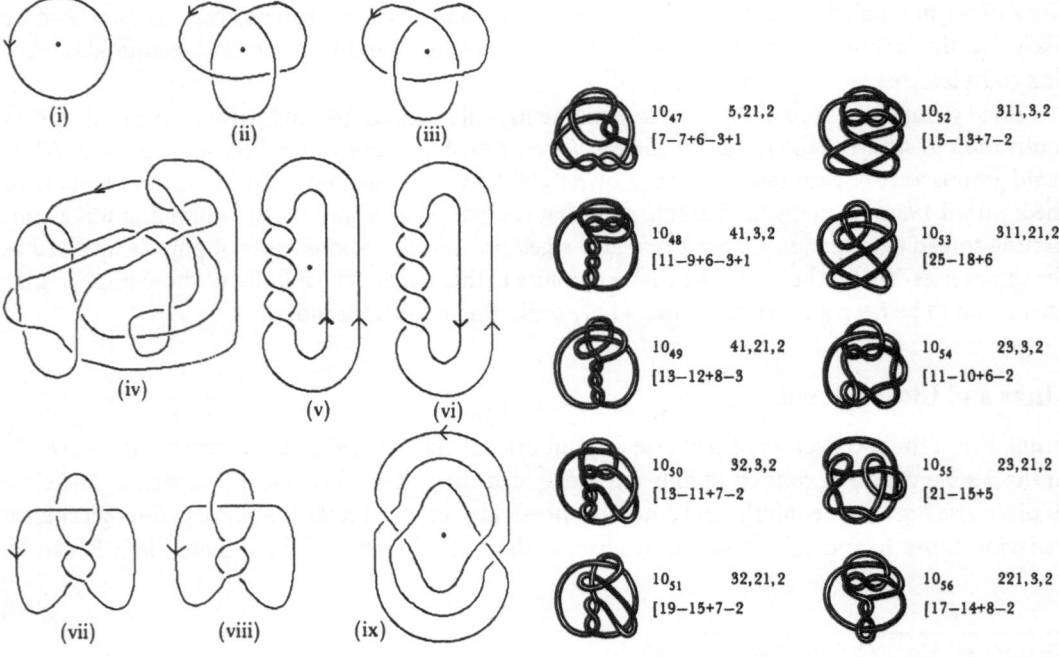

FIGURE 33-1 *Examples of knots and links.*

FIGURE 33-2 *A sample page from the knot tables.*

z-axis **A** is the *braid axis*. Examples can be seen in Figures 33-1(i)–(iii), (v), and (ix). The braid axis is orthogonal to the plane of the paper and its intersection with that plane is indicated by a black dot.

THEOREM 1. [1] Every link type can be represented by a closed braid.

PROOF: If **K** is not already a closed braid, we show how to change it to one. It will be convenient to assume that (possibly after a small deformation) **K** is polygonal, with edges $e_1, \ldots, e_m$, and that $d\theta/dt \neq 0$ on the interior of each edge. Call an edge e_i *bad* if $d\theta/dt < 0$ on e_i. By subdividing the bad edges if necessary, we can arrange that each bad edge e_i is an edge of a planar triangle τ_i such that $\mathbf{K} \cap \tau_i = e_i$ and $\mathbf{A} \cap \tau_i$ contains exactly one point. We can then replace e_i by $\partial\tau_i - e_i$ to remove the bad edge. After finitely many such replacements we'll have a closed braid representative of K. $\square$

Different proofs of Theorem 1 are given in [20], [28]; each brings new insights to the geometry.

Closed braids provide an immediate answer to a question that may have occurred to the reader. How can we describe a favorite link so that, for example, the description could be communicated by telephone to a colleague in another city? To obtain such a description from a closed braid representative **K**, let $\pi : \mathbb{R}^3 \to \mathbb{R}^2$ be orthogonal projection onto the plane $z = 0$. We may assume (if necessary after a small isotopy of **K**) that the singularities of $\pi \mid \mathbf{K}$ are at most a finite number of transverse double points at, say, $\theta = \theta_1 < \theta_2 < \cdots < \theta_k$. The intersections of **K** with the half-planes $\mathbf{H}(\theta)$ defined by $\theta = \theta_j - \epsilon$, ϵ small and $\neq 0$, will then be n points which have distinct r-coordinates $r_1(\theta) < r_2(\theta) < \cdots < r_n(\theta)$. Thus there will be a unique pair, say $r_i(\theta)$ and $r_{i+1}(\theta)$, that exchange r-order at the jth double point θ_j. We can then associate to the jth double point the symbol σ_i (respectively, σ_i^{-1}) according as the z-coordinate of $r_i(\theta_j)$ is bigger (or smaller) than that of $r_{i+1}(\theta_j)$. In this way we obtain a cyclic word of length k in the symbols $\sigma_1, \ldots, \sigma_{n-1}$ and their inverses, which describes **K** as an oriented closed braid. This word communicated by telephone, will give our friend precise instructions for reconstructing our picture of **K** as a closed braid. It also shows that the set of all link types is countable.

A Group of Braids

Closed braids lead naturally to open braids, as follows: Cut $\mathbb{R}^3 \setminus \mathbf{A}$ open along any half-plane $\mathbf{H}(\theta)$ to obtain an open solid cylinder $\mathbf{D} \times \mathbf{I}$, where **D** is an open 2-disc and **I** an interval. The closed braid **K** then goes over to a union of n disjoint arcs in $\mathbf{D} \times \mathbf{I}$, which intersect each disc $\mathbf{D} \times \{t\}$ in n points. The union of these arcs is an (open) braid. There is an equivalence relation on open braids induced by link equivalence. Figure 33-3 shows an open braid associated to the closed braid of Figure 33-1(ix).

A further standardization is now in order: we can choose n distinguished points $\mathbf{z}^0 = (z_1, \ldots, z_n)$ on **D**, and assume without loss of generality that the initial (resp., terminal) endpoints of our braided arcs are at $\mathbf{z}^0 \times \{0\}$ (resp., $\mathbf{z}^0 \times \{1\}$). This has an immediate bonus, because now we have a way to "multiply" two n-braids, namely by pasting together the associated $\mathbf{D} \times \mathbf{I}$'s, matching the $\mathbf{D} \times \{1\}$ face of the first with the $\mathbf{D} \times \{0\}$ face of the second, and rescaling. This multiplication can be seen to be associative. Moreover $\mathbf{z}^0 \times \mathbf{I}$ represents the identity element, and each braid has an inverse, obtained by reflection through the disc $\mathbf{D} \times \{1\}$. In short, the n-braids form a group, the *n-strand braid group B_n*, discovered by Artin in 1925 [2].

We now show that there is a beautiful and simple way to redefine the group B_n, which will at one and the same time make precise all we have just said, reveal the structure of B_n, and lead to generalizations and applications. With all those goals in mind, define the *configuration space* $\Sigma_n(\mathbf{M})$ of a manifold **M** to be the space $\{(z_1, \ldots, z_n) \mid z_i \in \mathbf{M}$ and $z_i \neq z_j$ if $i \neq j\}$ of all n-tuples of distinct points of **M**. Note that even though $\Sigma_n(\mathbf{M})$ has n times the dimension of **M**, we can think of its points as being a set of n distinct

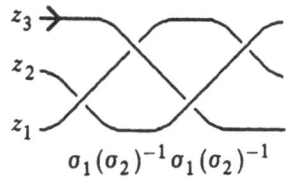

$$\sigma_1(\sigma_2)^{-1}\sigma_1(\sigma_2)^{-1}$$

FIGURE 33-3 *An open braid.*

points on a single copy of $\mathbf{M}$. Now, the permutation group S_n acts freely on $\Sigma_n(\mathbf{M})$, permuting coordinates, and we define a second and closely related configuration space to be the orbit space $\Omega(\mathbf{M})$ $= \Sigma_n(\mathbf{M})/S_n$. The new way to look at the braid group B_n, due to Fadell and Neuwirth [11], is to think of B_n as the fundamental group $\pi_1(\Omega_n(\mathbf{D}), \mathbf{z}^0)$. There is also a related *colored braid group* $P_n = \pi_1(\Sigma_n(\mathbf{D}), \mathbf{z}^0)$, so called because there is a well-defined assignment of colors to each of the n strands, which is preserved by group multiplication.

We can recover our intuitive definition of B_n as follows: An element $\beta \in \pi_1(\Omega_n(\mathbf{D}), \mathbf{z}^0)$ is represented by a $\mathbf{z}^0$-based loop in the space $\Omega_n(\mathbf{D})$, i.e., by n coordinate functions $\beta_1, \dots, \beta_n$, whose graphs are the n arcs that join $\mathbf{z}^0 \times \{0\}$ to $\mathbf{z}^0 \times \{1\}$ in $\mathbf{D} \times \mathbf{I}$. These arcs intersect each intermediate plane $\mathbf{D} \times \{t\}$ in exactly n distinct points. The fact that we are using Ω_n rather than Σ_n allows the ith braid strand to begin at z_i and end at some other z_j. The equivalence relation on the various geometric representatives of an element $\beta \in B_n$ is homotopy in the configuration space, which means that braid strands can be deformed by level-preserving deformations, arbitrary except that two strands cannot pass through one another. This is of course the essence of the phenomenon of knotting and linking.

The Algebraic Structure of B_n

We promised to uncover the structure of B_n, and shall do so. The first observation is that Σ_n is a (regular) covering space of Ω_n, with S_n as its group of covering translations. This reveals immediately that P_n is a normal subgroup of B_n, with quotient group S_n; i.e., we have a short-exact sequence:

$$\{1\} \to P_n \to B_n \to S_n \to \{1\}. \tag{1}$$

We can say more. There is a natural map $f_n : \Sigma_n \to \Sigma_{n-1}$, defined by forgetting the last coordinate. A straightforward construction (try it for an exercise!) proves that Σ_n is a fiber-space over the base Σ_{n-1}, with projection f_n, the fiber being $\mathbf{D} \setminus (z_1, \dots, z_{n-1})$, the $(n-1)$-times punctured plane $\mathbf{D}$. The latter group is a free group F_{n-1} of rank $n-1$. A deeper study of the long exact homotopy sequence of the fibration shows that groups that precede and follow the ones we have identified, i.e., F_{n-1}, P_n and P_{n-1}, are trivial. Thus one obtains a short exact sequence:

$$\{1\} \to F_{n-1} \to P_n \to P_{n-1} \to \{1\}. \tag{2}$$

We can say more. The group P_{n-1} clearly occurs as a *subgroup* of P_n. It is the subgroup of pure braids on the first $n-1$ strands. In fact, the short exact sequence of (2) splits, i.e., P_n is a semi-direct product of F_{n-1} and P_{n-1}.

The decomposition we just described can be repeated for P_{n-1}, and so on down to P_2, which is isomorphic to the infinite cyclic group F_1. In this way one sees that P_n is built up from the free groups

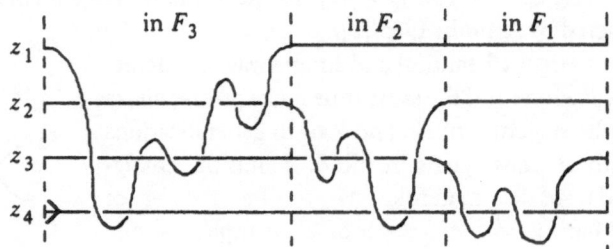

FIGURE 33-4 *Factorization of a pure 4-braid.*

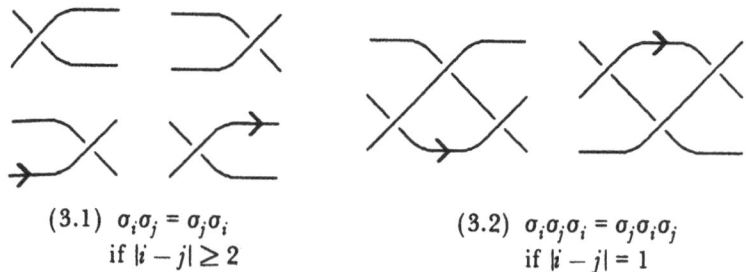

$(3.1)\ \sigma_i\sigma_j = \sigma_j\sigma_i$
if $|i - j| \geq 2$

$(3.2)\ \sigma_i\sigma_j\sigma_i = \sigma_j\sigma_i\sigma_j$
if $|i - j| = 1$

FIGURE 33-5 *Defining relations in the braid group.*

$F_{n-1}, F_{n-2}, \ldots, F_1$ by a sequence of semi-direct products. An example of a pure 4-braid that is a product of braids in F_3, F_2, and F_1 is given in Figure 33-4. The point is that *every* pure braid admits a unique factorization of this type. For details of the proof, see [11] or [5].

The approach just described yields, with a little more work, a presentation for the group B_n, with generators the elementary braids $\sigma_1, \ldots, \sigma_{n-1}$ described earlier. Defining relations can be shown to be:

$$(3.1) \qquad \sigma_i\sigma_j = \sigma_j\sigma_i \qquad\qquad \text{if } |i - j| \geq 2 \qquad\qquad (3)$$
$$(3.2) \qquad \sigma_i\sigma_j\sigma_i = \sigma_j\sigma_i\sigma_j \qquad\qquad \text{if } |i - j| = 1.$$

These relations are illustrated in Figure 33-5.

We make one final remark. There are two symmetries of special interest in knot and link theory, and both have simple meanings from the braid point of view. The first is a change in the orientation of **K**, and the second in the orientation of the ambient space S^3. Assume that **K** is the closure of an n-braid β, and that β is expressed as a word in the generators $\sigma_1, \ldots, \sigma_{n-1}$. Changing the orientation of **K** corresponds to reading the braid word backwards, and changing the orientation of S_3 corresponds to replacing each σ_i by its inverse. Replacing a braid word by its inverse thus corresponds to simultaneously changing the orientations on **K** and S^3.

Markov's Theorem

Markov's Theorem concerns the relationship between the various (open) braids whose closures define the same oriented link type. We would like to precede our discussion of this important theorem by backing up a little bit and returning to the looser concept of a link diagram, defined above. Let D and D' be oriented link diagrams. Call D and D' *Reidemeister-equivalent* if they define the same oriented link type. See Figure 33-6. Notice that relation (3.2) is a special case of R-III, whereas "free reduction" in B_n is essentially R-II.

THEOREM 2. [22] Reidemeister equivalence is generated by the three moves R-I, R-II, R-III that are illustrated in Figure 33-6.

A small amount of experimentation (or a proof based upon the use of polygonal representatives) should convince the reader of the reasonableness of Theorem

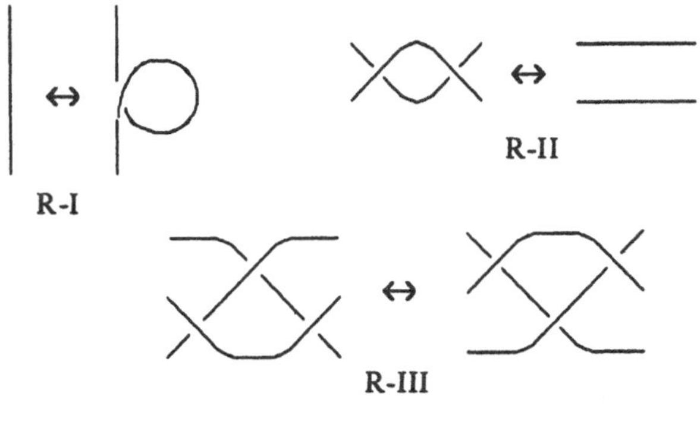

R-I

R-II

R-III

FIGURE 33-6 *Reidemeister's moves.*

2. A proof can be found in [9], but the reader who has succeeded in working out our earlier exercises might wish to construct one with the help of polygonal representatives. The proof proceeds in much the same way as the proof we sketched for Theorem 1, beginning with two polygonal representatives of a link, and assumes they are equivalent under a finite sequence of moves, each of which is an interchange that replaces one side of a planar triangle with the other two, or two sides with one.

Now, Markov's Theorem will be seen to be similar to Reidemeister's, except that the diagrams in question come from closed braids, and the equivalence is via a sequence of closed braid diagrams. The proof is harder because of these requirements. To state the theorem, let B_∞ denote the disjoint union of braid groups $B_1, B_2, B_3, \ldots$. Call braids $\beta \in B_n$ and $\beta \in B_m$ *Markov equivalent* if the closed braids they determine have the same oriented link type.

THEOREM 3.   Markov equivalence is generated by:

(i) Conjugacy
(ii) The maps $\mu^\pm: B_k \to B_{k+1}$, *with* $\beta \to \beta\sigma_k^{\pm 1}$.

Theorem 3 was announced by Markov in 1935 with a working outline for a proof. Proofs are in [5], [4], and [20]. Versions of Theorem 3 that replace the "stabilization" move (ii) with braid index-preserving moves are given in [6], [7].

Markov's Theorem has an immediate consequence. A function $f: B_\infty \to K$, where K is a ring, is a *Markov trace* if it is a class invariant in each B_k and satisfies $f(\beta) = f(\mu^+(\beta)) = f(\mu^-(\beta))$ for all $\beta \in B_\infty$. The following corollary of Markov's theorem is immediate:

COROLLARY 4.   *Any Markov trace is a link-type invariant.*

Indeed, most of the known link invariants have been interpreted as Markov traces on $B\infty$.

The Symmetric Group and the Braid Group

A head-on attack on Markov equivalence in B_∞ is hopelessly difficult, but one might hope that something could be done by passing to quotients of the B_n's. In view of the short exact sequence (1) above, a logical place to begin might be with the symmetric groups $S_1, S_2, S_3, \ldots$. Accordingly, let $\pi: B_n \to S_n$ be the homomorphism defined by the exact sequence (1). There is an obvious candidate for a Markov trace which factors through π, i.e., set $f(\beta) = $ the number of cycles in $\pi(\beta)$. The link invariant it determines is the number of components.

Encouraged by success, we look to representation theory for more subtle invariants. A remarkable phenomenon emerges. To illustrate this in a non-trivial case, we first note that the group S_n is generated by the transpositions $s_i = (i, i+1), i = 1, \ldots, n-1$, and in terms of these generators it has defining relations:

(4.1)	$s_1 s_j = s_j s_i$	if $|i - j| > 2$	
(4.2)	$s_i s_j s_i = s_j s_i s_j$	if $|i - j| = 1$	(4)
(4.3)	$s_i^2 = 1$	$i = 1, \ldots, n-1.$	

In view of the similarity between relations (3) and (4), one might wonder how the representations of B_n are related to those of S_n. We investigate an example. There is an $(n+1) \times (n+1)$-dimensional matrix representation of S_n over the integers, defined by mapping s_i to a matrix S_i that differs from the identity matrix only in the $(i-1)^{st}$, i^{th}, and $(i+1)^{st}$ entries in the ith row, which are

$(1, -1, 1)$ instead of $(0, 1, 0)$. Let t be a real number close to 1. We can "deform" this representation by changing these three entries to $(t, -t, 1)$ or $(1, -t^{-1}, t^{-1})$, the two choices giving a deformed matrix $S_i(t)$ or its inverse. A short calculation shows that the matrices $S_i(t)$ satisfy relations (4.1) and (4.2), but not (4.3); in fact, $S_i(t)$ has infinite order if $t \neq 1$. In view of the defining relations for B_n given by (3), the deformed matrices yield a one-parameter family of $(n + 1)$-dimensional representations of B_n. Alternatively, we can think of them as representations of B_n by matrices with entries in the ring $\mathbb{Z}[t, t^{-1}]$.

Label the rows and columns of the matrices $0, 1, \ldots, n$. Our representations are obviously reducible, because the 0th row and the nth row of each $S_i(t)$ are unit vectors. Therefore, the submatrix obtained by deleting the 0th and n^{th} rows and columns multiplies independently of the remaining entries and so yields an $(n - 1)$-dimensional representation $\rho_n : B_n \to M_n(\mathbb{Z}[t, t^{-1}])$. It is irreducible, because the representation of S_n obtained from it by setting $t = 1$ is known to be irreducible.

The representations ρ_n were discovered by Werner Burau in 1938 and have been the object of intensive investigations ever since. They yield Markov traces $\Delta : B_n \to \mathbb{Z}[t, t^{-1}]$, defined by the formula:

$$\Delta_\beta(t) = \frac{t^{n-1-\omega(\beta)} \det\left(1 - \rho_n(\beta)\right)}{1 + t + t^2 + \cdots + t^{n-1}} \tag{5}$$

where $\Delta_\beta(t)$ is the Laurent polynomial in t determined by the image $\rho_n(\beta)$ of $\beta \in B_n$ and $\omega(\beta)$ is the sum of the exponents of β when written as a product of the σ_i's. The invariant $\Delta_\beta(t)$ is the Alexander polynomial of the link defined by the closed braid. (Alexander's original methods, however, had nothing to do with braids.) To see that this is indeed a Markov trace, one must verify that it is invariant under the two changes described in Theorem 3. Invariance under (3.1) is immediate, because the characteristic polynomial of $\rho_n(\beta)$ is a class invariant; also $\omega(\beta)$ is invariant under the braid relations (3) and conjugacy. See [19] for a braid-theoretic proof of invariance under (3.2), or see Burau's proof, presented as Theorem 3.11 in [5].

The fact that we could deform one particular matrix representation of S_n to a parametrized family of representations of B_n was not an isolated phenomenon. Much more is true. The irreducible representations of S_n are well known. They are classified by Young diagrams and can be given by matrices whose entries are all 0s and 1s. *Every* irreducible representation of S_n deforms to a parametrized family of irreducible representation of B_n. In fact the entire group algebra of S_n deforms to an algebra H_n (the Hecke algebra of the symmetric group) that is a quotient of the group algebra of B_n. See [15] for details. The algebras H_n support a Markov trace that is a weighted sum of matrix traces on their irreducible summands. The link invariant determined by this Markov trace is the "Homfly" or 2-variable Jones polynomial of [12].

The picture does not end there. Kauffman has discovered in [17] yet another 2-variable polynomial-invariant of oriented link type, which is independent of all of the ones just described. Unlike the 1-variable Jones polynomial, the *Kauffman polynomial* was discovered by purely combinatorial techniques and seemed at first glance to be completely unrelated to braids. However, algebras we denote by W_n were constructed in [8]. They are quotients of the complex group algebra $\mathbb{C}B_n$ of the braid group, and support a 2-parameter family of Markov traces whose associated link invariant is the Kauffman polynomial. Each W_n contains H_n as a direct summand, and the Markov trace that defines the Homfly polynomial is the restriction to H_n of the Markov trace on W_n that defines the Kauffman polynomial. The algebras W_n are, moreover, deformations of a generalization of the complex group algebra $\mathbb{C}S_n$, much as H_n is a deformation of $\mathbb{C}S_n$.

If the irreducible representations of S_n are restricted to ones belonging to Young diagrams with at most two rows, the deformed algebra is the Jones algebra A_n, studied in [14]. That algebra is shown in [15] to support a Markov trace. The link invariant it determines is the Jones polynomial. We will learn how to compute it in the next section.

We summarize the situation. Let R_n denote the algebra (over the complex numbers) generated by the Burau matrices. Because R_n occurs as an irreducible summand of A_n, we have a sequence of algebra homomorphisms:

$$\mathbb{C}B_n \to W_n \to H_n \to A_n \to R_n \to S_n. \tag{6}$$

Each algebra supports a Markov trace and so determines a link-type invariant. In this way a uniform picture of the new and old link invariants is emerging, with the representation theory of B_n an important part of the picture.

Combinatorics and Link Theory

Among the new polynomials, the 1-variable Jones polynomial plays a very special role as the simplest example, after the connectivity, of a link invariant that arises as a Markov trace. We turn to Louis Kauffman's work in [16] for an exceptionally simple proof that the Jones polynomial is an invariant of oriented link type in oriented S^3. At the same time it will also show us how to compute the Jones polynomial from a link diagram.

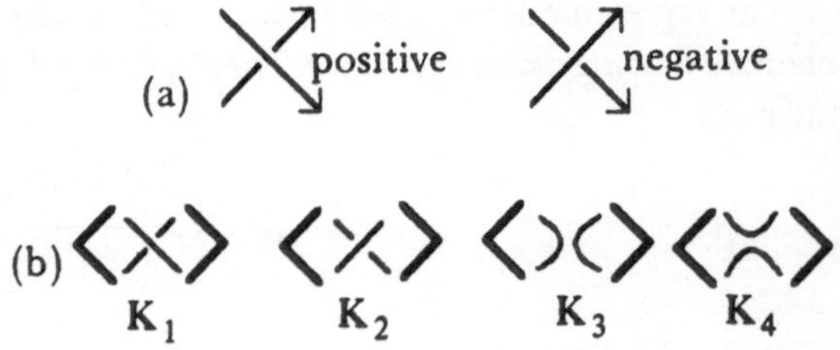

FIGURE 33-7 (a) Signed crossings; (b) Four related link diagrams.

Kauffman's work begins with Reidemeister's theorem, stated earlier as Theorem 2. The changes in a link diagram, which we depicted in Figure 33-6, are the *Reidemeister moves* of types I, II, and III. Kauffman's method is to use Reidemeister's moves to deduce the existence and invariance of the Jones polynomial. We begin with an oriented link defined by a diagram denoted **K**. Our diagram is *not*, in general, a braid diagram. The diagram determines an algebraic crossing number $\omega(\mathbf{K})$, the sign conventions being given in Figure 33-7a. Kauffman's version of the Jones polynomial takes the form:

$$F_{\mathbf{K}}(a) = (-a)^{-3\omega(\mathbf{K})}\langle \mathbf{K} \rangle, \tag{7}$$

where $\langle \mathbf{K} \rangle$ is a polynomial in the variable a which will be computed from the link diagram, ignoring the orientation. It is known as the "bracket polynomial."

We describe Kauffman's method for computing $\langle \mathbf{K} \rangle$. It rests on known properties of the Jones polynomial. Let **O** denote a simple closed curve in the plane. Let $\mathbf{K} \cup \mathbf{O}$ be the disjoint union of a non-empty diagram **K** and the diagram **O**. Consider four links, all defined by unoriented diagrams, the diagrams being identical except near a single crossing, where they differ in the manner illustrated in Figure 33-7b. We call our four diagrams $\mathbf{K}_1$, $\mathbf{K}_2$, $\mathbf{K}_3$, and $\mathbf{K}_4$. In general, they determine four distinct link types. The properties that will be seen to characterize $F_{\mathbf{K}}(a)$ are:

$$\langle \mathbf{O} \rangle = 1, \tag{P1}$$

$$\langle \mathbf{K} \cup \mathbf{O} \rangle = (-a^2 - a^{-2})\langle \mathbf{K} \rangle, \tag{P2}$$

$$\langle \mathbf{K}_1 \rangle = a^{-1}\langle \mathbf{K}_3 \rangle + a\langle \mathbf{K}_4 \rangle, \tag{P3}$$

$$\langle \mathbf{K}_2 \rangle = a\langle \mathbf{K}_3 \rangle + a^{-1}\langle \mathbf{K}_4 \rangle. \tag{P3'}$$

Note that (P3) implies (P3′) because if we rotate our pictures clockwise by 90°, we interchange $\mathbf{K}_1$ and $\mathbf{K}_2$, and $\mathbf{K}_3$ and $\mathbf{K}_4$. This would not be true if the diagrams were oriented.

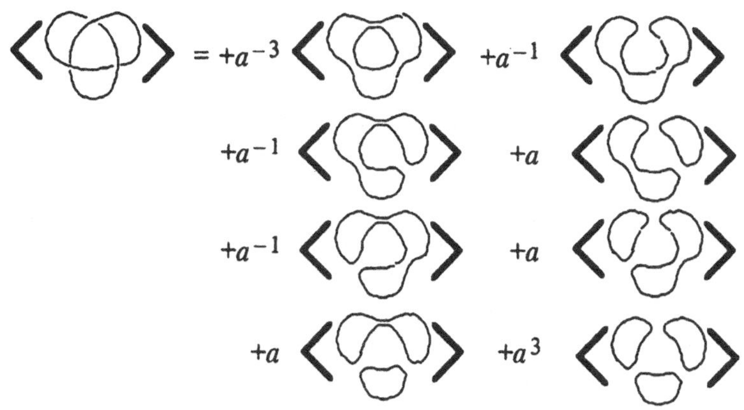

$$F_{\mathbf{K}}(a) = -a^9(-a^{-5} - a^3 + a^7) = a^4 + a^{12} - a^{16}$$

FIGURE 33-8 *Kauffman's version of the Jones polynomial of the trefoil.*

Because repeated applications of (P3) and (P3′) yield diagrams that have no crossings, and which therefore can only be a collection of disjoint circles, $\langle \mathbf{K} \rangle$ is determined unambiguously on all link diagrams and is a Laurent polynomial over the integers in the indeterminate a. Even more, if $\mathbf{K}$ had r crossings, it would follow that $\langle \mathbf{K} \rangle$ would be a sum of 2^r terms. We illustrate this with two examples:

Example 1 Let $\mathbf{K}$ be the r-component "unlink", i.e., a link that can be represented as the disjoint union of r planar circles. Applying (P2) $r - 1$ times we find that $F_{\mathbf{K}}(a) = (-a^2 - a^{-2})^{r-1} \langle \mathbf{O} \rangle$. It then follows from (P1) that the polynomial of the r-component unlink is $(-a^2 - a^{-2})^{r-1}$.

Example 2 is more complicated. We compute the polynomial of the trefoil knot shown earlier in Figure 33-1(ii). See Figure 33-8 for the steps in the calculation, which involve the repeated use of properties (P3) and (P3′). If we can prove that $F_{\mathbf{K}}(a)$ depends only on K we will have proved that this trefoil is not equivalent to the unknot. Fortunately, there is a very easy method, based upon Theorem 2. See Figure 33-9 for a picture proof that $f_{\mathbf{K}}(a)$ is invariant under R-II. We leave it as a simple exercise for the tireless reader to check invariance under R-I and R-III.

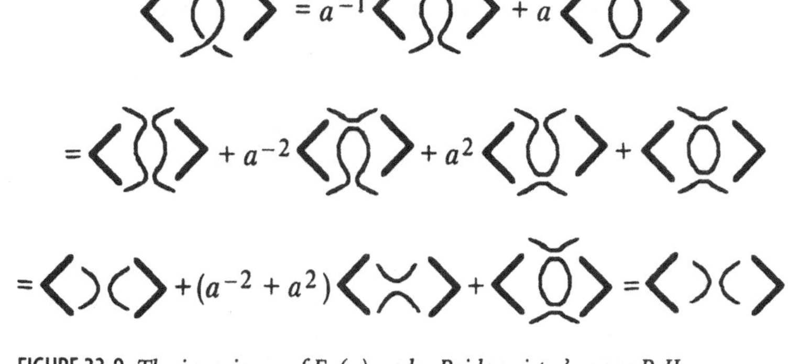

FIGURE 33-9 *The invariance of $F_{\mathbf{K}}(a)$ under Reidemeister's move R-II.*

The polynomial $F_K(a)$ is, after a change in variables, the Jones polynomial. See [16] for a proof of this assertion and other related ideas. It does a very good job of telling knots and links apart (although there are distinct links with the same Jones polynomial) and would have enabled Tait and his co-workers to reduce years of work into a few days of calculation.

The Yang-Baxter Equation

A head-on attack on the problem of finding invariants of Markov equivalence would be hopelessly difficult, but there are presents from the physicists. The Yang-Baxter equation and its solutions play a fundamental role in two physical problems; the theory of exactly solvable models in statistical mechanics (see [3]) and the theory of completely integrable systems. In statistical mechanics one studies systems of interacting particles and attempts to predict properties of the system that depend upon averages over all possible configurations or states of the system. As an example one might study a two-dimensional array of atoms located at the vertices of a lattice and interpret a "state" of the lattice to mean the assignment of a spin (which can take on $q \geq 2$ possible values) at each vertex. The total energy $E(\sigma)$ depends upon the state σ; the partition function, which is the object of ultimate interest, is a sum over all possible σs of the function $\exp(-kE(\sigma))$, where k is an appropriate constant.

The algebraic difficulties in computing the partition function are formidable; however, under certain conditions the problem is in fact solvable. The conditions are that certain matrices that describe the states of the system satisfy what is known as the Yang-Baxter equation. This has a curious geometric meaning. The matrices satisfy the Yang-Baxter equation if and only if they define a representation of the braid group B_n. To make this precise, we follow the development in [26] and let V be a free module with free basis $v_1, \ldots, v_m$ over a commutative ring K. Let $V^{\otimes n}$ be the n-fold tensor product of V with itself. We define elements $\{R_i, i = 1, \ldots, n-1\}$ of $\mathrm{Aut}(V^{\otimes n})$ by setting R_i equal to the identity on all but the ith and $(i+1)^{st}$ factors, while R_i restricted to those two factors is a fixed K-linear isomorphism:

$$R: V \otimes V \to V \otimes V. \tag{8.0}$$

Note that it follows immediately that $R_1, \ldots, R_{n-1}$ satisfy the condition:

$$R_i R_j = R_j R_i \quad \text{if } |i - j| > 1. \tag{8.1}$$

The automorphism R satisfies the *Yang-Baxter equation* if, in addition, $R_1 R_2 R_1 = R_2 R_1 R_2$. In view of the way in which we defined the R_i's, this implies that

$$R_i R_j R_i = R_j R_i R_j \quad \text{if } |i - j| = 1. \tag{8.2}$$

Comparing equations (3.1) and (3.2) with (8.1) and (8.2), we see that each solution to the Yang-Baxter equation determines, for each n, a representation of the braid group B_n in $\mathrm{Aut}(V^{\otimes n})$, with $\sigma_i \to R_i$. This representation in turn determines a finite-dimensional matrix representation of B_n. See [18] for explicit examples. Before 1984 it was not understood that the Yang-Baxter equation had anything to do with braids, and at this writing it is still unclear how braiding arises in the physical problem.

There are general methods [13] for finding solutions to the Yang-Baxter equations. (Solutions are actually associated to each representation of a simple Lie algebra.) In [26], further conditions are placed on the automorphism R, which ensure that the solutions so obtained support a Markov trace with values in the ring K. It turns out that *every* solution to the Yang-Baxter equation satisfies

Turaev's extra conditions. Thus, there is a machine ready to go, to produce further link invariants, all initially quite mysterious. Reshetiken has now shown in [23] that they can in fact all be obtained from the more basic polynomials described earlier via the algebra homomorphisms of (6), if one replaces a link by an appropriate "(p, q) cable" on it (see [24] or [9]). Thus order is emerging from chaos. The order, however, appears to be part of an even larger order, which involves the physics of conformal field theory and leads to further invariants, now in arbitrary 3-manifolds ([27],[10]).

Concluding Remarks

One of the great puzzles about the Jones polynomial and its various relatives is that, at this writing, we do not have any real understanding of their topological meaning. We know that they are invariants of oriented link type in oriented 3-space, but our proofs do not yield interpretations in terms of the link complement or link group or in terms of covering spaces or surfaces that the link bounds or, indeed, in terms of any of the familiar machinery of geometric and algebraic topology. There also appears to be little understanding, at this writing, of the "braiding" that occurs in the various areas of mathematics and physics that have been related to our story. These are hints of the enormous amount of work still to be done.

In conclusion, we challenge the reader to use the methods described above to distinguish the links in Figure 33-1, by computing their Jones and or Alexander polynomials, with the help of relations (5), or (P1)–(P3'). Examples were chosen to include a link whose type changes and one whose type does not change when one reverses the orientation of S^3, also to illustrate that a link type can change when one reverses the orientation on one of the components. Note that, by the derivation we gave above, the Jones polynomial is necessarily invariant under a reversal of orientation of *all* of the components.

If two diagrams determine links with district invariants, their link types will have been proved to be distinct. However, if they have the same invariants, they may or may not be distinct. Our examples illustrate this too. If one suspects they are not, one may then attempt to deform one diagram into the other to complete the proof.

References

1. J. W. Alexander, A lemma on systems of knotted curves, *Proc. Nat. Acad. Sciences USA* 9 (1923), 93–95.
2. E. Artin, Theorie der Zöpfe, *Hamburg. Abh.* 4 (1925), 47–72.
3. R. J. Baxter, *Exactly Solved Models in Statistical Mechanics.* London: Academic Press (1982).
4. D. Bennequin, Entrelacements et equations de Pfaff, *Astérisque* 107–108 (1983), 87–161.
5. J. S. Birman, Braids, links and mapping class groups. *Ann. of Math. Studies* No. 82, Princeton Univ. Press (1974).
6. J. Birman and W. Menasco, Closed braid representatives of the unlink, preprint, 1989.
7. J. Birman and W. Menasco, On the classification of links that are closed 3-braids, preprint (1989).
8. J. S. Birman and H. Wenzl, Braids, link polynomials and a new algebra. *Trans. AMS* 313 (1989), 249–273.
9. G. Burde and H. Zieschgang, *Knots.* Berlin; de Gruyter (1986).
10. L. Crane, Topology of 3-manifolds and conformal field theories, preprint, Yale Univ. (1989).
11. E. Fadell and L. Neuwirth, Configuration spaces, *Math. Scand.* 10 (1962), 111–118.
12. P. Freyd, J. Hoste, W. Lickorish, K. Millett, A. Ocneanu and D. Yetter, A new polynomial invariant of knots and links, *Bull Amer. Math. Soc.* (2)12 (1985), 257–267.

13. M. Jimbo, Quantum R-matrix related to the generalized Toda system: an algebraic approach, *Lecture Notes in Physics* 246 (1986), 335–361.

14. V. Jones, Braid groups, Hecke algebras and type II_1 factors, *Proc. US Japan Seminar Kyoto* (Araki and Effros, eds.). New York: John Wiley (1973).

15. ———. Hecke algebra representation of braid groups and link polynomials, *Ann. of Math.* (2)126 (1987), 335–388.

16. L. Kauffman, States models and the Jones polynomial, *Topology* 26 (1987), 395–407.

17. ———. An invariant of regular isotopy, *Trans. Amer. Math. Soc.* 318 (1990), 417–471.

18. T. Kohno, Linear representations of braid groups and classical Yang-Baxter equations, *BRAIDS*, *Contemp. Math.* 78, 339–364, *Amer. Math. Soc.* (1988).

19. A. King and M. Rocek, The Burau representation and the Alexander polynomial, preprint, Stony Brook: SUNY (1988).

20. H. Morton, Threading knot diagrams, *Math Proc. Camb. Phil. Soc.* 99 (1986), 246–260.

21. J. Prztycki and P. Traczyk, Invariants of links of Conway type, *Kobe J. Math.* 4 (1987), 115–139.

22. K. Reidemeister, *Knotentheorie*. New York: Chelsea Pub. Co. (1948). English translation: *Knot Theory*, BSC Associates, Moscow: ID (1983).

23. N. Reshetiken, Quantized universal enveloping algebras, the Yang-Baxter equation, and invariants of links I and II, preprint, Leningrad: Steklov Institute of Math. (1987).

24. D. Rolfsen, *Knots and Links*, Berkeley: Publish or Perish (1976).

25. P. G. Tait, On Knots I, II, III. *Scientific papers I*, London: Camb. Univ Press (1898).

26. V. Turaev, The Yang-Baxter equation and invariants of links, *Invent. Math.* 92 (1988), 527–553.

27. E. Witten, Quantum field theory and the Jones polynomial, *Comm. Math. Phys.* 121 (1989), 351–399.

28. S. Yamada. The minimum number of Seifert circles equals the braid index of a link, *Invent. Math.* 89 (1987), 347–356.

34

Hyperbolic Geometry and Spaces of Riemann Surfaces

Linda Keen

Introduction

Classifying Riemann surfaces is a problem that has fascinated mathematicians for more than a century. Real analytic, complex analytic, and geometric solutions have been found using a variety of techniques. In this article I shall examine several approaches; I shall restrict myself to the situation where the surface is a torus or a punctured torus and make the description very explicit.

Moduli Spaces for Riemann Surfaces

A Riemann surface is a topological surface with a complex analytic structure on it; that is, the surface is covered by a set of charts so that the relation between the maps defined on overlapping neighborhoods is complex analytic. If S_1 and S_2 are two Riemann surfaces, it can happen that there exist homeomorphisms from S_1 to S_2, and yet none of these homeomorphisms is complex analytic. In other words, S_1 and S_2 have the same underlying topological surface but are distinct as Riemann surfaces. It turns out that, unless the underlying surface is the 2-sphere or the 2-sphere minus 1, 2, or 3 points, there is a continuum of distinct Riemann surfaces with the same underlying surface. How then might we characterize the set of all distinct Riemann surfaces for a given topological surface S? This set is known as the *moduli space* of the surface and is denoted Mod(S). To put the characterization problem more concretely:

- Can we realize Mod(S) as some natural geometric object (e.g., as a real analytic manifold, or perhaps even as a complex analytic manifold)?
- Can we find parameters (these are the "moduli"), at least for some large open subset of this manifold, so that as we vary the parameters there is some aspect of the complex structure of the corresponding Riemann surfaces that is visibly varying with the parameters?

This article looks at several examples that illustrate what the problem is about and some of the methods that have been used to attack it. The geometric key is that a complex analytic homeomorphism is *conformal;* that is, it preserves angles locally. It is an easy exercise in calculus to show that if a map

Volume 16, No. 3 (Summer 1994), 11-19

is complex analytic and invertible at a point, then the angle between two curves intersecting transversally at that point, measured as the angle between their tangents, is equal to the angle between the image curves at the image point. Maps that distort angles cannot be complex analytic.

First Simple Example.   Let λ be a real number in the unit interval $I = \{0 < \lambda < 1\}$, and consider the cyclic group

$$G_\lambda = \{g_n : z \to \lambda^n z, n \in \mathbf{Z}\}$$

of conformal homeomorphisms of the punctured plane, $\mathbf{C}^* = \mathbf{C} - \{0\}$, to itself. The natural map $\mathbf{C}^* \to \mathbf{C}^*/G_\lambda \cong S_\lambda$ maps the half-open annulus $A_\lambda = \{|\lambda| \le |z| < 1\}$ one-to-one onto the quotient S_λ. Because $g_1 : z \to \lambda z$ maps the unit circle one-to-one onto the inner boundary of A_λ, we see that S_λ is topologically a torus. The image α of the unit circle and the image β of the real axis are a pair of generators for its homology. Projecting the complex structure from $\mathbf{C}$ onto S_λ makes it a Riemann surface.

When λ is close to 1, β is very short, so we get a very "skinny" torus, and as λ decreases, the torus gets "fatter." Because complex analytic maps preserve conformal geometry, this distortion is reflected in the complex structure, so it is plausible that we get a whole continuum of different complex structures as λ varies in the interval.

Now suppose that λ is no longer real but is a complex number $re^{i\theta}$ in the punctured unit disk, $D^* = \{z : 0 < |z| < 1\}$. Define the group G_λ, the quotient $\mathbf{C}^*/G_\lambda \cong S_\lambda$, and the annulus A_λ as above. The element $g_1 \in G_\lambda$ still identifies the inner boundary of A_λ with the outer boundary, but now the inner circle is twisted by the angle θ before it is glued. This twisting distorts the complex structure of the quotient, so for fixed r and varying θ, there is another whole continuum of different structures. In fact, classical theorems from elliptic function theory tell us that every possible complex structure on the torus is obtained from some $\lambda \in D^*$. Thus, D^* is a good candidate for out natural realization of $\text{Mod}(S)$ and λ is a natural parameter. However, D^* is *not* quite $\text{Mod}(S)$ because many different λ's may give rise to the same structure.

We shall return to this question after the next section. We shall see that the parameter space D^* is, in fact, a covering space of $\text{Mod}(S)$. It is typical that the moduli space is difficult to find; one often has to settle for a covering space.

Second Simple Example.   A more usual representation of the torus is obtained by considering the group $G_\tau = \{g_{m,n}(z) = z + m + n\tau : m, n \in \mathbf{Z}\}$, where τ is in the upper half-plane U, and forming the quotient $\mathbf{C}^* \to \mathbf{C}^*/G_\tau \cong S_\tau$. The parallelogram P_τ spanned by 1 and τ, with its opposite sides glued, is the analogue of the annulus. The complex structure on the torus is inherited from $\mathbf{C}$ and depends on G_τ. As uniform stretching doesn't change the complex structure, the quotient of $\mathbf{C}$ by the group $\{z \to z + rm + rn\tau : m, n \in \mathbf{Z}\}$, where r is any positive scalar, determines a torus equivalent to S_τ. The space of moduli is the collection of groups G_τ, $\tau \in U$. The parameter space U is, therefore, another covering space of $\text{Mod}(S)$ that is easy to find and to work with. To find $\text{Mod}(S)$ one has to see how different choices of generators for the groups G_τ are related. For these groups one knows how to do this. The classical modular group $PSL(2, \mathbf{Z})$ relates pairs of generators.

The plane $\mathbf{C}$ is simply connected and is the universal cover of the torus. The exponential maps the parallelogram P_τ onto the annulus A_λ for $\lambda = \exp(2\pi i \tau)$. Because the parameters τ and $-1/\tau$ give rise to equivalent tori, so do the parameters $\lambda = \exp(2\pi i \tau)$ and $\lambda' = \exp(-2\pi i/\tau)$. If $\tau = it$, for $t > 0$ real and large, λ is real and very small, so the underlying torus is fat. On the other hand, λ' is real and close to 1, so the underlying torus is skinny. Thus, a torus that is fat from one perspective is skinny from another.

The relation between τ and λ also shows that the parameter space D^* is an intermediate covering space between U and $\text{Mod}(S)$.

Boundary Behavior.   In our examples, the plane domains in which the parameter spaces are embedded have boundaries. This means that if our parameter reaches the boundary, something has happened—the construction of the torus no longer works.

Let us look at what happens when $\lambda = re^{i\theta}$ tends to the boundary of D^*. The absolute value of the parameter, $r = |\lambda|$, is measuring the size of the annulus. The open arc from 1 to λ projects to a closed curve β on the torus S_λ. If $\theta = 0$, so that $\lambda = r$, the length of β on S_λ is $|\log r|$; as $r \to 0$, it becomes infinite. On the other hand, suppose $\theta = 2\pi i p/q$ for p/q rational, and consider the collection of arcs

$$\{(r, 1), (re^{i\theta}, e^{i\theta}), (re^{2i\theta}, e^{2i\theta}), \dots, (re^{(q-1)i\theta}, e^{(q-1)i\theta})\}.$$

They project to a closed curve on S_λ that I again call β. Now if $r \to 1$, the arcs get short, β becomes "pinched," and its length on S_λ goes to zero. In either case, there is no longer a torus; it has become a doubly infinite cylinder.

Now let us look at the parameter space U. What happens as τ approaches the rational points on the boundary of U? Suppose $\tau = p/q + it$, $p, q \in \mathbf{Z}$, $t \in \mathbf{R}^+$. Draw the parallelogram P_τ spanned by 1 and τ; it contains the vertical line joining the origin and $-p + q\tau = qit$, which projects to a closed curve β on S_τ. As $t \to 0$, β is "pinched," and when $t = 0$, S_τ has degenerated to a doubly infinite cylinder again.

It is much more difficult to describe what is degenerating on S_τ as we approach the irrational points on the boundary. If we write $\tau = r + it$, where r is irrational and $t > 0$, the projection β of the vertical line in the parallelogram joining 0 and $-r + \tau = it$ never closes up on S_τ and, hence, is an open curve of infinite length. If we call α the projection of the generator joining 0 and 1, we see that what is getting shorter is the length of the segment of β between its successive intersections with α.

Third Example: A First Taste of Hyperbolic Geometry.   In our first example, we can think of the annulus A_λ as the torus cut open along a curve. The domain $\mathbf{C}^*$ is "tiled" by more annuli $A_n = g_n(A_\lambda)$; the annuli A_n don't overlap and together fill out all of $\mathbf{C}^*$. The group G_λ is a discrete group of conformal self-maps of $\mathbf{C}^*$.

In our second example, we can think of the parallelogram P_τ as the torus cut along a pair of simple curves that intersect exactly once. The group of translations G_τ determines a collection of parallelogram tiles, $P_{m,n} = g_{m,n}(P_\tau)$, that do not overlap and fill out all of $\mathbf{C}$. Again G_τ is a discrete group of conformal self-maps of $\mathbf{C}$.

We can apply this idea of "tiling" to obtain techniques that work not only on tori but also on more complicated Riemann surfaces. We cut the surface up to obtain a piece P of complex plane; we then try to find a group G of conformal maps to obtain a collection $\{gP\}$ of images of P that fill up some simply connected domain Ω in the plane without overlap; each element of G should be a conformal self-map of Ω.

I illustrate again with a simple example. Start with a torus, and take a pair of simple closed curves on it intersecting in the point η. Now remove the point η to

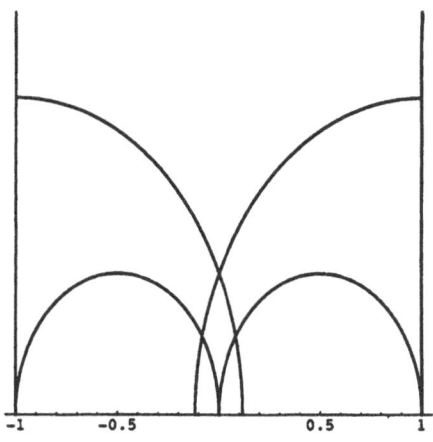

FIGURE 34-1 *The hyperbolic quadrilateral.*

obtain a *punctured* torus. Cutting along the curves gives us a quadrilateral with its corners removed. As we try to make our tiling, we see that because η is simply connected, the removed corners will have to be on the boundary of the domain η we are tiling. It follows that η will have to have at least four boundary points, and hence, by the Riemann mapping theorem, cannot be complex analytically equivalent to either $\mathbf{C}$ or $\mathbf{C}^*$.

Let us get very specific. Suppose that P (see Figure 34-1) is the region inside the upper half-plane U bounded by the semi-circles

$$C_1 = \left\{ \left| z + \frac{1}{2} \right| = \frac{1}{2} \right\} \cap U \quad \text{and} \quad C_2 = \left\{ \left| z - \frac{1}{2} \right| = \frac{1}{2} \right\}$$

and by the semi-infinite vertical lines

$$I_1 = \{ \Re z = 1 \} \cap U \quad \text{and} \quad I_2 = \{ \Re z = -1 \} \cap U.$$

Now consider the linear fractional transformations

$$g(z) = \frac{2z + 1}{z + 1}, \quad h(z) = \frac{z + 1}{z + 2},$$

and let $G = \langle g, h \rangle$ be the group they generate.

We easily compute

$$g(-1) = \infty, \quad g(0) = 1 \quad \text{and} \quad g((-1 + i)/2) = 1 + i,$$

so g maps the semi-circle C_1 onto the vertical line I_1. Moreover, it maps P onto a quadrilateral gP adjacent to P along I_1. It does not overlap P, and its vertices are again on $\mathbf{R}$. Similarly we see that h maps the vertical line I_2 to the semi-circle C_2; and that hP is a quadrilateral with vertices on $\mathbf{R}$, not overlapping P, and adjacent to P along the semi-circle. The group G is a discrete group of conformal self-maps of U, and because P has zero angles, one may show that the images of P under G do, in fact, tile U.

This example gives us a once-punctured torus, and it has a complex structure inherited from U. Now we can puncture any torus (the torus being homogeneous, it doesn't matter *where* we puncture it), so there is again a whole family of possible complex structures for the punctured torus. How can we introduce parameters into the group we just constructed to vary it and obtain these other punctured tori?

In the late nineteenth century, Poincaré [1] discovered a technique which he, Fricke [2], and others used on this problem. It was used again by a number of people in the mid-1960s, including Ahlfors [3], Bers [4], Fenchel [5], Maskit [6], and the author [7-9]; in the 1970s, it was enlarged and developed further by Thurston, Sullivan, and Gromov [10]. What Poincaré remarked on was that the group of linear transformations

$$(az + b)/(cz + d), \quad a, b, c, d \in \mathbf{R}, \quad ad - bc > 0,$$

are not only conformal homeomorphisms of U but are also isometries with respect to the *hyperbolic metric* on U.

The hyperbolic metric is defined by $ds = |dz|/\Im z$. Geodesics are circles orthogonal to the real axis (and vertical lines). The distance from any point inside U to a point on $\mathbf{R} \cup \{\infty\}$ is infinite.

In our second example, where we tile the plane by parallelograms, we may convince ourselves that we can choose the basic parallelogram in any shape by choosing the lengths of the sides and the angle between them. These lengths and the angle determine generating translations for the group. Because rescaling doesn't change the complex structure of the quotient, we may always assume one of the lengths is 1. Then, as the angles of a parallelogram add up to 2π, four copies fit around each corner, and we can tile the plane.

In our punctured torus example, the quadrilateral P is bounded by hyperbolic geodesics, but they have infinite length. Moreover, they meet at 0 angles at the boundary. Are there hyperbolic geometric invariants sitting inside P somewhere? Does it have a "hyperbolic shape"? The answer is yes!

The hyperbolic isometry $g(z)$ fixes exactly two points on $\mathbf{R}$ and leaves the hyperbolic geodesic A_g joining them invariant. This geodesic is called the *axis* of g. Unlike the Euclidean case, the hyperbolic distance between z and $g(z)$, $d_U(z, g(z))$, is not the same for all $z \in U$. This distance is minimal for any $z \in A_g$; the minimum distance l_g is called the *translation length* of g. Similarly the isometric h has an axis A_h and a translation length l_h; one sees that A_g and A_h intersect in exactly one point.

Suppose now that we try to construct an arbitrary hyperbolic quadrilateral P with four infinite sides, meeting in vertices on the real axis, and such that there are hyperbolic isometrics g and h identifying the pairs of opposite sides. It is a theorem, certainly known to Fricke and Fenchel, but first published by the author [8], that

- the "shape" of such a hyperbolic P is determined by the translation length of either isometric, l_g or l_h, and the angle θ between the axes A_g and A_h, and
- there is a P and a group for any given shape.

Only one length is necessary in this case because there are no isometries that change scale.

In sum, we have constructed a simply connected covering of the moduli space of a punctured torus parametrized by two real variables, $\{(l_g, \theta) \in \mathbf{R}^+ \times (0, \pi)\}$. These parameters have a geometric interpretation on the surface. There is also a simple way to write these parameters as real analytic functions of the coefficients of the generators of the group.

An important point here is that the methods of Examples 1 and 2 do *not* generalize to surfaces of higher genus, but these methods *do*.

Complex Moduli Spaces

The parameters for conformal structures on Riemann surfaces that we found above using hyperbolic geometric methods have many desirable properties. They are intrinsically defined. We can explicitly compute the polygonal tile and, hence, the group they determine; they work for arbitrary Riemann surfaces.

In the first two examples, we see how the complex structure on the torus depends on the parameter as a complex variable, so these parameter spaces have a rich structure. The methods, however, depend on elliptic function theory and work only for tori. In the third example, where the methods do generalize, the complex structure on the punctured torus depends on the parameters as independent real variables, so the parameter space has less structure. Ideally one would like to find a method for constructing parameter spaces for general Riemann surfaces so that the parameters are complex and the dependence of the geometry of the Riemann surface on these complex variables can be understood.

The Punctured torus revisited.   Here is another complex representation of the moduli space of a punctured torus that will generalize.

For $\mu \in U$, consider the group $G_\mu \langle g, h_\mu \rangle$ where

$$g(z) = z + 2, \quad h_\mu(z) = \frac{1}{z} + \mu.$$

Using techniques originated by Maskit (e.g., [6], VII), one can show that for appropriately chosen μ, there is a simply connected domain $\Omega(G_\mu)$ such that the group G_μ is a discrete group of conformal automorphisms and $\Omega(G_\mu)/G_\mu$ is a punctured torus.

To get an indication of how this works, choose $\mu = 3i$ and let P be the region (see Fig. 34-2)

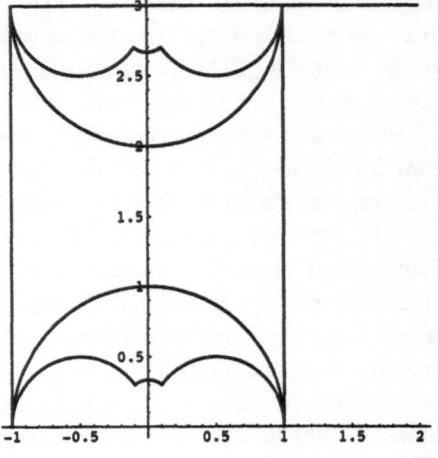

FIGURE 34-2 *The tile P for G_{3i}.*

- between the vertical lines $\Re z = -1$, $\Re z = 1$ and the circles $|z| = 1$ and $|z - 3i| = 1$,
- with vertices $-1, 1, 1 + 3i$ and $-1 + 3i$.

The map $g(z)$ takes the left side of P to the right side and maps P to a translate adjacent along the right side. The map $h(z)$ takes the lower semi-circular boundary onto the upper one and maps P to a quadrilateral adjacent along the upper semi-circle. This is the start of our tiling. It is not obvious, but it follows from Maskit's theory that the images of P under G_{3i} do not overlap. As we generate these images of P, they fill out some domain $\Omega(G_{3i})$ in $\mathbf{C}$, which, by construction, is invariant under G_{3i}.

Unlike Example 3, where the images of the tile P filled out the recognizable upper half-plane, the domain $\Omega(G_{3i})$ is not easily described; in fact, $\Omega(G_\mu)$ is different for different choices of μ. To get some idea of what the domains $\Omega(G_\mu)$ can look like, in Figures 34-3, 34-4, and 34-5, I show the computer pictures made by Ian Redfern at Warwick University for the groups G_μ with $\mu = 3i$, $\mu = 0.0533 + 1.9i$, and $\mu = 0.5001 + 1.667i$. The domain $\Omega(G_\mu)$ is the complement of the closed circles; its boundary is quite intricate.

The Maskit parameter space. To show that the specific group G_{3i} gives us another way to represent a particular punctured torus by forming the quotient $\Omega(G_{3i})/G_{3i}$ requires Maskit's combinatorial theory for groups that represent Riemann surfaces. To show that every punctured torus can be

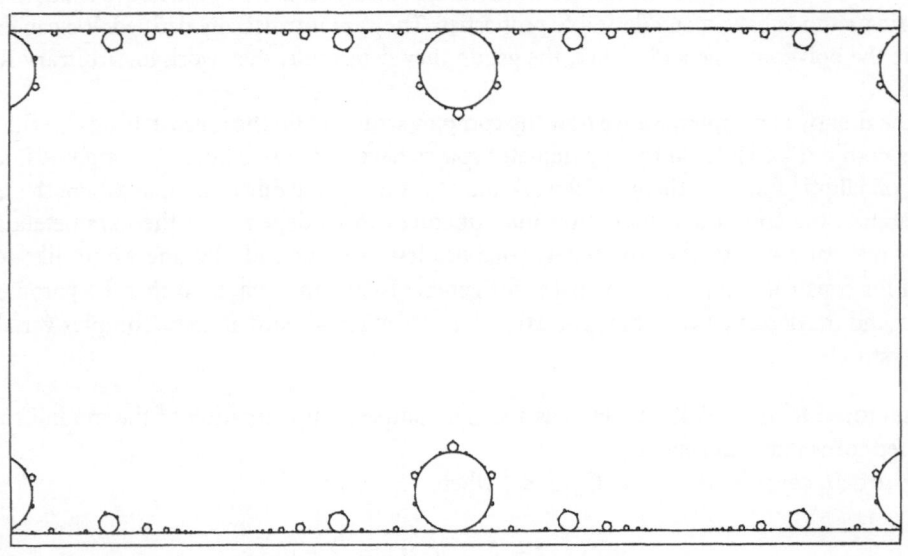

FIGURE 34-3 $\Omega(G_{3i})$

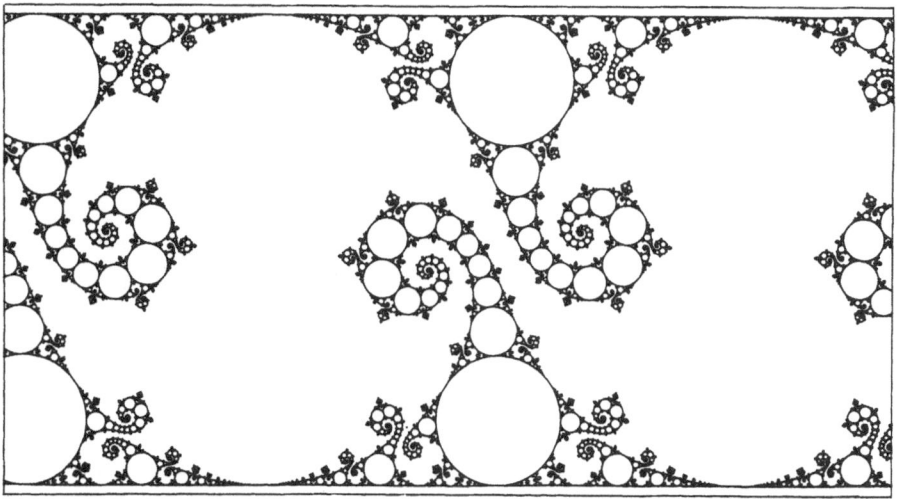

FIGURE 34-4 $\Omega(G_{0.0533+1.9i})$.

represented by some group in the family $\{G_\mu\}$ takes a different set of techniques, from partial differential equations and the theory of quasi-conformal mappings developed by Ahlfors and Bers in the early 1960s (see, e.g., [3, 4, 11]). I will not explain these theories but just report that for the punctured torus they tell us that:

- there is a simply connected domain $\mathcal{M} \subseteq U$ such that for $\mu \in \mathcal{M}$ there is a domain $\Omega(G_\mu)$ on which the group G_μ acts as a group of conformal homeomorphisms and such that $\Omega(G_\mu)/G_\mu$ represents a punctured torus, and
- every punctured torus is represented by some $\mu \in \mathcal{M}$.

In the first example, the cyclic group depending on a parameter $\lambda \in D^*$, we saw that the boundary of the disk was a natural boundary because the tori had degenerated. Similarly, the Maskit parameter space $\mathcal{M}$ is embedded as a domain inside the upper half-plane and the tori must degenerate as we approach its boundary. How can we find or describe this boundary? Using Maskit's

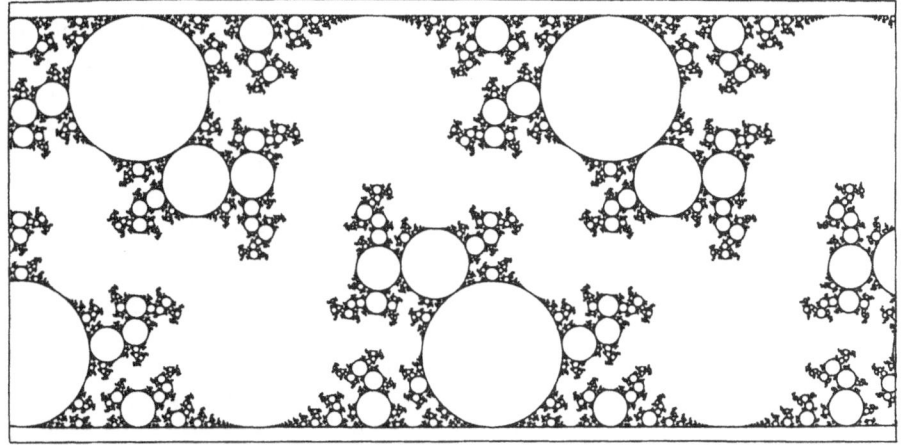

FIGURE 34-5 $\Omega(G_{0.5001+1.667i})$.

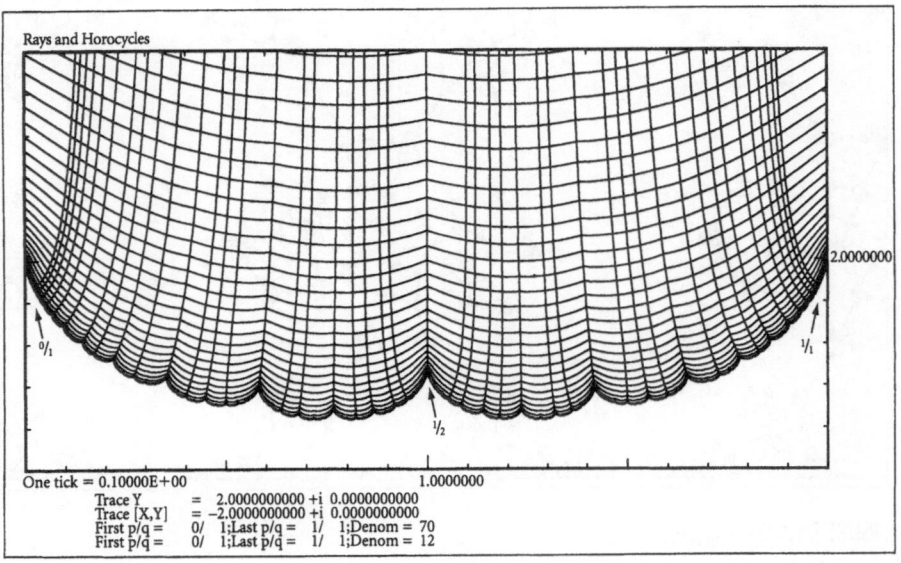

FIGURE 34-6 *The Maskit embedding with pleating coordinates.*

techniques one can prove that the half-plane $\Im\mu > 2$ is contained in $\mathcal{M}$. Therefore, the boundary $\partial\mathcal{M}$ in $\mathbf{C}$ is in the horizontal strip $0 < \Im\mu \leq 2$. Wright [12] used experimental techniques to compute this boundary and came up with the picture in Figure 34-6. This picture was the jumping-off point for the author's ongoing collaboration with Caroline Series on complex moduli spaces [13-17].

Hyperbolic geometry again—this time in three dimensions.   When Poincaré realized that linear fractional transformations with real coefficients were isometries of the hyperbolic plane, he also realized that linear fractional transformations with complex coefficients were isometries of hyperbolic 3-space. Hyperbolic 3-space can be modeled by the upper half-space $\mathbf{H}^3 = \{(z, t) : z \in \mathbf{C}, t \in \mathbf{R}^+\}$. The hyperbolic metric there is given by $ds = \sqrt{|dz|^2 + dt^2}/t$. Geodesics are circles orthogonal to the base $\mathbf{C}$, and hyperbolic planes are hemispheres orthogonal to the base $\mathbf{C}$.

Linear fractional transformations map circles and straight lines in $\mathbf{C}$ onto circles and straight lines; in fact, each is a product of an even number of reflections in lines and inversions in circles. Given a circle in $\mathbf{C}$, we can view it as the equator of a sphere in $\mathbf{R}^3$. An inversion in the circle extends naturally to an inversion in the sphere. Similarly, reflections in lines in $\mathbf{C}$ extend to reflections in the planes through them orthogonal to $\mathbf{C}$. The isometry of $\mathbf{H}^3$ corresponding to a linear transformation is a product of these extended inversions and reflections. One checks that since there are an even number, the upper half-space is preserved. It is an exercise to check that the metric is preserved.

Fenchel [5] and Greenberg [18], in the early 1960s, began to use techniques of 3-dimensional hyperbolic geometry to study groups representing Riemann surfaces. The idea was that because the group was discrete, one could look at the quotient 3-manifold, $\mathbf{H}^3/G$. It is a manifold with boundary, and the Riemann surfaces represented by the group are the boundary components.

Let us see how this works in our third example, the group $G = \langle g, h \rangle$. The action of G on U can be extended to $\mathbf{H}^3$ and to the lower half-plane L. Reflect the circles C_i and the lines I_i, $i = 1, 2$, in the real axis. These reflections determine a region $\bar{P}$ in L, and L/G is again a punctured torus. Note that the surfaces U/G and L/G are *anti*-holomorphically equivalent, for the maps on local neighborhoods are given by complex conjugation. The hemispheres over the circles C_i and the vertical half-planes over the lines I_i, $i = 1, 2$, bound a region $R \subset \mathbf{H}^3$; and g identifies the hemisphere over C_1 with the plane over I_1, while h identifies the plane over I_2 with the hemisphere over C_2. The polyhedron R is a tile for the group G acting on $\mathbf{H}^3$. The quotient $(U \cup L \cup \mathbf{H}^3)/G$ is a 3-manifold whose boundary consists of a pair of antiholomorphically equivalent punctured tori. As we shall see, not all groups of linear fractional transformations acting on $\mathbf{H}^3$ act so symmetrically with respect to the real line, nor are the relations among their boundary surfaces so easy to determine.

In the early 1970s, Marden [19] studied the relationship between groups of linear fractional transformations acting on $\mathbf{H}^3$ and the topological properties of their quotient 3-manifolds, and in the late 1970s, Thurston [10] introduced revolutionary new techniques involving this hyperbolic geometry to attack classification problems for both Riemann surfaces and 3-manifolds.

Convex hulls and pleated surfaces.   Let us return to the family of groups $\{G_\mu\}$ for $\mu \in \mathcal{M}$. For each group we have an open plane domain $\Omega(G_\mu)$ invariant under G_μ. The boundary, $\lambda(G_\mu) = \Omega(G_\mu)$, is a closed G_μ-invariant set called the *limit set of* G_μ. Let us turn our attention to it.

One of Thurston's ideas was to consider the (hyperbolic) *convex hull* C in $\mathbf{H}^3$ of the set $\lambda(G_\mu)$ and to study its boundary. This boundary is also G_μ-invariant, and one can prove that there is a G_μ-invariant component of this boundary, $\partial C(G_\mu)$, that is homeomorphic to $\Omega(G_\mu)$. The quotient, $\partial C(G_\mu)$, that is homeomorphic to $\Omega(G_\mu)$. The quotient, $\partial C(G_\mu)/G_\mu$, is, therefore, again a punctured torus.

Thurston saw that ∂C had certain geometric properties that were very useful. It is a surface in $\mathbf{H}^3$ made of pieces of hyperbolic plane joined along geodesic curves that, because of the convexity, can only meet on $\hat{\mathbf{C}} = \partial \mathbf{H}_3$ in points of $\lambda(G_\mu)$. The quotient surface $S_\mu = \partial C(G_\mu)/G_\mu$ is, therefore, also made up of pieces of hyperbolic plane joined along non-intersecting geodesics. Thurston called such surfaces *pleated*, and the geodesics along which they are pleated, the *pleating locus*.

Before we see how to extract information about the groups G_μ from these ideas, let us see why they don't give any new information about the groups of Example 3. There, the set Ω is always U and the limit set λ is the real line. The convex hull of λ is the vertical plane above λ and so is equal to its boundary. There is only one hyperbolic plane and so no geodesic where two planes are joined; the pleating locus is, therefore, empty.

Figure 34-4 is a picture of $\lambda(G_{0.0533+1.9i})$. This limit set is very intricate and its convex hull C has interior. To get a sense of what C looks like, note that by definition the hyperbolic geodesic joining any pair of points in λ belongs to C, as does the hyperbolic triangle spanned by any three points in λ. If four or more points of λ lie in a circle C, the convex polygon they span is in a plane and in C. If, moreover, there are no points of λ inside C, then the hyperbolic plane spanned by C intersects ∂C in a hyperbolic convex polygon.

If we look carefully at Figure 34-3, 34-4, or 34-5, we see a pattern of closed circles and other overlapping circles with missing boundary arcs. The interiors of these circles contain no points in λ. The boundary of the convex hull in $\mathbf{H}^3$ consists of the intersection of the hyperbolic planes spanned by all the circles that we see. The full planes spanned by the closed circles belong to ∂C is an infinite-sided convex polygon. When a pair of circles intersect, the planes spanning them intersect in a circular arc that is a boundary curve of the polygon on each plane. It is a geodesic with its endpoints in the limit set; ∂C is "bent" along this geodesic at an angle equal to the angle between the circles. The set of geodesics formed by the intersecting planes is the pleating locus.

In computer pictures for various groups $G_\mu \in \mathcal{M}$ made by Wright and Redfern, and particularly for those groups near the boundary, one could see patterns of circles in the limit sets $\lambda(G_\mu)$, and we thought that there should be meaning to the patterns. For example, note that the patterns in Figures 34-3, 34-4, and 34-5 are decidedly different. What Series and I realized is that whenever a pattern of circles appears in $\lambda(G_\mu)$, the quotient S_μ is pleated along some simple closed curve and the curve is determined by the pattern!

Enumerating simple closed curves on the punctured torus.   Consider the unpunctured torus with simple closed geodesics α and β intersecting once. We may assume α is the projection of the line joining 0 and τ. Every simple closed geodesic on the torus then has the form $p\alpha + q\beta$ and is the projection of a line joining 0 to $p + q\tau$ for relatively prime integers p and q. For each such pair, (p, q), we have a family of parallel lines projecting onto a family of parallel geodesics.

The fundamental group of the punctured torus, $\pi_1(S)$ is also generated by a pair of simple closed geodesics intersecting once, but it is a free group not an abelian group. We know from Maskit's theory that for $\mu \in \mathcal{M}$, the domain $\Omega(G_\mu)$ is simply connected; it follows that G_μ is isomorphic to $\pi_1(S)$. The "forgetful map" from S into the unpunctured torus, defined by forgetting the puncture, shows that each simple closed curve on S is also a simple closed curve on the unpunctured torus. It induces a projection on fundamental groups

$$\pi_1(S) \to \mathbf{Z} + \mathbf{Z},$$

from which we see that there are many elements in $\pi_1(S)$ that project to $(p\alpha + q\beta)$.

It is an interesting fact, proved by Series [20], that there is a unique simple closed geodesic $\gamma_{p/q}$ in the inverse image of $(p\alpha + q\beta)$. Moreover, there is a unique conjugacy class in G_μ containing a shortest cyclically reduced representative for that geodesic. Hence, there is a canonical word $W_{p/q}$ in G_μ associated to each simple closed geodesic on the punctured torus. These words may be enumerated recursively using continued fractions.

Pleating curves and moduli.   Given a linear fractional transformation, $(az + b)/(cz + d)$, we may assume without loss of generality that $ad - bc = 1$. The *trace* of the transformation, $a + d$, is then well defined, and conjugate transformations have the same trace.

If we look at words in G_μ they are compositions of the maps g and h_μ, so their coefficients and their traces are polynomials in μ with integral coefficients.

The crucial observation [13] that relates the complex parameter μ to the geometry of the hyperbolic 3-manifold $\mathbf{H}^3/G_\mu$ is

THEOREM 1.   Whenever the quotient of the convex hull boundary S_μ is pleated along the curve $\gamma_{p/q}$, the trace polynomial $\mathrm{Tr}\, W_{p/q}(\mu)$ is real-valued.

We also prove

THEOREM 2.   For any pair of relatively prime integers, (p, q), there is some $\mu \in \mathcal{M}$ such that S_μ is pleated along $\gamma_{p/q}$.

These theorems together give this picture of the parameter space:

THEOREM 3.   The space $\mathcal{M}$ is foliated by real analytic curves $\mathcal{P}_r$, $r \in \mathbf{R}$, a dense subfamily of which is defined by properly chosen branches of the curves defined by

$$\mathfrak{I}\, \mathrm{Tr}\, W_{p/q}(\mu) = 0$$

(the "vertical" curves in Fig. 6). These curves meet the boundary of $\mathcal{M}$ at points where the torus has degenerated because the curve $\gamma_{p/q}$ has been pinched.

The final piece of the relationship between the complex and geometric parameters is given by

THEOREM 4. There is a family, $\{l_r(\mu)\}_{r \in \mathbf{R}}$, of analytic maps from $\mathcal{M}$ to $\mathbf{C}$ that vary continuously with r. The value $l_r(\mu)$, for r rational and $\mu \in \mathcal{P}_r$, equals the appropriately normalized length of the pleating locus. The pairs $(r, l_r(\mu))$ define a new set of coordinates for $\mathcal{M}$.

The level curves $l_r(\mu) = $ const are the "horizontal" curves in Figure 34-6.

We have generalized these techniques to *twice*-punctured tori and expect them to generalize to arbitrary surfaces [15, 17].

References

1. H. Poincaré, *Papers on Fuchsian Functions*, translated by J. Stillwell, New York: Springer-Verlag (1985).
2. R. Fricke and F. Klein, *Vorlesungen über die Theorie der automorphen Funktionen*, New York: Johnson Reprint (1965), Vol. 2.
3. L. Ahlfors, *Lectures on Quasi-conformal Mappings*, New York: Van Nostrand (1966).
4. L. Bers, Finite dimensional Teichmüller spaces and generalizations, *Bull. AMS* (2) 5 (1972), 257–300.
5. W. Fenchel and J. Nielsen, Discrete groups, unpublished manuscript.
6. B. Maskit, *Kleinian Groups*, New York: Springer-Verlag (1987).
7. L. Keen, Canonical polygons for finitely generated Fuchsian groups, *Acta Math.* 115 (1966), 1–16.
8. L. Keen, Intrinsic moduli on Riemann surfaces, *Ann. Math.* 84(3) (1966), 404–420.
9. L. Keen, On Fricke moduli, *Advances in the Theory of Riemann Surfaces*, Princeton, NJ: Princeton University Press (1971), 205–224.
10. W P. Thurston, The geometry and topology of three-manifolds, unpublished manuscript.
11. F. Gardiner, *Teichmüller Theory and Quadratic Differentials*, New York: Wiley (1987).
12. D. J. Wright, The shape of the boundary of Maskit's embedding of the Teichmüller space of once punctured tori, preprint.
13. L. Keen and C. Series, Pleating coordinates for the Maskit embedding of the Teichmüller space of punctured tori, *Topology* 32(4) (1993), 719–749.
14. L. Keen and C. Series, Pleating coordinates for the Teichmüller space of punctured tori, *Bull. AMS.* (2) 26 (1992), 141–146.
15. L. Keen and C. Series, The Riley slice of Schottky space, *Proc. London Math. Soc.* (3) 69 (1994), 72–90.
16. L. Keen, B. Maskit, and C. Series, Geometric finiteness and uniqueness for Kleinian groups with circle-packing limit uniqueness for Kleinian groups with circle-packing limit sets, *J. Reine angew. Math.* 436 (1993), 209–219.
17. L. Keen, J. Parker, and C. Series, Pleating coordinates for the twice-punctured torus, preprint.
18. L. Greenberg, Fundamental polyhedra for Kleinian groups. *Ann. Math.* 84(2) (1966), 433–441.
19. A Marden, The geometry of finitely generated Kleinian groups, *Ann. Math.* 99 (1974), 383–462.
20. C. Series, The geometry of Markoff numbers, *Math Intelligencer* 7(3) (1985), 20–29.

Part Seven

HISTORY OF MATHEMATICS

35

Kurt Gödel in Sharper Focus

John W. Dawson Jr.

1. Introduction

The lives of great thinkers are sometimes overshadowed by their achievements—a phenomenon perhaps no better exemplified than by the life and work of Kurt Gödel, a reclusive genius whose incompleteness theorems and set-theoretic consistency proofs are among the most celebrated results of twentieth-century mathematics, yet whose life history has until recently remained almost unknown.

Several tributes to Gödel have appeared since his death in 1978, most notably the obituary memoirs of Curt Christian [2], Georg Kreisel [11], and Hao Wang [15].[1] But none of those authors developed a close personal acquaintance with Gödel before the 1950s, and there are discrepancies among their accounts. To resolve them, to substantiate or refute various rumors that have circulated, and to learn further details about Gödel's life and work, scholars must therefore turn to primary documentary sources. In the remainder of this article I shall highlight some aspects of Gödel's life that have been thrown into sharper focus as the result of my own explorations among such sources; I have drawn primarily on my experiences over the past two years in cataloguing Gödel's *Nachlass* at the Institute for Advanced Study in Princeton, supplemented by personal interviews, visits, and correspondence with various individuals here and abroad. Of necessity, I have repeated some biographical information already available in the memoirs cited. My aim, however, has been to amplify or correct details of those accounts, insofar as primary documentation is available.

I am grateful to the Institute for Advanced Study for permission to quote from, and to reproduce photographs of, unpublished items in the *Nachlass;* to Rudolf Gödel, Kurt's brother and only surviving close relative, for his gracious responses to my inquiries; and to H. Landshoff for assistance in preparing illustrations for this article.

2. The Gödel *Nachlass:* Provenance, Arrangement, and Disposition

The scientific *Nachlass* of Kurt Gödel, including correspondence, drafts, notebooks, unpublished manuscripts, books from his library, and all manner of loose notes and memoranda, was donated to the Institute for Advanced Study after Gödel's death by his widow. Gödel himself made no provision for the disposition of his papers, although correspondence in the *Nachlass* indicates that the Library

---

[1]Wang's later memoir [16] contains some interesting additional information, but on the whole it is less reliable, even though it alone was written before Gödel's death and was submitted to Gödel for his approval and correction. In particular, [16] is marred by several errors in references to dates and places.

Volume 6, No. 4 (Fall 1984), 9–17

Kurt Gödel, March 1939.

of Congress solicited them from him. Indeed, his attitude toward posthumous preservation or publication of his papers seems to have been ambivalent. Thus he spoke to Dana Scott of his desire to have certain papers published posthumously and even asked Scott to prepare typescripts of some of them. Yet on the other hand, he declined several invitations to consider publication of his collected works, maintaining that the most important of them were readily available and that the rest were only of historical and biographical interest. Ultimately, his papers were gathered into boxes and placed in a cage in the basement of the Institute's historical studies library (which has no archival facilities) to await further disposition.

My own involvement with Gödel's papers began in the fall of 1980. In an effort to track down Gödel's lesser-known publications, I consulted the bibliography in [1], prepared on the occasion of Gödel's sixtieth birthday. Though I presumed that that listing would be complete, a colleague soon called my attention to an item not cited there, and I later found two more. Subsequently, I wrote to the I.A.S. to inquire whether Gödel himself had ever prepared a bibliography of his own work; in response, I received a typewritten list identical to that in [1]. Concluding that no comprehensive bibliography had ever been attempted, I undertook to compile one myself, at the same time embarking on the more ambitious program of translating all of Gödel's previously untranslated works into English.[2] Eventually I was offered the opportunity of cataloguing the *Nachlass,* and in June 1982 I arrived in Princeton to begin work.

At the outset, there were three major problems to be faced: the sheer bulk of the materials comprising the *Nachlass;* my lack of training in archival technique; and the necessity of penetrating Gödel's Gabelsberger shorthand, an obsolete German script[3] he used extensively. It was at first somewhat daunting to find that the *Nachlass* occupied two large filing cabinets plus some *sixty*-odd cardboard packing boxes. The majority of the boxes, however, contained books from Gödel's library, back issues of journals he received, and preprints and offprints sent to him by others. The remaining, much more manageable, body of primary documentary material required about three months to survey in order to devise an appropriate arrangement scheme. (See [4] for further details.)

My ignorance of archival principles was remedied by consultations with archivists and by reference to Gracy's helpful manual [9]. Gödel's eminently logical mind, extremely methodical habits, and clear handwriting were also a boon to my efforts. In particular, one fundamental archival conflict—that of preserving the original order of a manuscript collection while facilitating scholarly access to it—has seldom arisen. My task has rather been that of restoring Gödel's order to papers gathered somewhat haphazardly after his death.

---

[2]The bibliography has now appeared [3]; note also [5]. The translations will be included in the first volume of our forthcoming edition of Gödel's collected works.

[3]Devised by Franz Xavier Gabelsberger, for whom it is named, the script was one of two competing German shorthand systems in widespread use during the early decades of this century. Eventually the two systems merged to form the modern *Einheitskurzschrift* ("unified shorthand"); alas, however, those trained only in the modern system can read *neither* the Gabelsberger nor Stolze-Schrey scripts from which it was derived. Yet there is a need for some younger scholars to learn to read these scripts, since many prominent intellectuals used them in their daily lives—not, as is often supposed, for reasons of secrecy, but as efficient means for rapid and concise recording of events and ideas.

The shorthand problem proved the most difficult. Early on, my wife volunteered to learn the script, provided suitable instruction manuals could be found. At the same time we began a search for "native" stenographers. Eventually, both approaches proved successful. In the *Nachlass* itself, Gödel's own shorthand textbook turned up, along with several "Rosetta stones"—vocabulary notebooks in several languages, with foreign words in longhand paired with their German synonyms in Gabelsberger—that allowed comparison of Gödel's individual "hand" with the textbook examples. Much later, we also located a German émigré, Hermann Landshoff, who had learned the Gabelsberger script as a youth and was willing to assist us.

Cataloguing of the *Nachlass* should be completed by the summer of 1984. After that, it is expected that the papers will be donated to the library of Princeton University, where they will be made available to scholars.

3. Gödel's Childhood and Youth

Kurt Friedrich Gödel was born April 28, 1906 in Brünn, Moravia (then part of the Austro-Hungarian empire; now Brno, Czechoslovakia), the second of the two children of Rudolf and Marianne Gödel. Then as now, Brünn was a major textile center, and Kurt's father worked as director of the Friedrich Redlich textile factory. (Redlich himself was Kurt's godfather, later killed by the Nazis. From him the infant's middle name was presumably taken. Gödel officially dropped his middle name when he became a U.S. citizen, but the initial "F." survives on his tombstone.) Patent correspondence preserved among Gödel's papers attests to his father's inventiveness in the textile field, and even Marianne Gödel's maiden name, Handschuh ("glove"), suggests the garment trade. (As noted in [2] and [11], her father was in fact a weaver.)

The bare facts about Gödel's birth are recorded on a copy of his *Taufschein* (baptismal certificate), preserved in the *Nachlass* along with his naturalization papers. That document shows that he was born at 5 Bäckergasse (now Pekařska) and baptized in the German Lutheran congregation of Brünn. Later the family moved to a villa (fig. 35-1) at 8A Spilberggasse (now Pellikova), a residence more befitting their moderately wealthy circumstances.

Gödel's ethnic heritage was thus neither Czech nor Jewish, as has sometimes been asserted. Though his parents were both born in Brünn, they were part of the German community there, and the children attended German-language schools. Kurt did not enroll in optional courses in the Czech language, and both he and his brother gave up their postwar Czech citizenship after they became students at the University of Vienna.

Other items in the *Nachlass* pertaining to Gödel's youth include report cards from both elementary and secondary schools and several of his school notebooks. Particularly quaint is his first arithmetic workbook (Figure 35-2), which contains but a single error in computation. The report cards are indicative of curricula of the time, which laid heavy stress on science and languages, in addition to such required courses as religion, drawing, and penmanship. Latin and French were required, and Gödel chose English as the second of his two elective subjects (after shorthand). In general, he seems to have been quite interested in languages. His

FIGURE 35-1 *The Gödel villa in Brno as it is today.*

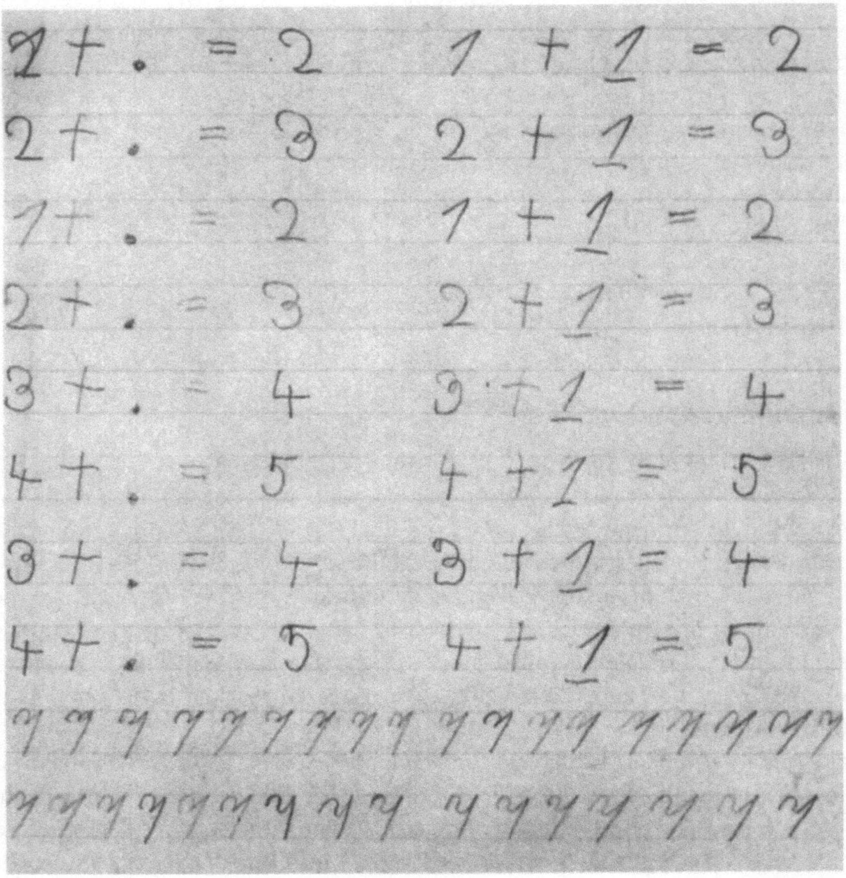

FIGURE 35-2 *A page from Gödel's first arithmetic workbook.* (H. Landshoff)

library contains many foreign-language dictionaries, and there are vocabulary and exercise note-books in Italian and Dutch in addition to the languages already mentioned.

All of Gödel's school records attest to his diligence and outstanding performance as a student. Indeed, only once did he receive less than the highest mark—in mathematics. (See Figure 35-3.) But the report cards also record a rather large number of excused absences, including exemptions during the years 1915–16 and 1917–18 from participation in physical education. The earlier of those exemptions probably corresponds to a childhood bout with rheumatic fever, an episode Rudolf Gödel believes to have been the source of his brother's later hypochondria.

Some later material in the *Nachlass* also relates to Gödel's youth. One especially valuable item is a questionnaire sent to Gödel in 1974 by the sociologist Burke D. Grandjean—a document Gödel dutifully filled out but never returned. In response to some of its queries, Gödel noted that his interest in mathematics began at about age 14, stimulated by an introductory calculus text in the well-known Göschen collection; that his family was little affected by World War I and the subsequent inflation; that he was never a member of any religious congregation, although he was a believer (describing himself as a theist rather than pantheist, "following Leibniz rather than Spinoza"); and that prior to his enrollment at the University of Vienna, he had little contact with Vienna's intellectual or cultural life except through the newspaper *Neue freie Presse*.

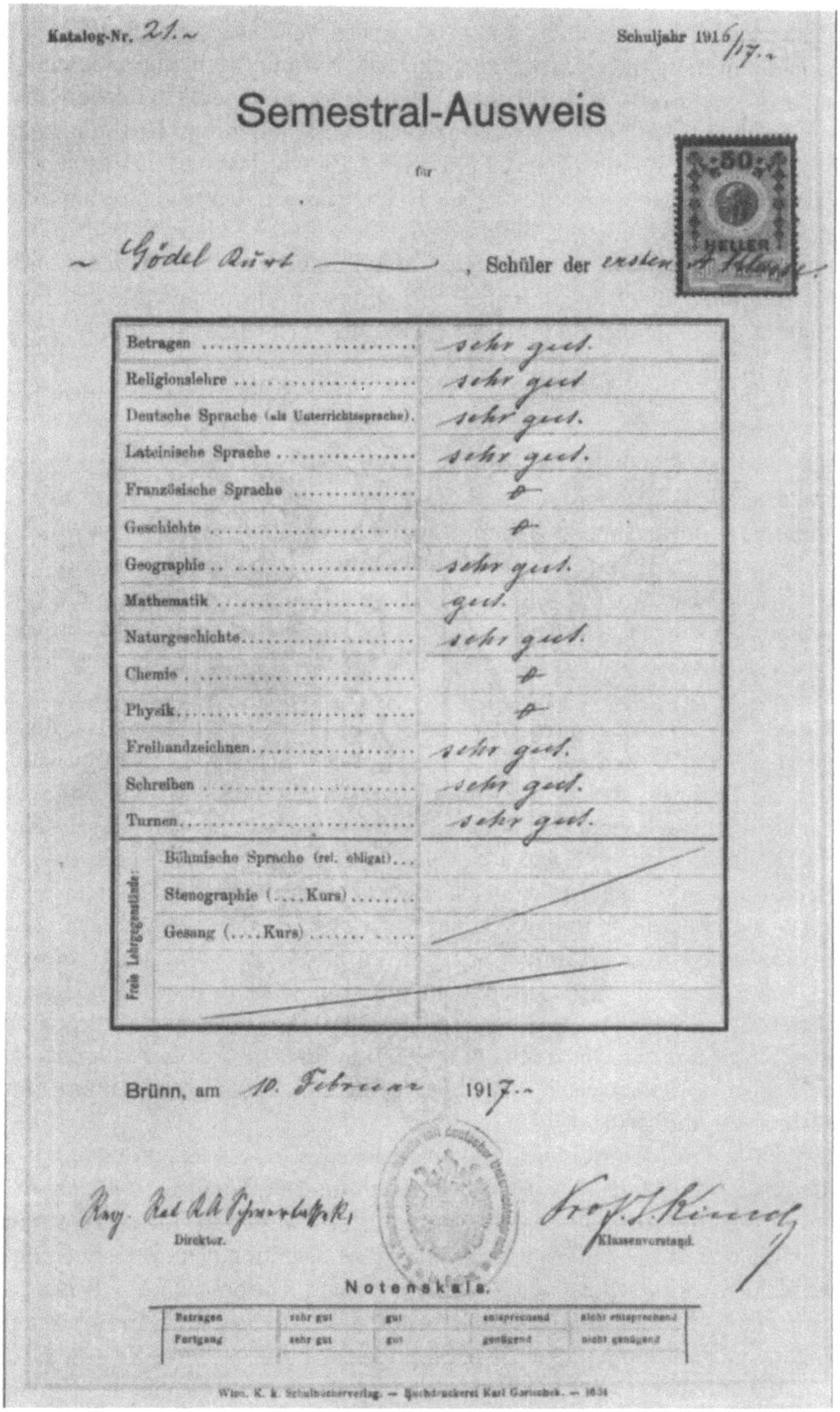

Katalog-Nr. 21.– Schuljahr 1916/17.–

Semestral-Ausweis

für

Gödel Kurt, Schüler der ersten A Klasse

Betragen	sehr gut.
Religionslehre	sehr gut
Deutsche Sprache (als Unterrichtssprache)	sehr gut.
Lateinische Sprache	sehr gut.
Französische Sprache	—
Geschichte	—
Geographie	sehr gut.
Mathematik	gut.
Naturgeschichte	sehr gut.
Chemie	—
Physik	—
Freihandzeichnen	sehr gut.
Schreiben	sehr gut.
Turnen	sehr gut.
Freie Lehrgegenstände: Böhmische Sprache (rel. obligat)	
Stenographie (....Kurs)	
Gesang (....Kurs)	

Brünn, am 10. Februar 1917.–

Direktor. Klassenvorstand.

Notenskala.

Betragen	sehr gut	gut	entsprechend	nicht entsprechend
Fortgang	sehr gut	gut	genügend	nicht genügend

FIGURE 35-3 *Semester report for Kurt Gödel, K. K. Staatsrealgymnasium in Brünn, February 1917. Note the grading scale at the bottom. (H. Landshoff)*

4. Vienna Years and Visits to Princeton

Aside from his doctoral diploma and some of his course notebooks, the *Nachlass* contains few records of Gödel's university career. According to his own account (again in response to the Grandjean questionnaire), he entered the University of Vienna in 1924 intending to major in physics. Once there, however, he was influenced by the mathematical lectures of Philipp Furtwängler and the lectures of Heinrich Gomperz on the history of philosophy. He switched into mathematics in 1926. At about the same time, under the guidance of Hans Hahn, he began to attend meetings of the Vienna Circle, with whose views, however, he disagreed, and from which he later took pains to dissociate himself (as revealed, for example, in several letters found in the *Nachlass*). Gödel submitted his dissertation in the autumn of 1929, a year marked not only by worldwide financial collapse but by the premature death of Gödel's father on February 23, five days before his fifty-fifth birthday. On February 6, 1930, Gödel's doctorate in mathematics was conferred by the University of Vienna (not, as E. T. Bell asserts in his *Mathematics, Queen and Servant of Science,* by "the University of Brno in engineering").

Shortly afterward, in an attempt to pursue the aims of Hilbert's program, Gödel sought to find an interpretation of analysis within arithmetic. In so doing, he came to realize that the concept of provability could be defined arithmetically. This led to his incompleteness proof, which, ironically, overturned Hilbert's program (at least as originally envisioned). Yet Gödel announced his momentous discovery almost casually, toward the end of a discussion on foundations[4] at a conference in Königsberg where only the day before he had lectured on his dissertation results (the completeness of the first-order predicate calculus). Reaction (not always accompanied by understanding) was immediate, ranging from profound appreciation by von Neumann (who two months later nearly anticipated Gödel's discovery of the *second* incompleteness theorem) to vigorous criticism by Zermelo (see [7] and [10]), and even to a claim to priority by Finsler (see van Heijenoort's note, pp. 438–440 in [14]), dismissed by Gödel with uncharacteristic disdain. The incompleteness paper appeared early in 1931. Later Gödel submitted it to the University of Vienna as his *Habilitationsschrift,* thereby earning the right to teach as *Privatdozent.* In the meantime he participated actively in Karl Menger's colloquium, where he presented nearly a dozen papers and collaborated in editing volumes 2–5 and 7–8 of the colloquium proceedings (*Ergebnisse eines mathematischen Kolloquiums*).

Officially, Gödel's tenure as *Privatdozent* extended from 1933 to 1938. In fact, however, his lecturing at the University of Vienna was repeatedly interrupted, both by visits to America and by episodes of ill health. Indeed, on the basis of enrollment slips saved by Gödel and records of the University of Vienna, it appears that Gödel actually taught only three courses there: foundations of arithmetic, in the summer of 1933; selected chapters in mathematical logic, in the summer of 1935; and axiomatic set theory, in the spring of 1937.

From the published memoirs it is difficult to piece together a coherent chronology of Gödel's visits to America; there were actually three such prior to his emigration in 1940. He first came in 1933–34 to lecture on his incompleteness theorems at the Institute for Advanced Study, where he spent the academic year. It was the Institute's first year of operation, without a building of its own and with titles for the visiting scholars yet to be decided upon. The official I.A.S. *Bulletin* of that year lists Gödel simply as a "worker." In April, Gödel also travelled to New York and Washington, where he lectured before the Philosophical Society of New York and the Washington Academy of Sciences.

After his return to Europe, Gödel suffered a nervous breakdown. He entered a sanatorium and was forced to postpone an invitation to return to the I.A.S. for the second term of 1934–35. In the meantime he began his investigations in set theory; and when he did return to the I.A.S. in October 1935, he told von Neumann of his consistency proof for the axiom of choice. A month later Gödel

---

[4]An abridged transcript of the discussion appeared in *Erkenntnis* 2 (1931), pp. 135–151; for an English translation and commentary, see [6].

suddenly resigned, suffering from depression and overwork. Veblen saw him aboard ship in New York and telegraphed ahead to Gödel's family. More time in sanatoria followed, and, as noted above, Gödel only resumed his teaching in Vienna in the spring of 1937.

Later that summer,[5] Gödel saw how to extend his consistency proof to the generalized continuum hypothesis. In the fall of 1938 he returned once more to America, spending the first term at the I.A.S. and the second, at Menger's invitation, at Notre Dame. (Not "Rotterdam", as stated in [16].) At both institutions he lectured on his consistency results, and at Notre Dame he and Menger offered a joint course on elementary logic. The *Nachlass* contains manuscripts of all those lectures—carefully written in English, except for a single page of examination questions, effectively concealed in Gabelsberger.

5. Emigration and American Career

Gödel intended to return again to Princeton in the fall of 1939, but personal and political events intervened. The previous September, only about two weeks before his departure for America, he had married Adele Nimbursky (née Porkert) in Vienna. Though he had known Adele for over a decade, their marriage had been delayed by opposition from Gödel's family; for she was not only a divorcee, older than Kurt, but she had worked as a dancer and was somewhat disfigured by a facial birthmark. It proved nevertheless to be an enduring union. Their first year of marriage, however, was spent apart, as Adele remained behind in Vienna during the academic year 1938–39.

After his return to Vienna to rejoin his bride in the summer of 1939, Gödel was called up for a military physical by the Nazi government. Writing to Veblen in November, Gödel reported that contrary to his expectation, "I was mustered and found fit for garrison duty."[6] At the same time, to retain his right to teach at the University of Vienna, he was obliged to apply to Nazi authorities for appointment as a *Dozent neuer Ordnung*, thereby subjecting himself to politicial and racial scrutiny; and though his mother and brother lived unmolested in Brno and Vienna throughout the Nazi occupation, Gödel was suspect because of his association with Jewish intellectuals such as Hahn. Eventually his application was approved, but only after his emigration to America.

Even there, however, he was thought by many to be Jewish. Thus Bertrand Russell declared in the second volume of his *Autobiography*:

I used to go to [Einstein's] house once a week to discuss with him and Gödel and Pauli. These discussions were in some ways disappointing, for, although all three of them were Jews and exiles and, in intention, cosmopolitans, I found that they all had a German bias toward metaphysics. . . . Gödel turned out to be an unadulterated Platonist, and apparently believed that an eternal "not" was laid up in heaven, where virtuous logicians might hope to meet it hereafter.

In 1971, Gödel's attention was called to this passage by Kenneth Blackwell, curator of the Russell archives at McMaster University. Gödel drafted a reply (never actually sent) that is preserved in the *Nachlass*:

As far as the passage about time [in Russell's autobiography] is concerned, I have to say *first* (for the sake of truth) that I am not a Jew (even though I don't think this question is of any impor-

---

[5]This date is based on Gödel's correspondence with von Neumann. On July 13, 1937, von Neumann wrote Gödel from Budapest, saying that he expected to visit Vienna in a few weeks and that while there he hoped to speak with Gödel and learn more about his plans. In the same letter he argued Gödel to consider publishing his work on the axiom of choice in the *Annals of Mathematics*. In his next letter, however, written September 14 from New York, von Neumann advised Gödel that the editors of the *Annals* were prepared to expedite publication of his work on the generalized continuum hypothesis. In the end, both consistency results were announced late the following year in the *Proceedings of the National Academy of Sciences*.
[6]By "mustered" Gödel apparently meant only that he had to report for the physical examination. It seems very unlikely that he was actually sworn in.

tance), 2.) that the passage gives the wrong impression that I had many discussions with Russell, which was by no means the case (I remember only one). 3.) Concerning my "unadulterated" Platonism, it is no more "unadulterated" than Russell's own in 1921 when in the *Introduction* [*to Mathematical Philosophy*] he said "[Logic is concerned with the real world just as truly as zoology, though with its more abstract and general features]". At that time evidently Russell had met the "not" even in this world, but later on under the influence of Wittgenstein he chose to overlook it.

(In Gödel's draft, only an ellipsis (. . .) appears between the quotation marks. The bracketed passage inserted here was quoted by Gödel in his 1944 essay "Russell's mathematical logic".)

Somehow, in the midst of the political turmoil, Gödel succeeded in obtaining exit visas: his passport, preserved in the *Nachlass*, testifies to his frantic efforts to obtain transit documents from consulates in Vienna and Berlin. By then it was too dangerous to risk crossing the Atlantic, so, in January 1940, he and Adele travelled through Lithuania and Latvia to board the trans-Siberian railway at Bigosovo. After crossing Russia and Manchuria they made their way to Yokohama and thence by ship to San Francisco, where they arrived March 4, 1940.

At the I.A.S. Gödel found an intellectual haven from which he was rarely to venture again. Reclusive by nature, he seems not to have minded (and perhaps even to have sought) his growing isolation. Several of Gödel's acquaintances, however, have remarked that Princeton society proved unreceptive (not to say hostile) toward Adele, and she appears to have led a very lonely life there.

Professionally, the I.A.S. offered Gödel relative job security. Yet he was appointed on a yearly basis until 1946, when he was finally made a permanent member. Only in 1953, five years after becoming a U.S. citizen and two years after sharing the first Einstein award,[7] was he promoted to professor. Gödel himself seems never to have expressed dissatisfaction about this long delay, but others have called for an explanation. In particular, Stanislaw Ulam ([13], p. 80) and Freeman Dyson ([8], p. 48) have brought the matter to public attention. Ulam, indeed, quoting remarks made to him by von Neumann, has suggested that Gödel's treatment was occasioned by the personal opposition of some unnamed I.A.S. colleague. The *Nachlass* itself sheds no light on the matter, but some "old-timers" at the Institute have suggested that a division of opinion prevailed among Gödel's colleagues. Some felt that Gödel would not welcome the administrative responsibilities entailed by faculty status, while others feared that if he were promoted, his sense of duty and legalistic habit of mind might impel him to undertake such responsibilities all too seriously, perhaps hindering efficient decision-making by the faculty. In the event, such fears seem to have been justified; but one should also note Gödel's own statement in 1946 in a letter to C. A. Baylis, in which he noted that "apprehension that cooperation of this kind [service in offices or on committees] would be expected" was "the very reason" he had so belatedly joined the Association for Symbolic Logic.

In any case, the Institute allowed Gödel freedom to pursue a broad range of intellectual interests. At first he labored to prove the independence of the axiom of choice and the continuum hypothesis, but the later especially proved unyielding, and eventually he gave up the attempt.[8] Instead, he turned

---

[7]With Julian Schwinger; not, as stated in [12], with von Neumann. Von Neumann was actually a member of the awards committee, and he may have introduced Gödel's name for consideration. In any case, there is evidence in von Neumann's papers that Schwinger alone was originally proposed for the award.

[8]Rumors have persisted that Gödel actually obtained the independence of the axiom of choice in the early 1940s but refused to publish his results. In particular, after Cohen's proof in 1964, Mostowski asserted, "Es ist seit 1938 bekannt, dass Gödel einen Unabhängigkeitsbeweis dieser Hypothesen besitzt; trotz vielen Anfragen verriet aber nie sein Geheimnis." (*Elemente der Mathematik* 19, p. 124.) But Gödel himself denied this. In a letter to Wolfgang Rautenberg (published in *Mathematick in der Schule* 6, p. 20) he stated explicitly:

Die Mostowskische Behauptung ist insofern unrichtig, als ich bloss in Besitzte gewissen Teilresultate war, nämlich von Beweisen fur die Unabhängigkeit der Konstruktibilitäts- und Auswahlaxioms in der Typentheorie. Auf Grund meiner höchst unvollständigen Aufzeichnungen von damals (d.h. 1942) könnte ich ohne Schwierigkeiten nur den ersten dieser

to philosophy. The transition is marked by his 1944 essay "Russell's mathematical logic", solicited by P. A. Schilpp for his *Library of Living Philosophers* series. Subsequently, Schilpp was to solicit essays for the Einstein, Carnap, and Popper volumes of the series as well. Gödel accepted all but the last, and he devoted great care to each—so much so that in every case his essay was among the last to be received. Schilpp displayed great patience and diplomacy, but Gödel could not be hurried; and when his Russell essay was received too late for Russell to reply, Gödel considered withdrawing it altogether. Eventually he yielded to Schilpp's entreaties, but when the situation recurred a few years later, Schilpp was forced to send the Carnap volume to press without Gödel's contribution. It remains unpublished in the *Nachlass*.

In contrast, Gödel's contribution to the Einstein volume was not only commented upon, but praised by Einstein as marking a significant advance in the physical, as well as philosophical, understanding of relativity; for Gödel had in fact discovered an unexpected solution to Einstein's field equations of gravitation, one permitting "time travel" into the past.[9] The contribution to Schilpp's anthology is quite brief, but Gödel prepared a much longer essay that has remained unpublished. It too is preserved in the *Nachlass*—in six different versions. It is worth noting that Gödel's interest in relativity went beyond the purely theoretical. In his essay he argues in favor of the possible relevance of his models to our own world, and in the *Nachlass* there are two notebooks devoted to tabulations of the angular orientations of galaxies (which Gödel hoped might exhibit a preferred direction). Freeman Dyson has remarked that even much later, Gödel maintained a keen interest in such observational data.

Yet another unpublished item in the *Nachlass* is the text of Gödel's Gibbs Lecture to the American Mathematical Society, delivered at Brown University on December 26, 1951. Titled "Some basic theorems on the foundations of mathematics and their philosophical implications," it is Gödel's contribution to the debate on mechanism in the philosophy of mind.

In addition to such relatively finished papers, there are a great many pages of notes by Gödel, including sixteen mathematical workbooks, fourteen philosophical notebooks, and voluminous shorthand notes on Leibniz. The latter appear to be partly bibliographic, but a recently discovered memorandum suggests that there are about one thousand pages of Gödel's own philosophical assertions as well.

Gödel's long-standing interest in Leibniz is also indicated by an extensive correspondence with archivists, conducted jointly with Oskar Morgenstern during the years 1949–53. The object of the correspondence was the microfilming of some of Leibniz's unpublished manuscripts in Hannover, with the aim not only of preserving them but of making them available to American scholars. Ultimately the attempt failed, but copies of the manuscripts were later deposited at the University of Pennsylvania through the independent efforts of Prof. Paul Schrecker.

6. Later Years

Gödel's last published paper appeared in 1958. Based on results obtained nearly eighteen years earlier (on which Gödel lectured at Yale, April 15, 1941), it presented a consistency proof for arithmetic by means of "a [t]hitherto unused extension" of principles formulated in intuitionistic mathematics. As such, it represented a return to earlier mathematical interests (and to the German language—it

---

beiden Beweise rekonstruieren. Meine Methode hat eine sehr nahe Verwandtschaft mit der neuerdings von Dana Scott entwickelten, weniger mit den Cohenschen.

It is clear from Gödel's correspondence that he had great respect for Cohen's work; indeed he described Cohen's achievements as "the greatest advance in abstract set theory since its foundation by Georg Cantor." Nevertheless, the method by which Gödel obtained his partial results may still prove to be of interest.

[9]Technical details were published that same year (1949) in *Reviews of Modern Physics*, and a year later Gödel spoke on his results at the International Congress of Mathematicians in Cambridge, Massachusetts.

was the only paper Gödel published in German after his emigration); nevertheless, it is decidedly philosophical in character and was published in the philosophical journal *Dialectica*. Unfortunately, the paper is notoriously difficult to translate. In the early 1970s Gödel himself prepared a revised and expanded English version that reached the stage of galley proofs but never appeared.

After 1958, Gödel devoted himself to revisions of earlier papers, to a search for new axioms to settle the continuum hypothesis (in the wake of Cohen's independence proofs), and to a study of the philosophy of Edmund Husserl. Honorary degrees, academy memberships, and awards (including the National Medal of Science in 1975) were bestowed upon him from many quarters, but he became increasingly withdrawn and preoccupied with his health. He consulted doctors but distrusted their advice, which he often failed to heed. Thus, in the late 1960s, he refused recommended surgery for a prostate condition, despite the urgings of concerned colleagues; and earlier, in the 1940s, he delayed treatment of a bleeding duodenal ulcer until life-saving blood transfusions had to be administered. Afterward he observed a strict dietary regimen, and as the years wore on his figure became even more gaunt.

During the final decade of Gödel's life, his wife underwent surgery and suffered two strokes that led to her placement in a convalescent home. By all accounts, Gödel attended her devotedly, but he soon began to display signs of depression and paranoia. Correspondence, even with his brother, virtually ceased during the last two years of Gödel's life. In the end, his paranoia conformed to a classic syndrome: fear of poisoning leading to self-starvation. After a relatively brief hospitalization, Gödel died January 14, 1978, of "malnutrition and inanition" caused by "personality disturbance". (His death certificate, from which these phrases are quoted, is on file in the Mercer County courthouse, Trenton, N.J.) He is buried beside his wife and mother-in-law in the historic Princeton Cemetery.

7. Prospects

To what extent can study of Gödel's *Nachlass* be expected to reveal hitherto unknown discoveries? It seems unlikely that major new mathematical results will be found in Gödel's notebooks, although the reputed "general" consistency proof for the axiom of choice mentioned in [15] and [16] may prove to be an exception—if, indeed, it can be reconstructed. Gödel was certainly cautious and overly fastidious in submitting his work for publication, but there is no evidence that he actively withheld important mathematical discoveries; and though details of his researches remain largely concealed behind his shorthand, the topics of his investigations can nonetheless be determined to a large extent from (longhand) headings in the notebooks. On that basis, it seems safe to predict that some lesser results of mathematical interest will be found there, along with some anticipations of, or alternative approaches to, results of others (such as Gödel's early recognition of errors in Herbrand's work, cited in [11], or his partial independence results in set theory). Of course, details of Gödel's explorations should be of great interest to mathematical historians. Beyond that, I would venture to predict that of all the unpublished material in the *Nachlass*, Gödel's philosophical investigations will turn out to be of greatest interest. Certainly they figure most prominently among the items he left in relatively finished form and that he himself considered to be potentially publishable.

Plans for publication of Gödel's collected works are now well under way.[10] Two volumes are presently envisioned, under the editorship of Solomon Feferman (editor-in-chief), myself, Stephen C. Kleene, Gregory H. Moore, Robert M. Solovay, and Jean van Heijenoort. The first volume, now in preparation, will contain all of Gödel's published articles and reviews, together with his doctoral dissertation (in its original unpublished form) and his revised English version of the *Dialectica* paper, as well as three short notes appended to galley proofs of the latter. Papers in German will be accompanied by English translations on facing pages, and each article will be preceded by introductory

---

[10]A nearly complete edition of Gödel's published works, excluding reviews, has already appeared in Spanish translation; see Jesús Mosterín, ed., *Kurt Gödel, Obras Completas*, Alianza Editorial, Madrid, 1981.

commentary. Textual notes, a short biographical essay, and an extensive bibliography will round out the volume.

Detailed contents of volume two remain to be determined, subject both to the success of our decipherment efforts and to our ability to obtain necessary funding and copyright permissions. We hope, however, to include all the relatively finished papers mentioned previously, plus other lecture texts, excerpts from the mathematical notebooks, and selected correspondence, including not only extensive exchanges with other mathematicians but individual letters of interest. Should there be enough material to warrant it, further volumes may also be considered. The editors will welcome contributions of correspondence with Gödel or recollections of him.

References

1. Bulloff, Jack, T. C. Holyoke and S. W. Hahn (eds.) (1969) *Foundations of Mathematics: Symposium Papers Commemorating the Sixtieth Birthday of Kurt Gödel*, New York: Springer-Verlag.

2. Christian, Curt (1980) "Leben und Wirken Kurt Gödels," *Monatshefte für Mathematik 89*, 261–273.

3. Dawson, J. W., Jr. (1983) "The published work of Kurt Gödel: an annotated bibliography." *Notre Dame Journal of Formal Logic 24*, 255–284.

4. Dawson, J. W., Jr. (1983) "Cataloguing the Gödel *Nachlass* at the Institute for Advanced Study," *Abstracts of the 7th International Congress of Logic, Methodology, and Philosophy of Science 6*, 59–61.

5. Dawson, J. W., Jr. (1984) "Addenda and corrigenda to 'The published work of Kurt Gödel,' " *Notre Dame Journal of Formal Logic 25*, 283–287.

6. Dawson, J. W., Jr. (1984) "Discussion on the foundation of mathematics, *History and Philosophy of Logic 5*, 111–129.

7. Dawson, J. W., Jr. (1985) "Completing the Gödel-Zermelo correspondence," *Historia Mathematica 12*, 66–70.

8. Dyson, Freeman (1983) "Unfashionable pursuits," *The Mathematical Intelligencer 5:3*, 47–54.

9. Gracy, David B., II (1977) *Archives & Manuscripts: Arrangement & Description* (Basic Manual Series), Chicago: Society of American Archivists.

10. Grattan-Guinness, Ivor (1979) "In memoriam Kurt Gödel: his 1931 correspondence with Zermelo on his incompletability theorem," *Historia Mathematica 6*, 294–304.

11. Kreisel, Georg (1980) "Kurt Gödel, 1906–1978, elected For. Mem. R.S. 1968," *Biographical Memoirs of Fellows of the Royal Society 26*, 148–224; (1981) corrigenda, *27*, 697; (1982) further corrigenda, *28*, 718.

12. Quine, Willard V. (1978) "Kurt Gödel (1906–1978)," *Year Book of the American Philosophical Society*, 81–84.

13. Ulam, Stanislaw, (1976) *Adventures of a Mathematician*, New York: Scribner's.

14. van Heijenoort, Jean (ed.) (1967) *From Frege to Gödel, A Source Book in Mathematical Logic, 1879–1931*, Cambridge: Harvard.

15. Wang, Hao, (1978) "Kurt Gödel's intellectual development," *The Mathematical Intelligencer 1*, 182–184.

16. Wang, Hao, (1981) "Some facts about Kurt Gödel," *The Journal of Symbolic Logic 46*, 653–659.

36

Who Would Have Won the Fields Medals a Hundred Years Ago?

Jeremy Gray

Every four years the mathematical community honors a select few of its younger members by awarding them Fields Medals. On the most recent occasion (1982) I fell to wondering, as an historian, who would have won the medals a hundred years ago. Since the medals were first awarded in 1936, the question is necessarily speculative, but it seemed to me to admit some tentative answers. The more I thought about it, the more clearly it seemed possible to find out who were the bright young mathematicians of the day, and, as importantly, by whom they were so regarded. I began to follow a hypothetical "Fields," perhaps an American or, like the real one, a Canadian, on his journey round Europe, as he talked to distinguished older mathematicians, formed his Prize Committee, and then watched them making their difficult decisions.

What emerged is, possibly, surprising, and, I think, necessary for us to consider when we think about the dynamic of mathematics and seek to explain its development at that time. I shall draw out one or two more serious points at the end, but first to New York, sometime in 1878, where Fields is even now taking his leave of the irrepressible J. J. Sylvester, Professor of Mathematics at Johns Hopkins University, and preparing to leave for England. Fields is anxious to see mathematics, then greatly underrepresented in American intellectual life, given higher status, and hopes that by establishing an international prize in mathematics, young Americans will be drawn to the subject. Sylvester, in America partly to proselytize for mathematics himself, will have advised him to start by visiting his old friend, Arthur Cayley, in Cambridge.

Cayley (b. 1821) was then Sadlerian Professor of Mathematics at Cambridge and a member of twelve foreign scientific academies including those of Berlin, Boston, Göttingen, and Rome. On the occasion of his election to the Presidency of the British Association for the Advancement of Science in 1883, his colleague Salmon was to praise him for "the amount and universality of his reading," for his fairness, and his mastery of several languages, writing that "as was said of Moltke, there are few European languages in which he does not know how to hold his tongue." At a time when women were still not admitted to the University, he was chairman of the Association for Promoting the Higher Education of Women. Moreover, Cayley was still a highly active mathematician; in 1878 alone he published forty papers. Many of these were surveys, aimed at acquainting an English audience with recent continental work. He would be an ideal choice for the Fields Medals Committee.

Fields might also have considered H.J.S. Smith (b. 1826), the Savilian Professor of Geometry at Oxford. A number theorist, and the author of an excellent survey of that subject, he was also much

Volume 7, No. 3 (Summer 1985), 10–19

FIGURE 36-1

respected for his administrative skills, and he served on the Royal Committee on Scientific Education. I must suppose, however, that Fields would not choose Smith, if only because Smith's health was poor. He was to fall seriously ill in 1881 and his last public engagement was to speak at Oxford in favor of extending the franchise to agricultural laborers. He died in 1883. It is also true that Smith was not well known on the Continent. The Paris Académie des Sciences was to be very embarrassed when it announced a prize in 1881 for the theory of the decomposition of integers as a sum of five squares, only to find that Smith had successfully treated this problem in 1859. Worse, by the time they announced the award (which went to Smith and the young Hermann Minkowski), Smith was dead.

Cayley would doubtless send Fields to Paris, with a view to finding a Frenchman for his committee. The choice would lie between Charles Hermite (b. 1822) and C. Bertrand (b. 1822) Although Bertrand was then the Perpetual Secretary to the Académie des Sciences, a post he had held since 1874, and a Professor at both the Ecole Polytechnique and the Collège de France, Hermite's greater range as a mathematician would commend him more strongly to Cayley and to Fields. Hermite had been converted by Cauchy to Roman Catholicism in 1856 after a very serious attack of smallpox and was henceforth a man of the clerical right, yet he corresponded though the 70s with the German mathematician Lazarus Fuchs, so he was not affected by the poisonous nationalism engendered by the Franco-Prussian war. Felix Klein said of him in his *Entwicklung der Mathematik* [1922, 292] that he was "not temperamentally a leader of men," a remark which must be contrasted with Klein's view of Klein. In the late 1870s Hermite was pioneering the use of elliptic functions in applied mathematics, via the theory of Lamé's equation.

Fields would then undoubtedly go to Berlin, the acknowledged center of the mathematical world. The three professors there were Ernst Edouard Kummer (b. 1810), Leopold Kronecker (b. 1823), and

Georg Cantor

FIGURE 36-2

Carl Weierstrass (b. 1815). Kummer's chief contribution to mathematics was, of course, his algebraic theory of numbers, but he had also worked on differential equations and elliptic functions in the 1830s, and more recently on geometry. In 1864 he discovered the quartic surface with sixteen nodal points that now bears his name, and it continued to attract quite a lot of interest. He even described how to make a gypsum model of it in 1866. At Berlin he took on the task of lecturing on the elementary branches of mathematics and giving them firm foundations. He drew huge audiences to these lectures, up to 150 students. Apparently he took great care of his students, and in a curious phrase it has been said that in return "their devotion sometimes approached enthusiasm." In 1878 he resigned as Perpetual Secretary of the Berlin Academy after fifteen years.

Kronecker was Kummer's best student, an independently wealthy man who lectured at the University in virtue of his status as a member of the Berlin Academy. He proposed fifteen people for membership of the Academy between 1863 and 1886, including Riemann, Sylvester, Smith, Dedekind, Betti, Brioschi, Hermite, Fuchs, and Casorati, which suggests something of his range as a mathematician. Frobenius, in his eulogy of Kronecker in 1891, regarded Kronecker as too broad, so that he was only the second-best at everything he did. His main interests were, of course, elliptic functions and number theory, but he had published little between 1860 and 1880, confining himself to the research seminars at Berlin.

Like Kronecker, with whom he shared the running of the seminars, Weierstrass had published little by 1878. He lectured on all branches of analysis, most importantly on elliptic and abelian functions. Like Kummer he was a modest, moderate, and kind man. Unhappily, Fields would enter a deteriorating situation at Berlin. Weierstrass was appalled by Kronecker's hardening views on arithmetization, most memorably expressed in his famous dislike of Cantor's work, and was afraid that Kronecker, as the youngest of the triumvirate, would eventually dominate the department. Nor could he look for much support to Kummer, who tended to side with his old student. Although strictly this anecdote lies outside the period 1878–1882, one may suppose the signs were building up. In 1884–85 Schwarz wrote to invite Kronecker to a New Year celebration, in these terms: "He who does not honour the Smaller is not worthy of the Greater"—a reference to the respective physical sizes of Kronecker and Weierstrass. Kronecker absurdly took this to refer to intellectual size, although "the Smaller" was his nickname in Berlin, and broke with Schwarz completely!

In such a delicate situation the practical businesslike Fields would surely choose Kummer, as much for his position as a compromise candidate as for his wide grasp of mathematics. But he would be disappointed. Kummer was withdrawing from academic life and was to startle his colleagues by resigning altogether from the university in 1883. When Kummer finally declined to take on this new commitment, Fields could make only one choice if he was to avoid Schwarz's mistake. He would choose both Kronecker and Weierstrass for his committee, making Weierstrass the president. Two Berlin mathematicians would seem only a fair reflection of the size and importance of their school of mathematics, which every young German mathematician and many foreign ones had attended.

Next, on to Italy. Brioschi (b. 1824) would be the obvious choice. Still active as a mathematician, and interested in invariant theory, cubic surfaces, and linear differential equation theory, the editor of *Annali di Matematiche,* on the executive council of the Ministry of Education, and since 1863 the head of the Istituto Tecnico Superiore, he was also to be elected President of the Accademia dei Lincei in 1884. His only rival, Enrico Betti (b. 1823), was by contrast shy and retiring.

For a sixth member, Fields would be strongly advised to choose the Swiss Ludwig Schläfli (b. 1814). Cayley had been much taken with Schläfli's earliest work on invariant theory, saying it contained "a very beautiful theorem on resultants," and had corresponded with him about it; Brioschi also praised this work. For his study of the twenty-seven lines on a cubic surface, in which he described the thirty-six "double-six" configurations they form, he had won the Steiner prize in 1870. This would have further endeared him to Cayley, who had originally discovered the lines themselves. His work on the theory of modular equations would have pleased Hermite; one in particular cleared up a mysterious table of transformations published without explanation by Hermite in 1854. Schläfli was to be elected to the Accademia dei Lincei in 1883. In choosing Schläfli, Fields would also discretely inform himself of the Göttingen school of German mathematics which was already somewhat different from Berlin's, for Schläfli also belonged to the Göttingen Academy of Sciences.

So we may suppose that Fields drew around him a committee consisting of Brioschi, Cayley, Hermite, Kronecker, and Schläfli, with Weierstrass as president. All of these men were active mathematically, although Schläfli was finding no takers for his immense study of n-dimensional geometry. Who would they consider for the Fields medals of 1882, as they began their leisurely deliberations?

It is not a difficult task to draw up a complete list of candidates. We can do so today from the pages of the annual abstracting journal *Jahrbuch über die Fortschritte der Mathematik*, but the mathematical community was smaller then than now, and the scholars would have little trouble in recollecting everyone personally. In 1879 they would, I imagine, draw up the following list if the requirement of being not more than forty in 1882 was imposed: Cantor (who would be 37 in 1882), Clifford (37), Darboux (40), Dini (37), Frobenius (33), Halphen (38), Klein (33), Lie (40), M. Noether (38), Schwarz (39), Weber (40).

If they decided to look just a bit beyond the forty mark, they would then find Casorati (who would be 47 in 1882), Cremona (52), Dedekind (52), Fuchs (49), Gordan (45), Jordan (44), Mathieu (47), Sylow (50).

By 1880 Cantor had over twenty publications to his name. He had worked first on the convergence of trigonometric series, where he showed that there is at most one trigonometric series which converges pointwise to a given function for all x. This work led him to embark on his theory of derived sets, to propose a theory of the reals (as equivalence classes of Cauchy sequences of rationals), and to show that the reals are uncountable whereas the algebraic numbers are countable. In 1878 he had established the invariance of dimension with a proof that was generally considered valid until 1898. Publication in the *Journal für Mathematik* was, however, delayed by Kronecker, but Weierstrass argued successfully for it. As a result Cantor took his work thereafter to the rival *Mathematische Annalen*, and, albeit briefly, to Mittag-Leffler's *Acta Mathematica* (founded 1882), which argues for Weierstrass' continuing support. By 1882 three in his series of papers "Ueber unedliche, lineare Punktmannigfaltigkeiten" had appeared, and he had proceeded as far as the definition of the power of a set and the introduction of infinite symbols (to denote limits of limits of limits of . . .).

Kronecker's hostility to Cantor is well known. Indeed, Cantor believed that the entire Berlin school was against him except for Weierstrass, as several of his letters quoted in Dauben's recent biography of him [1979] testify, and he singled out Kronecker and Schwarz. So Kronecker would argue against him, and Weierstrass on the other hand would support him. The other members of the committee would find it difficult to take sides, because Cantor's work was little known outside Germany at this time, and Fields

Gaston Darboux

FIGURE 36-3

Ulisse Dini

Georg Frobenius

Georges Halphen

FIGURE 36-4

would surely find the discussions drifting towards a compromise amid talk of safer choices, other young mathematicians, and so forth. Cantor would be quietly passed over.

Clifford's untimely death from tuberculosis would remove the only English candidate from the list and so prevent Cayley from facing the embarrassment of a curious squabble in English intellectual life. Clifford, a friend of Huxley's, had interpreted Riemann's ideas on the foundations of geometry in a strictly empiricist fashion, thus challenging the defenders of scientific and moral a priori reasoning. Cayley had sided with the latter group, and passions were running high on the issue. Clifford's death would unhappily prevent the committee from considering his more strictly mathematical accomplishments.

If Cantor's treatment seems short shrift to us now, it is history that has dealt unkindly with Darboux. He was publishing prolifically throughout the 1870s—an average of nine papers a year, many of them very long. He studied second and higher order partial differential equations, such as arise, for example, in the study of curves of specified kinds on various surfaces; his cyclides; and many other questions of a differential geometric nature. In 1874 he had proved this version of the fundamental theorem of the calculus: if a function f has a bounded Riemann-integrable derivative f' on $[0, 1]$ then $\int_0^x f'(t)dt = f(x) - f(0)$, for all $x \in [0, 1]$. In 1878 Dini observed that Darboux's theorem suggests that there might be functions f whose derivatives f' vanish on some point in every interval, in which case if f' is bounded it is either constant or not Riemann-integrable, and the existence of such surprising functions was confirmed by Volterra in 1881. If the committee took as part of their brief the terms of the real committee, that the award should be: "Not alone because of the outstanding character of the achievement, but also with a view to encouraging further development along these lines," then Darboux's theorem might strike them as exactly right. Since he had also studied shockwaves, the Kummer surface, and proved that line and cross-ratio-preserving maps in real protective geometry are automatically continuous, Fields would have heard a strong French case with support from Cayley and Schläfli at the very least. Darboux's name would go forward, and indeed I note that he was to be nominated to the Académie des Sciences in Paris in 1884 on the death of Puiseux.

Dini presents a curious case. He had studied under Hermite in 1865 and worked originally on differential geometry, publishing important theorems on conformal representation and maps of one surface onto another which preserve geodesics. But then in 1880 he gave up mathematics for a career in Italian politics, only returning to mathematics ten years later. Fields might have thought this did not present the example he wanted, so I imagine Dini would not be chosen.

Frobenius, then in Zurich, had not really begun to do his best work and was still reworking and improving the work of others. His papers on Pfaff's equation and on quasi-elliptic functions would not, I think, merit the prize and his great work on group representations was still over a decade away, whereas Halphen actually won two prizes in 1882: the Grand Prix of the Académie des Sciences, and the Steiner prize of the Berlin Academy. The Grand Prix was awarded, by a panel headed by Hermite, for his work on differential equations having algebraic solutions. Hermite praised it for its skillful use of the idea of the genus of a curve together with the ideas of invariant theory. The Steiner prize was awarded to Halphen for a long paper in which he classified all algebraic space curves of degree less than or equal to 20. To Hermite's quiet satisfaction, one supposes, the judges would readily agree to acknowledge Halphen's claim on a prize.

Felix Klein would present the committee with its strongest disagreements. The author of nearly eighty papers by 1882, he had written on non-Euclidean geometry, line geometry, and algebraic curves and had recently begun to bring the tools of group theory, invariant theory, and Riemann surface theory to bear on problems to do with modular functions. In 1882 the fruit of his collaboration with Poincaré, an account of Riemann surfaces defined by the action of discrete subgroups of $SL(2; \mathbf{R})$ on the upper-half plane, appeared in the *Mathematische Annalen,* of which he had been an editor for ten years. As a very young man he had studied with Plücker and Clebsch, then passed through Berlin, and had learned group theory in Paris with Jordan. But in Berlin, although he had won the annual prize for the best student of the year, he had made enemies. Weierstrass was not persuaded that non-Euclidean geometry should be seen as a species of projective geometry, as Klein would have it, because it was a metrical geometry, and Klein to the end of his life recalled the hostility he had met in Berlin. From its inception in 1869 the *Mathematische Annalen* was perceived as an unwelcome rival for the *Journal für Mathematik,* and Klein energetically poached articles by people who, like Cantor, were out of favor in Berlin. The startling view of Klein expressed by the selection committee for Weierstrass' successor in 1891, that Klein was a mere organizer,

Felix Klein

Sophus Lie

Max Noether

FIGURE 36-5

Hermann Amandus Schwarz

FIGURE 36-6

incapable of original work, perhaps reflects the altered status Klein possessed after the breakdown of his health in September, 1882, but it surely says something about how he had been seen in Berlin all along. I imagine Kronecker and Weierstrass would find him insufficiently profound, and even at times too imprecise (Lipschitz's judgment on Klein's doctoral thesis). On the other hand, Cayley might have spoken up for him, for they had similar interests, and Schläfli would have pointed to their joint work on the non-orientability of the real projective plane. Brioschi's opinion I cannot determine, but I feel that Hermite would not have liked Klein. Klein was unsparing in public criticism of Hermite's friend Fuchs and may well have struck Hermite as being too pushy and ambitious. Moreover, although Hermite was impressed by Riemann's theory of functions, he preferred to avoid it personally and may not have been enthusiastic about Klein's striving to don the Riemannian mantle.

If we imagine that Fields saw in Klein an energetic man keen to advance mathematics and pressed his committee to rise above narrow personal judgments, then I think the tide would still run Berlin's way. The projective interpretation of non-Euclidean geometry, the yoking together of geometry, group theory, and invariant theory are good pieces of work, but not perhaps outstanding. The newer work extending the definition of modular functions and clarifying their transformations is certainly better, but once again I hear the murmurs—"a young man ... other opportunities ... results not completely established. . . ."

Moreover, Klein had a rival whose claims would be preferred. This was not his friend Sophus Lie, whose theory of contact transformations with applications to lines of curvature on a surface and the theory of first-order partial differential equations had appeared in the 1870s. Lie was now beginning to develop his ambitious theory of differential invariants and transformation groups, but almost nobody read it. The *Fortschritte* reports on Lie's work at this time are written by Lie himself, for example, which is very unusual. A little later, in 1886, Klein was to be instrumental in bringing Lie to Leipzig, which greatly helped to publicize his work, and Klein (and later Poincaré) sent students to Leipzig to learn the new mathematics. But in 1882 Lie's name would scarcely be mentioned. No, Klein's rival for the title of Germany's leading geometer was his former colleague in Clebsch's school of projective geometry, Max Noether (b. 1844).

With Noether one can trace the rise to respectability, even in Berlin, of algebraic geometry, culminating in his award of the Steiner prize, jointly with Halphen, for his study of algebraic space curves in 1882. His paper with Brill in 1874 marks the start of the rigorous, purely algebraic theory of singular curves, by means of his theory of divisors, and includes a sketch of how "bad" singularities can be reduced to ordinary ones (those having all tangent directions distinct). He had also begun to study algebraic surfaces, about which very little was known, and most recently shown how to incorporate the theory of Weier-

Heinrich Weber

FIGURE 36-7

strass points into his own approach. In his day, Noether would have outshone Klein.

The claims of Schwarz (b. 1843) and H. Weber (b. 1842) would not have been very strongly pressed. Between 1877 and 1882 Schwarz only published five papers, and in the whole decade from 1871 he had only managed two striking results. One was his enumeration of those hypergeometric equations all of whose solutions are algebraic, thereby giving a partial answer to a question of Fuchs. The other was his alternating method, which solved Dirichlet's problem for a large class of boundaries of disklike regions. His important study of the calculus of variations lay in the future and the contemporary judgment was that he squandered his abilities by failing to distinguish important from unimportant problems.

Felice Casorati

FIGURE 36-8

Weber's claim is also weak. His first truly important work was his joint paper with Dedekind on algebraic functions. Submitted to the *Journal für Mathematik* in 1880, Kronecker delayed its publication until 1882 for reasons which are not clear, so the profundity of its arithmetical approach would not have had time to sink in.

I feel sure that as Fields, perhaps spending Christmas 1881 in Paris with Hermite, discussed how things stood, three frontrunners would be well established: Darboux, Halphen, and Noether. But equally the celebrations in Berlin would be marred by the knowledge that none of their younger generation was quite of that caliber. Could matters be allowed to rest that way? Suddenly a solution would suggest itself. It is obvious that mathematicians do not do their best work until they are about 40; Weierstrass' own fame only came with his solution to the Jacobi inversion problem for hyperelliptic integrals done when he was that age, for example. He would propose to the committee that one looks among the slightly older men, who have, after all, only his one chance of getting a Medal. His colleagues, innocent of the twentieth-century glorification of youth, would agree.

Once this is done, four more men must be considered. Casorati (b. 1835), an Italian function-theorist who had visited Berlin in 1864 and remained in touch with many German mathematicians; Fuchs (b. 1833), a former student of Kummer and Weierstrass; Gordan (b. 1837), the so-called King of Invariants; and Jordan (b. 1838). Of those, Jordan stands out. The leader, with Kronecker, among those who had pioneered the understanding of Galois' ideas, he had not only written the first major book on abstract algebra, the *Traité des substitutions et des équations algébriques* (1872), he had throughout the 1870s shown how to use group theory to illuminate problems in other branches of mathematics. As an example, he had taken up Fuchs' problem of algebraic solutions to an nth order linear ordinary differential equation and had shown how to reduce it to the problem of enumerating the finite subgroups of $SL(n; \mathbf{C})$ which can arise as monodromy groups of the equation. This problem he then solved explicitly for $n = 2$ and 3, and for general n he proved his finiteness theorem, publishing his results in the *Journal für Mathematik* and, in an improved form, in Brioschi's *Annali di Matematiche*. Cayley would have

Sofia Kovalevskaia

FIGURE 36-9

Richard Dedekind

FIGURE 36-10

approved, as, one supposes would Weierstrass, whose student Hamburger had shown how to use the Jordan canonical form of a matrix to simplify the general solution to a linear differential equation.

The 1870s had not been a good decade for invariant theory. Very frequently Gordan had been reduced to solving problems already solved in *ad hoc* ways by others, and on occasion following men like Jordan and Klein to their results. The committee would pass him over, if only to land on Fuchs. Now the Berlin professors would press their candidate home. The architect of the rigorous modern theory of complex linear ordinary differential equations, he represented the school of Weierstrassian analysts. But since that work in the 1860s he had raised and solved the problem of when a differential equation has algebraic solutions. His work was more general than that of Schwarz, to which Brioschi had also contributed, more explicit than that of Klein, and if it was not as general as Jordan's, then it certainly was a stimulus for it. Now he was discussing an interesting generalization of Jacobi inversion theory to differential equations and pushing well towards a solution. For all these reasons, Fuchs was soon to be chosen as Kummer's successor in Berlin and elected to the Academy of Sciences there. Hermite would have joined in, praising his friend's study of modular functions, which, as he had written to Fuchs, also shed light on Kronecker's difficult theory of complex multiplication. Cayley, who certainly liked Schwarz's work on differential equations, would doubtless concur, sharing with Hermite and Fuchs a preference for the traditional in the theory of elliptic functions over and above the newer ideas advanced by Klein. So Fuchs would be home, Berlin's honor would be saved, Weierstrass could report to Fields that the committee was nearing unamimity. Unless, of course, five names was too many. . . ?

Fields also had another worry. Nearly four years had gone by, and perhaps there were some young mathematicians who had just emerged but whose claims to attention were just as strong. It would not do for an American to go all the way with the European preference for seniority. He had heard good things, for example, about a young Frenchman, Picard (b. 1856).

Indeed, by the end of 1881 Picard had published thirty-four papers, developing the ideas of Hermite on Lamé's equation into a theory of quasi-elliptic functions. He was certainly a man to watch. There was even someone else, a slightly older man but suddenly publishing profusely: Poincaré (b. 1854). There was little doubt that his work on differential equations and Riemann surfaces was more profound than Klein's, or so Hermite would argue, but it was visionary, very imprecise, and while it suggested many important paths to follow, it still left a lot that needed to be explored. For this reason the Paris Académie had placed his preliminary essays on the subject behind those of Halphen when awarding their Grand Prix. Kronecker's opinion was more trenchant. He told Mittag-Leffler in 1882 that to publish Poincaré's long papers on automorphic functions would kill his new journal, *Acta Mathematica*. No, these brilliant young Frenchmen could wait and mature a little.

Weierstrass might then be tempered to advance the claim of his friend Sofia Kovalevskaia (b. 1850). Her existence theorem for analytic partial differential equations had come up with an unexpected subtlety which had surprised and pleased him. Unhappily, her best work lay in the future. Her paper on Saturn's rings was published in 1883, and her actually monstrous paper on hyperelliptic integrals, which was highly regarded in its day, only came out in 1884. Her theory of the rotation of a solid body was awarded the Bordin prize in 1888; it seems indeed that Hermite devised the competition that year with that aim in mind. In 1889 she became the first woman to be a professor of anything since the Renaissance, so her canditature would have increased with the years, but 1882 was too early.

A quick look round to make sure no one has been missed. Lindemann (b. 1852) was to excite the mathematical world with his proof that π was transcendental, submitted to the *Mathematische Annalen* in June 1882, promptly published there and publicly verified by Weierstrass in the same month. This work was based on Hermite's earlier proof of the transcendence of e. Just too late for Fields and his committee, but surely a certainty for 1886. Could there be anyone else? Temporarily there were no strong Russian candidates except Kovalevskaia herself, and no, none anywhere else. No young Americans? Precisely the reason for my prize, Fields would reply (curiously not mentioning Gibbs, out of ignorance, I imagine). So, who would they choose? It would take the joke too far to press it home this far. Of the five, I imagine Darboux, Fuchs, Halphen, and Jordan are home, perhaps Noether would lapse, but I have already protested a little too much. Let me leave the game of historical competitions to your imaginations (there are after all many other years for which it can be played) and make a few more substantial points.

Lazarus Fuchs

FIGURE 36-11

My first point is methodological. It is necessary to conduct an analysis like this if we are ever to understand the way in which mathematics has developed. To explain how it attracts students from, or loses them to, the rival disciplines of theology, philology, physics, or engineering, we must consider how mathematics has been considered at a given time. We must ask the same question if we see to understand why one department of mathematics rather than another is being favored. Great ideas may win through in the end, deep problems may halt the most promising advance, but the truly historical story must deal with how that greatness or difficulty was originally perceived.

When we do that here, we find some interesting novelties. I do not think I have had to strain the historical record unduly to exclude Cantor or Klein from consideration in 1882, and it is certain that Lie was not considered a major mathematician until at least a decade later. It follows that histories of mathematics which concentrate on these men and downplay Darboux or Halphen risk distorting the historical development. The selection, on the other hand, of Fuchs is, I must admit, deliberately contrived. He is best remembered now for Fuchsian functions and groups, which he did not create, but in his day he represented the future of Berlin University. His concerns were the important concerns of the day, which have perhaps been undervalued in our search for the nineteenth-century origins of twentieth-century structural mathematics (a valid search in its own way, of course). Conversely, Dedekind owes his modern reputation mostly to Hilbert and his school; his omission above is deliberate.

Other, more specific, points can be made more briefly. It becomes clear that Berlin failed to provide for its own future. The triumvirate of Weierstrass, Kummer, and Kronecker was succeeded by that of Frobenius, Fuchs, and Schwarz, of whom only Frobenius was really of equal caliber. The largest university of its day was soon overtaken by Göttingen, with lessons any administrator or ambitious head of department might like to ponder. The foundations of analysis does not emerge as the central topic in mathematics that one might think it was from histories of mathematics. The central topics were arguably the theory of differential equations and a variety of topics in geometry, including the theory of algebraic curves. On the other hand the modern identification of mathematics with pure mathematics seems well established by 1882; I believe it was visible even fifty years before that. What we do see is a fashionable dallying with mathematical physics (e.g., Lamé's equation) accompanied by a vague feeling that applications are good for mathematics. It would be interesting to reconstruct debates on the borderline between mathematics and physics to see how the two disciplines saw each other and evaluated their ideas.

Finally, I wish to say that I have tried to make Fields and his committee reach decisions that genuinely reflect well on them. I believe that they chose five of the very best mathematicians of their generation. Then, as now, others were *not* chosen who were neither better nor worse but simply incomparable. The diversity of mathematics is one of its charms, the value judgments of mathematicians no less important for being, inevitably, partial.

Bibliographical Comments and Sources

I have not burdened this essay with the usual historical fortifications, so I must now add a few words concerning my sources. Almost everything about Berlin mathematics has been taken from K.-R. Biermann's fascinating *Die Mathematik und ihre Dozenten an der Berliner Universität, 1810–1920* (Academie Verlag, Berlin, 1973). Other personal information can be found in J. W. Dauben's biography of Cantor. The various collected works of Fuchs, Hermite, and Klein are full of interesting remarks, and the reviewers in *Fortschritte* provide further glimpses of contemporary opinion. P. Dugac's study of Dedekind, *Richard Dedekind et les fondements des mathématiques,* Vrin, Paris, 1976, contains many contemporary letters which further illuminate the scene. The most important history of nineteenth-century mathematics is Klein's *Entwicklung der Mathematik im 19th Jahrhundert,* Chelsea reprint, 1976, which historians are only now learning to handle with care. In almost all cases the short biographies of mathematics in the *Dictionary of Scientific Biography* (Scribners, 15 vols., 1970–1977) proved to be very useful; they provide a source of information many mathematicians could profitably consult on many historical questions.

37

The Last 100 Days of the Bieberbach Conjecture

O. M. Fomenko and G. V. Kuz'mina

1. Short Historical Excursus

In mathematics there exist many beautiful hypotheses, but few of them have had profound influence on the development of the subject in the large. One of these hypotheses, however, is the Bieberbach conjecture.

This conjecture arose at the very outset of geometric function theory (GFT). The essential part in this theory belongs to univalent functions which are regular (holomorphic) or meromorphic functions that determine one-to-one mappings. Univalent functions may be considered in various domains of definition, even on a Riemann surface, but attention is usually directed to certain specific classes. Two of the most important classes are S and Σ. (For a description of class Σ, see Section 4.) The class S consists of functions

$$f(z) = z + \sum_{n=2}^{\infty} c_n z^n \tag{1}$$

that are regular and univalent in the disk $|z| < 1$.

The investigation of the class S and of all the theory of univalent functions had its beginning in results of Koebe in 1907–09 (see [1, 2, 3]). Somewhat later the "area method" arose—the first consistent method in the theory of univalent functions. This method is based on the following simple fact: the area enclosed by the image under $1/f(z)$, $f \in S$, of the circle $|z| = r$ $(r < 1)$ is positive. The area method established a number of remarkable properties of the Koebe functions

$$K_\epsilon(z) = \frac{z}{(1 + \epsilon z)^2}, \ |\epsilon| = 1. \tag{2}$$

In particular, the functions $K_\epsilon(z)$, and only these functions, realize the maximum and the minimum of $|f(re^{i\theta})|$ and $|f'(re^{i\theta})|$ by fixed r, $0 < r < 1$, in the class S. There is a general principle of GFT which asserts that the extremal case is also a symmetric case. The Koebe functions $K_\epsilon(z)$ are the most symmetric mappings of all slit mappings in S. These functions map the disk $|z| < 1$ onto the whole plane with a radial slit.

In 1916 L. Bieberbach [4] conjectured that for any function $f \in S$ the inequality

$$|c_n| \leq n, \quad n \geq 2, \tag{B}$$

Volume 8, No. 1 (Winter 1986), 40–47

holds and that the equality in (B) occurs for the Koebe functions only. Because of its elegant simplicity of formulation, the Bieberbach Conjecture has stimulated the work of many analysts and has inspired, directly or indirectly, the development of various methods in GFT. Thus, in 1923 K. Löwner [5] created the parametric method and proved the inequality (B) in the case $n = 3$ using it.

In the course of the 1930s to 1940s, Grötzsch's method of strips, the method of contour integration, and the variational method were given. Later the method of symmetrization and the method of extremal metric arose. Just these methods, which are essentially geometric, have determined the face of modern GFT. They allowed one to obtain many deep results for general classes of functions, especially from the knowledge of the values assumed by such a function or its derivatives. In particular, great attention was devoted to the coefficients in power series expansions of functions. Nevertheless, the Bieberbach conjecture continued to present a challenge to all methods in complex analysis.

The inequality (B) for $n = 4$ was first proved in 1955 by Garabedian and Schiffer[1] in an extremely complicated way. In 1960 Charzynski and Schiffer gave an elementary and simple proof that $|c_4| \le 4$, aided by the following Grunsky result in 1939 [11]. For $f \in S$ let

$$\log \frac{f(z) - f(\zeta)}{z - \zeta} = \sum_{n=0}^{\infty} \sum_{k=0}^{\infty} c_{nk} z^n \zeta_k, \quad |z| < 1, \quad |\zeta| < 1,$$

where the coefficients c_{nk} are polynomials in the coefficients c_n of f. Then for each integer N and for all complex numbers $\lambda_1, \lambda_2, \ldots, \lambda_N$ the inequalities

$$\left| \sum_{n=1}^{\infty} \sum_{k=1}^{\infty} c_{nk} \lambda_n \lambda_k \right| \le \sum_{n=1}^{\infty} \frac{1}{n} |\lambda_n|^2 \tag{3}$$

hold.

The Grunsky inequalities (3) are of great importance in GFT. After $n = 4$ came $n = 6$. Also using the inequalities (3), Pederson and Ozawa each independently proved in 1968–69 that $|c_6| \le 6$, and then in 1972–73 Ozawa and Kubota proved that $|c_8| \le 8$. The inequality $|c_5| \le 5$ was obtained only in 1972 by Pederson and Schiffer. These authors used the Garabedian-Schiffer inequalities of 1967, which are interesting generalizations of the Grunsky inequalities. Later, new proofs were obtained for some values of n cited above. As a rule, the inequality (B) was proved indirectly: at first some "averaging" inequality was established from which the desired inequality $|c_n| \le n$ was obtained by more elementary means. The indicated proof always used deep techniques that were of independent interest.

There is another hypothesis which is directly connected with the Bieberbach Conjecture. Let $S^{(2)}$ be the class of odd functions in S; that is, the class of functions

$$f_2(z) = \{f(z^2)\}^{\frac{1}{2}} = \sum_{k=1}^{\infty} c_k^{(2)} z^{2k-1},$$

where $f \in S$. The functions

$$K_\epsilon^{(2)}(z) = \frac{z}{1 + \epsilon z^2}, \, |\epsilon| = 1, \tag{4}$$

play the role of the Koebe functions in the class $S^{(2)}$. However the analogue of the Bieberbach Conjecture is false for this class: as early as 1933 Fekete and Szegö obtained the exact inequality

---

[1]Omitted bibliographic references can be found in [2, 6, 7, 8]. See also the recent articles of C. FitzGerald [9] and Ch. Pommerenke [10].

$|c_3^{(2)}| \le \frac{1}{2} + e^{-\frac{2}{3}} = 1.013. \ldots$. Nevertheless, in 1936 M. Robertson conjectured [12] that the inequality

$$\sum_{k=1}^{n} |c_k^{(2)}|^2 \le n, \quad n \ge 2, \tag{R}$$

holds for every $f \in S^{(2)}$. The Robertson Conjecture implies the Bieberbach Conjecture, since

$$c_n = \sum_{k=1}^{n} c_k^{(2)} c_{n+1-k}^{(2)}$$

and hence

$$|c_n| \le \sum_{k=1}^{n} |c_k^{(2)}|^2,$$

by the Cauchy-Schwarz inequality. The Robertson Conjecture was given much less attention, however. Before 1984 the inequality (R) was proved only in the cases $n = 3, 4$.

Naturally, the attack on the Bieberbach Conjecture also took place in other directions. In 1955 W. K. Hayman proved that for any $f \in S$ there is an integer $n = n_f$ such that $|c_n| \le n$ for all $n \ge n_f$. The estimates of the form $|c_n| \le Cn$ for all $n \ge 2$ were obtained by various authors and by different methods. Before 1984 the best estimates were proved by FitzGerald in 1972 ($C = \sqrt{7/6} < 1.081$) and then by Horowitz in 1976 ($C = (209/140)^{1/6} < 1.0691$).

Such was the situation with the Bierberbach Conjecture when in the beginning of 1984 the American mathematician Louis de Branges from Purdue University announced that he had obtained the proof of inequality (B) simultaneously for all n.

In connection with de Branges' proof, we must say something about one further result. Let γ_k be the logarithmic coefficients of a function $f \in S$; that is,

$$\log \frac{f(z)}{z} = \sum_{k=1}^{\infty} 2\gamma_k z^k.$$

A natural way for estimating the logarithmic coefficients provides the Grunsky inequalities (3). In 1965–67 N. A. Lebedev and I. M. Milin [13, 14] established the following inequality:

$$\left(|c_{n+1}| \le \right) \sum_{k=1}^{n+1} |c_k^{(2)}|^2 \le (n+1) \exp\left\{ \frac{1}{n+1} \sum_{k=1}^{n} (n+1-k) \left(k|\gamma_k|^2 - \frac{1}{k} \right) \right\}. \tag{5}$$

The equality in the second inequality (5) holds only in the case

$$\gamma_k = \frac{1}{k} \eta^k, \quad k = 1, \ldots, n, \quad |\eta| = 1.$$

For the Koebe function

$$|\gamma_k| = \frac{1}{k} \text{ for all } k \ge 1,$$

and in (5) the equalities hold:

$$|c_{n+1}| = \sum_{k=1}^{n+1} |c_k^{(2)}|^2 = n + 1.$$

It follows immediately from (5) that, if for $f \in S$ the inequality

$$\sum_{k=1}^{n}(n+1-k)k|\gamma_k|^2 \leq \sum_{k=1}^{n}(n+1-k)\frac{1}{k} \tag{L}$$

holds, then the inequalities (R) and (B) also hold. The conjecture that the inequality (L) is true for every function $f \in S$ we will call the Lebedev-Milin Conjecture, and it is precisely this fact that was proved by de Branges. Moreover, de Branges obtained a stronger result; he proved some inequalities for bounded univalent functions from which the inequality (L) is obtained as a simple limiting case.

2. The Classical Version of de Branges' Proof

Below we give an account of the classical version of de Branges' proof. This version was published in the series *LOMI Preprints* [16] and arose as a result of the creative contact of de Branges with the members of the Leningrad Seminar in GFT. This seminar has existed for more than forty years. The leaders of the seminar were G. M. Goluzin (until 1952) and N. A. Lebedev (1952 to 1982). At present the seminar works in LOMI and its leader is the second author of this article. During de Branges' visit, the seminar consisted of E. G. Emeljanov, S. I. Fedorov, E. G. Goluzina, A. Z. Grinspan, V. I. Kamozkii, V. O. Kuznetzov, I. A. Lebedev, I. M. Milin, V. I. Milin, and N. A. Schirokov.

Before formulating de Branges' results, we need to give some definitions. Let B_S be the class of functions $b(z)$ that are regular and univalent in the disk $|z| < 1$ and satisfy the conditions $b(0) = 0$, $b'(0) > 0$ and $|b(z)| < 1$ for $|z| < 1$. Let

$$b_{\tau,x}(z) = K_x^{-1}(\tau K_x(z)), \quad 0 < \tau \leq 1, \quad |x| = 1, \tag{6}$$

where $K_x(z)$ is the Koebe function (2). The function $b_{\tau,x}(z)$ is in B_S and maps $|z| < 1$ onto the disk $|w| < 1$ with a radial slit.

The following system of time-dependent weight functions plays the key role in de Branges' proof. Let $\{\sigma_k(t)\}_1^{n+1}$ be the sequence of functions, defined on $[1, \infty)$, which are obtained as solutions of the recurrent system of equations

$$\sigma_k(t) + \frac{t}{k}\sigma_k'(t) = \sigma_{k+1}(t) - \frac{t}{k+1}\sigma_{k+1}'(t) \tag{7}$$

with initial conditions

$$\sigma_k(1) = n + 1 - k \quad (k = 1, \ldots, n+1). \tag{8}$$

De Branges essentially uses the following property of these functions: $\sigma_k(t)$, $k = 1, \ldots, n + 1$, are non-negative and non-increasing functions for all $t \in [1, \infty)$. Establishing this property of the functions $\sigma_k(t)$ later, de Branges first proves the following theorem.

THEOREM 1 (The main theorem of de Branges) Let $n \geq 1$. Assume that the above determined system of functions $\{\sigma_k(t)\}_1^{n+1}$ satisfies the conditions $\sigma_k(t) \geq 0$, $\sigma_k(t) \leq 0$ for all $t \in [1, \infty)$ ($k = 1, \ldots, n + 1$). Then the inequality

$$\sum_{k=1}^{n}\left|\left\{\log\frac{b(z)}{b'(0)z} + p \circ b(z)\right\}_k\right|^2 k\sigma_k(\alpha) \leq \sum_{k=1}^{n}|p_k|^2 k\sigma_k(\beta) + \sum_{k=1}^{n}\frac{4}{k}(\sigma_k(\alpha) - \sigma_k(\beta)) \tag{*}$$

holds for every $b(z) \in B_S$ and for every function

$$p(z) = \sum_{k=1}^{\infty} p_k z^k,$$

regular for $|z| < 1$, and all numbers α and β, $1 \leq \alpha < \beta < \infty$, such that $\alpha = b'(0)\beta$. (The notation $\{F(z)\}_k$ is used for the coefficient of z^k in the expansion of $F(z)$ in powers of z.) For the functions $b(z) = b_{\tau,x}(z)$, where $\tau = \alpha/\beta$, $|z| = 1$, and $p(z) = -2\log(1 + xz)$, the equality holds in (∗).

The properties of functions $\sigma_k(t)$, used in Theorem 1, are established by the following theorem of de Branges.

THEOREM 2:   Let $\{\sigma_k(t)\}_1^{n+1}$ be a system of functions satisfying the system of equations (7) with initial conditions (8). Then the inequalities

$$\sigma_k(t) \geq 0, \sigma_k'(t) \leq 0, \quad k = 1, \dots, n+1, \tag{9}$$

hold for all $t \in [1, \infty)$.

One obtains directly from Theorem 1 that the Lebedev-Milin Conjecture, the Robertson Conjecture, and the Bieberbach Conjecture are all true for all $n \geq 2$. Also, the equality in (L) occurs for the Koebe functions (2); the equalities in (R) and (B) occur only for the functions (4) and (2) correspondingly.

Indeed, assume that $f(z) = z + c_2 z^2 + \cdots \in S$ and that $|f(z)| < M$ ($M > 1$), in the disk $|z| < 1$. Then $f_1(z) = M^{-1} f(z) \in B_S$ and inequality (∗) applies to $f_1(z)$ with $\beta = M\alpha$. Letting $M \to \infty$, one obtains the inequality (L). Then applying the Lebedev-Milin inequalities (5) (and remembering that equality holds in the second inequality (5) only in the case $\gamma_k = (1/k)\eta^k$, $|\eta| = 1, k = 1, \dots, n$, and therefore only in the case $f(z) = K_x(z)$, $|x| = 1$) one obtains that the inequalities (R) and (B) are true, and that the equalities in (R) and (B) hold only in the indicated cases.

We now give the details of the classical version of de Branges' proof which was published in the series *LOMI Preprint* [16]. Later de Branges gave another variant of the same version of his proof [18].[2] The last proof uses the most general form of the Löwner equation. In the present article we concentrate on the first classical version of de Branges' proof since the first variant, in our opinion, shows more clearly the possibilities of the Löwner method and the virtuoso skill with which de Branges adapted it.

The Proof of Theorem 1

There are three steps in the proof.
  1.   Löwner's theorem (see page 436) shows that it is sufficient to prove the inequality (∗) for the functions in B_S. It is possible in turn to restrict the family of functions for which it is sufficient to prove the inequality. Indeed, the following lemma is true.

LEMMA.   Every function $b(z) \in B_S$, $b'(0) = \tau, 0 < \tau < 1$, can be approximated by compositions of a finite number of functions of the form (6) uniformly in every closed disk $|z| \leq r < 1$.

---

[2]See also the joint work of FitzGerald and Pommerenke [19].

The proof of this lemma is based on the fact that the function $x(t)$ in the Löwner equation (10), as a continuous function in the closed interval $[\tau, 1]$, can be approximated by a sequence of piecewise constant functions $\{x_m(t)\}$, $|x_m(t)| = 1$, which converges uniformly to $x(t)$. The solution $b(z) = b(z, \tau)$ of the Löwner equation (10) for the function $x_m(t)$ is a composition of the mappings of the form (6).

This lemma is not an essentially new result but it is just Löwner theory applied by de Branges.

A composition of such mappings can be described as "forking the slit":

$$\bigodot_{\cdot} \underset{\rightarrow}{b_{\tau_1, x_1}(z)} \bigodot_{\diagup} \underset{\rightarrow}{b_{\tau_n, x_n} \circ \ldots \circ b_{\tau_1, x_1}(z)} \bigodot_{\cdot}$$

2. Thus it is sufficient to prove the inequality $(*)$ for a composition $b(z)$ of a finite number of the mappings of the form (6). It is possible to restrict the family of such functions still more. De Branges does this by the following highly elegant reasonings. We shall consider the difference between the left and right sides of $(*)$:

$$\Phi(b, p, \beta) = \sum_{k=1}^{n} \left| \left\{ \log \frac{b(z)}{\tau z} + p \circ b(z) \right\}_k \right|^2 k \sigma_k(\beta \tau) - \sum_{k=1}^{n} |p_k|^2 k \sigma_k(\tau) - I(\beta \tau, \beta).$$

Here

$$\tau = b'(0), \, I(\alpha, \beta) = \sum_{k=1}^{n} \frac{4}{k} (\sigma_k(\alpha) - \sigma_k(\beta)) = 2 \int_{\alpha}^{\beta} \left[\sigma_1(t) - t\sigma_1'(t) \right] \frac{dt}{t}.$$

Let $b(z)$ be the composition of $m \geq 2$ mappings of the form (6):

$$b(z) = b_2 \circ b_1(z)$$

where $b_2(z)$ is a mapping of the form (6), and $b_1(z)$ is the composition of $m - 1$ such mappings. We have $b'(0) = \tau = \tau_2 \tau_1$ where $\tau_j = b_j'(0), j = 1, 2$. Using the evident identities

$$\log \frac{b(z)}{\tau z} + p \circ b(z) = \log \frac{b_1(z)}{\tau_1 z} + \left(\log \frac{b_2}{\tau_2 z} + p \circ b_2 \right) \circ b_1(z)$$

and

$$I(\beta \tau, \beta) = I(\beta \tau, \beta \tau_2) + I(\beta \tau_2, \beta),$$

one obtains

$$\Phi(b_2 \circ b_1, p, \beta) = \phi(b_2, p, \beta) + \Phi \left(b_1, \log \frac{b_2}{\tau_2 z} + p \circ b_2, \beta \tau_2 \right). \tag{11}$$

It shows that it is sufficient to prove the inequality $(*)$ only for the mappings $b_{\tau, x}(z)$ themselves and for arbitrary functions $p(z), p(0) = 0$, regular in the disk $|z| < 1$. The identity

$$\Phi(b_{\tau, x}, p, \beta) = \Phi(b_{\tau, 1}, \bar{p}, \beta) \quad \text{where } \bar{p}(z) = p(\bar{x}z)$$

shows that it is sufficient to obtain inequality $(*)$ for the functions $b_{\tau, 1}(z)$.

3.   It is clear that for $\tau = 1$ we have $b_{\tau,1}(z) = z$ and (*) is true: $\Phi(b_{1,1}, p, \beta) = 0$. Therefore it is sufficient to prove that

$$\tau\frac{\partial}{\partial\tau}\Phi(b_{\tau,1}, p, \beta) \geq 0 \text{ for } 1/\beta < \tau < 1.$$

On applying the identity (11) to the functions $b_{\tau(1-\epsilon),1}(z) = b_{\tau,1} \circ b_{1-\epsilon,1}(z)$ one obtains

$$\tau\frac{\partial}{\partial\tau}\Phi(b_{\tau,1}, p, \beta) = \lim_{\epsilon\to 0}\left\{\frac{1}{-\epsilon}\left[\Phi(b_{1-\epsilon,1}, q, \beta\tau) - \Phi(b_{1,1}, q, \beta\tau)\right]\right\},$$

where

$$q = q(z, \tau) = \log\frac{b_{\tau,1}(z)}{\tau z} + p \circ b_{\tau,1}(z) = \sum_{k=1}^{\infty}q_k(\tau)z^k, q(z, 1) = p(z).$$

Consequently it is sufficient to show that

$$\Phi_0(p, \beta) = \lim_{\tau\to 1-0}\tau\frac{\partial}{\partial\tau}\Phi(b_{\tau,1}, p, \beta) \geq 0,$$

for every considered function $p(z)$ and for every $\beta > 1$. Observe that

$$\tau\frac{\partial}{\partial\tau}\Phi(b_{\tau,1}, p, \beta) = 2\,\mathrm{Re}\,\sum_{k=1}^{n}\overline{q_k(\tau)}\tau\frac{\partial q_k(\tau)}{\partial\tau}k\sigma_k(\beta\tau)$$

$$+ \sum_{k=1}^{n}|q_k(\tau)|^2k\beta\tau\sigma_k'(\beta\tau) + 2(\sigma_1(\beta\tau) - \beta\tau\sigma_1'(\beta\tau)).$$

From the definition of the functions $b_{\tau,1}(z)$:

$$\frac{b_{\tau,1}(z)}{(1 + b_{\tau,1}(z))^2} = \frac{\tau z}{(1 + z)^2},$$

and one obtains the identity

$$\tau\frac{\partial}{\partial\tau}q(z, \tau) = \frac{2b_{\tau,1}(z)}{1 - b_{\tau,1}(z)} + b_{\tau,1}(z)\frac{1 + b_{\tau,1}(z)}{1 - b_{\tau,1}(z)}p'(b_{\tau,1}(z)).$$

Write

$$\frac{2z}{1 - z} + z\frac{1 + z}{1 - z}p'(z) = \sum_{k=1}^{\infty}\psi_k z^k.$$

Then

$$\Phi_0(p, \beta) = 2\,\mathrm{Re}\,\sum_{k=1}^{n}\bar{p}_k\psi_k k\sigma_k(\beta) + \sum_{k=1}^{n}|p_k|^2k\beta\sigma_k'(\beta) + 2(\sigma_1(\beta) - \beta\sigma_1'(\beta))$$

Thus, using the determination of the functions $\sigma_k(t)$, one obtains the desired inequality by elementary calculations:

$$\Phi_0(p, \beta) = -\sum_{k=1}^{n}\frac{\beta}{k}\sigma_k'(\beta)|\psi_k|^2 \geq 0.$$

This completes the proof of Theorem 1.

The Magical Löwner Method

De Branges' proof uses the Löwner method, well known in the theory of univalent functions. In application to the class B_S this method is based on the fact that the family B'_S of those functions from the class B_S which map $|z| < 1$ onto a domain obtained from $|w| < 1$ by producing a Jordan arc is dense in the entire class B_S. Controlling the growth of the arc by a parameter t, it is shown that the corresponding mapping function satisfies a certain differential equation. Such a parametrization is very useful for the solution of many extremal problems.

Later, essential development of the Löwner method was given. But de Branges avoids the general case of the Löwner equation: the Löwner method in its classical form of 1923 is sufficient for his goal.

LÖWNER'S THEOREM   For every function $b(z) \in B_S$ there exists a sequence of functions $\{b_n(z)\}$ in B_S which converges to $b(z)$ uniformly in the disk $|z| < 1$. Every function $b_n(z)$ of this sequence is obtained as a solution of the equation

$$t \frac{\partial}{\partial t} b(z, t) = b(z, t) \frac{1 + x(t)b(z, t)}{1 - x(t)b(z, t)}$$

$$\text{for } b'_n(0) \leq t \leq 1 \tag{10}$$

with a function $x(t)$, $|x(t)| = 1$, continuous for $b'_n(0) \leq t \leq 1$ and with the initial condition $b(z, 1) = z$.

A solution of the Löwner equation (10) can be described as "rolling up the slit"

The Proof of Theorem 2

We come now to the proof of Theorem 2. From the second condition (9) and the equations (7) one obtains the first condition (9): $\sigma_{n+1}(t) = 0, \sigma_k(t) \geq \sigma_n(t) = t^{-n}$ for $k = 1, \ldots, n$. It remains to show that the second condition (9) is true. The notation $_{p+1}F_p$ is used for the hypergeometric series:

$$_{p+1}F_p \begin{pmatrix} a_1, \ldots, a_{p+1}; \\ b_1, \ldots, b_p; \end{pmatrix} = \sum_{k=0}^{\infty} \frac{(a_1)_n \ldots (a_{p+1})_n}{(b_1)_n \ldots (b_p)_n} \frac{z^n}{n!}, \quad |z| < 1,$$

where $(a)_0 = 1, (a)_n = a(a + 1) \ldots (a + n - 1)$ for $n \geq 1$.

The solution of the system (7) is of the form

$$\frac{\sigma_k(t)}{k} = \sum_{v=0}^{n-k} \frac{(2k + v + 1)_v}{(-1)^v v!} \Delta_{k+v}(t) \, (k = 1, \ldots, n), \quad \text{and } \sigma_{n+1}(t) = 0,$$

where $\Delta_k(t)$ satisfies the elementary differential equation

$$t\Delta_k'(t) = -k\Delta_k(t)$$

with solution

$$\Delta_k(t) = \Delta_k(1)t^{-k}.$$

The conditions (8) lead to the equality

$$(n - k)\,\Delta_{n-k}(1) = (k + 1)\,_2F_1\left(-k, 2n - k; -k - 1; 1\right).$$

From this de Branges obtains, in a natural way, the representation for $\sigma_{n-k}'(t)$ in the form of the hypergeometric series:

$$-\frac{t\sigma_{n-k}'(t)}{n - k} = \frac{(2n - 2k + 2)_k}{k!}\,t^{-n+k}\,_3F_2\left(\begin{matrix}-k, 2n - k + 2, n - k + \frac{1}{2};\\ 2n - 2k + 1, n - k + \frac{3}{2};\end{matrix}t^{-1}\right).$$

Now the following result is used. Askey and Gasper [17] show that, when $\alpha > -2$ and $k = 0, 1, 2, \ldots,$

$$_3F_2\left(\begin{matrix}-k, k + \alpha + 2, (\alpha + 1)/2;\\ \alpha + 1, (\alpha + 3)/2;\end{matrix}z\right) \geq 0 \text{ for } z \in [0, 1].$$

On applying the result with $\alpha = 2n - 2k$, one obtains the desired inequality:

$$\sigma_k'(t) \leq 0 \text{ for all } t \in [1, \infty], k = 1, \ldots, n.$$

3. The "Last 100 Days" of the Bieberbach Conjecture

The history of such an outstanding mathematical discovery as the proof of the Bieberbach Conjecture is undoubtedly interesting to readers, so we would like to describe the "last 100 days."

It was the end of January 1984 when de Branges phoned Leningrad. He said that he was going to come to the city in April 1984 for two months under the terms of the exchange agreement between the Soviet and U.S. Academies of Sciences; he had the manuscript of his new book. The book was devoted to the theory of square summable power series, and he added that it contained the proof of the Bieberbach Conjecture as the final chapter.

The last announcement was received with distrust. Could one prove the conjecture by the methods of functional analysis? There was general skepticism. Some remembered as well that, using the same concepts, de Branges had earlier obtained an erroneous proof of the Ramanujan-Peterson Conjecture. However, the Leningrad Seminar in GFT decided to prepare for the discussions seriously, and for two months preceding the arrival of de Branges, the seminar studied his previous work thoroughly and informally. Besides the theory of complementary spaces, this work used the methods of univalent function theory, yet it was far from a proof of the Bieberbach Conjecture. At that stage of the seminar a great deal of work was done by Kamozkii and Emeljanov.

In April 1984 de Branges arrived in Leningrad with a voluminous manuscript (385 typewritten pages!) of his book. During the first preliminary meeting with the seminar members, de Branges briefly outlined the basic ideas of his proof. We learned that de Branges used not only his concepts of functional analysis but also the Löwner method and the Lebedev-Milin inequalities. The seminar leader pointed out immediately to de Branges that if his proof was correct, then there should exist a classical version of the proof, that is, a version using no functional analysis. De Branges replied that he had been questioned about that many times, and he believed it was impossible to find a classical version of his proof. It was strange to hear such a statement, but maybe it was not supposed to be taken too literally. In any event, it was clear that for our seminar the simplest way to verify the proof was to translate it into the language of GFT.

FIGURE 37-1 *Leningrad Seminar. From left: S. I. Fedorov, E. G. Goluzina, G. V. Kuz'mina, A. Z. Grinspan, A. Grinspan, V. I. Milin, I. M. Milin, P. N. Pronin.*

The seminar members asked de Branges to give them the manuscript to study at home. De Branges made several seminar reports; simultaneously the seminar members were examining and analyzing de Branges' manuscript (Emeljanov, Fedorov, Goluzina, and Schirokov took active part in this work). It was the general expectation that some serious error would soon be found. However, little by little it became evident that de Branges actually had a correct method of proceeding towards a proof: his proof of the main theorem (see Section 2, Theorem 1) was essentially correct! The exception was the argument related to approximation by functions of the special form, but it was easy to substitute a correct proof. Some of de Branges' formulations and arguments also were translated into the language of GFT.

The culmination was the following episode. During a discussion following one of de Branges' reports, Emeljanov wrote on a blackboard the main inequality (*). It should be noted that in the manuscript, the inequality (*) has a different form: it was an inequality containing norms in some functional spaces. De Branges agreed that his inequality could be rewritten in the form of the inequality (*). At this time he understood that the classical version of his proof existed and that the seminar was working toward this version. It remained to convince de Branges to publish the proof of the classical version in the near future. The seminar leader explained to de Branges that, being a closed account of the proof of the Bieberbach Conjecture in classical terms of GFT, this version would undoubtedly rouse great interest and suggested to de Branges that he publish the proof as an LOMI preprint. De Branges' first answer was categorically negative. His main goal was to publish the book, so he didn't want to lose such a brilliant example of the effectiveness of his theory. But de Branges was impressed by the arguments in favor of the publication of the classical version, and finally he agreed.

To complete the verification of de Branges' proof, it was necessary to finish the work on the classical version. An essentially completed variant of the proof of the main theorem of de Branges (see Theorem 1) was prepared by Emeljanov. The text of this proof was slightly modified by I. M. Milin, who prepared a report to acquaint mathematicians of Leningrad with the proof. The second author had then to complete the course of the seminar. It was necessary to verify the proof of de Branges' second theorem (see Theorem 2) to be sure that he had indeed established the inequality $|c_n| \le n$ for all n. It seems strange now, but then it was difficult to believe that this had really been accomplished; it was considered likely that de Branges obtained the proof of the Bieberbach Conjecture only for some coefficients. However, de Branges' Theorem 2 turned out to be true!

Almost at the same time, Theorem 2 was verified in another way. A. N. Kirillov (from the laboratory of algebraic methods of LOMI) obtained a simple and short proof of the inequalities de Branges needed. This proof was based on the classical theorem of Clausen in the theory of hypergeometric series. This proof was shown to de Branges, but earlier he had received the same proof from Gasper. So the production of the classical version of the proof was complete.

FIGURE 37-2 *N. A. Lebedev.*

Then the work on the preparation of the proof for publication began. It was not easy since de Branges was still doubtful about the publication of the classical version, and the seminar had to put pressure on him. This work was in progress during all of June 1984. At this time de Branges and the seminar leader discussed the proof in detail, improving separate parts of the account in de Branges' preprint. A project involving a more detailed exposition of the proof, which de Branges primarily hoped to publish in the "Mathematical Sbornik," was also discussed.

FIGURE 37-3 *A. N. Kirillov.*

At the end of June, de Branges departed for Europe and then for the U.S.A. "I return to America, and nobody will believe me there again," he said during his last conversation.[3]

In July 1984 the preprint of de Branges was published. Following de Branges' request, we mailed a part of the edition to him, and then congratulated him on the first publication of the proof. The remaining part of the edition was mailed mostly to Soviet mathematicians, so that the specialists could be convinced that the conjecture was indeed proved.

Remembering the "100 days," the second author asked herself repeatedly if the Leningrad Seminar in GFT had acted correctly toward de Branges; perhaps de Branges had to wait for us, but undoubtedly the activity of the seminar played a useful role. It broke the psychological barrier which would naturally arise before anybody undertook the study of the primary 385-page manuscript.

4. Some Controversial Remarks

With the appearance of de Branges' proof there arose many questions. Should the Bieberbach Conjecture have been solved considerably earlier? Our answer is yes. (FitzGerald [9] is of another opinion.) In fact, the Löwner theory used by de Branges is a classical fact of univalent function theory. The Lebedev-Milin inequalities were obtained in 1965–67, but they could possibly have been established long before (but not earlier than 1939 when the Grunsky work [11] had been done). It should be noted that logarithmic coefficients directly connected with the Grunsky inequalities were outside the basic current of investigations for some time. In our opinion it is wrong to attribute a crucial role

---

[3]As it turned out, these fears of de Branges were incorrect. In March 1985 Purdue University hosted an International Symposium on the occasion of de Branges' proof of the Bieberbach Conjecture.

in de Branges' proof to the result of Askey and Gasper of 1976 [17]. Indeed, although not acquainted with [17], Kirillov obtained the required result, and this fact confirms our standpoint.

De Branges' brilliant idea is to use the system of functions $\sigma_k(t)$. Undoubtedly this idea was motivated by a complementation concept in functional analysis. On the other hand, there is a direct connection between the functions $\sigma_k(t)$ and the functions $b_{\tau,x}(z)$ which are analogous to the Koebe functions in the class of bounded univalent functions. Just this connection determined the essential role that the functions $\sigma_k(t)$ play in de Branges' proof.

Someone might puzzle over how it could happen that from all methods developed in the effort to solve the Bieberbach Conjecture, only the Löwner method was used for its proof. Of course, the circumstances mentioned above should not discredit the other methods in GFT in the least degree. Each of the methods of GFT has its own range of action and perhaps no universal method can exist.

Apparently a number of analysts are sorry that the Bieberbach Conjecture is finally proved. However, it should bring great satisfaction to know that there is such a beautiful proof as de Branges'.

In GFT there are many interesting unsolved problems and the list of these problems increases constantly. For example, even the class S, which is one of the most studied classes of functions, has not been investigated from the point of view of the regions of values for basic functionals. As to a question of estimating the coefficients, this problem for the class Σ of functions $f(z) = z + \alpha_0 + \alpha_1 z^{-1} + \ldots$, which are meromorphic and equivalent in the exterior of the unit disc, is considerably more difficult than the Bieberbach conjecture. The following question arises: is the maximum of $|\alpha_n|$ in the class Σ realized by functions with real coefficients $\alpha_1, \alpha_2, \ldots$, that is, by symmetric elements in Σ? This question is still open for every $n \geq 4$.

References

1. Goluzin, G. M. *Geometric Theory of Functions of a Complex Variable.* Moscow, 1952; German trans., Deutsher Verlag: Berlin, 1957; 2nd ed., Izdat. "Nauka": Moscow, 1966; English trans., Amer. Math. Soc., 1969.
2. Jenkins, J. A. *Univalent Functions and Conformal Mapping.* 2nd ed. Springer-Verlag: Berlin, 1965.
3. Lebedev, N. A. *The Area Principle in the Theory of Univalent Functions.* Izdat. "Nauka": Moscow, 1975 (in Russian).
4. Bieberbach, L. Über die Koeffizienten derjenigen Potenzreihen, welche eine schlichte Abbildung des Einheitskreises vermitteln. *S.-B. Preuss. Akad. Wiss.,* 1916, 940–955.
5. Löwner, K. Untersuchungen über schlichte konforme Abbildungen des Einheitskreises, I. *Math. Ann.,* 89 (1923), 103–121.
6. Pommerenke, Ch. *Univalent functions* (with a chapter on quadratic differentials by G. Jensen). Vandenhoeck and Ruprecht: Göttingen, 1975.
7. Duren, P. L. *Univalent functions.* Springer-Verlag, 1983.
8. Duren, P. L. Coefficients of univalent functions. *Bull. Amer. Math. Soc.,* 83 (1977), 891–911.
9. FitzGerald, C. H. The Bieberbach Conjecture: Retrospective. *Notices Amer. Math. Soc.,* (1985), 2–6.
10. Pommerenke, Ch. The Bieberbach Conjecture. *The Math. Intelligencer,* 7, no. 1 (1985), Springer-Verlag.
11. Grunsky, H. Koeffizientenbedingungen für schlicht abbildende meromorphe Funktionen. *Math. Z.,* 45 (1939), 29–61.
12. Robertson, M. S. A remark on the odd schlicht functions. *Bull. Amer. Math. Soc.,* 42 (1936), 366–370.
13. Lebedev, N. A. and Milin, I. M. An inequality. *Vestnik Leningrad. Univ.,* 20 (1965), no. 19, 157–158 (in Russian).

14. Milin, I. M. On the coefficients of univalent functions. *Dokl. Akad. Nauk SSSR*, 176 (1967), 1015–1018 (in Russian) = *Soviet Math. Dokl.*, 8 (1967), 1255–1258.

15. Milin, I. M. *Univalent Functions and Orthonormal Systems.* Izdat. "Nauka": Moscow, 1971 (in Russian); English trans., Amer. Math. Soc., Providence, R.I., 1977.

16. de Branges, L. *A proof of the Bieberbach Conjecture.* Steklov Math. Inst. Leningrad Department, LOMI Preprints, E-5-84 (1984).

17. Askey, R. and Gasper, G. Positive Jacobi polynomial sums, II. *Amer. J. Math.*, 98 (1976), 709–731.

18. de Branges, L. A proof of the Bieberbach Conjecture. *Acta Math.*, 154 (1985), 137–152.

19. FitzGerald, C. H. and Pommerenke, Ch. The de Branges theorem on univalent functions. *Trans. Amer. Math. Soc.*, 290 (1985), 683–690.

38

A Little-Known Chapter in the History of German Mathematics

Walter Kaufmann-Bühler

Walter Kaufmann-Bühler enjoyed writing imaginary history. This article, along with a few biographies of non-existent mathematicians, was found among his papers.

The creation of a Max-Planck-Institut for Mathematics in Bonn (*Mathematical Intelligencer*, Vol. 7 (1985), No. 2, 41–52) reminds us of earlier efforts to set up a similar establishment devoted to the development of German mathematics.

Just prior to the first convention of the Deutsche Mathematikervereinigung in 1892, Heinrich Schröter, the algebraist from Karlsruhe, sent a confidential circular to several of his friends and colleagues at other universities, expressing his fear that German mathematics, only twenty years after the victorious war against France ("and a world of enemies"), might fall back behind England and France, its natural rivals as "Kulturvölker." The older generation, men like Lifschitz, Knierim, or Pilz, could remember the time when the young German mathematicians had to go to Paris to sit at the feet of first-rate teachers—an altogether humiliating and degrading experience. Now, after the tragic breakdown of Felix Klein in 1882 (a carefully kept domestic secret), the field again seemed to belong to the genius of Poincaré and his younger colleagues Darboux, Hadamard, and Grosser-Pointillist.

A few months after Schröter's letter, sixty-five university professors sent a formal petition to Herr Althoff, the ministerial director in the department of education in the Prussian government and well-known benefactor of mathematics. This petition discussed the situation of mathematics in Germany, comparing it, in the flowery language of the time, to a young virgin in need of protection "lest she will never grow to be the chaste mother of a new heroic age," and culminating in the proposal to establish a unique new research facility, an institute, provisionally to be called Kaiser-Friedrich-Gedächtnis-Institut. The choice of the name was particularly fortunate because Emperor Friedrich III, who had died of cancer of the throat in 1888 after a reign of only three months, had been popular among the Germans because of his interest in science and his love for mathematics. He often referred to his famous (and devastating) bombardment of the French village St. Marie-sur-Ouse in 1870 as his *Gesellenstück* in ballistics, the most important and interesting branch of applied mathematics.

Shortly after the professors sent their petition to Althoff, the apl. Professoren, the ao. Professoren, and the Privatdocenten duly filed their petitions, respectfully addressed to Althoff's assistant, his first deputy assistant, and his secretary, Frl. Anna von Wolkenstein-Drachenfels.

Volume 9, No. 4 (Fall 1987), 20–22

FIGURE 38-1A *Leopold Kronecker.*

FIGURE 38-1B *Georg Cantor.*

Althoff, forward-looking and enthusiastic as ever, immediately devised a plan to reach a decision about the project. Althoff's efficiency was proverbial (he is said to have inspired Hilbert's *Basissatz*); right away he knew what to do. Eventually, his committee of capable and right-minded mathematicians consisted of Kronecker, Heine, Cantor (M. and G. for a joint opinion), H. A. Schwarz, General von Pöckeritz, a disciple of von Müffling, and the Generalpostmeister Heinrich von Stephan.

Stephan's and Pöckeritz's opinions are among the treasures of the Stiftung Preussicher Kulturbesitz in Berlin. Both are positive, but Stephan points out that one should not expect quick improvement of the postal service once the institute was set up. Pöckeritz quotes from a letter Müffling had written to von Lindenau when the former tried to convince Gauss to move to Berlin. Pöckeritz's opinion concludes climactically with: "If we don't have a Gauss, we should at least have a research institute."

Of the professional mathematicians, Heine responded first, arguing strongly against the idea of a central institute. Instead, he proposed to establish 150 small institutes, evenly distributed over the German Reich. Heine was confident that this would lead to a rise in the level of mathematical cul-

FIGURE 38-1C *Felix Klein.*

FIGURE 38-1D *Hermann Amandus Schwarz.*

ture in Germany. He did not see much point in the development of advanced and specialized research topics and suggested concentration on elementary mathematics. His impassioned memorandum concludes with a philosophical passage, explaining the unity between higher and lower mathematics: "Only if we have a firm base, will we be able to erect a lofty structure, the envy of our neighbors and the mirror of our culture."

The Cantors' opinion need not be discussed here. It is extensively presented in numerous textbooks on set theory. Most undergraduates have, in one way or another, been exposed to it. As usual, Schwarz was straightforward. He came out in favor of an institute to be established in the small provincial town of Bonn (Rhein), not far from Düsseldorf, Felix Klein's birthplace. Klein and Schwarz were not friends, but Klein, according to Schwarz, was the natural choice for director. Yet one had to proceed with caution because Klein had recently betrayed a most unnatural and dangerous interest in educational questions. It was important to make sure that he would not lose himself in a maze of abstruse pedagogical controversies. Klein, as was well known, had his decisive ideas in the theory of automorphic forms in the course of a sleepless night, suffering from a severe attack of asthma. This prompted Schwarz to suggest a special room for Klein, equipped with an abundance of the then-fashionable heavy furniture: a dusty sofa, a secondhand imitation of an empire chaise lounge, and velvet draperies. Schwarz's main concern was the furtherance of mathematical knowledge but he was not unkind; asthma attacks are rarely dangerous, though hardly a pleasant experience. There can be no doubt that Klein was prepared to make sacrifices but, ironically, Schwarz came forth with his proposal exactly at the time when Klein became a convinced internationalist, without any interest in competing with the French or English. "Ach—too late," Schwarz is said to have exclaimed when Klein's conversion became known.

Kronecker was against the institute but died before the final decision was reached. Apparently, he felt that sufficient opportunities existed for the few gifted researchers without independent income. There was the royal war research group in Charlottenburg with its focus on geometry and ballistics; the existence of a new institute, he felt, might mislead people and even lure them into mathematics. Even 100 new institutes and hundreds of reichsmarks in funds would not, to vary a theme of Jacobi, pave a royal road for mathematicians.

On his deathbed, Kronecker suggested Frege as his replacement, probably to annoy G. Cantor. This was an unusual choice and led, much to Althoff's chagrin, to a delay of three years because Frege developed a specific symbolic calculus to assess the proposal. Frege's essay, though in symbolic language, is quite spirited; we quote a few decisive sentences (in the translation from symbolic calculus by L. von Patzer): "Mathematics is a homeless orphan, without a bank account or a cheque book. In what coin will she [!] ever pay her bills and how would she do it? Would a cheque be written by the hand that guides the pen, by the pen, or by its ink? Is the cheque, drawn up by mathematics, a mathematical cheque and what do the numbers on the cheque mean? Is money a function of these numbers or are the numbers a function of the money?"

Convincingly, even from today's perspective, Frege shows how the establishment of an institute leads to so many theoretical complications that it is practically impossible. However, and this will not surprise any student of Frege's work, there is a constructive alternative—a symbolic, ideal institute. Full of enthusiasm, Frege proposes the establishment of innumerable such symbolic institutes. Every mathematician, every German should harbor such an institute in his breast!

Obviously, Frege's proposal was appealing to the generally parsimonious German authorities. The decision to implement Frege's plan in 1895 marks the final decline of the influence of the Berlin group on Althoff. In several conversations, Althoff openly called them materialistic, money-hungry riffraff, without any appreciation of ideal concepts.

When the Kaiser-Wilhelm-Institut was set up by Adolph von Harnack, clearly there was no need to consider mathematics. History has borne out Frege. Never before or since has German mathematics been as strong as between 1895 and 1933.

39

The War of the Frogs and the Mice, or the Crisis of the *Mathematische Annalen*

D. van Dalen

Will no one rid me of this turbulent priest?

Henry II

On 27 October 1928, a curious telegram was delivered to L. E. J. Brouwer, a telegram that was to plunge him into a conflict that for some months threatened to split the German mathematical community. This telegram set into motion a train of events that was to lead to the end of Brouwer's involvement in the affairs of German mathematicians and indirectly to the conclusion of the *Grundlagenstreit*.

The story of the ensuing conflict that upset the mathematical world is not a pleasant one; it tells of the foolishness of great men, of loyalty, and of tragedy. There must have been an enormous correspondence relating to the subject. Only a part of that was available to me, but I believe that enough of the significant material could be consulted so as to warrant a fairly accurate picture.

The telegram was dispatched in Berlin, and it read:[1]

Professor Brouwer, Laren N.H. Please do not undertake anything before you have talked to Carathéodory who must inform you of an unknown fact of the greatest consequence. The matter is totally different from what you might believe on the grounds of the letters received. Carathéodory is coming to Amsterdam on Monday.

Erhard Schmidt.

[1]All the correspondence in the *Annalen* affair was in German; in the translations I have exercised some freedom in those cases where a literal translation would have resulted in overly awkward English.

Volume 12, No. 4 (Fall 1990), 17–31

A message of this kind cold hardly be called reassuring. Brouwer duly collected two registered letters from Göttingen and waited for the arrival of Constantin Carathéodory. The letters were still unopened when Carathéodory arrived in Laren[2] on the thirtieth of October.

Bearer of Bad News

Carathéodory's visit figures prominently in the history that is to follow.

In order to appreciate the tragic quality of the following history, one must be aware that Brouwer was on friendly terms with all the actors in this small drama, with the exception of David Hilbert; some of them were even intimate friends, for example Carathéodory and Otto Blumenthal.

Carathéodory found himself in the embarrassing position of being the messenger of disturbing, even offensive, news, and at the same time disagreeing with its contents. It was regrettable, he said, that the two unopened letters had been written. The first letter contained a statement that should have carried more signatures, or at least Blumenthal's signature. Carathéodory's name was used in a manner not in accordance with the facts, although he would not disown the letter should Brouwer open it. Finally, the sender of the letter would probably seriously deplore his action within a couple of weeks. The second letter was written by Carathéodory himself, although Blumenthal's name was on the envelope. He, Carathéodory, regretted the contents of the letter.

Thereupon, Brouwer handed the second letter over to Carathéodory, who proceeded to relate the theme of the letters. The contents of the second can only be guessed, but the first letter can be quoted verbatim. It was written by Hilbert, and copies were sent to the other actors in the tragedy that was about to fill the stage for almost half a year.

Hilbert's letter was a short note:

Dear Colleague,

   Because it is not possible for me to cooperate with you, given the incompatibility of our views on fundamental matters, I have asked the members of the board of managing editors of the Mathematische Annalen *for the authorization, which was given to me by Blumenthal and Carathéodory, to inform you that henceforth we will forgo your cooperation in the editing of the* Annalen *and thus delete your name from the title page. And at the same time I thank you in the name of the editors of the* Annalen *for your past activities in the interest of our journal.*

Respectfully yours,
D. Hilbert

The meeting of the old friends was painful and stormy; it broke up in confusion. Carathéodory left in despondency and Brouwer was dealt one of the roughest blows of his career.

The *Annalen*

The *Mathematische Annalen* was the most prestigious mathematics journal at that time. It was founded in 1868 by A. Clebsch and C. Neumann. In 1920 it was taken over from the first publisher, Teubner, by Springer.

For a long period the name of Felix Klein and the *Mathematische Annalen* were inseparable. The authority of the journal was mostly, if not exclusively, based on his mathematical fame and management abilities. The success of Klein in building up the reputation of the *Annalen* was largely the result of his choice of editors. The journal was run, on Klein's instigation, on a rather unusual basis; the editors formed a small exclusive society with a remarkably democratic practice. The board of editors

[2]Brouwer lived in a small town, Laren, some distance from Amsterdam. He also owned a house in Blaricum.

met regularly to discuss the affairs of the journal and to talk mathematics. Klein did not use his immense status to give orders, but the editors implicitly recognized his authority.

Being an editor of the *Mathematische Annalen* was considered a token of recognition and an honor. Through the close connection of Klein—and after his resignation, of Hilbert—with the *Annalen*, the journal was considered, sometimes fondly, sometimes less than fondly, to be "owned" by the Göttingen mathematicians.

Brouwer's association with the *Annalen* went back to 1915 and before, and was based on his expertise in geometry and topology. In 1915 his name appeared under the heading "With the cooperation of" (*Unter Mitwirkung der Herren*). Brouwer was an active editor indeed; he spent a great deal of time refereeing papers in a most meticulous way.

FIGURE 39-1 *Constantin Carathéodory.*

The status of the editorial board, in the sense of bylaws, was vague. The front page of the *Annalen* listed two groups of editors, one under the head *Unter Mitwirkung von* (with cooperation of) and one under the head *Gegenwärtig herausgegeben von* (at present published by).

I will refer to the members of those groups as associate editors and chief editors. The contract that was concluded between the publisher, Springer, and the *Herausgeber* Felix Klein, David Hilbert, Albert Einstein, and Otto Blumenthal (25 February 1920) speaks of *Redakteure* but does not specify any details except that Blumenthal is designated as managing editor. The loose formulation of the contract would prove to be a stumbling block in settling the conflict that was triggered by Hilbert's letter.

At the time of Hilbert's letter the journal was published by David Hilbert, Albert Einstein, Otto Blumenthal, and Constantin Carathéodory, with the cooperation of (*unter Mitwirkung von*) L. Bieberbach, H. Bohr, L. E. J. Brouwer, R. Courant, W. v. Dyck, O. Hölder, T. von Kármán, and A. Sommerfeld. The daily affairs of the *Annalen* were managed by Blumenthal, but the chief authority undeniably was Hilbert.

Brouwer and Hilbert

Nowadays the names of Brouwer and Hilbert are automatically associated as the chief antagonists in the most prominent conflict in the mathematical world of this century, the notorious *Grundlagenstreit*. But things had not always been like that; some twenty years earlier Brouwer had met Hilbert, who was nineteen years his senior, in the fashionable seaside resort of Scheveningen and had instantly admired "the first mathematician of the world."[3] Hilbert obviously recognized the genius of the young man and on the whole accepted and respected him. Brouwer's letters to Hilbert for a prolonged period were written in a warm and friendly tone.

Already in his dissertation of 1907 Brouwer was markedly critical of Hilbert's formalism; this caused, however, no observable friction, probably because the dissertation was written in Dutch and thus escaped Hilbert's attention. The relationship remained friendly for a long time; Göttingen was Brouwer's second scientific home, and Hilbert wrote a warm letter of recommendation in 1912 when Brouwer was considered for a chair at the University of Amsterdam. In 1919 Hilbert went so far as to offer Brouwer a chair in Göttingen, an offer that Brouwer turned down.

The initially warm relationship between Hilbert and Brouwer began to cool in the twenties, when Brouwer started to campaign for his foundational views. Hilbert accepted the challenge—he took the

---

[3]From a letter of Brouwer to the Dutch poet C. S. Adama van Scheltema (9 November 1909): "This summer the first mathematician of the world was in Scheveningen, I was already acquainted with him through my work; now I have repeatedly walked with him, and talked to him as a young apostle to a prophet. He was 46, but young in heart and body, he swam vigorously and enjoyed climbing over walls and fences with barbed wire" [2, p. 100].

FIGURE 39-2 *L. E. J. Brouwer.*

threat of an intuitionistic revolution seriously. Brouwer lectured successfully at meetings of the German Mathematical Society. His series of Berlin lectures in 1927 caused a considerable stir; there was even some popular reference to a *Putsch* in mathematics. In March 1928 Brouwer gave talks of a mainly philosophical nature in Vienna (tradition has it that these talks were instrumental in Wittgenstein's return to philosophy). On the whole the future of intuitionism looked rosy.

Gradually the scientific differences between the two adversaries turned into a

FIGURE 39-3 *David Hilbert.*

personal animosity. The *Grundlagenstreit* is in part the collision of two strong characters, both convinced that they were under a personal obligation to save mathematics from destruction.

Brouwer's involvement in the national affairs of the German mathematicians also played a role. In so far as Brouwer had any political views, they could not be called sophisticated. From the end of the First World War, Brouwer had taken up the cause of the German mathematicians, subjected as they were to harsh measures and an international boycott.[4] For example, he forcefully opposed the participation of certain French mathematicians in the Riemann memorial volume of the *Mathematische Annalen,* much to the chagrin of Hilbert. His latest exploit in this area was his campaign against the participation of German mathematicians in the International Congress of Mathematicians at Bologna in August 1928. Hilbert put the full weight of his authority to bear on this matter, with the result that a sizable delegation followed Hilbert to Bologna [4, p. 188].[5]

Hilbert's Decision

The stage was set for the final act, and the letter of dismissal was the signal to raise the curtain. It is hard to imagine what Hilbert had expected; he could not have counted on a calm, resigned acquiescence from the highly strung emotional Brouwer. In Brouwer's eyes (and quite a few colleagues would have taken the same view) a dismissal from the *Annalen* board was a gross insult.

Carathéodory must have revealed some of the underlying motive to Brouwer, who wrote in his letter of 2 November to Blumenthal:

> *Furthermore Carathéodory informed me that the* Hauptredaktion *of the* Mathematische Annalen *intended (and felt legally competent) to remove me from the* Annalenredaktion. *And only for the reason that Hilbert wished to remove me, and that the state of his health required giving in to him. Carathéodory begged me, out of compassion for Hilbert, who was in such a state that one could not hold him responsible for his behavior, to accept this shocking injury in resignation and without resistance.*

Hilbert himself was explicit; in a letter of 15 October he asked Einstein for his permission (as a *Mitherausgeber*) to send a letter of dismissal (the draft to the chief editors did not contain any explanation) and added

[4]Brouwer's views and actions in this area can easily be (and have been) misrepresented; they deserve a more detailed treatment. The matter will be covered in a forthcoming biography.
[5]It was felt by a number of Germans, and by Brouwer, that the Germans were tolerated only as second-rate participants at the Bologna conference. Rather than suffer such an insult, they advocated a boycott of the conference. This topic has also received some degree of notoriety and is in need of a more balanced treatment. It will find a place in the forthcoming biography.

Just to forestall misunderstandings and further ado, which are totally superfluous under the present circumstances, I would like to point out that my decision—to belong under no circumstances to the same board of editors as Brouwer—is firm and unalterable. To explain my request I would like to put forward, briefly, the following:

1. *Brouwer had, in particular by means of his final circular letter to German mathematicians before Bologna, insulted me and, as I believe, the majority of German mathematicians.*
2. *In particular because of his strikingly hostile position vis-à-vis sympathetic foreign mathematicians, he is, in particular in the present time, unsuitable to participate in the editing of the* Mathematische Annalen.
3. *I would like to keep, in the spirit of the founders of the* Mathematische Annalen, *Göttingen as the chief base of the* Mathematische Annalen—*Klein, who earlier than any of us realized the overall detrimental activity of Brouwer, would also agree with me.*

In a postscript he added: "I myself have for three years been afflicted by a grave illness (pernicious anemia); even though the deadly sting of this disease has been taken by an American invention,[6] I have been suffering badly from its symptoms."

Clearly, Hilbert's position was that the *Herausgeber* (chief editors) could appoint or dismiss the *Mitarbeiter* (associate editors). As such he needed the approval of Blumenthal, Carathéodory, and Einstein. Blumenthal had complied with Hilbert's wishes, but for Carathéodory, consent was problematic; apparently he did not wish to upset Hilbert by contradicting him, but neither did he want to authorize him to dismiss Brouwer. Hilbert may easily have mistaken Carathéodory's evasive attitude for an implicit approval. Carathéodory had landed in an awkward conflict between loyalty and fairness. He obviously tried hard to reach a compromise. In view of Hilbert's firmly fixed conviction, he accepted the unavoidable conclusion that Brouwer had to go; but at least Brouwer should go with honor.

Einstein's Neutrality

Being caught in the middle, Carathéodory sought Einstein's advice. In a letter of 16 October he wrote "It is my opinion that a letter, as conceived by Hilbert, cannot possibly be sent off." He proposed instead to send a letter to Brouwer explaining the situation and suggesting that Brouwer should voluntarily hand in his resignation. Thus a conflict would be avoided and one could do Brouwer's work justice: "Brouwer is one of the foremost mathematicians of our time and of all the editors he has done most for the M.A."

The second letter we mentioned above must have been the concrete result of Carathéodory's plan. Einstein answered: "It would be best to ignore this Brouwer-affair. I would not have thought that Hilbert was prone to such emotional outbursts" (19 October 1928).

The managing editor, Blumenthal, must have been in an even greater conflict of loyalties, being a close, personal friend of Brouwer and the first Ph.D. student (1898) of Hilbert, whom he revered.

Einstein did not give in to Hilbert's request. In his answer to Hilbert (19 October 1928) he wrote:

I consider him [Brouwer], with all due respect for his mind, a psychopath and it is my opinion that it is neither objectively justified nor appropriate to undertake anything against him. I would say: "Sire, give him the liberty of a jester (Narrenfreiheit)!" If you cannot bring yourself to this, because his behavior gets too much on your nerves, for God's sake do what you have to do. I, myself, for the above reasons cannot sign such a letter.

---

[6]The work of G. H. Whipple. F. S. Robscheit-Robbins and of G. R. Minot, cf. [4, p. 179].

Carathéodory, however, was seriously troubled and could not let the matter rest. He again turned to Einstein (20 October 1928):

FIGURE 39-4 *Albert Einstein.*

> *... Your opinion would be the most sensible, if the situation would not be so hopelessly muddled. The fight over Bologna ... seems to me a pretext for Hilbert's action. The true grounds are deeper—in part they go back for almost ten years. Hilbert is of the opinion that after his death Brouwer will constitute a danger for the continued existence of the M.A. The worst thing is that while Hilbert imagines that he does not have much longer to live ... he concentrates all his energy on this one matter. ... This stubbornness, which is connected with his illness, is confronted by Brouwer's unpredictability. ... If Hilbert were in good health, one could find ways and means, but what should one do if one knows that every excitement is harmful and dangerous? Until now I got along very well with Brouwer; the picture you sketch of him seems to me a bit distorted, but it would lead too far to discuss this here.*

This letter made Einstein, who in all public matters practised a high standard of moral behavior, realize that these were deep waters indeed (23 October 1928):

> *I thought it was a matter of mutual quirk, not a planned action. Now I fear to become an accomplice to a proceeding that I cannot approve of, nor justify, because my name—by the way, totally unjustifiedly—has found its way to the title page of the* Annalen. *... My opinion, that Brouwer has a weakness, which is wholly reminiscent of the* Prozessbauern,[7] *is based on many isolated incidents. For the rest I not only respect him as an extra-ordinarily clear visioned mind, but also as an honest man, and a man of character.*

From these letters, even before the real fight had started, it clearly appears that Einstein was firmly resolved to reserve his neutrality. Einstein called Brouwer "an involuntary proponent of Lombroso's theory of the close relation between genius and insanity," but Einstein was well aware of Brouwer's greatness and did not wish him to be victimized. It is not clear whether Einstein's opinion was based on personal observation or on hearsay; there are no reports of personal contacts between Brouwer and Einstein, but one may conjecture that they had met at one of the many meetings of the *Naturforscherverein* or in Holland during one of Einstein's visits to Lorentz.

Unsound Mind

It did not take Brouwer long to react. Brouwer was a man of great sensitivity, and when emotionally excited, he was frequently subject to nervous fits. According to one report (a letter from Dr. Irmgard Gawehn to von Mises), Brouwer was ill and feverish for some days following Carathéodory's visit.

---

[7]This probably refers to the troubles in Schleswig-Holstein during roughly the same period, when farmers resisted the tax policies of the government. Hans Fallada has sketched the episode in his *Bauern, Bonzen und Bomben.*

On 2 November Brouwer sent letters to Blumenthal and Carathéodory, from which only the copy of the first one is in the Brouwer archive—it contained a report of Carathéodory's visit. The letter stated that "in calm deliberation a decision on Carathéodory's request was reached."

The answer to Carathéodory, as reproduced in the letter to Blumenthal, was short:

Dear Colleague,

After close consideration and extensive consultation I have to take the position that the request from you to me, to behave with respect to Hilbert as to one of unsound mind, qualifies for compliance only if it should reach me in writing from Mrs. Hilbert and Hilbert's physician.

Yours
L. E. J. Brouwer

This solution, although perhaps a clever move in a political game of chess, was of course totally unacceptable—even worse, it was a misjudgment of the matter. In a more or less formal indictment, Blumenthal declares concerning "this frightful and repulsive letter" that apparently Brouwer had picked from Carathéodory's statements and entreatments the ugliest interpretation. "I must confess, and Cara has written me likewise, that I have been thoroughly deceived in Brouwer's character and that Hilbert has known and judged him better than we did."

So Brouwer's first action only served to rob him of his potential support.

The conflict had presented itself so suddenly and so totally unexpectedly to Brouwer that he failed to realize to what extent Hilbert saw him as a deadly danger for mathematics and as the bane of the *Mathematische Annalen*. His belief that the announced dismissal was the whim of a sick and temporarily deranged man emerges from a letter he dispatched to Mrs. Hilbert three days later:

I beg you, use your influence on your husband, so that he does not pursue what he has undertaken against me. Not because it is going to hurt him and me, but in the first place because it is wrong, and because in his heart he is too good for this. For the time being I have, of course, to defend myself, but I hope that it will be restricted to an incident within the board of editors of the Annalen, and that the outer world will not notice anything.

A copy of this letter went to Courant with a friendly note, asking him (among other things) to keep an eye on the matter: "As a matter of course, I count especially on you to bring Hilbert to reason, and to make sure that a scandal will be avoided" (6 November 1928).

Courant, after visiting Mrs. Hilbert, replied to Brouwer (10 November 1928) that Hilbert was in this matter under nobody's influence and that it was impossible to exert any influence on him.

Apart from Einstein, who kept a strict neutrality, all the editors (mostly reluctantly) did take sides—the majority with Hilbert, but Hilbert himself no longer took part in the conflict. His position was fixed once and for all, and in view of his illness the developments were as far as possible kept from him (e.g., Blumenthal to Courant on 4 November 1928: "Hilbert must not find out about Cara's trip to Brouwer").

One might wonder whether Brouwer, as a relative outsider (one of the three non-Germans among the editors), stood a chance from the beginning; his letter of 2 November to Carathéodory, however, definitely lost him a good deal of sympathy and proved a weapon to his opponents.

The Ripples Spread

In a circular letter of 5 November 1928, Brouwer appealed directly to the publishers and editors, thus widening the circle of persons involved:

To the publisher and the editors of the Mathematische Annalen.

From information communicated to me by one of the chief editors of the Mathematische Annalen *at the occasion of a visit on 30-10-1928 I gather the following:*

1. *That during the last years, as a consequence of differences between my opinion and that of Hilbert, which had nothing to do with the edition of the* Mathematische Annalen *(my turning down the offer of a chair in Göttingen, conflict between formalism and intuitionism, difference in opinion concerning the moral position of the Bologna congress), Hilbert had developed a continuously increasing anger against me.*

2. *That lately Hilbert had repeatedly announced his intention to remove me from the board of editors of the* Mathematische Annalen, *and this with the argument that he could not longer "cooperate" (zusammenarbeiten) with me.*

3. *That this argument was only a pretext, because in the editorial board of the* Mathematische Annalen *there has never been a cooperation between Hilbert and me (just as there has been no cooperation between me and various other editors). That I have not even exchanged any letters with Hilbert for many years and that I have only superficially talked to him (the last time in July 1926).*

4. *That the real grounds lie in the wish, dictated by Hilbert's anger, to harm and damage me in some way.*

5. *That the equal rights among the editors (repeatedly stressed by the editorial board within and outside the board[8]) allow a fulfillment of Hilbert's will only in so far that from the total board a majority should vote for my expulsion. That such a majority is scarcely to be thought of, since I belong among the most active members of the editorial board of the* Mathematische Annalen, *since no editor ever had the slightest objection against the manner in which I fulfill my editorial activities, and since my departure from the board, both for the future contents and for the future status of the* Annalen, *would mean a definite loss.*

6. *That, however, the often proclaimed equal rights, from the point of view of the chief editors, was only a mask, now to be thrown down. That as a matter of fact the chief editors wanted (and considered themselves legally competent) to take it upon themselves to remove me from the editorial board.*

7. *That Carathéodory and Blumenthal explain their cooperation in this undertaking by the fact that they estimate the advantages of it for Hilbert's state of health higher than my rights and honor and freedom of practice (Wirkungsmöglichkeiten) and than the moral prestige and scientific contents of the* Mathematische Annalen *that are to be sacrificed.*

I now appeal to your sense of chivalry and most of all to your respect for Felix Klein's memory and I beg you to act in such a way, that either the chief editors abandon this undertaking, or that the remaining editors separate themselves [from the chief editors, v.D.] and carry on the tradition of Klein in the managing of the journal by themselves.

Laren, 5. November 1928
L. E. J. Brouwer

---

[8]From the editorial obituary of Felix Klein, written by Carathéodory: "He (Klein) has taken care that the various schools of mathematics were represented in the editorial board and that the editors operated with equal rights alongside of himself—He has . . . never heeded his own person, always had kept in view the goal to be achieved."

From a letter from Blumenthal to me, 13-9-1927: "I believe that you overestimate the meaning of the distinction between editors in large and small print. It seems to me that we all have equal rights. In particular we can speak of the *Annalenredaktion* if and only if we have made sure of the approval of the editors interested in the matter under consideration.—Although I too take the distinction between the two kinds of editors more to be typographical than factual (I make an exception for myself as managing editor), I very well understand your wish for a better typographical make-up. You know that I personally warmly support it. However, we can for the time being, as long as the state of Hilbert's health is as shaky as it is now, change nothing in the editorial board. I thus cordially beg you to put aside your wish. In good time I will gladly bring it out."

The above circular letter was dispatched on the same day as Brouwer's plea to Mrs. Hilbert; the two letters are in striking contrast. One letter is written on a conciliatory note; the other is a determined defence and closes with an unmistakable incitement to mutiny.

Blumenthal immediately took the matter in hand; he wrote to the publisher and the editors (16 November) to ignore the letter until he had prepared a rejoinder. The draft of the rejoinder was sent off to Courant on 12 November, with instructions to wait for Carathéodory's approval and to send copies to Bieberbach, Hölder, von Dyck, Einstein, and Springer. It appears from the accompanying letter that Carathéodory had already handed in his resignation, although he had given Blumenthal permission to postpone its announcement, so that it would not give food to the rumor that Carathéodory had turned against Hilbert.

In the meantime Brouwer had travelled to Berlin to talk the matter over with Erhard Schmidt and to explain his position to the publisher, Ferdinand Springer. Brouwer, accompanied by Bieberbach, called at the Berlin office of Springer, who reported the discussion in an *Aktennotiz* "Unannounced and surprising visit of Professor Bieberbach and Professor Brouwer" (13 November 1928). As Springer wrote, his first idea was to refuse to receive the gentlemen, but he then realized that a refusal would provide propaganda material for the opposition.

Springer opened the discussion with the remark that he was firmly resolved not to mix in the skirmishes and that he did not consider the *Annalen* the sole property of the Company (like other journals) but that the proper *Herausgeber*, Klein and Hilbert, had been in a sense in charge. Moreover, he would choose Hilbert's side out of friendship and admiration, if he would be forced to choose sides.

The unwelcome visitors then proceeded to inquire into the legal position of Hilbert, a topic that Springer was not eager to discuss without the advice of his friends and which he could not enter into without consulting the contract. Thence the two gentlemen proceeded to "threaten to damage the *Annalen* and my business interests. Attacks on the publishing house, which could get the reputation of lack of national feeling among German mathematicians, could be expected."

This threat was definitely in bad taste, not in the last place because the Springer family had Jewish ancestry. Bieberbach's later political views have gained a good measure of notoriety (cf. [3]); it certainly is true that already before the arrival of the Third Reich he held extreme nationalistic views. Brouwer's position in this matter of *Nationalgefühl* was rather complex; it was not based on a political ideology but rather on his moral indignation at the boycott of German science.

Be that as it may, this particular approach was not likely to mollify Springer, who calmly answered that he would deplore damage resulting from this quarrel but that he would bear it without complaints under the present circumstances.

Thus rejected, Bieberbach and Brouwer asked if Springer could suggest a mediator, upon which Springer answered that he was not sufficiently familiar with the personal features involved, but that two *deutschfreundliche* foreigners like Harald Bohr and G. H. Hardy might do.[9]

FIGURE 39-5 *Ferdinand Springer. This photograph was taken on Hilbert's sixtieth birthday, 23 January 1922.*

[9]This suggestion of the publisher encouraged the impression that the conflict had a political origin. Blumenthal complained to Courant (letter of 18 November 1928) ". . . the bad thing is, that Brouwer managed to move everything on to the political plane, just what Carathéodory thought he had prevented." The idea of mediation was not pursued.

Before leaving, Brouwer threatened to found a new journal with De Gruyter, and Bieberbach declared that he would resign from the board of editors if it definitively came to the exclusion of Brouwer.

In a letter to Courant (13 November 1928) Springer dryly commented "On the whole the founding of a new journal, wholly under Brouwer's supervision, would be the best solution out of all difficulties.[10] He also conveyed his impression of the visit: "I would like to add that Brouwer, as a matter of fact, does make a scarcely pleasant (*unerfreulich*) impression. It seems, moreover, that he will carry the fight to the bitter end (*der Kampf bis aufs Messer führen wird*).

The Case for the Prosecution

In Aachen Blumenthal was preparing his defense of the intended dismissal of Brouwer and, following an old strategic tradition, he took to the attack. After consulting Courant, Carathéodory, and Bohr he drew up a kind of indictment. I have not seen the draft of 12 November, but from a letter from Bohr and Courant to Blumenthal (14 November 1928), one may infer that it was harder in tone and more comprehensive than the final version. There is mention of a detailed criticism of Brouwer's editorial activities and of matters of formulation ("... leave out *Schrullenhaftigkeit* [capriciousness] ...."). Moreover Bohr and Courant warned Blumenthal:

> To what extent Brouwer exploits without consideration every tactical advantage that is offered to him, and how dangerous his personal influence is (Bieberbach), can be seen from the enclosed notice which Springer has just sent us [the above-mentioned Aktennotiz].

The correspondence of Blumenthal, Bohr, and Courant shows an unlimited loyalty to Hilbert, which it would be unjust to ascribe to Hilbert's state of health alone. There is no doubt that Hilbert as a man and a scientist inspired a great deal of loyalty in others, let alone in his students. Sentences like "We don't particularly have to stress that we are, like you, wholly on Hilbert's side, and also, when necessary, prepared for action" (same letter), illustrate the feeling among Hilbert's students.

A revised version of Blumenthal's letter is dated 16 November, and it is this version that was in Brouwer's possession. It incorporated remarks of Bohr and Courant but not yet those (at least not all of them) of Carathéodory. It contained a concise *résumé* of the affair so far and proceeds to answer Brouwer's points (from the letter of 5 November 1928).

As Blumenthal put it, he partly based his handling of the matter on letters from Hilbert, Carathéodory, and Brouwer, partly on an extensive conversation with Hilbert in Bologna. The contents of the latter conversation remain a matter of conjecture, but it may be guessed that in August at the conference Hilbert had made clear his objections to Brouwer—in particular after Brouwer's opposition to the German participation in the conference.

From Blumenthal's circular letter, the editors—and also Brouwer—learned the contents of Hilbert's letter of 25 October. In answering Brouwer's points Blumenthal quite correctly stressed that Brouwer interpreted "cooperation" in a too narrow fashion. Hilbert, he said, found it impossible to justify his sharing responsibilities in an editorial board together with Brouwer. As to point 4 of Brouwer's letter, "the motivation is ugly and thus needs no answer." The scientific opposition in foundational matters had not played a role, according to Blumenthal. Even Brouwer's circular letter concerning the Bologna Congress "by which statements Hilbert felt insulted" had, according to Blumenthal, only cited in a catalytic way on his decision: "The motives lie much deeper."

Concerning Klein's position, Blumenthal remarked that Klein always acted as a kind of higher authority, to which one could appeal. After Klein's death Hilbert felt obliged to take on Klein's role.

[10]Brouwer indeed founded a new journal, the *Compositio Mathematica*, with the Dutch publisher Noordhoff.

"Hilbert has recognized in Brouwer a stubborn, unpredictable, and ambitious (*herrschsüchtig*) character. He has feared that when he should eventually resign from the editorial board, Brouwer would bend the editorial board to his will and he had considered this such a danger for the *Annalen* that he wanted to oppose him as long as he still could do so."

How strongly Blumenthal and Carathéodory wished to spare Brouwer, while complying with Hilbert's wishes, can be seen from the following paragraph:

> *Cara and I, who were associated with Brouwer in a long-standing friendship, had objectively to recognize Hilbert's objections to Brouwer's editorial activity.*
>
> *True, Brouwer was a very conscientious and active editor, but he was quite difficult in his dealings with the managing editor and he subjected the authors to hardships that were hard to bear.*
>
> *E.g., manuscripts that were submitted for refereeing to him lay around for months, while in principle he had prepared a copy of each submitted paper (I recently had an example of this practice). Above all there is no doubt that Klein's premature resignation from the editorial board is to be traced back to Brouwer's rude behaviour (in a matter in which Brouwer was formally right). The further course of events has shown that Hilbert was even far more right than we thought at the time.*
>
> *Since we could not reject the objective justification of Hilbert's point of view, and were confronted by his immutable will, we have given our permission for the removal of Brouwer from the editorial board [at this point it should be made clear that Carathéodory had not given his permission, as he wrote to Courant (14 November) in his letter with corrections to the draft]. We only wished—unjustified, as I now realize—a milder form, in the sense that Brouwer should be prevailed upon to resign. Hilbert could not be induced to this procedure, so we finally, though reluctantly, have decided to give in to him* (den Weg freigeben). *Mr. Einstein did not comply, with the argument that one should not take Brouwer's peculiarities seriously.*

To what the reader already knows about Carathéodory's trip to Brouwer, Blumenthal's letter adds the following: In Göttingen on 26 and 27 October, Blumenthal and Carathodory discussed the situation. In a last attempt to bring the matter to a good end through a mitigation of the categorical form of the statement of notice, Carathéodory travelled to Berlin and discussed the matter, as it appears, with Erhard Schmidt. The result was, as we know, the request to Brouwer not to take action before Carathéodory's arrival.

Finally, Blumenthal proceeded to reproduce the text of Brouwer's letter to Carathéodory of 2 November, with the "unsound mind" phrase, concluding that "I have thoroughly misjudged Brouwer's character and that Hilbert has known and judged him better than we have." The letter ended with the request to the editors for permission to delete Brouwer's name from the title pages of the *Annalen*.

Defence of the Underdog

A few editors responded to Blumenthal's letter in writing, but the majority remained silent. Only von Dyck, Hölder, and Bieberbach sent their comments. Von Dyck could "neither justify Brouwer's views nor Hilbert's action" and he hoped that a peaceful solution could be found. Hölder was of the opinion that he could not approve of a removal of Brouwer by force (27 November).

Bieberbach's letter showed a thorough appreciation of the situation. And he at least was willing to take up the case of the underdog. In view of his later political extremism one might be inclined to question the purity of his motives; however, in the present letter there is no reason not to take his arguments at their face value. Like Brouwer, and probably the majority if not the whole of the editorial board, Bieberbach contested the right of the *Herausgeber* to decide matters without the support of the majority of all the editors, let alone without consultation. Indeed, this seems to be a shaky point in the whole procedure.

FIGURE 39-6 *Otto Blumenthal (left) and L. E. J. Brouwer (right) in happier days (1920?).*

As a matter of fact, the contract between Springer and the *Herausgeber* (25 February 1920) is not very concrete in this particular point. It states: "Changes in the membership of the editorial board require the approval of the publisher." The correspondence does not lead me to believe that Hilbert observed this rule.

Bieberbach observed that a delay in handling papers cannot be taken seriously as grounds for dismissal; such things ought to be discussed in the annual meetings of the board. Bieberbach's comments on Hilbert's annoyance (to say the least) with Brouwer's actions in the matter of the Bologna conference are of a rather scholastic nature and border on nit-picking: the objectionable statements of Brouwer concerned Germans who were to attend the *Unionskongress* in Bologna, and since Hilbert denied that the meeting in Bologna was a *Unionskongress,* the statements did not apply to him.

Bieberbach correctly spotted a serious flaw in Blumenthal's charge involving Brouwer's "terrifying and repulsive" letter:

> *Finally, I hold it totally unjustified to forge material against Brouwer from letters that he wrote after learning about the action that was mounted against himself. For it is morally impossible to use actions, to which a person is driven in a fully understandable emotion over an injustice that is inflicted on him, afterwards as a justification of this injustice itself.*

The point is well taken. It does not exonerate Brouwer from hitting below the belt, but it at least makes clear that to use it against Brouwer is distasteful.

Bieberbach explicitly stated that he would not support Brouwer's dismissal; on the contrary, he strongly sided with Brouwer, without, however, attacking Hilbert.

The publisher reacted in a cautious way. Springer thought that Brouwer was "an embittered and malicious adversary" and that he should not receive a copy of the circular letter without the permission of the lawyer of the firm. Springer also concluded that the publisher should not state in writing that he officially agreed to Brouwer's dismissal, because it would imply a recognition of Brouwer's membership on the board of editors in the sense of the contract. In short, Springer abstained from voting on Blumenthal's proposal.

Froschen–Mäusekrieg

At this point the whole action against Brouwer seemed to reach a climax. One may surmise a good deal of activity in the camp of the *Göttinger*, as the sympathizers of Hilbert were called.

A certain amount of animosity between Göttingen and Berlin mathematicians was a generally acknowledged fact. The Berlin faction had suffered a setback in the matter of the Bologna boycott, where Hilbert had undeniably carried the day. Born, in his letter to Einstein (20 November) quotes von Mises (a *Berliner*), "the *Göttinger* simply run after Hilbert, who is not completely responsible for his actions (*sei wohl nicht mehr ganz zurechnungsfähig*)." The friction between Berlin and Göttingen was a weighty reason to settle the *Annalen* conflict as speedily and quietly as possible. If there was any risk of a rift in the German mathematical community, it was here.

A key figure, in view of his immense scientific and moral prestige, was Albert Einstein. If he could be persuaded to side with Hilbert the battle would be half won. In spite of personal pressure from Born (20 November 1928) on behalf of Hilbert, Einstein remained stubbornly neutral. In his letters to Born and to Brouwer and Blumenthal one may recognize a measure of disgust behind a facade of raillery.

In the letter to Born (27 November) the apt characterization of "*Frosch-Mäusekrieg*" (war of the frogs and the mice)[11] was introduced. After declaring his strict neutrality Einstein continued:

> *If Hilbert's illness did not lend a tragic feature, this ink war would for me be one of the most funny and successful farces performed by that people who take themselves deadly seriously.*
>
> *Objectively I might briefly point out that in my opinion there would have been more painless remedies against an overly large influence on the managing of the* Annalen *by the somewhat mad (verrückt) Brouwer, than eviction from the editorial board.*
>
> *This, however, I only say to you in private and I do not intend to plunge as a champion into this frog-mice battle with another paper lance.*

Einstein's letter to Brouwer and Blumenthal (25 November) is even more cutting and reproving.

> *I am sorry that I got into this mathematical wolf-pack (Wolfsherde) like an innocent lamb. The sight of the scientific deeds of the men under consideration here impresses me with such cunning of the mind, and I cannot hope, in this extra-scientific matter, to reach a somewhat correct judgment of them. Please, allow me therefore, to persist in my "booh-nor-bah" (Muh-noch-Mäh) position and allow me to stick to my role of astounded contemporary.*
>
> *With best wishes for an ample continuation of this equally noble and important battle, I remain*
>
> *Yours truly,*
> *A. Einstein*

Deadlock

The whole affair now rapidly reached a deadlock. A week before, Springer, who had at Blumenthal's urging sought legal advice, had optimistically written to Courant (17 November 1928) that the legal adviser of the firm, E. Kalisher, was of the opinion that it would suffice that those of the four chief editors who did not want to advocate Brouwer's dismissal actively would abstain from voting, thus giving the remaining chief editors a free hand. Apparently Springer did not realize that since two edi-

---

[11]War of the frogs and the mice—a Greek play of unknown authorship; a late medieval German version, *Froschmeuseler*, is from the hand of Rollenhagen.

tors with a high reputation had already decided not to support Hilbert, the solution, even if it was legally valid, would lack moral credibility.

If this solution should turn out to raise difficulties within the editorial board, the publisher could still fire the whole editorial board and reappoint Hilbert and his supporters, so the advice ran. In the opinion of the legal adviser the publishing house was contractually bound to the chief editors (*Herausgeber*) only; there was no contract with the remaining editors.

Bieberbach's letter, mentioned above, apparently worried Carathéodory to the extent that he decided to ask a colleague from the law faculty for advice. This advice from Müller-Erzbach (Munich) plainly contradicted the advice from the Springer lawyer. It made clear that

FIGURE 39-7 *Richard Courant.*

1. Brouwer and Springer-Verlag were contractually bound since Brouwer had obtained a fee.
2. Hilbert's letter was not legally binding.

Müller-Erzbach sketched three solutions to the problem:

1. Springer dismisses Brouwer. A letter of dismissal should, however, contain appropriate grounds.
2. The four chief editors and the publisher form a company (*Gesellschaft*) and dismiss Brouwer.
3. A court of law could count the "*Mitarbeiter*" as editors. In that case the only way out would be to dissolve the total editorial board and to form a new one.

Carathéodory considered the first two suggestions inappropriate because it would not be fair to saddle Springer with the internal problems of the editors. Hence he recommended the third solution (letter to Blumenthal, 27 November) Here, for the first time, appeared the suggestion that was to be the basis of the eventual outcome of the dispute.

Hilbert, the main contestant in the *Annalen* affair, had quite sensibly withdrawn from the stage. The developments, had he known them, would certainly have harmed his still precarious health. In a short notice he had empowered Harald Bohr and Richard Courant to represent him legally in matters concerning the *Mathematische Annalen*. Thus the whole matter became more and more a shadow fight between Brouwer and an absentee.

At this point the dispute had reached an impasse. Although Springer upheld in a letter to Bieberbach the principle that the chief editors could dismiss any of the other editors, the impetus of the attack on Brouwer seemed to ebb. A meeting between Carathéodory, Courant, Blumenthal, and Springer had repeatedly to be postponed and finally had been cancelled.

Courant agreed with Carathéodory that the dissolution of the complete board would be a good solution (30 November 1928); however, it would require a voluntary action from the editors and the ultimate organization of the editorial board should not have the character of a legal trick with the sole purpose of rendering Brouwer's opposition illusory.

Carathéodory, who, on the basis of Müller-Erzbach's information, had come to the conclusion that the original plan of Hilbert, even in a modified form, would not stand up in a court of law, expressed his willingness to assist "out of devotion to Hilbert" in the liquidation of the affair but quite firmly refused to be involved in the future organization of the *Annalen*.

Dissolution

The reluctance of Carathéodory to be involved in the matter beyond the bare minimal efforts to satisfy Hilbert and spare Brouwer (his friend) is throughout understandable. As far as we can judge from the correspondence, only Blumenthal exhibited an unbroken fighting spirit. He realized, however, that his circular had not furthered an acceptable solution (letter to Courant and Bohr, 4 December), and he leaned towards alternative solutions. In particular, Blumenthal wrote, the time was favorable to Carathéodory's plan. The *Annalen* were completing their hundredth volume, and it would present a nice occasion to open with volume 101 a "new series" or "second series" with a different organization of the editorial board. But at the present time he was facing a dilemma. Because Hilbert's letter clearly had no legal status, Brouwer was still a *Mitarbeiter* and his name should appear on the cover of the issue that was to appear—this, however, conflicted with Hilbert's wishes. Could Bohr and Courant, as proxies of Hilbert, authorize him to print Brouwer's name on the cover? Otherwise the publication would have to be postponed. The authorization probably was given.

It seems that Bohr had also put forward a solution to the affair. From the correspondence of Carathéodory and Bohr with Blumenthal, one gets the impression that Bohr's proposal was a slight variant of Carathéodory's suggestion. The main difference was that Bohr advocated a total reorganization of the editorial board. In his proposal there would only remain *Herausgeber* and no *Mitarbeiter*. So the solution would look like a fundamental change of policy, and hence it would no longer be recognizable as an act levelled against Brouwer.

Apparently Bohr envisaged Hilbert, Blumenthal, Hecke, and Weyl as the members of a new board. And should Weyl decline, one might invite Toeplitz. Blumenthal questioned the wisdom of reinstating himself as an editor; it could easily be viewed as the old board of *Herausgeber* in disguise (letter to Bohr, 5 December). In his letter to Courant, the next day, he considered the dissolution of the editorial board at large as necessary, and he fully agreed that Hilbert should choose the new editors.

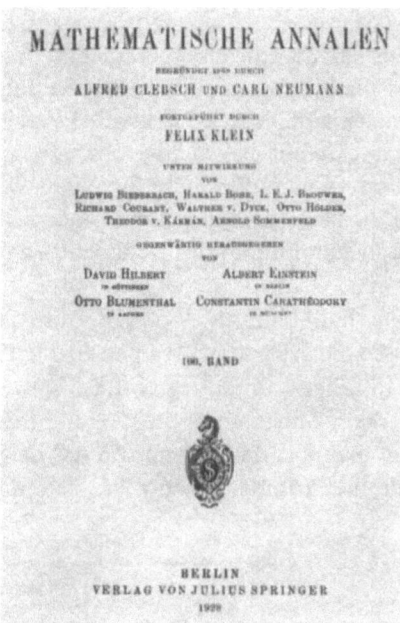 

FIGURE 39-8 *Mathematische Annalen, volume 100 (1928) and volume 101 (1929). Notice the change in editors from volume 100 to volume 101.*

From then on things moved smoothly; Springer accepted the dissolution of the editorial board and agreed to enter into a contract with Hilbert on the subject of the reorganized *Annalen*. By and large only matters of formulation and legal points remained to be solved.

One might wonder where Brouwer was in all this—he was completely ignored. In a letter of 30 November to the editors and the publisher he confirms the receipt of Blumenthal's indictment, which had only just reached him. In a surprisingly mild reaction he merely asked the editors to reserve their judgment—blissfully unaware that nobody was going to take a vote—for the composition of a defence would take some days.

Because the dissolution of the editorial board had to be a voluntary act, it was a matter of importance to get Einstein's concurrence. The contract of 1920 presented an elegant loophole that would allow both parties to settle the matter without breaking the rules. In Section 5 the clauses for termination of the contract were listed, and one of them stipulated that if the editors (*Redaktion*) renounced the contract, without a violation from the side of the publisher, the latter could continue the *Mathematische Annalen* at will.

Possibly Einstein's agreement could be dispensed with, but it is likely that a decision to ignore Einstein's vote would influence general opinion adversely; moreover, it would be wise to opt for a watertight procedure, as Brouwer would not hesitate to test the outcome in court.

So pressure was brought to bear on Einstein. James Franck, a physicist and a friend of Born, begged him to listen to the new plan. He stressed the political side of the issue, "At this time, . . ., whether the mathematicians split into factions, or whether the affair is arranged smoothly, depends on your decision. It would almost be an ill-chosen joke (*ein nicht all zu guter Witz*) if in this case you would be claimed for the nationalistic side" (undated).

Franck was not the only person to discover a (real or imaginary) political aspect in the controversy at hand. Blumenthal had already complained to Courant (18 November) that Brouwer had managed to introduce the political element into the matter. Born also, in his letter to Einstein of 11 November, tied the conflict to the political issue of the German nationalists and the animosity of Berlin vs. Göttingen.

The successful conclusion of the undertaking was conveyed to Springer by Courant. In his letter of 15 December he announced the cooperation of Einstein, Carathéodory, Blumenthal, and Hilbert. At the same time he proposed that a new contract be made between Hilbert and the publisher and that Hilbert get *carte blanche* for organizing the editorial board. Blumenthal should be invited to continue his activity as managing editor and, according to Courant, he would probably accept. Also— and this is a surprising misjudgment of Einstein's mood—Courant thought that there was a 50 percent chance that Einstein would join the new board. As far as he himself was concerned, Courant thought it wiser to postpone his own introduction as an editor until the dust had settled (the matter apparently had been discussed earlier).

Finally Courant suggested that the publisher alone should inform all present editors of the collective resignation. With respect to Brouwer, he advised Springer to write a personal letter explaining the solution to the conflict and to stress that he [Springer] would regret it if Brouwer were left with the impression that the whole affair would restrict his freedom of practice and that the publishing house would be at his disposal should he wish to report on his foundational views. It is not known whether this letter was ever written, but Courant's attitude certainly was statesmanlike and conciliatory.

"No Personal Motives"

Once the decision was taken, no time was wasted; after the routine legal consultations the publisher carried out the reorganization and the editors were informed of the outcome (27 December). In spite of Courant's considerations mentioned above, the letter was signed by Hilbert and Springer.

Brouwer, like everybody else, was thanked for his work and was given the right to a free copy of the future *Annalen* issues. The matter would have been over, were it not for some rumblings among the former editors and for a desperate but hopeless rearguard action by Brouwer.

Carathéodory had been considerably distressed during the whole affair; from the beginning he had been torn between his loyalty to Hilbert and his abhorrence of the injustice of Brouwer's dismissal. His efforts to mediate had only worsened the matter and the final solution was an immense relief to him. In a fit of despondency he wrote to Courant (12 December) "You cannot imagine how deeply worried I was during the last weeks. I envisioned the possibility that, after I had parted with Brouwer, the same thing would happen with all my other friends." He had even considered accepting a chair at Stanford that was offered to him.

In his answer (15 December) Courant tried to set Carathéodory's mind at ease: he believed that he had succeeded in convincing Hilbert that Carathéodory, in his position, could not have acted differently; the matter was settled now "without fears of a residue of resentment on Hilbert's part."

Two days later Courant wrote that the night before he had discussed the whole matter with Hilbert, who asked Courant to tell Carathéodory that "he thinks that you would have done everything for him, as far as possible." Hilbert was completely satisfied with the result of the undertaking, and in his opinion the *Annalen* were even better protected now than through his original dismissal of Brouwer ". . . and by and by it has become completely clear to me that in fact no personal motives have inspired Hilbert's first step. . . ."

Carathéodory expressed his pleasure with Hilbert's attitude (9 December) but he was not wholly satisfied with Courant's evaluation of the motives behind Hilbert's move. "Now, he himself has given as the exclusive motive for his decision that he felt insulted by Brouwer; I would find it unworthy of him, to construe after the fact, that only impersonal motives had guided him."

The last remark could hardly be left unanswered by Courant. He had worked hard to pacify the participants in the affair, and here one of the former *Herausgeber* was lending support to the rumor that Hilbert was not completely devoid of some personal feelings of revenge. In an attempt to quench this source of dissent, he and Bohr admonished Carathéodory. Courant calmly repeated his view (23 December) and referred to Hilbert's personal statements that he "fostered no personal feelings of hate, anger or insult against Brouwer." Even a bit of subtle pressure was brought to bear on Carathéodory: "Our responsibility to Hilbert at this point is even greater, as he is not yet filled in on the development of the conflict; in particular he does not surmise your visit to Laren and the disconcerting report of it by Brouwer."

Bohr was less subtle in his approach (same letter); if Carathéodory were not convinced of Hilbert's impersonal motives, he should ask Hilbert himself. "For, that Hilbert—without being aware of it and without being able to defend himself—should first be considered 'of unsound mind,' and then 'not to the point' (*unzurechnungsfähig . . . unsachlich*), that is a situation that I as a representative of Hilbert cannot in the long run witness without action."

In spite of Bohr's saber rattling, Carathéodory stuck to his guns: "To judge Hilbert's motives is a very complicated matter; I believe that I see through his motives because I have known his way of thinking for more than 25 years. It is true that the motivations that you indicate, and which H. also expounded in Bologna in discussion with Blumenthal, were there. The total complex of thoughts that caused the explosion of feeling of 15 October [cf. letter to Einstein, 15 October] was much more complicated."

Who was right, Courant and Bohr or Carathéodory? The matter will probably never be completely settled. There is no doubt that the question of "how to safeguard the *Annalen* from Brouwer's negative influence (real or imagined)" was uppermost in Hilbert's mind. But who is to say that no personal motives were involved? There are Hilbert's own statements (e.g., to Blumenthal and Courant) to the effect that no personal grudge led to his action, but how much weight can be attached to them? In any case they contradict the letter of 15 October.

Last Ditch

The whole problem seemed to have been settled satisfactorily. Hilbert, who was only partially informed of the goings on, wrote to Blumenthal (Blumenthal to Courant, 31 December) "a triumphant letter, that everything was glorious." Courant had written a conciliatory letter to Brouwer (23 December) in which he expressed the hope that the solution to the matter satisfied Brouwer. He also wished to convince Brouwer that no personal motives had played a role in Hilbert's action and definitely no motives "whose existence were in conflict with the respect for your scientific or moral personality." Little did he know Brouwer!

To begin with, Brouwer had not yet received the letter from Springer and Hilbert, so he was unaware that the matter had been settled (unless he was informed by one of the other editors).

As a matter of fact Brouwer launched another appeal to the publisher and the editors the same day Courant was offering Brouwer the "forgive-and-forget" advice. Brouwer insisted that in the interest of mathematics the total editorial board of the *Mathematische Annalen* should remain in function; as he realized that a written defense from his hand would inevitably wreck the unity of the editors, he was willing to postpone such a letter; moreover, Carathéodory, in a letter of 3 December, had promised him to do his utmost to find an acceptable solution and had begged him to be patient for a couple of more weeks. Sommerfeld had also pressed Brouwer to wait for Carathéodory's intervention.

The final solution, as formulated in the Hilbert-Springer letter, did not satisfy Brouwer. He recognized that the reorganization of the *Annalen* was mostly, if not wholly, designed to get rid of him. Also, Brouwer had explicit views on the ideal organization of the *Annalen*.

In a circular letter (23 January 1929) to the editors, Blumenthal and Hilbert excluded, Brouwer rejected the final solution. According to him, the *Mathematische Annalen* was a spiritual heritage, a collective property of the total editorial board. The chief editors were, so to speak, appointed by free election and they were merely representatives *vis-à-vis* the mathematical world. Thus, Brouwer argued, the contractual rights of the chief editors were not a personal but an endowed good. Hilbert and Blumenthal, in his view, had abstracted this good from their principals, and hence were guilty of embezzlement, even if this could by sheer accident not be dealt with by law (the reader may hear a faint echo of Brouwer's objections to the principle of the excluded third [1]).

Brouwer then proceeded to attack Blumenthal's role in the *Annalen*. He repeated Blumenthal's earlier views on the equal rights of all editors and referred to certain irregularities in the management of the *Annalen* in 1925 resulting in Blumenthal's promise to resign after the appearance of volume 100.

Carathéodory also deplored the end of the old régime. When confronted with Hecke's comments on the practice of the past (letter from Courant to Carathéodory, 17 December): ". . . that Hecke, when he learned about the organization of the editorial board and the competency of the *Beirat* [the advisory editors] grasped his head and judged a revision and a more strict organization absolutely necessary," Carathéodory heartily disagreed (to Courant, 19 December 1928):

> For, Klein had organized the board of editors of the Mathematische Annalen *in such a way that it formed really a kind of Academy, in which each member had the* same rights as the others. That was in my opinion the main reason why Annalen *could claim to be the first mathematics journal in the world. Now it will become a journal like all other ones.*

The wisdom of severely restricting the size of the editorial board was questioned. Already on 2 February 1929 Blumenthal sent out a note on the future organization of the *Annalen,* in which he drew the attention to the decline of the journal compared to other journals. Since the *Nebenredaktion* had been eliminated (*ausgeschaltet*) one simply needed a larger staff: "the increasing necessity of scientific advisers follows inevitably from the increasing specialization." In short Blumenthal proposed to reinstate something like the old *Mitarbeiter* under a different name. In the same letter he broached the question of the successor of Hilbert, should he step down. One finds it difficult to reconcile this letter with the arguments that were put forward in favor of the solution to the conflict.

Parting Shot

The *Annalen* settled down under the new regime. Due to tactful handling of all publicity, the excitement in Germany died out, even, as Courant wrote to Hecke, among the colleagues in Berlin—and Brouwer was completely ignored. After waiting for months—and probably realizing that the battle was over and that everybody had gone home—Brouwer fired his parting shot, the letter of defense against Blumenthal's indictment of 16 November 1928. The letter is three-and-a-half folio sheets long and contains a report of the events mentioned above as experienced by Brouwer.

In the first place he denied Blumenthal's claim that Brouwer had substituted his own interpretation for Carathéodory's version of the developments leading to, and including, Hilbert's action. The views, he wrote, were not mine, but "views that during the aforementioned visit, came up between Carathéodory and me in mutual agreement, i.e., that were successively uttered by one of us and accepted by the other." He also elaborated the grounds for not acquiescing in the dismissal. He had told Carathéodory that

> he would consider a possible dismissal from the editorial board not only a revolting injustice, but also a serious damage to my freedom to act (Wirkungsmöglichkeit) and, in the face of public opinion, as an offending insult; that, if it really came to this unbelievable event, my honour and freedom of practice could only be restored by the most extensive flight into public opinion.

At the end of the otherwise friendly visit of 30 October, Carathéodory had once more returned to the matter. At Brouwer's exclamations Carathéodory could only answer "What can one do?" and "I don't want to kill a person." The final farewell was accompanied by Brouwer's bitter "I don't understand you any more," "I consider this visit as a farewell," and "I am sorry for you."

After attacking Blumenthal for his desire to remove Brouwer from the board of editors, Brouwer went on to answer Blumenthal's points. Without repeating the argument *verbatim*, some points may be taken from it to represent Brouwer's side in the discussion. Blumenthal accused Brouwer of rudeness; the latter answered that if Blumenthal meant by "rude" the "desire for integrity (duty of every human) increased by the will for clarity (the destiny of the mathematician)," there could have been cases of rudeness, in which—in Brouwer's words—neither the vanity of the author nor the wish of Blumenthal to appear pleasant could be spared. These cases, moreover, were entrusted by Blumenthal to Brouwer as a trouble shooter, and thus Blumenthal could not possibly find support among his fellow editors if and when he complained.

The matter of the resignation of Klein was, according to Brouwer, misrepresented; an author had, after his paper had been turned down by Brouwer, appealed to Klein and made the contents plausible. When Brouwer afterwards showed Klein that the author was ("not formally, but materially") wrong, Klein saw that he could not fulfill his promise to the author. In the discussion with Brouwer, Klein then uttered the opinion that the public was misled by the lists of editors on the cover of the journal and that, as far as he was concerned, he could no longer carry the responsibility for this impression. He retired soon afterwards.

The reproach concerning the long delay of papers at Brouwer's desk was dismissed by Brouwer as nonsense. Papers with lots of mistakes take time—and never a paper got lost, as happened with Hilbert, he said. In any case, Blumenthal's reproaches had never been uttered before.

The battle being lost, Brouwer no longer attempted to reverse the reorganization of the *Annalen*. He merely challenged Blumenthal to open the archive of the *Annalen*, claiming the correspondence would fully vindicate Brouwer.

Not Just Another Battle

The whole history of the *Mathematische Annalen* conflict was quietly incorporated into the oral tradition of European mathematics. Little is known of the aftermath; the *Göttinger* had won the battle, and they may have been tempted to pick a bone or two with some of the minor actors. For instance, Harald Bohr drafted a letter to the effect that "Schmidt for once realizes that he is vulnerable and that it is dangerous just to make a telephone call to Brouwer" (letter to Courant, 31 December 1928). After some reflection the letter to Schmidt was never sent.

From the gossip generated by the *Annalen* affair, a few rumors have surfaced in print. Only in one case could some evidence be unearthed, to wit the claim that Brouwer's dismissal was partly motivated by the fact that he had reserved the right to handle all papers from Dutch mathematicians [4, p. 187]. Professor Freudenthal told me that this was indeed commonly believed at the time of the conflict. By chance this particular rumor was confirmed in the draft of a letter from Felix Klein to the Dutch mathematician Schouten (13 March 1920). Klein wrote that "Prof. Brouwer . . . who at his entry in the editorial board of the *Annalen* has reserved the right to decide, in particular about Dutch papers. . . ." In general not much is known about the actual use Brouwer made of this prerogative; the letter of Klein dealt with a paper of Schouten that had received a negative evaluation from Brouwer.

Looking back, without the emotions of the contemporaries, we can only say that the whole affair was a tragedy of errors. Hilbert's annoyance with Brouwer was understandable. There had been a long series of conflicts, the *Grundlagenstreit*, the Göttingen chair that was turned down, the Riemann volume of the *Annalen*, and finally the Bologna affair. In a sense there had been an ongoing battle and each antagonist was firmly convinced that the survival of mathematics depended on him. Hilbert's illness, with the real danger of a fatal outcome, must have influenced his power of judgment. I do not see how Brouwer could have marched the *Annalen* to its doom. One has to agree with Einstein: if Brouwer was a menace of some sort, there were other ways to safeguard the *Annalen*. The question of the real motives behind Hilbert's action remains a matter of conjecture. Most likely the letter to Einstein shows an unguarded Hilbert with personal motives after all.

For Brouwer the matter had, in my opinion, far more serious consequences. His mental state could, under severe stress, easily come dangerously close to instability. Hilbert's attack, the lack of support from old friends, the (real or imagined) shame of his dismissal, the cynical ignoring of his undeniable efforts for the *Annalen*; each and all of these factors drove Brouwer to a self-chosen isolation.

Although it is most unlikely that intuitionism would have become the dominant doctrine of mathematics, there was a real possibility that it would develop into a recognized, although limited, activity. As it was, history took another turn, the development of intuitionistic (or constructive) mathematics suffered a setback, from which it recovered only some forty years later.

After the *Annalen* affair, little zest for the propagation of intuitionism was left in Brouwer; he continued to work in the field, but on a very limited scale with only a couple of followers. Actually, his whole mathematical activity became rather marginal for a prolonged period. During the thirties Brouwer hardly published at all (only two small papers on topology); he undertook all kinds of projects that had nothing to do with mathematics or its foundations. For all practical purposes, 1928 marks the end of the *Grundlagenstreit*.

Acknowledgments. The material used for this paper comes from various sources; the letters of Einstein are published with the permission of the Department of Manuscripts and Archives of the Jewish National and University Library at Jerusalem; the Niedersächsiche Staats- und Universitäts-bibliothek Göttingen gave permission for publication of the quotation from Klein's letter to Schouten; Professor Freudenthal kindly made some of the correspondence available, and material from the Brouwer Archive has been used. I have received advice and help from a great number of people and institutions to whom I express my gratitude. I would like to thank in particular P. Forman, H. Freudenthal, H. Mehrtens, D. Rowe, and C. Smorynski.

References

1. L. E. J. Brouwer, Über die Bedeutung des Satzes vom ausgeschlossenen Dritten in der Mathematik, insbesondere in der Funktionentheorie. *Journal für die reine und angewandte Mathematik* 154 (1924), 1–7. Also in *Collected Works*, Vol. 1 (A. Heyting, ed.), Amsterdam: North-Holland Publ. Co. (1975), 268–274.
2. D. van Dalen (ed.), *L. E. J. Brouwer, C. S. Adama van Scheltema. Droeve snaar, vriend van mij.* Amsterdam: Arbeiderspers (1984).
3. H. Mehrtens, Ludwig Bieberbach and "Deutsche Mathematik," *Studies in History of Mathematics* (E. R. Philips. ed.). The Mathematical Association of America (1987), 195–241.
4. C. Reid, *Hilbert-Courant.* Berlin: Springer-Verlag (1986). (Originally two separate vols. 1970, 1976).

40

Hilbert's Problems and Their Sequels

Jean-Michel Kantor

Il n'y a pas de problèmes résolus, il n'y a que des problèmes plus ou moins résolus.[1]

Henri Poincaré

"To speak of the future of mathematics is an exercise in fantasy which can not be discouraged too strongly; it is really a pure absurdity ... [The state of mathematics in the year 2000] will be just fine if the atomic physicists or some peace conferences don't suddenly interrupt the course of progress" (R. Godement, 1948, Gen. Ref. [LeL]).[2] Today the historical development of the subject can no longer be treated more than partially, and even then it takes a team such as was assembled at the conference in the United States in 1974 on the problems David Hilbert had posed in 1900. Poincaré and Hilbert were no doubt the last mathematicians to have an overall conception of their science; it is no longer within the scope of one mathematician to handle the mathematical legacy of Hilbert. This may serve to excuse in part such omissions and errors as may be found in the following, which are my responsibility alone, despite the kind help of V. I. Arnold, M. Berger, M. Berry, H. Brézis, P. Cartier, J.-L. Colliot-Thélène, J. Dixmier, M. Hindry, J.-P. Kahane, V. Kharlamov, B. Malgrange, Y. Matijasevich, B. Mazur, J. Milnor, C. Sabbagh, J.-P. Serre, G. Tenenbaum, and M. Waldschmidt. My warm thanks to them, to the staff of the interuniversity science library of Jussieu, and to Alberto Arabia.

The Context[3]

Poincaré marked the first International Congress of Mathematicians in 1897 by a lecture on the relations between mathematics and physics. When Hilbert was invited to speak to the next International Congress (Paris, 6–10 August 1900), he hesitated between an answer to Poincaré and a list of problems to stimulate the research of the new century. He asked Minkowski, "Might I point to probable

---

[1]"There are no solved problems, there are only more-or-less solved problems."
[2]References are indicated by authors' initials. Some are general (Gen. Ref.) and some pertain to particular problems.
[3]For the entire section, see Gen. Refs. [H] and [M].

Volume 18, No. 1 (Winter 1996), 21–30

directions of the mathematics of the new century . . . a look to the future?" Minkowski was enthusiastic. At Paris, Hilbert had a rapt audience:

> *Who would not lift the veil that hides the future from us, to have a look at the progress of our Science and the secrets of its further development in future centuries?* [H, p. 58]

He lectured on 10 of the 23 problems later published in the Proceedings of the Congress:

On foundations—**Problems 1, 2, and 6.**
Four problems on arithmetic and algebra—**Problem 7,** Irrationality and transcendence; **Problem 8,** The Riemann hypothesis; **Problem 13,** Superposition of two functions; and **Problem 16,** Ovals and limit cycles.
Three more on theory of functions—**Problem 19,** Calculus of variations; **Problem 21,** Fuchsian differential equations and monodromy; and **Problem 22,** Uniformization.

To Hilbert, the choice of problems was supposed to show the deep unity of mathematics and support his declaration of faith in a glorious future (in mathematics there is no "ignorabimus"—Du Bois–Reymond). The Hilbert problems went on to have the success we all know; their history is rich in instructive anecdotes, but also in blind alleys. There are also some surprises waiting in their present status. Let me try to bring the problems up to date; I emphasize the progress made between 1975 and 1992, because a serious work was produced by the American Mathematical Society seminar in 1974. For an introduction to the situation in number theory (before the "Wiles bomb"), see [Ma].

The Present Status

PROBLEMS 1 AND 2: The continuum hypothesis; consistency of arithmetic.

Associated with the first two problems are the names of Kurt Gödel and Paul Cohen.
The continuum hypothesis, the subject of Problem 1, may be stated as

Every uncountable subset of **R** has the same cardinality as **R**,

or

$$2^{\aleph_0} = \aleph_1.$$

The consistency of arithmetic was treated by Gödel, who showed (1931) the incompleteness and the undecidability of arithmetic. Gödel showed in 1938 that if set theory (according to the Zermelo–Fraenkel axioms) is consistent, then it remains consistent if augmented by adjoining the axiom of choice and/or the continuum hypothesis. Finally, in 1963 Paul Cohen showed that if the Zermelo–Fraenkel axioms are consistent, then the negation of the axiom of choice or even the negation of the continuum hypothesis[4] can be adjoined and the theory will remain consistent!
Gödel's impact was felt especially outside of mathematical circles: it did not change the way of working of mathematicians—with the obvious exception of logicians—and it gave rise to a huge

---

[4]The generalized continuum hypothesis implies the axiom of choice.

trove of citations eligible for the next edition of the *Dictionary of Stupidity*.[5] J.Y. Girard sums up Gödel's discovery by this image:

There are things in mathematical thought which are not part of the mechanism. [G]

PROBLEM 3: Congruence of polyhedra. "On the equality of the volumes of two tetrahedra with equal base and altitude."

> *Two pyramids with triangular base and the same altitude are in the same ratio as their bases. The proof (Euclid) uses the method of exhaustion, whereas for plane figures the calculation of the area of any polygon is reducible by decomposition ("scissors geometry") to that of the square. To show that such a reduction is impossible for volumes.*

Actually the problem was already solved by Bricard (1896) and Hilbert's student Dehn (1900), based on the notion of invariant; two solids which can be cut up into congruent pieces necessarily have the same Dehn invariant. Call "scissors congruence invariant" an element $D(P)$ (of a group) associated to every polytope P in such a way that

$$D(P \cap P') + D(P \cup P') = D(P) + D(P')$$

if P, P', and $P \cup P'$ are polytopes. Further, $D(P) = 0$ if the polytope is planar; and for g a motion of space, $D(g(P)) = D(P)$. The Dehn invariant is defined as follows. The group of angles between lines in the plane is identified with $\mathbf{R}/2\pi\mathbf{Z}$, and one sets

$$D(P) = \sum_i |L_i| \otimes \delta_i, \quad D(P) \in \mathbf{R} \otimes_{\mathbf{Z}} (\mathbf{R}/2\pi\mathbf{Z}),$$

where L_i is an edge and δ_i the dihedral angle at L_i. Then Dehn shows that the cube and the regular tetrahedron of the same volume do not have the same Dehn invariant. The problem was considered as solved. However, Sydler in 1965 showed that two polytopes are equivalent if and only if they have the same volume and the same Dehn invariant (see [B], [C], and [S]). On the other hand, no analogous results are known in higher dimensions or in non-Euclidean geometry in three dimensions or more.

More recently, other invariants, the Hadwiger invariants, were introduced in arbitrary dimensions, with respect to the group of translations, and it was shown that for two polytopes to be equivalent when only translations are allowed, it is necessary and sufficient that they have the same Hadwiger invariants. We refer to [C] for the connection to the calculation of Eilenberg–MacLane homology groups and for current developments.

PROBLEM 5: Topological groups and Lie groups. Is a locally Euclidean group a Lie group?

This problem was in fashion in the 1950s. It was solved in 1953 by Gleason and by Montgomery and Zippin. The following question still remains open: Given a locally compact topological group acting faithfully on a topological manifold, is it a Lie group? For example (actually, equivalently), can

---

[5]Example: R. Debray (*Le Scribe*, 1980): "From the day that Gödel showed that there is no consistency of Peano arithmetic formalizable within that theory (1931), political scientists were in a position to understand why Lenin had to be mummified and displayed . . . in a mausoleum."

the p-adic integers act faithfully on a compact topological manifold? We observe that the Montgomery-Zippin theorem plays a crucial role in a recent important work of M. Gromov [G].

PROBLEM 6:   Axiomatization of physics.

There are two main reasons why all relation has disappeared between this problem as Hilbert stated it and as it would be stated today:

1.   The revolution in modern physics: relativity, general relativity, quantum physics.
2.   The introduction of new mathematical tools, first of all Hilbert space. Hilbert was starting work on it in winter 1900, in the following spirit:

> *It appears to me of outstanding interest to undertake an investigation of the convergence conditions which serve for the erection of a given analytic discipline so that we can set up a system of the simplest fundamental facts which require for their proofs a specific convergence condition. Then by the use of that one convergence condition alone—without the addition of any other convergence condition whatsoever—the totality of the theorems of the particular discipline can be established.* (Gen. Ref. [R, p. 85])

Stimulated by his reading of Fredholm's work, Hilbert introduced the space which bears his name. One notes with astonishment that he makes no mention of this research in the problems stated at the Paris Congress. The notion of Hilbert space completely transformed the mathematical formalism in use in theoretical physics.

Skipping forward a century, if we try to indicate the most active directions of the present, these seem to stand out:

General relativity and global differential geometry. Construction of manifolds which could be effective cosmological models (S. Hawking, R. Penrose).

Mathematics of quantum field theory; gauge theory; introduction of quantum groups in the formalism of quantum physics.

PROBLEM 7:   Irrationality and transcendence of α^β for α algebraic and β algebraic irrational (for example, $2^{\sqrt{2}}$).

The problem was solved by Gel'fond (1935) and Th. Schneider. Euler's constant remains as mysterious as ever.

Let us note some recent progress in this area: A. Baker (1966, Fields Medal in 1970): If the a_i are non-zero algebraic numbers whose logarithms are linearly independent over $\mathbf{Q}$, then

$$1, \log a_1, \ldots, \log a_n$$

are linearly independent over the field of algebraic numbers.

In the last twenty years tools have been brought into play from various mathematical domains. The theory of functions of several complex variables enabled Enrico Bombieri (Fields Medal 1974) to resolve a conjecture of Nagata and enabled W. D. Brownawell in 1985 to give the first effective version of the Hilbert *Nullstellensatz*. The latter theorem is useful in problems of algebraic independence; the method initiated by A. O. Gel'fond in 1949 and developed by G. V. Chudnovsky in the 1970s enabled him to prove the transcendence of numbers like $\Gamma(1/4)$ or $\Gamma(1/3)$ (where Γ is the Euler gamma function). Commutative algebra, with Chow forms, was introduced in this context with great success by Yu. V. Nesterenko. Results on transcendence involving the usual exponential function or elliptic functions (Th. Schneider) were extended to (commutative) algebraic groups by S. Lang in the 1960s; this theme was abundantly developed especially by D. W. Masser, first alone and then in collaboration with G. Wüstholz, in connection with the work of Faltings on the Mordell conjecture.

This result of G. Faltings implies that, for a fixed $p \geq 3$, the equation

$$x^p + y^p = z^p$$

has a finite number of solutions without a common factor. What remained of the Fermat problem was to show that this number is zero. K. Ribet showed that the Fermat conjecture is implied by a deep conjecture (Taniyama–Weil) related to number theory and also to algebraic geometry and group theory. This got the Fermat problem out of its isolation.[6]

PROBLEM 8:   The Riemann hypothesis.

Surely the most celebrated problem in the history of mathematics; according to Hilbert, the most important problem of mathematics, and he even went so far as to say in conversation that it was the most important problem for humanity! (*loc. cit.*) It is now known that more than 40 percent of the zeros are on the fateful line (Gen. Ref. [C] 1989), and the fluctuations among the zeros have been studied. It seems, as was suggested very early, that the distribution of zeros resembles that of the eigenvalues of a self-adjoint operator (Hilbert and Pólya, about 1915). About 1973 Dyson suggested that the operator could be a random hermitian operator. For several decades, progress on the hypothesis has been essentially technical, and the most powerful computers have been tried on it. Some of the achievements have been real *tours de force* ([B-I], cf. [G-K]).

On the other hand, there have been attempts, following a suggestion by Hilbert, to generalize the Riemann hypothesis by replacing the field of rationals by a field of functions on a projective curve defined over $\mathbf{F}_q$. This was a motivation of the founders of modern algebraic geometry, A. Weil and O. Zariski. The "Weil conjectures," stated in 1949, which extended the earlier conjectures to higher-dimensional varieties, were proved in 1973 by P. Deligne (Fields Medal 1978) (Gen. Ref. [K]).

PROBLEM 10:   "On the possibility of solving a diophantine equation."

The problem is to give an algorithm which tests the solubility of diophantine equations (polynomial equations with integer coefficients whose integer solutions are sought). As the statement indicates, the problem is at the border between logic—recursion theory—and number theory. It was settled negatively by Yu. Matijasevich in 1970. This negative result was obtained by deepening our knowledge of recursive sets (recursively enumerable or "listable") and diophantine sets; that is, sets of integer parameters a_i for which

$$P(a_1, a_2, \ldots, a_n, z_1, z_2, \ldots, z_m) = 0$$

---

[6]Having already cited Gen. Ref. [Ma] for number theory through the 1980s, let me draw on the introduction to the 1993 English edition for the recent developments: ". . . the famous problem 'Fermat's Last Theorem,' together with the Taniyama–Weil conjecture for semi-stable elliptic curves, seem to have been completely proved by A. Wiles. Wiles used various sophisticated techniques and ideas due to himself and a number of other mathematicians (K. Ribet, G. Frey, Y. Hellegouarch, J.-M. Fontaine, H. Hida, J.-P. Serre, J. Tunnel, . . .). This genuinely historic event concludes a whole epoch in number theory, and opens at the same time a new period which could be closely involved with implementing the general Langlands program. . . . One of the characteristic features of the new methods and ideas is the intensive use of p-adic L-functions and Galois representations. Another striking example of this feature is K. Rubin's construction of rational points on elliptic curves using special values of p-adic L-functions and their derivatives."

has solutions (z) in integers. Matijasevich's theorem asserts that these two families of sets of integers are the same. Astonishing application: There exists an effectively computable polynomial (in 10 integer variables) with integer coefficients whose positive values are precisely the set of prime numbers!

Other problems of the same sort remain unsolved; for example, does there exist a system

$$f(x_1, x_2, \ldots, x_n, t) = 0 \tag{1}$$

of equations with integer coefficients which has rational solutions $(x_1, \ldots, x_n)$ if and only if the parameter t is an integer? Otherwise stated, does there exist a morphism of varieties over $\mathbf{Q}$

$$V \to \text{Affine line}$$

such that the fiber at a is non-empty if and only if a is an integer?

PROBLEM 11:   Classification of quadratic forms with coefficients in rings of algebraic integers.

See Gen. Ref. [K].

PROBLEM 12:   This concerns generalizations of Gauss' quadratic reciprocity law, the law from which has flowed the modern algebraic theory of numbers.

Consider the equation (in integers)

$$x^2 + 1 \equiv 0 (\text{mod } p).$$

This equation has solutions if and only if the prime p is congruent to 1 modulo 4. This rule is at the origin both of group theory and of modern number theory.

To sum up the ramifications of this problem, one would cite class field theory (Hilbert, and around 1930 Furtwangler, Takagi, E. Artin); the introduction of methods of cohomology of groups; the study of L-series; and the vast "Langlands program" aiming to extend the quadratic reciprocity law to the non-abelian case, on which numerous mathematicians around the world are laboring (whereas Langlands himself has turned to percolation!). See Gen. Ref. [Ma].

In the abelian case, one would like to describe explicitly the abelian extensions of a number field K which is a finite extension of $\mathbf{Q}$, in terms of the values of certain special functions (exponential, elliptic) and the action of the Galois group. For $\mathbf{Q}$, the Kronecker–Weber theorem gives an explicit description of the abelian extensions by means of the action of the Galois group on the roots of unity. The work of Shimura and Taniyama [S-T] concerning abelian varieties with complex multiplication provides an almost complete answer for "CM fields" (totally imaginary quadratic extensions of a totally real field).

PROBLEM 13:   "Impossibility of solving the general equation of degree 7 in terms solely of functions of two arguments": superposition of functions.

The equation of third degree can be reduced by translation to

$$X^3 + pX + q = 0,$$

which has the solution (Scipione del Ferro, 16th century)

$$X = \left(-\frac{q}{2} + \sqrt{\frac{4p^3 + 27q^2}{4(27)}}\right)^{1/3} + \left(-\frac{q}{2} - \sqrt{\frac{4p^3 + 27q^2}{4(27)}}\right)^{1/3}.$$

The equation of fourth degree can be solved by superposition (composition) of addition, multiplication, square roots, cube roots, and fourth roots.

To try to solve algebraic equations of higher degree (a vain hope according to N. Abel and E. Galois), the idea of Tschirnhaus (1683) was to adjoin a new equation:
to

$$P(X) = 0$$

one adjoins

$$Y = Q(X),$$

where Q is a polynomial of degree strictly less than that of P, chosen expediently. In this way one can show that the roots of an equation of degree 5 can be expressed via the usual arithmetic operations in terms of radicals and of the solution $\phi(x)$ of the quintic equation

$$X^5 + xX + 1 = 0,$$

depending on the parameter x. Similarly for the equation of degree 6, the roots are expressible in the same way if we include also a function $\theta(x, y)$, a solution of a 6th-degree equation depending on two parameters x and y.

For degree 7 we would have to include also a function $\sigma(x, y, z)$, solution of the equation

$$X^7 + xX^3 + yX^2 + zX + 1 = 0.$$

Hence the natural question: can $\sigma(x, y, z)$ be expressed by superposition of algebraic functions of two variables? V. I. Arnold asked whether σ can be represented using as "irreducible branches" of the superposition mappings topologically equivalent to algebraic functions of two complex variables. Despite results of V. Lin, this question, along with the original question of D. Hilbert, remains unanswered.

Let us note one more strange result in this circle of ideas. The solution of

$$X^5 + xX^2 + yX + 1 = 0$$

cannot be expressed using algebraic functions of one variable, addition and multiplication (division is excluded!). (A. Khovanskii, see [R]; historical articles [JPK] and [D].)

A. N. Kolmogorov's important notions of complexity (created around 1955) were the subject of interesting comments by Yuri I. Manin at the International Congress at Kyoto (Gen. Ref. [C]). Hilbert had had another motivation for this problem: nomography, the method of solving equations by drawing a one-parameter family of curves. This problem, rising in the methods of computation of Hilbert's time, inspired the development of Kolmogorov's notion of ϵ-entropy; among its applications is its crucial role in theories of approximation now used in computer science. Let us briefly recall the history.

Contrary to the expectations of Hilbert and of contemporary mathematicians, in 1957 V. I. Arnold (then a student of Kolmogorov) showed that any continuous function f of three variables can be written

$$f(x, y, z) = \sum_i f_i(\phi_i(x, y), z)$$

with functions f_i and ϕ_i continuous. A few weeks later Kolmogorov showed that any continuous function f of n variables can be written in the form

$$f(x_1, x_2, \ldots, x_n) = \sum_{i=1}^{2n+1} f_i\left(\sum_j \phi_j(x_j)\right),$$

where the ϕ_j are continuous, monotone, and independent of f.

In the opposite direction are the results of Vitushkin (1954). When we deal with superpositions of formal series, or for analytic functions, or even of infinitely differentiable functions, we can show by an elementary enumeration technique followed by a Baire argument that, for example, almost every entire function has at an arbitrary point of $\mathbf{C}^3$ a germ which is not expressible by superposition of series of two variables. So there are many more entire functions of three variables than of two. The result of Vitushkin is that for functions of smoothness α (with α larger than 1, not necessarily an integer), representability depends on n/α.

PROBLEM 15:   "Rigorous foundation of the enumerative geometry of Schubert." Schubert calculus.

Schubert's study of the number of points of intersection of varieties—for example, of the number of lines meeting four given lines in 3-space—was taken up by the Italian school (Severi). The non-rigorous arguments—"principle of conservation of numbers," that is, heuristic arguments of continuity—grew into the modern theory of multiplicities of Pierre Samuel and Alexander Grothendieck (Fields Medal 1966). A topological point of view was taken by René Thom (Fields Medal 1958): Thom–Bordman polynomials, enumerative theory of singularities. Problem 15 can not yet be regarded as solved, despite some recent progress (Demazure, Fulton, Kleiman, R. MacPherson).

PROBLEM 16:   "Problems of the topology of curves and of algebraic surfaces."

This is the only problem directly about topology. It falls into two distinct parts.

A: Consider a non-singular curve of degree m in $\mathbf{P}^2(\mathbf{R})$. From a topological point of view such a curve is composed of ovals of which there are at most $\frac{1}{2}(m-1)(m-2)+1$, and this bound is the best possible. Curves attaining the maximum are called M-curves. *The problem is to study what configurations of the ovals are possible.* The problem is non-trivial beginning with $m=6$, the case raised by Hilbert; this was solved by Gudkov in the 1970s. (See Figure 40-1).

In 1933 Petrovskii proved inequalities concerning the topological invariants of

$$B = \{x \mid f(x) \geq 0\},$$

where f determines a curve of even degree; general inequalities were proved by Petrovskii and Oleinik (1949–1951), mainly concerning surfaces in $\mathbf{R}^3$, and similar results were obtained by Thom and Milnor

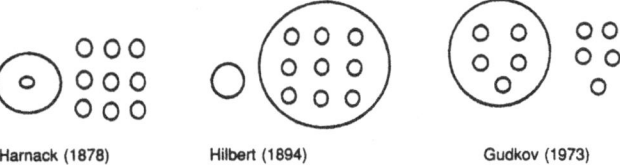

Harnack (1878)   Hilbert (1894)   Gudkov (1973)

FIGURE 40-1 *Possible arrangements for $m = 6$.*

(1964), see [R]. In 1971 V. Arnold proved congruence relations connecting the number P of even ovals (contained in an even number of ovals) and the number I of odd ovals. For a curve of even degree m, $P - I$ is congruent to $(m/2)^2$ modulo 8. For degree 8 the problem is still not solved. An account of the work, from O. Oleinik and I. Petrovskii (1949) to O. Viro (1979), is in [R]. The Leningrad school (Kharlamov, Gudkov, Rokhlin, Viro) turned up subtle relations between a non-singular complex algebraic variety and its real part [V]. The current work is very technical. The results have applications in analysis (partial differential equations, completely integrable systems).

B: *Problem of finiteness of limit cycles.*

Given a vector field V in the plane, a cycle is a periodic trajectory, and a limit cycle is a cycle on which non-periodic trajectories accumulate. When V is analytic, a limit cycle is isolated in the set of cycles. But it is possible for cycles to accumulate at a point or more generally on a "polycycle" (concept due to H. Poincaré) formed of pieces of integral curves ending at singular points of V.

Hence the natural question posed by Hilbert: Given a vector field $V = (X, Y)$, X and Y being polynomials of degree $\leq n$, find the maximum possible number of limit cycles. To begin with, one would want to try to prove

CONJECTURE: A polynomial vector field in the plane has only finitely many limit cycles or polycycles.

In 1923, Dulac proved this conjecture, but his proof was insufficient, as was recognized in the 1970s by Yuli Il'yashenko. However, some of Dulac's ideas were revived by Il'yashenko and Ecalle, in different directions. Il'yashenko (1984) proved that the finiteness conjecture holds on the complement of a proper algebraic subset of the vector space of vector polynomials of degree n.

Actually, the proper setting is that of singular foliations **F** (defined locally by analytic vector fields with isolated singular points). The return-map is defined in the neighborhood of a polycycle like C (Figure 40-2).

There exists an analytic curve $T : [0, 1] \to M$ with $T(0) \in C$ such that T is transversal to **F** and the leaf through $T(t)$—for t sufficiently small—cuts T again for the first time at $T(f(t))$. The mapping f leaves 0 fixed, preserves orientation, and is analytic on a punctured neighborhood of 0, but need not be analytic at 0. Thus it may have a sequence of isolated fixed points (t_n) which converge to 0. C is then the limit of the limit cycles through the $T(t_n)$. It is to be shown that a polycycle cannot be the limit of limit cycles, which is to say that if f is not the identity, then 0 is an isolated fixed point. This would imply the finiteness by the Poincaré–Bendixson theorem. Il'yashenko's proof of the finiteness conjecture and Ecalle's are given in [I] and [E]; Ecalle pursues his investigation, which uses his theory of generalized asymptotic developments and resurgence) to study the return-map f.

Let me mention finally that there is a "linearized" version of Problem **B** [A-I].

PROBLEM 17: Sums of squares: "Representation of definite forms by sums of squares."

Here "definite" means that the forms (homogeneous polynomials) in n variables with real coefficients have non-negative values for all real values of the variables.

Can they always be represented as quotients of sums of squares of forms?

This has been a motivating problem in real algebraic geometry.

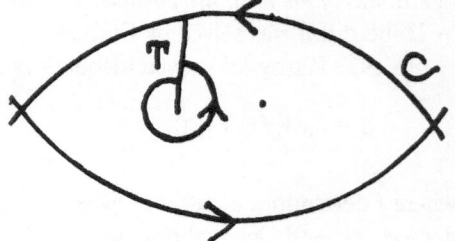

FIGURE 40-2 *Polycycle (of a vector field) and return-map.*

THEOREM OF EMIL ARTIN (1927):   Let X be an irreducible algebraic variety over **R** or **Q**, let f be a rational function on X, and assume f is positive at the real points of X where it is defined. Then f is a sum of squares in the function field of X (functions with coefficients in **Q**, if X is defined over **Q**).

THEOREM OF PFISTER (circa 1970):   Let X be an irreducible algebraic variety over **R** of dimension d, and assume f in the function field of X is positive (in the sense above). Then f can be written as the sum of at most 2^d squares.

Hilbert had already studied the conditions under which a homogeneous polynomial of degree m in n variables can be written as a sum of squares of homogeneous polynomials. Recently, this has been studied for analytic functions, for differentiable functions, and for Nash functions; see [B]. Real algebraic geometry is attracting renewed interest, by virtue of its relations with logic (elimination principle of Tarski and Seidenberg, model theory) and with industrial applications (robotics), but it is intrinsically more complicated than complex analytic geometry.

PROBLEM 18:   "To rebuild space with congruent polyhedra."

Hilbert's problem is divided into three distinct parts.
A: "*In Euclidean space of n dimensions, show that there are only finitely many different kinds of groups of displacements with a (compact) fundamental domain.*" In other words, one looks for discrete subgroups with compact quotient of the group $E(n)$ of isometries of $\mathbf{R}^n$. The result was proved by Bieberbach in 1910. The classification of these groups is important in crystallography and it generalizes to questions about lattices in Lie groups.
B: "*Tiling of space by a single polyhedron which is not a fundamental domain as in* **A**."
J. Milnor's remarks in Gen. Ref. [M] still apply: This is a very lively topic today [G-S1]. Important developments have been the theory of Penrose tilings [P] and the closely related physical problem of anomalous crystal structure (see, for example, [J]), as well as Thurston's theory of self-similar fractal tilings [T].
C: "*Packing of spheres. How should spheres of the same radius be arranged in space (of any dimensionality) so as to achieve the greatest density of packing?*"
Hilbert's text gives the impression that he did not anticipate the success and the developments this problem would have. The hexagonal packing in the plane is the densest (proof by Thue in 1882, completed by Fejes in 1940). In space, the problem is still not solved. There is very recent progress by Hales. For spheres whose centers lie on a lattice, the problem is solved in up to eight dimensions. The subject has various ramifications: application to the geometry of numbers, deep relations between coding theory and sphere-packing theory, the very rich geometry of the densest known lattices.
Before entering on analysis, Hilbert poses, in general, the question of choosing the class of functions with which to work: differentiable, analytic. . . ? There are other, intermediate classes which in my view have not been looked at sufficiently—for example, classes of functions defined by inequalities on their derivatives (Gevrey classes) or functions which are solutions of certain types of differential equations (differentiably algebraic functions).

PROBLEMS 19, 20, AND 23:   Partial differential equations: calculus of variations and the Dirichlet problem.

The developments in calculus of variations are so diverse (finite-element method, control theory) that it is impossible to survey them here. They play a central role in the study of non-linear phenomena (Plateau problem, equations of minimal surfaces) and, in the form of optimal control, in industrial applications of mathematics. Taking for granted the exposition by James Serrin (Gen. Ref. [M]), I may indicate for Problem 20 (solvability of boundary-value problems) developments involving elliptic systems, both linear with non-regular coefficients and non-linear: study of harmonic mappings; questions of regularity motivated by global geometry, as in the work of Schoen and Uhlenbeck (cf. Gen. Ref. [C], Karen Uhlenbeck's article); mechanics (elasticity), or more recently the study of liquid crystals (Gen. Ref. [I], 1988 and 1990). The study of extensions of the calculus of variations (Problem 23) has had a very wide development in recent years. Take for example the work of Bahri on the existence of periodic solutions of the three-body problem.

PROBLEM 21:   Monodromy of Fuchsian equations. "Does there always exist a linear differential equation of the Fuchs class having given critical points and a given monodromy group?"

One might try to prove the existence on the Riemann sphere of the following objects with given singular points and given monodromy groups:

A: Fuchsian equations (the Fuchs class): a linear first-order differential equation whose monodromy matrix has only simple poles at the given singular points;
B: one requires all the solutions to have regular singular points, that is, with moderate growth in angular sectors (this is the problem treated by Deligne [D]);
C: Fuchsian systems.

"Fuchsian" implies singular regular but not conversely. Plemelj (1930) claimed a solution to **A**. However, there was an error in his proof (see Gen. Ref. [E], vol. 1), and counter-examples have been found (see exposition in [B]).

PROBLEM 22:   Uniformization of analytic curves.

This is a fine example of a crossroads problem: topology, complex function theory, group theory, partial differential equations, ... Recent developments concern the extension to higher dimensions (Griffiths).

Some Reflections

One must not overlook little problems although they seem like dead ends without much future: there are many such in Hilbert's list, and they gave rise to important developments (particularly in geometry).

The phenomenon keeps occurring since then:
  the Henon attractor;
  non-periodic tilings of Penrose, used and studied by physicists, by Connes . . .
  . . .

If we look at the whole list of Hilbert problems and the works that have grown out of them, it is clear that they connect to an important part of this century's mathematics. They stimulated major efforts of numerous mathematicians. Who in this century has not dreamed of solving a question posed by Hilbert? However, let us look at the topics locally, undistracted by the impressive figure of their author. Even aside from the absence of certain important questions (functional analysis, measure theory) and the minor part given to others (topology—no doubt Poincaré has something to do with this), some mathematicians find the interest excited by Hilbert for certain problems to have been disproportionate. J. Dieudonné [LeL] judges a good portion of these problems "isolated," but this reproach can be invalidated, as we see, for example, from the recent work that "uncloisters" Fermat's conjecture. Some problems not presented at Paris (Problems 3, 5 and 15) and their sequels also took on luster from the great mathematician.

Problems, program?   As his correspondence with Minkowski shows, Hilbert must have harbored the dream of writing the program of action for the mathematicians of the twentieth century—at least the first stages. This was in line with his structured, centralized vision of mathematics: Göttingen at the center, number theory at the heart of mathematical science . . . Didn't Ostrowski even find a curious resemblance to another great advocate of programs of action, V. I. Lenin? Rather than a program, which would inevitably be lacking in means of implementation, it was more a matter of a *project*, a projection toward the future, bringing out ideas dear to Hilbert:

mathematicians' power to resolve crises (there are no problems without solution—contrast the dialectical point of view of Poincaré);
the deep unity of mathematics, around number theory;
mathematics as independent of the physical sciences.

The other subject that Hilbert had thought of treating in his Paris lecture, the relationship between pure mathematics and physics, he would take up in 1930, after his retirement, when he was made honorary citizen of the great city of Koenigsberg–Kaliningrad, the city of Kant and Jacobi. In his talk, which he summarized in an interview on the local radio, the defense of pure mathematics is linked, by way of a reading of Kant, to his profound conviction—which he asserted as if to prove— of the omnipotence of mathematics:

We must know. We will know.

And on these words one hears a great burst of laughter from Hilbert. What could it mean? An intuition of what was happening? Two months later the *Monatshefte für Mathematik* received an article from the 25-year-old mathematician Kurt Gödel . . .

Yesterday, today, tomorrow.   With the year 2000 coming up, it is tempting—but dangerous—to imagine an analogous effort. Indeed, other mathematicians since Hilbert have dared to envisage partial programs (A. Weil for one [LeL], or the team assembled by F. Browder [M], or finally the "sketch of a program" of A. Grothendieck [G]).

Lucky chance, or sign of the end of an era? Klein's *Encyclopedia of Mathematical Sciences* marked the end of the century; Hilbert's lecture marked the start of a formidable development (say 1900–1970). The appearance today of the *Encyclopaedia of Mathematical Sciences* [E] could fill a role analogous to the former. But how could we imagine it followed by a "new Hilbert program"?

In the last twenty years the progress of mathematics has been considerably transformed. Whereas mathematics in the first part of the century, up to the sixties, was motivated essentially by an internal logic and thus came under the sponsorship of Hilbert (and his disciple Bourbaki), today "external needs"—in a global sense—are the principal motor driving mathematics as a science. Poincaré's revenge. The laboration of electronic processing of information allows a "positive feedback" between theory and applications. This is indeed what the editors of the *Encyclopaedia of Mathematical Sciences* say (Vol. 1), speaking in their introduction of "the industrialization of mathematics." Let us take a few examples:

the motivating role of computer science in research in logic [I];
problems of coding, cryptography, and signal theory in number theory; "experimental" research in this area using the most powerful computers;

rapprochement of theoretical physics and mathematics (Fields Medals of 1990);
topology and formal calculus (F. Sergeraert [I, 1990]);
nonlinear systems (chaos, Korteweg–de Vries equation, . . .);
minimal surfaces of constant mean curvature (applications to interfaces of polymers).

This new nature of the evolution of mathematics (which merits a deeper study comparing this era to others) is summed up by Jacques-Louis Lions in reference to optimal control theory [L]: "Hermès, Hilbert program (Problem 23) in practice" (the Kalman filter already has an important role in controlling the Airbus).

This transformation naturally carries with it sociological consequences, bringing the ways of the mathematical tribe closer to those of other subcultures: the competition for contracts, the growing weight of military contracts together with a romantic tendency in the practice of research (one dreams, "but perhaps one ought not to dream too much" [F]).

In another direction, the relation between philosophy and mathematics, I would refer to [K2] but especially to [C'].

The "we must know" of Hilbert should today be accompanied by a "we want to act." One could imagine beginning the "program of twenty-first-century mathematics" with a study, done jointly by mathematicians and other scientists, which would list the important scientific problems deserving of mathematicians' efforts—new or renewed—(Feynman integrals, turbulence, complex systems not yet studied from this point of view, theoretical biology, cognitive sciences . . .) and would extract choices corresponding to a real social demand.

Might a program of this sort, putting the power of mathematics to work, make David Hilbert laugh—but with delight?

Acknowledgment

Thanks to Barry Mazur for these remarks on reading an early version of this article:

Looking over the Hilbert problems as you reviewed them, for me, is *quaintness*. One thinks of the mathematical developments in the latter part of the century (Grothendieck's, Langlands', Beilinson's conjectures, classification of finite simple groups, of differential structures, singularity theory, conformal and topological field theories with their concomitant theories of quantum groups, balanced tensor categories, invariants of three-manifolds, classification of three-manifolds à la Thurston, . . .) and one squints at a stock of old daguerrotypes, charming mementos of the period.

General References

[C]   *International Congress of Mathematicians, Kyoto, 1990. Proceedings*, Mathematical Society of Japan/Springer-Verlag, Tokyo, 1992.

[C']   P. Cartier, La pratique—et les pratiques—des mathématiques, in *Encyclopédie philosophique universelle*, vol. 1, 1991.

[D]   *Ein Jahrhundert Mathematik 1890–1990*, Vieweg & S., 1990. Review in *Mathematical Intelligencer* 13(1991), no. 4, 70–74.

[E]   *Encyclopaedia of Mathematical Sciences*, Heidelberg: Springer-Verlag (1990) (Russian original Moscow, 1985).

[F]   J. M. Fontaine, Valeurs spéciales des fonctions L des motifs, *Sémin. Bourbaki* (1992), Exposé 751. *Astérisque* 206, (1992).

[G]   A. Grothendieck, Esquisse d'un programme, Dossier de candidature au C.N.R.S., 1985.

[H]   D. Hilbert, Sur les problèmes futurs des mathématiques, in *Proceedings of the Second International Congress of Mathematicians*, Paris: Gauthier-Villars (1902), pp. 58–114.

[I]   Images des mathématiques, Annual Supplement to *Courrier du CNRS*, CNRS, Paris.

1985

J. F. Boutot and L. Moret-Bailly, Equations diophantiennes: la conjecture de Mordell.
Images des nombres transcendants, d'après P. Philippon.

1988

M. Parigot, Preuves et programmes; les mathématiques comme langage de programmation.
J. M. Ghidaglia and J. C. Saut, Equations de Navier–Stokes, turbulence et dimension des attracteurs.

1990

F. Sergeraert, Infini et effectivité: le point de vue fonctionnel.
F. Bethuel, H. Brezis, J. M. Coron, and F. Helein, Problèmes mathématiques des cristaux liquides.

[K]   J.-M. Kantor, Hilbert (problèmes de), in *Encyclopaedia Universalis* (1989), vol. 11.

[K2]   J.-M. Kantor, L'intuition en équations, Report on the Kyoto Congress, *Le Monde* (August 1990).

[LeL]   A. Blanchard, *Les grands courants de la pensée mathématique* (F. LeLionnais, ed.), (1948), new edition 1962, Paris; in particular the articles by J. Dieudonné, R. Godement, A. Weil.

[L]   J.-L. Lions, L'Ordinateur, nouveau Dédale, Daedalon Gold Medal Lecture, 1991.

[M]   *Mathematical Developments Arising from Hilbert Problems* (F. Browder, ed.), Providence, RI: American Mathematical Society (1976).

[Ma]   Yu. I. Manin and A. A. Panchishkin, Number Theory I: Introduction to Number Theory, in *Encyclopaedia of Mathematical Sciences*, Heidelberg: Springer-Verlag (1993).

[P]   H. Poincaré, L'avenir des mathématiques, International Congress of Mathematicians, Rome, 1908.

[R]   C. Reid, *Hilbert*, New York: Springer-Verlag (1970).

References Specific to the Problems

Problems 1 and 2

[B]   J. Y. Bouleau and A. Louveau Girard, *Cinq conférences sur l'indécidabilité*, Paris: Presses de l'Ecole nationale des Ponts et Chaussées (1982).

[G]   E. Nagel, J. Newman, K. Gödel, and J. Y. Girard, *Le théorème de Gödel*, Paris: Seuil (1989).

Problem 3

[A]   Agrégation de mathématiques, problèmes de mathématiques générales, *Revue de mathématiques spéciales*, 1985–1986, Paris, pp. 139–150. Vuibert.

[B]   V. Boltianski, *Hilbert's Third Problem*, New York: Wiley (1978).

[C]   P. Cartier, Décomposition des polyèdres: le point sur le troisième problème de Hilbert, *Sémin. Bourbaki* (1984–1985), no. 646.

[S]   G. Sah, *Hilbert's Third Problem: Scissors Congruence*, London: Pitman (1979).

Problem 5

[A]   J. Aczél, The state of the second part of Hilbert's fifth problem, *Bull. Am. Math. Soc.* 20 (1982).

[G]   M. Gromov, Groups with polynomial growth and expanding maps, *Publ. I.H.E.S.*, no. 53.

[S]   I. Sillmann, Every proper smooth action of a Lie group is equivalent to a real analytic action: a contribution to Hilbert's fifth problem, *M.P.I. Mathematik*, Bonn, 1993.

Problem 6

[B]   *P. Bohl's Fourth Thesis and Hilbert's Sixth Problem*, Moscow: Nauka (1986) [in Russian].

Problem 7

[B-M] A. Baker and D. W. Masser (eds.), *Transcendence Theory, Advances and Applications*, New York: Academic Press (1977).

[B-W] D. Bertrand and M. Waldschmidt (eds.), *Approximations diophantiennes et nombres transcendants*, Basel: Birkhäuser (1983).

[B]   A. Baker (ed.), *New Advances in Transcendence Theory*, Cambridge: Cambridge University Press (1988).

[P]   P. Philippen (ed.), *Approximations diophantiennes et nombres transcendants*, de Gruyter (1992).

Problems 8 and 9

[B]   M. V. Berry, Semi-classical formula for the number variance of the Riemann zeroes, *Nonlinearity* 1 (1988), 399–407.

[B-I] E. Bombieri and H. Iwaniec, On the order of $\zeta(1/2 + it)$, *Ann. Scuola Norm. Sup. Pisa* 13 (1986), 449–472.

[G-K] S. W. Graham and G. Kolesnik, *Van der Corput's Method of Exponential Sums*, Heidelberg: Springer-Verlag (1991).

[I]   A. Ivic, *The Riemann Zeta Function*, New York: Wiley (1985).

[T]   E. C. Titchmarsh, *The Riemann Zeta Function*, 2d. edn. (revised by D. R. Heath-Brown), Oxford: Oxford University Press (1986).

Problem 10

[M]   Yu. Matijasevich, *The Tenth Problem of Hilbert*, Moscow: Nauka (1993) [in Russian].

[M1]   Yu. Matijasevich, My collaboration with Julia Robinson, *Mathematical Intelligencer* 14 (1992), no. 4, 38–45.

Problem 12

[K-S] K. Kato and S. Saito, Global class field theory of arithmetical schemes, *Contemp. Math.* 55 (Part I) (1986), 5–331.

[S-T] G. Shimura and Y. Taniyama, Complex multiplication of algebraic varieties and its application to number theory, *Publ. Math. Soc. Japan* 6 (1961).

Problem 13

[D]   J. Dixmier, Histoire du treizième problème de Hilbert, *Séminaire d'historie des mathématiques de l'Institut Henri Poincaré*, 1991–1992.

[JPK] J.-P. Kahane, Le treizième problème de Hilbert: un carrefour de l'analyse, de l'algèbre et de la géométrie, *Cahiers Sémin. Hist. Math. Sér.* 1, 3 (1982).

[R]   J. J. Risler, Complexité et géométrie réelle, d'après A. Khovanskii, *Sémin. Bourbaki* (1984–1985), no. 637.

[L]   V. Y. Lin, Superposition of algebraic functions, *Funct. Anal. ego Prim.* 10 (1976), 32–38.

Problem 15

[L]   A. Lascoux, Anneaux de Grothendieck de la variété des drapeaux, in *The Grothendieck Festschrift*, Vol. III, Basel: Birkhäuser (1993), pp. 1–34.

[S]   P. Samuel, Sur l'histoire du quinzième problème de Hilbert, *Gazette des Mathématiciens* (1975), (Oct. 1974), 22–32.

Problem 16

[A-I]   V. Arnold and Yu. Il'yashenko, Ordinary differential equations, in *Dynamical Systems—I, Encyclopaedia of Mathematical Sciences*, Heidelberg: Springer-Verlag (1988).

[B]   J. Bochnak, M. Coste, and M. F. Roy, *Géométrie algébrique réelle*, Heidelberg: Springer-Verlag (1986).

[E]   J. Ecalle, *Introduction aux fonctions analysables et preuve constructive de la conjecture de Dulac*, Paris: Hermann (1992); and Six lectures on trousseries, analysable functions, and the constructive proof of Dulac's conjecture, in *Bifurcations and Periodic Orbits of Vector Fields* (D. Schlominck, ed.), Kluwer Acad. Publ. (1993), pp. 75–184.

[I]   Yu. Il'yashenko, *Finiteness Theorems for Limit Cycles*, American Mathematical Society Providence, RI: (1991).

[K]   V. Kharlamov and I. Itenberg, Towards the maximal number of components of a nonsingular surface of degree 5 in $\mathbf{R}P3$, preprint (1993).

[M]   R. Moussu, Le problème de la finitude du nombre de cycles, d'après R. Bamon et Y. S. Il'yashenko, *Sémin. Bourbaki* (1985–1986), no. 655. *Astérisque*, 145–146 (1987).

[R]   J. J. Risler, Les nombres de Betti des ensembles algébriques réels, une mise au point, *Gazette des Mathématiciens* (1992).

[V]   O. Viro, Progress in the topology of real algebraic manifolds, in *Proceedings of the International Congress of Mathematicians, Warsaw, 1983*, pp. 595–611.

[Y]   J.-C. Yoccoz, Non-accumulation des cycles limites, *Sémin. Bourbaki* (1987–1988), no. 690.

Problem 17

[BS]   H. Benis-Sinaceur, De D. Hilbert à E. Artin: Les différents aspects du dix-septième problème de Hilbert et les filiations conceptuelles de la théorie des corps réels clos, *Arch. Hist. Exact Sci.* 29(3) (1984), 267–286.

[K]   A. Khovanskii, *Fewnomials*, Providence, RI: American Mathematical Society (1991).

Problem 18, Parts A and B

[D-G]   L. Danzer, B. Grünbaum, and G. C. Shephard, Does every type of polyhedron tile three-space? *Topol Struct.* 8 (1983).

[D-K]   M. Duneau and A. Katz, *Cristaux apériodiques et groupe de l'icosaèdre*, Palaiseau, France: Publications du Centre de Physique théorique, Ecole polytechnique.

[G-S]   B. Grünbaum and G. C. Shephard, Tiling with congruent tiles, *Bull. Am. Math. Soc.* 3 (1980), 951–973.

[G-S1]   B. Grünbaum and G. C. Shephard, *Tilings and Patterns*, New York: W.H. Freeman (1987).

[J]   M. Jaric, *Introduction to the Mathematics of Quasi-crystals*, New York: Academic Press (1989).

[P]   R. Penrose, Pentaplexity, *Eureka* 39 (1978), 16–22; Pentaplexity: a class of non-periodic tilings of the plane, *Mathematical Intelligencer* 2 (1979), no. 1, 32–37.

[T]   W. Thurston, *Groups, Tilings, and Finite State Automata*, Providence, RI: American Mathematical Society (1989); Research Report GCG-1, Geometry Supercomputer Project, University of Minnesota, Minneapolis (1989).

Problem 18, Part C

[C-S]   J. H. Conway and N. J. Sloane, *Sphere Packings, Lattices and Groups*, New York: Springer-Verlag (1988).

[O]   J. Oesterlé, Empilement de sphères, *Sémin. Bourbaki* (1989–1990), No. 727; "Les sphères de Kepler," video (series "Mosaïque mathématique"), Paris: Productions "Les films d'ici."

[S]   F. Sigrist, Sphere packing, *Mathematical Intelligencer* 5 (1983), 34–38.

Problems 19, 20, and 23

[V]   C. Viterbo, Orbites périodiques dans le problème des trois corps, *Sémin. Bourbaki* (1992–1993), no. 774.

Problem 21

[B]   A. Beauville, Equations différentielles à points singuliers réguliers d'après Bolybrukh, *Sémin. Bourbaki* (1992–1993), no. 765. *Astérisque*, 216 (1993).

[D]   P. Deligne, *Equations différentielles à points singuliers réguliers*, Heidelberg: Springer-Verlag (1970), p. 136.

Index of Names